Electrician's Exam Preparation

Electrical Theory
National Electrical Code
NEC® *Calculations*
Contains 2,400 Practice Questions

Mike Holt

Delmar Publishers

an International Thomson Publishing company I(T)P®

Albany • Bonn • Boston • Cincinnati • Detroit • London • Madrid
Melbourne • Mexico City • New York • Pacific Grove • Paris • San Francisco
Singapore • Tokyo • Toronto • Washington

NOTICE TO THE READER

Cover Design: Courtesy of Brucie Rosch

Delmar Staff
Publisher: Alar Elken
Acquisitions Editor: Mark Huth
Project Editor: Barbara Diaz
Production Coordinator: Toni Hansen
Art/Design Coordinator: Cheri Plasse
Editorial Assistant: Dawn Daugherty
Production Manager: Mary Ellen Black
Marketing Manager: Mona Caron

Online Services

Delmar Online
To access a wide variety of Delmar products and services on the World Wide Web, point your browser to:
http://www.delmar.com
or email: info@delmar.com

A service of I(T)P®

COPYRIGHT © 1999
by Delmar Publishers

an International Thomson Publishing company I(T)P®

The ITP logo is a trademark under license
Printed in the United States of America

Design Layout and Typesetting: Paul Wright

Graphic Illustrations: Mike Culbreath

COPYRIGHT © 1999 Charles Michael Holt Sr.

For more information contact:

Delmar Publishers
3 Columbia Circle, Box 15015
Albany, New York 12212-5015

International Thomson Editores
Seneca 53
Colonia Polanco
11560 Mexico D. F. Mexico

International Thomson Publishing Europe
Berkshire House
168 - 173 High Holborn
London WC1V 7AA
United Kingdon

International Thomson Publishing GmbH
Königswinterer Straße 418
53227 Bonn
Germany

Nelson ITP, Australia
102 Dodds Street
South Melbourne
Victoria, 3205 Australia

International Thomson Publishing Asia
60 Albert Street
#15 - 01 Albert Complex
Singapore 189969

Nelson Canada
1120 Birchmount Road
Scarborough, Ontario
M1K 5G4, Canada

International Thomson Publishing Japan
Hirakawa-cho Kyowa Building, 3F
2-2-1 Hirakawa-cho, Chiyoda-ku
Tokyo 102, Japan

International Thomson Publishing France
Tour Maine-Montparnasse
33 Avenue du Maine
75755 Paris Cedex 15, France

ITE Spain/Paraninfo
Calle Magallanes, 25
28015-Madrid, Espana

5 6 7 8 9 10 XXX 03 02 01 00

Library of Congress Cataloging-in-Publication Data
Holt, Charles Michael.
    Electrician's exam preparation : electrical theory, National
electrical code, NEC calculations : contains 2,400 practice
questions / Mike Holt.
        p.   cm.
    ISBN 0-7668-0376-7 (alk. paper)
    1. Electric engineering—United States Examinations, questions,
etc.   2. Electricians—Licenses—United States.   3. Electric
engineering Problems, exercises, etc.   I. Title.
TK169.H598   1999
621.319'24'076—dc21
                                        99-13241
                                        CIP

I dedicate this book to God, who has given me the ability to write it.

Delmar Publishers Is Your Electrical Book Source!

Whether you're a beginning student or a master electrician, Delmar Publishers has the right book for you. Our complete selection of proven best-sellers and all-new titles is designed to bring you the most up-to-date, technically-accurate information available.

NATIONAL ELECTRICAL CODE

National Electrical Code® 1999/NFPA
Revised every three years, the *National Electric Code®* is the basis of all U.S. electrical codes.
Order # 0-8776-5432-8
Loose-leaf version in binder
Order # 0-8776-5433-6

National Electrical Code® Handbook 1999/NFPA
This essential resource pulls together all the extra facts, figures, and explanations you need to interpret the 1999 *NEC®*. It includes the entire text of the Code, plus expert commentary, real-world examples, diagrams, and illustrations that clarify requirements.
Order # 0-8776-5437-9

Illustrated Changes in the 1999 National Electrical Code®/O'Riley
This book provides an abundantly-illustrated and easy-to-understand analysis of the changes made to the 1999 *NEC®*.
Order # 0-7668-0763-0

Understanding the National Electrical Code®, 3E/Holt
This book gives users at every level the ability to understand what the *NEC®* requires, and simplifies this sometimes intimidating and confusing code.
Order # 0-7668-0350-3

Illustrated Guide to the National Electrical Code®/Miller
Highly-detailed illustrations offer insight into Code requirements, and are further enhanced through clearly-written, concise blocks of text that can be read very quickly and understood with ease. Organized by classes of occupancy.
Order # 0-7668-0529-8

Interpreting the National Electrical Code®, 5E/Surbrook
This updated resource provides a process for understanding and applying the *National Electrical Code®* to electrical contracting, plan development, and review.
Order # 0-7668-0187-X

Electrical Grounding, 5E/O'Riley
Electrical Grounding is a highly illustrated, systematic approach for understanding grounding principles and their application to the 1999 *NEC®*.
Order # 0-7668-0486-0

ELECTRICAL WIRING

Electrical Raceways and Other Wiring Methods, 3E/Loyd
The most authoritive resource on metallic and nonmetallic raceways, provides users with a concise, easy-to-understand guide to the specific design criteria and wiring methods and materials required by the 1999 *NEC®*.
Order # 0-7668-0266-3

Electrical Wiring Residential, 13E/Mullin
Now in full color! Users can learn all aspects of residential wiring and how to apply them to the wiring of a typical house from this, the most widely-used residential wiring book in the country.
Softcover Order # 0-8273-8607-9
Hardcover Order # 0-8273-8610-9

House Wiring with the NEC®/Mullin
The focus of this new book is the applications of the *NEC®* to house wiring.
Order # 0-8273-8350-9

Electrical Wiring Commercial, 10E/Mullin and Smith
Users can learn commercial wiring in accordance with the *NEC®* from this comprehensive guide to applying the newly revised 1999 *NEC®*.
Order # 0-7668-0179-9

Electrical Wiring Industrial, 10E/Smith and Herman
This practical resource has users work their way through an entire industrial building—wiring the branch-circuits, feeders, service entrances, and many of the electrical appliances and sub-systems found in commercial buildings.
Order # 0-7668-0193-4

Cables and Wiring, 2E/AVO
This concise, easy-to-use book is your single-source guide to electrical cables—it's a "must-have" reference for journeyman electricians, contractors, inspectors, and designers.
Order # 0-7668-0270-1

ELECTRICAL MACHINES AND CONTROLS

Industrial Motor Control, 4E/Herman and Alerich
This newly-revised and expanded book, now in full color, provides easy-to-follow instructions and essential information for controlling industrial motors. Also available are a new lab manual and an interactive CD-ROM.
Order # 0-8273-8640-0

Electric Motor Control, 6E/Alerich and Herman
Fully updated in this new sixth edition, this book has been a long-standing leader in the area of electric motor controls.
Order # 0-8273-8456-4

Introduction to Programmable Logic Controllers/Dunning
This book offers an introduction to Programmable Logic Controllers.
Order # 0-8273-7866-1

Technician's Guide to Programmable Controllers, 3E/Cox
Uses a plain, easy-to-understand approach and covers the basics of programmable controllers.
Order # 0-8273-6238-2

Programmable Controller Circuits/Bertrand
This book is a project manual designed to provide practical laboratory experience for one studying industrial controls.
Order # 0-8273-7066-0

Electronic Variable Speed Drives/Brumbach
Aimed squarely at maintenance and troubleshooting, *Electronic Variable Speed Drives* is the only book devoted exclusively to this topic.
Order # 0-8273-6937-9

Electrical Controls for Machines, 5E/Rexford
State-of-the-art process and machine control devices, circuits, and systems for all types of industries are explained in detail in this comprehensive resource.
Order # 0-8273-7644-8

Electrical Transformers and Rotating Machines/Herman
This new book is an excellent resource for electrical students and professionals in the electrical trade.
Order # 0-7668-0579-4

Delmar's Standard Guide to Transformers/Herman
Delmar's Standard Guide to Transformers was developed from the best-seller *Standard Textbook of Electricity* with expanded transformer coverage not found in any other book.
Order # 0-8273-7209-4

DATA AND VOICE COMMUNICATION CABLING AND FIBER OPTICS

Complete Guide to Fiber Optic Cable System Installation/Pearson
This book offers comprehensive, unbiased, state-of-the-art information and procedures for installing fiber optic cable systems.
Order # 0-8273-7318-X

Fiber Optics Technician's Manual/Hayes
Here's an indispensible tool for all technicians and electricians who need to learn about optimal fiber optic design and installation as well as the latest troubleshooting tips and techniques.
Order # 0-8273-7426-7

A Guide for Telecommunications Cable Splicing/Highhouse
A "how-to" guide for splicing all types of telecommunications cables.
Order # 0-8273-8066-6

Premises Cabling/Sterling
This reference is ideal for electricians, electrical contractors, and inspectors needing specific information on the principles of structured wiring systems.
Order # 0-8273-7244-2

ELECTRICAL THEORY

Delmar's Standard Textbook of Electricity, 2E/Herman
This exciting full-color book is the most comprehensive book on DC/AC circuits and machines for those learning the electrical trades.
Order # 0-8273-8550-1

Industrial Electricity, 6E/Nadon, Gelmine, and Brumbach
This revised, illustrated book offers broad coverage of the basics of electrical theory and industrial applications. It is perfect for those who wish to be industrial maintenance technicians.
Order # 0-7668-0101-2

EXAM PREPARATION

Journeyman Electrician's Exam Preparation, 2E/Holt
This comprehensive exam prep guide includes all of the topics on journeyman electrician competency exams.
Order # 0-7668-0375-9

Master Electrician's Exam Preparation, 2E/Holt
This comprehensive exam prep guide includes all of the topics on master electrician's competency exams.
Order # 0-7668-0376-7

REFERENCE

ELECTRICAL REFERENCE SERIES
This series of technical reference books is written by experts and designed to provide the electrician, electrical contractor, industrial maintenance technician, and other electrical workers with a source of reference information about virtually all of the electrical topics that they encounter.

Electrician's Technical Reference—Motor Controls/Carpenter
Electrician's Technical Reference—Motor Controls is a source of comprehensive information on understanding the controls that start, stop, and regulate the speed of motors.
Order # 0-8273-8514-5

Electrician's Technical Reference—Motors/Carpenter
Electrician's Technical Reference—Motors builds an understanding of the operation, theory, and applications of motors.
Order # 0-8273-8513-7

Electrician's Technical Reference—Theory and Calculations/Herman
Electrician's Technical Reference—Theory and Calculations provides detailed examples of problem-solving for different kinds of DC and AC circuits.
Order # 0-8273-7885-8

Electrician's Technical Reference—Transformers/Herman
Electrician's Technical Reference—Transformers focuses on the theoretical and practical aspects of single-phase and 3-phase transformers and transformer connections.
Order # 0-8273-8496-3

Electrician's Technical Reference—Hazardous Locations/Loyd
Electrician's Technical Reference—Hazardous Locations cover electrical wiring methods and basic electrical design considerations for hazardous locations.
Order # 0-8273-8380-0

Electrician's Technical Reference—Wiring Methods/Loyd
Electrician's Technical Reference—Wiring Methods covers electrical wiring methods and basic electrical design considerations for all locations, and shows how to provide efficient, safe, and economical applications of various types of available wiring methods.
Order # 0-8273-8379-7

Electrician's Technical Reference—Industrial Electronics/Herman
Electrician's Technical Reference—Industrial Electronics covers components most used in heavy industry, such as silicon control rectifiers, triacs, and more. It also includes examples of common rectifiers and phase-shifting circuits.
Order # 0-7668-0347-3

RELATED TITLES

Common Sense Conduit Bending and Cable Tray Techniques/Simpson
Now geared especially for students, this manual remains the only complete treatment of the topic in the electrical field.
Order # 0-8273-7110-1

Practical Problems in Mathematics for Electricians, 5E/Herman
This book details the mathematics principles needed by electricians.
Order # 0-8273-6708-2

Electrical Estimating/Holt
This book provides a comprehensive look at how to estimate electrical wiring for residential and commercial buildings with extensive discussion of manual versus computer-assisted estimating.
Order # 0-8273-8100-X

Electrical Studies for Trades/Herman
Based on *Delmar's Standard Textbook of Electricity,* this new book provides non-electrical trades students with the basic information they need to understand electrical systems.
Order # 0-8273-7845-9

Preface

INTRODUCTION

Passing the Journeyman or Master Electrician's exam is the dream of every electrician; unfortunately, many electricians don't pass it the first time. The primary reasons that people fail their exam are because they are not prepared on the technical material and/or on how to take and pass an exam.

Typically, an electrical exam contains 25 percent Electrical Theory/Basic Calculations, 40 percent *National Electrical Code®,* and 35 percent *National Electrical Code®* Calculations. This book contains hundreds of explanations with illustrations, examples, and 2,400 practice questions covering these subjects.

The writing style of this book is informal and relaxed, and the book contains clear graphics and examples that apply to the electrical exam.

To get the most out of this book, you should answer the 200 questions at the end of each unit. The 2,400 questions contained in this book are typical questions from electrician exams across the country. After you have read each unit, take about ten hours to complete the two hundred unit practice questions. If you have difficulty with a question, skip it and go back to it later. You will find that the answer key contains detailed explanations for each question.

HOW TO USE THIS BOOK

Each unit of this book contains objectives, explanations with graphics, examples, steps for calculations, formulas and practice calculations, and 1999 *NEC®* questions. As you read this book, review the author's comments, graphics, and examples with your 1999 Code book.

Note: This book contains many cross-references to other related Code rules. Please take the time to review the cross-references.

As you progress through this book, you will find some formulas, rules, or comments that you don't understand. Don't get frustrated. Highlight the section in the book that you are having a problem with. Discuss it with your boss, inspector, co-worker, etc.; maybe they'll have some additional feedback. If necessary, just skip the difficult points. Once you have completed the book, review those highlighted sections again and see if you understand.

Note: Some words are italicized to bring them to your attention. Be sure that you understand the terms before you continue with each unit.

Note: An answer key is located at the end of the book.

DIFFERENT INTERPRETATIONS

Electricians, contractors, some inspectors, and others love arguing Code interpretations and discussing Code requirements. As a matter of fact, discussing the *NEC®* and its application is a great way to increase your knowledge of the Code and its intended use. The best way to discuss Code requirements with others is by referring to a specific Section in the Code, rather than by talking in vague generalities.

I have taken great care in researching the Code rules in this book, but I'm not perfect. If you feel that I have made an error, please let me know by contacting Delmar Publishers by mail addressed to Electrical Editor, Delmar Publishers, 3 Columbia Circle, Box 15015, Albany, NY 12212. You can also send e-mail to info@delmar.com.

THE *NATIONAL ELECTRICAL CODE*®

The *NEC*® is intended to be used by experienced persons having an understanding of electrical terms, theory, and trade practices. Such persons include electrical contractors, electrical inspectors, electrical engineers, and qualified electricians. The Code was not written to serve as an instructive or teaching manual [*Section 90-1(c)*] for untrained persons.

Learning to use the *NEC*® is like learning to play the game of chess. You first must learn the terms used to identify the game pieces, the concepts of how each piece moves, and the layout of how the pieces are placed on the board. Once you have this basic understanding of the game you're ready to start playing it, but all you can do right now is make crude moves because you really don't understand what you're doing. To play the game well, you'll need to study the rules, understand the subtle and complicated strategies, and then practice, practice, practice.

The same with the Code. Learning the terms, concepts, and layout of the *NEC*® gives you just enough knowledge to be dangerous. Perhaps most difficult are the subtle meanings within the Code rules.

There are thousands of different applications of electrical installations and there is not a specific Code rule for every application. To properly apply the *NEC*®, you must understand the safety-related issue of the rule and then apply common sense.

Note: The following pages should help you get started learning how to use the *National Electrical Code*®. Once you have completed tabbing your *Code* book, you will be ready to answer the *NEC*® questions at the end of each unit.

NEC® TERMS AND CONCEPTS

The *NEC*® uses many technical terms and expressions. It's crucial that you understand the meanings of basic words like ground, grounded, grounding, and neutral. If you don't understand basic terms used in the Code, you won't understand the rule itself.

It is not only the technical words that require close attention in the *NEC*®; even the simplest words can make a big difference. The word "or" can imply alternate choices for equipment, wiring methods, and other requirements. Sometimes "or" can mean any item in a group. The word "and" can be an additional requirement, or any item in a group.

Note: Electricians, engineers, and other trade-related professionals have created their own terms and phrases (slang or jargon). One of the problems with the use of slang terms is that the words mean different things to different people.

Understanding the safety-related concepts behind *NEC*® rules means understanding how and why things work the way they do (electrical theory). How does a bird sit on an energized power line without getting fried? Why, when we install a lot of wires close together, do we reduce the amount of current that each conductor can carry? Why can't a single current-carrying conductor be installed within a metal raceway? Why does the *NEC*® permit a 40 ampere circuit breaker to protect motor circuit conductors that are only rated 20 ampere? Why are bonding jumpers sometimes required for metal raceways containing 480Y/277-volt circuits, but not for 120/240-volt circuits?

If you understand why or how things work, you have a better chance of understanding the *NEC*® rules.

THE *NEC*® STYLE

Contrary to popular belief, the *NEC*® is a fairly well organized document, although parts of it are somewhat vague. Understanding the NEC structure and writing style is extremely important to understand and use the *Code* book effectively.

The *National Electrical Code*® is organized into 12 components.

1. Chapters (major categories)

2. Articles (individual subjects)

3. Parts (divisions of an Article)

4. Sections, Lists, and Tables (*Code* rules)

5. Exceptions (*Code* rules)

6. Fine Print Notes (explanatory material, not mandatory *Code* language)

7. Definitions (*Code* rules)

8. Superscript Letter ^{X}

9. Marginal Notations, *Code* changes (|) and deletions (•)

10. Table of Contents

11. Index

12. Appendices

1. Chapters. There are nine chapters and each chapter contains Articles. The nine chapters fall into four groupings:

• General Rules: Chapters 1 through 4

• Specific Rules (Hazardous locations, signs, control wiring): Chapters 5 through 7

• Communication Systems (Telephone, Radio/Television, and Cable TV Systems): Chapter 8

• Tables: Chapter 9

2. Articles. The *NEC®* contains approximately 125 Articles. An Article covers a specific subject, as in the following examples:

Article 110 – General Requirements

Article 250 – Grounding

Article 300 – Wiring Methods

Article 430 – Motors

Article 500 – Hazardous (classified) Locations

Article 680 – Swimming Pools

Article 725 – Control Wiring

Article 800 – Communication Wiring

3. Parts. When an Article is sufficiently large, the Article is subdivided into Parts. For example, *Article 250* contains nine parts, including:

• *Part A.* General

• *Part B.* Circuit and System Grounding

• *Part C.* Grounding Electrode System

CAUTION: The "Parts" of a *Code* Article are not included in the Section numbers. Because of this, we have a tendency to forget what "Part" the *Code* rule is relating to. For example, *Table 110-34* gives the dimensions of working space clearances in front of electrical equipment. If we are not careful, we might think that this table applies to all electrical installations. But *Section 110-34* is located in *Part C Over-600 volt Systems of Article 110!* The working clearance rule for under-600 volt systems is located in *Part A* of *Article 110,* in *Table 110-26.*

---

4. Sections, Lists, and Tables.

Sections. Each actual *Code* rule is called a Section and is identified with numbers, such as Section 225-26. A *Code* Section may be broken down into subsections by letters in parentheses, and numbers in parentheses may further break down each subsection. For example, the rule that requires all receptacles in a bathroom to be GFCI protected is contained in Section 210-8(a)(1).

Note: Many in the electrical industry incorrectly use the term "Article" when referring to a *Code* Section.

Lists. The 1999 *NEC®* has changed the layout of some Sections that contain lists of items. If a list is part of a numeric subsection, such as Section 210-52(a)(2), then the items are listed as a., b., c., etc. However, if a list is part of a Section, then the items are identified as (1), (2), (3), (4), etc.

Tables. Many *Code* requirements are contained within Tables which are a systematic list of *Code* rules in an orderly arrangement. For example, Table 300-15 lists the burial depths of cables and raceways.

5. Exceptions. Exceptions are italicized and provide an alternative to a specific rule. There are two types of exceptions: mandatory and permissive. When a rule has several exceptions, those exceptions with mandatory requirements are listed before those written in permissive language.

(a) Mandatory Exception. A mandatory exception uses the words "shall" or "shall not." The word "shall" in an exception means that if you are using the exception, you are required to do it in a particular way. The term "shall not" means that you cannot do something.

(b) Permissive Exception. A permissive exception uses such words as "shall be permitted," which means that it is accepted to do it in this way.

6. Fine Print Note, (FPN) [90-5]. A Fine Print Note contains explanatory material intended to clarify a rule or give assistance, but it is not *Code* requirement. FPNs often use the term "may," but never "shall."

7. Definitions. Definitions are listed in Article 100 and throughout the *NEC®.* In general, the definitions listed in Article 100 apply to more than one *Code* Article, such as "branch circuit," which is used in many Articles.

Definitions at the beginning of a specific Article apply only to that Article. For example, the definition of a "Swimming Pool" is contained in Section 680-4 because this term applies only to the requirements of Article 680 – Swimming Pools.

Definitions located in a Part of an Article apply only to that Part of the Article. For example, the definition of "motor control circuit" applies only to Article 430, Part F.

Definitions located in a *Code* Section apply only to that *Code* Section. For example, the definition of "Festoon Lighting" located in Section 225-6(b) applies only to the requirements contained in Section 225-6.

8. Superscript Letter ^X. The superscript letter ^X indicates that the material was extracted from other technical standards published by the NFPA. Appendix A, at the back of the *Code* book, identifies the NFPA documents and the Section(s) from which the material was extracted.

9. Changes and Deletions. Changes and deletions to the *NEC®* are identified in the margins of the 1999 *NEC®* in the following manner: A vertical line (|) marks changes and a bullet (•) identifies deletion of a *Code* rule. Many rules in the 1999 *NEC®* were relocated. The place from which the *Code* rule was removed has a bullet (•) in the margin, and the place where the rule was inserted has a vertical line (|) in the margin.

10. Table of Contents. The Table of Contents located in the front of the *Code* book displays the layout of the Chapters, Articles, and Parts as well as their location in the *Code* book.

11. Index. We all know the purpose of an index, but it's not that easy to use. You really need to know the correct term. Often it's much easier to use the Table of Contents.

12. Appendices. There are four appendices in the 1999 *NEC*®:

• Appendix A – Extract Information

• Appendix B – Ampacity Engineering Supervision

• Appendix C – Conduit and Tubing Fill Tables

• Appendix D – Electrical Calculation Examples

HOW TO FIND THINGS IN THE CODE

How fast you find things in the *NEC*® depends on your experience. Experienced *Code* users often use the Table of Contents instead of the index.

For example, what *Code* rule indicates the maximum number of disconnects permitted for a service?

Answer. You need to know that Article 230 is for Services and that it contains a Part F, Disconnection Means. If you know this, using the Table of Contents, you'll see that the answer is contained at page 66.

People frequently use the Index which lists subjects in alphabetical order. It's usually the best place to start for specific information. Unlike most books, the *NEC*® Index does not list page numbers; it lists Sections, Tables, Articles, Parts, and Appendices by their Section number.

Note: Many people say the *Code* takes them in circles, and sometimes it does. However, this complaint is often heard from inexperienced persons who don't understand electrical theory, electrical terms, and electrical practices.

CUSTOMIZING YOUR CODE BOOK

One way for you to get comfortable with your *Code* book is to customize it to meet your needs. This you can do by highlighting, underlining *Code* rules, and using convenient tabs.

Highlighting and Underlining. As you read through this book, highlight in the *NEC*® book those *Code* rules that are important to you such as yellow for general interest, and orange for rules you want to find quickly. As you use the Index and the Table of Contents, highlight terms in those areas as well. Underline or circle key words and phrases in

the NEC with a red pen (not a lead pencil) and use a 6-inch ruler to keep lines straight and neat.

Because of the new format of the 1999 *NEC*® ($8^1/_2 \times 11$), I highly recommend that you highlight in green the parts of at least the following Articles.

Article 230 – Services

Article 250 – Grounding

Article 410 – Fixtures

Article 430 – Motors

Author's Comment: Trust me; you'll be glad you did.

Tabbing the *NEC*®. Tabbing the *NEC*® permits you to quickly access *Code* Article, Section, or Tables. However, too many tabs will defeat the purpose.

Experience has shown that the best way to tab the *Code* book is to start by placing the last tab first and the first tab last (start at the back of the book and work your way toward the front). Install the first tab, then place each following tab so that they do not overlap the information of the previous tab.

The following is a list of Articles and Sections I most commonly refer to. Place a tab only on the Sections or Articles that are important to you.

Tab#	Description	Page
1	Index	621
2	Examples: Appendix D	609
3	Raceway Fill Tables: Appendix C	585
4	Conductor Area: Table 5 of Chapter 9	564
5	Raceway Area: Table 4 of Chapter 9	562
6	Satellites and Antennas: Article 820	545
7	Communication: Article 800	533
8	Fiber Optic Cables: Article 770	527
9	Fire Alarms: Article 760	519
10	Control Circuits: Article 725	510
11	Emergency Circuits: Article 700	501
12	Pools, Spas, and Fountains: Article 680	476
13	Electric Signs: Article 600	433
14	Marinas: Article 555	431
15	Mobile/Manufactured: Article 550	398
16	Carnivals, Circuses, and Fairs: Article 525	387
17	Health-Care Facilities: Article 517	359
18	Gasoline Dispensers: Article 514	346
19	Hazardous Locations: Article 500	309
20	Transformers: Article 450	290
21	Generators: Article 445	298
22	Air-Conditioning: Article 440	283
23	Motors: Article 430	256
24	Electric Space Heating: Article 424	241
25	Appliances: Article 422	236
26	Lighting Fixtures: Article 410	225

ABOUT THE AUTHOR

I have worked my way up through the electrical trade from an apprentice electrician to a master electrician and electrical inspector. I did not complete high school due to circumstances and dropped out after completing 11th grade. Realizing that success depends on one's education, I immediately attained my GED, and ten years later I attended the University of Miami's Graduate School for a Master's in Business Administration (MBA).

I am nationally recognized as one of America's most knowledgeable electrical trainers and have touched the lives of thousands of electricians, inspectors, contractors, and engineers.

I reside in Central Florida, am the father of seven children, and have many outside interests and activities. I am a National Barefoot Waterskiing Champion, have set five barefoot waterski records and am currently ranked No. 2. I enjoy white water rafting, racquetball, playing my guitar, and spending time with my family. My commitment to God has helped me develop a lifestyle that balances family, career, and self.

ACKNOWLEDGMENTS

I would like to say thank you to all the people in my life who believed in me, and even to those who didn't. There are many people who played a role in the development and production of this book. I will start with Mike Culbreath (Master Electrician), who helped me transform my words and visions into lifelike graphics. I could not have produced such a fine book without his help.

Next, Paul Wright of Digital Design Group, for the electronic production and typesetting. A very special thanks goes to Brooke Stauffer from NECA, Ray Cotter from North Tech Education Center, Elzy R. Williams, P.E., Adjunct Professor at John Tyler Community College, and Craig H. Matthews, P.E., from Austin Brockenbrough and Associates, L.L.P, for their incredible job of proofing this textbook for technical correctness.

To my beautiful wife, Linda, and my seven children, Belynda, Melissa, Autumn, Steven, Michael, Meghan, and Brittney, thank you for loving me.

Also thanks to all those who helped me in the electrical industry, *Electrical Construction* and *Maintenance* magazines for my first "big break," and to Joe McPartland, my "mentor," who was there to help and encourage me. Joe, I'll never forget to help others as you helped me.

In addition, I would like to thank Joe Salimando, the former publisher of *Electrical Contractor,* and Dan Walters of the National Electrical Contractors Association (NECA) for my second "big break," and for putting up with all of my crazy ideas.

I would also like to thank James Stallcup, Dick Lloyd, Mark Ode, D.J. Clements, Joe Ross, John Calloggero, Tony Selvestri, and Marvin Weiss for being special people in my life.

The final personal thank you goes to Sarina, my long-time friend and office manager. Thank you for covering the office for me while I spend so much time writing books, doing seminars, and producing videos and software. Your love and concern for me has helped me through many difficult times.

Delmar Publishers and the author would also like to thank those individuals who reviewed the manuscript and offered invaluable suggestions and feedback. Their assistance is greatly appreciated.

Robert Blakely
Mississippi Gulf Coast Community College

Ray Cotter, Electrical Instructor
North Tech Education Center, Palm Beach, Florida

John P. Cox
Brevard Community College

Tim DeDonder
Kaw Area Vocational-Technical School

William Elarton
Los Angeles Trade and Technical College

David Gehlauf
Tri-County Vocational Technical

Larry Killebrew
Mid-America Vocational-Technical School

Craig H. Matthews, P.E.
Austin Brockenbrough and Associates, L.L.

John Mills, Master Electrician, Instructor
Dade County, Florida

John Penley
Albert Lea Technical College

Larry C. Phillips
Minneapolis Technical College

Charles Plimpton
New Hampshire Technical College

Rodney Stanley
Morehead State University

Brooke Stauffer
National Electrical Contractors Association (NECA)

Kurt A. Stout, Electrical Inspector
Plantation, Florida

Elzy R. Williams, P.E.
Adjunct Professor, John Tyler Community College

GETTING STARTED

THE EMOTIONAL ASPECT OF LEARNING

To learn effectively, you must develop an attitude that learning is a process that will help you grow both personally and professionally. The learning process has an emotional as well as an intellectual component that we must recognize. To understand what affects our learning, consider the following:

Positive Image. Many feel disturbed by the expectation of being treated like children, and we often feel threatened by the learning experience.

Uniqueness. Each of us will understand the subject matter from different perspectives, and we all have some unique learning problems and needs.

Resistance to Change. People tend to resist change and information that appear to threaten their comfort level of knowledge. However, we often support new ideas that support our existing beliefs.

Dependence and Independence. The dependent person is afraid of disapproval, often will not participate in class discussion, and will tend to wrestle alone. The independent person spends too much time asserting differences and too little time trying to understand others' views.

Fear. Most of us feel insecure and afraid of learning until we understand the process. We fear that our performance will not match the standard set by us or by others.

Egocentricity. Our ego tendency is to prove someone is wrong, feeling a victorious surge of pride. Learning together without a win/lose attitude can be an exhilarating experience.

Emotion. It is difficult to discard our cherished ideas in the face of contrary facts when overpowered by the logic of others.

HOW TO GET THE BEST GRADE ON YOUR EXAM

Studies have concluded that for students to get their best grades, they must learn to get the most from their natural abilities. It's not how long you study or how high your IQ is, it's what you do and how you study that counts the most. To get your best grade, you must make a decision to do your best and follow as many of the following techniques as possible.

Reality. These instructions are a basic guide to help you get the maximum grade. It is unreasonable to think that all of the instructions can be followed to the letter all of the time. Day-to-day events and unexpected situations must be taken into consideration.

Support. You need encouragement in your studies and support from your loved ones and employer. To properly prepare for your exam, you need to study ten to fifteen hours per week for about three to six months.

Communication with Your Family. Good communication with your family members is very important. Studying every night and on weekends may cause tension. Try to get their support, cooperation, and encouragement during this trying time. Let them know the benefits. Be sure to plan some special time with them during this preparation period; don't go overboard and leave them alone too long.

Stress. Stress can really take the wind out of you. It takes practice, but get into the habit of relaxing before you begin your studies. Stretch; do a few sit-ups and push-ups; take a twenty-minute walk or a few slow, deep breaths. Close your eyes for a couple of minutes; deliberately relax the muscle groups that are associated with tension, such as the shoulders, back, neck, and jaw.

Attitude. Maintaining a positive attitude is important. It helps keep you going and helps keep you from getting discouraged.

Training. Preparing for the exam is the same as training for any event. Get plenty of rest and avoid intoxicating drugs, including alcohol. Stretch or exercise each day for at least ten minutes. Eat light meals such as pasta, chicken, fish, vegetables, fruit, etc. Try to avoid heavy foods, such as red meats, butter, and other high-fat foods. They slow you down and make you tired and sleepy.

Eye Care. It is very important to have your eyes checked! Human beings were not designed to do constant seeing less than arm's length away. Our eyes were designed for survival and for spotting food and enemies at a distance. Your eyes will be under tremendous stress because of prolonged, near-vision reading, which can result in headaches, fatigue, nausea, squinting, or eyes that burn, ache, water, or tire easily.

Be sure to tell the eye doctor that you are studying to pass an exam (bring this book and the Code Book), and you expect to do a tremendous amount of reading and writing. Reading glasses can reduce eye discomfort.

Reducing Eye Strain. Be sure to look up occasionally from near tasks to focus on distant objects. Your work area should be three times brighter than the rest of the room. Don't read under a single lamp in a dark room. Try to eliminate glare. Mixing of fluorescent and incandescent lighting can be helpful.

Sit up straight with your chest up and shoulders back so that both eyes are an equal distance from what you are viewing.

Getting Organized. Our lives are so busy that simply making time for homework and exam preparation is almost impossible. You can't waste time looking for a pencil or missing paper. Keep everything you need together. Maintain folders, one for notes, one for exams and answer keys, and one for miscellaneous items.

It is very important that you have a private study area available at all times. Keep your materials there. The dining room table is not a good spot.

Time Management. Time management and planning is very important. There simply are not enough hours in the day to get everything done. Make a schedule that allows time for work, rest, study, meals, family, and recreation. Establish a schedule that is consistent from day to day.

Have a calendar and immediately plan your exam preparation schedule. Try to follow the same routine each week, and try not to become overtired. Learn to pace yourself to accomplish as much as you can without the need for cramming.

Learn How to Read. Review the book's contents and graphics. This will help you develop a sense of the material.

Clean up Your Act. Keep all of your papers neat, clean, and organized. Now is not the time to be sloppy. If you are not neat, now is an excellent time to begin.

Speak up in Class. If you are in a classroom setting, the most important part of the learning process is class participation. If you don't understand the instructor's point, ask for clarification. Don't try to get attention by asking questions you already know the answer to.

Study with a Friend. Studying with a friend can make learning more enjoyable. You can push and encourage each other. You are more likely to study if someone else is depending on you.

Students who study together perform above average because they try different approaches and explain their solutions to each other. Those who study alone spend most

of their time reading and rereading the text and trying the same approach time after time even though it is unsuccessful.

Study Anywhere/Anytime. To make the most of your limited time, always keep a copy of the book(s) with you. Any time you get a minute free, study. Continue to study any chance you get. You can study at the supply house while waiting for your material; you can study during your coffee break, or even while you are at the doctor's office. Become creative!

You need to find your best study time. For some it could be late at night when the house is quiet. For others, it's the first thing in the morning before things get going.

Set Priorities. Once you begin your study, stop all phone calls, TV shows, radio, snacks, and other interruptions. You can always take care of it later.

HOW TO TAKE THE EXAM

Being prepared for an exam means more than just knowing electrical concepts, the Code, and the calculations. Have you felt prepared for an exam, then choke when actually taking it? Many good and knowledgeable electricians couldn't pass their exam because they did not know how to take an exam.

Taking exams is a learned process that takes practice and involves strategies. The following suggestions are designed to help you learn these methods.

Relax. This is easier said than done, but it is one of the most important factors in passing your exam. Stress and tension cause us to choke or forget. Everyone has had experiences where they get tense and couldn't think straight. The first step is to become aware of the tension, and the second step is to make a deliberate effort to relax. Make sure you're comfortable; remove clothes if you are hot, or put on a jacket if you are cold.

There are many ways to relax, and you have to find a method that works for you. Two of the easiest methods that work very well for many people follow:

Breathing Technique: Take some slow, deep breaths every few minutes. Do not confuse this with hyperventilation, which is abnormally fast breathing.

Single-Muscle Relaxation: When we are tense or stressful, many of us do things like clench our jaw, squint our eyes, or tense our shoulders without even being aware of it. If you find a muscle group that does this, deliberately relax that one group. The rest of the muscles will automatically relax also. Try to repeat this every few minutes, and it will help you stay more relaxed during the exam.

Have the Proper Supplies. First of all, make sure you have everything you need several days before the exam. The night before the exam is not the time to be out buying pencils, calculators, and batteries. The night before the exam, you should have a checklist (prepared in advance) of everything you could possibly need.

The following is a sample checklist to get you started.
- Six sharpened #2H pencils or two mechanical pens with extra #2H leads. The type with the larger lead is better for filling in the answer key.
- Two calculators. Most examining boards require quiet, paperless calculators. Solar calculators are great, but there may not be enough light to operate them.
- Spare batteries. Two sets of extra batteries are necessary. It's very unlikely you'll need them, but it's better to be prepared.
- Extra glasses, if you use them.
- A watch for timing questions.
- All your reference materials, even the ones not on the list. Let the proctors tell you which ones are not permitted.
- A Thermos of something you like to drink. Coffee is excellent.
- Some fruit, nuts, candy, aspirin, analgesic, etc.
- Know where the exam is going to take place and how long it takes to get there. Arrive at least thirty minutes early.

Note: It is also a good idea to pack a lunch rather than go out. It can give you a little time to review the material for the afternoon portion of the exam, and it reduces the chance of coming back late.

Understand the Question. To answer a question correctly, you must first understand the question. One word in a question can totally change the meaning of it. Carefully read every word of every question. Underlining key words in the question will help you focus.

Skip the Difficult Questions. Contrary to popular belief, you do not have to answer one question before going on to the next one. The irony is that the question you get stuck on is one that you'll probably get wrong. This will result in not having enough time to answer the easy questions. You will get all stressed-out, and a chain reaction will start. More people fail their exams for this reason than for any other.

The following strategy should be used to avoid getting into this situation.
- **First Pass:** Answer the questions you know. Give yourself about thirty seconds for each question. If you can't find the answer in your reference book within the thirty seconds, go on to the next question. Chances are that you'll come across the answers while looking up another question. The total time for the first pass should be 25 percent of the exam time.
- **Second Pass:** This pass is done the same way as the first pass except that you allow a little more time for

each question, about sixty seconds. If you still can't find the answer, go on to the next one. Don't get stuck. Total time for the second pass should be about 30 percent of the exam time.
- **Third Pass:** See how much time is left and subtract thirty minutes. Spend the remaining time equally on each question. If you still haven't answered the question, it's time to make an educated guess. Never leave a question unanswered.
- **Fourth pass:** Use the last thirty minutes of the exam to transfer your answers from the exam booklet to the answer key. Read each question and verify that you selected the correct answer on the test book. Transfer the answers carefully to the answer key. With the remaining time, see if you can find the answer to those questions you guessed at.

Guessing. When time is running out and you still have questions remaining, GUESS! Never leave a question unanswered.

You can improve your chances of getting a question correct by the process of elimination. When one of the choices is "none of these," or "none of the above," it is usually not the correct answer. This improves your chances from one-out-of-four (25 percent), to one-out-of-three (33 percent). Guess "All of these" or "All of the Above," and don't select the high or low number.

How do you pick one of the remaining answers? Some people toss a coin, others will count up how many of the answers were As, Bs, Cs, and Ds and use the one with the most tallied as the basis for their guess.

CHECKING YOUR WORK

The first thing to check (and you should be watching out for this during the whole exam) is to make sure you mark the answer in the correct spot. People have failed the exam by $1/2$ of a point. When they reviewed their exam, they found they correctly answered several questions on the test booklet but marked the wrong spot on the exam answer sheet. They knew the answer was "(b) False" but marked "(d)" in error.

Another thing to be very careful of is marking the answer for, let's say, question 7, in the spot reserved for question 8.

CHANGING ANSWERS

When re-reading the question and checking the answers during the fourth pass, resist the urge to change an answer. In most cases, your first choice is best. If you aren't sure, stick with the first choice. Only change answers if you are sure you made a mistake. Multiple choice exams are graded electronically, so be sure to thoroughly erase any answer that you changed. Also erase any stray pencil marks from the answer sheet.

ROUNDING OFF

You should always round your answers to the same number of places as the exam's answers. Numbers below "5" are rounded down, while numbers "5" and above are rounded up.

Example: If an exam has multiple choice of:

(a) 2.2   (b) 2.1

(c) 2.3   (d) none of these

And your calculation comes out to 2.16, do not choose the answer "(d) none of these." The correct answer is (a) 2.2, because the answers in this case are rounded off to the tenth.

Example: It could be rounded to tens, such as:

(a) 50   (b) 60

(c) 70   (d) none of these.

For this group, an answer such as 67 would be (c) 70, while an answer of 63 would be (b) 60. The general rule is to check the question's choice of answers, then round off your answer to match it.

SUMMARY

- Make sure everything is ready and packed the night before the exam.
- Don't try to cram the night before the exam if you don't know the material by then.
- Have a good breakfast. Get the Thermos and energy snacks ready.
- Take all of your reference books. Let the proctors tell you what you can't use.
- Know where the exam is to be held, and be there early.
- Bring ID and your confirmation papers from the license board, if there are any.

- Review your *NEC*® while you wait for your exam to begin.
- Try to stay relaxed.
- Determine the time per question for each pass, and don't forget to save thirty minutes for transferring your answers to the answer key.
- Remember, in the first pass, answer only the easy questions. In the second pass, spend a little more time per question, but don't get stuck. In the third pass, use the remainder of the time minus thirty minutes. In the fourth pass, check your work, and transfer the answers to the answer key.

THINGS TO BE CAREFUL OF

- Don't get stuck on any one question.
- Read each question carefully.
- Be sure you mark the answer in the correct spot on the answer sheet.
- Don't get flustered or extremely tense.

EXAMINATION COPIES

To request examination copies of this book, call or write to:

Delmar Publishers
3 Columbia Circle
P.O. Box 15015
Albany, NY 12212-5015
Phone: 1-800-347-7707 • 1-518-464-3500 •
Fax: 1-518-464-0301

CHAPTER 1
Electrical Theory and Code Questions

Scope of Chapter 1

1

Unit 1

Electrician's Math and Basic Electrical Formulas

OBJECTIVES

After reading this unit, the student should be able to briefly explain the following concepts:

Part A – Electrician's Math	Reciprocals	Electrical circuit values
Fractions	Square root	Ohm's law
Kilo	Squaring	PIE circle formula
Knowing your answer	Transposing formulas	Power changes with
Multiplier	**Part B – Basic Electrical Formulas**	the square of the
Parentheses	Conductance and resistance	voltage
Percent increase	Electric meters	Power source
Percentage R	Electrical circuit	Power wheel

After reading this unit, the student should be able to briefly explain the following terms:

Part A – Electrician's Math	Ampere	Megohmeter
Fractions	Armature	Ohmmeter
Kilo	Clamp-on ammeters	Ohms
Multiplier	Conductance	P – Power
Parentheses	Conductors	Perpendicular
Percentage	Current	Polarity
Ratio	Direct current	Polarized
Reciprocals	Directly proportional	Power
Rounding off	E – Electromotive Force	Power source
Square root	Electric meters	Resistance
Squaring a number	Electromagnetic	Shunt bar
Transposing	Electromagnetic field	Shunt meter
Part B – Basic Electrical Formulas	Electron pressure	Solenoid
A – Ampere	Helically wound	V – Voltage
Alternating current	Intensity	Voltmeter
Ammeter	Inversely proportional	W – Watt

PART A – ELECTRICIAN'S MATH

1–1 FRACTIONS

Fractions represent parts of numbers. To change a fraction to a decimal form, divide the numerator (top number of the fraction) by the denominator (bottom number of the fraction).

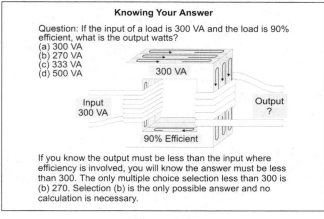

Figure 1–1
Knowing Your Answer

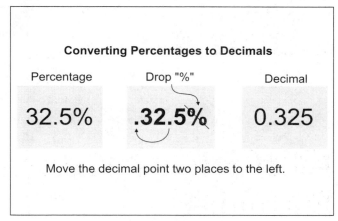

Figure 1–2
Converting Percentages to Decimals

❏ **Fractions to Decimal**

Convert the following fractions to a decimal.

$^1/_6$ = one divided by six = 0.167

$^5/_4$ = five divided by four = 1.25

$^7/_2$ = seven divided by two = 3.5

1–2 KILO

The letter k is the abbreviation of *kilo,* which represents 1,000 in the metric system of measurement.

❏ **Kilo**

What is the wattage for an 8kW-rated range?

(a) 8 watts   (b) 8,000 watts   (c) 4,000 watts   (d) none of these

• Answer: (b) 8,000 watts

Wattage = kW $\times$ 1,000. In this case 8 kW $\times$ 1,000 = 8,000 watts.

❏ **Kilo**

What is the kVA rating of a 300 VA load?

(a) 300 kVA   (b) 3,000 kVA   (c) 30 kVA   (d) 0.3 kVA

• Answer: (d) 0.3 kVA, kVA = VA divided by 1,000.

In this case 300 VA/1,000 = 0.3 kVA

Note: The use of k is not limited to kW or kVA; it is also used to express wire size, such as 250 kcmil.

1–3 KNOWING YOUR ANSWER

When working with mathematical calculations, you should know if the answer is greater than or less than the values given.

❏ **Knowing Your Answer**

If the input of a load is 300 watts and the load is 90 percent efficient, what is the output watts?

Note. Because efficiency is always less than 100 percent, the output is always less than the input (Figure 1–1).

(a) 300 VA   (b) 270 VA   (c) 333 VA   (d) 500 VA

• Answer: (b) 270 VA

Since the use of efficiency in the question implied that the output had to be less than the input, the answer must be less than 300 watts. The only choice that is less than 300 watts is (b) 270 VA.

1–4 MULTIPLIER

Often a number is required to be increased or decreased by a percentage. When a percentage or fraction is used as a multiplier, follow these steps:

Step 1: ➥ Convert the multiplier to a decimal form, then

Step 2: ➥ Multiply the number by the decimal value from Step 1.

❏ Increase by 125 Percent

An overcurrent protection device (breaker or fuse) must be sized no less than 125 percent of the continuous load. If the load is 80 ampere, the overcurrent protection device would have to be sized no less than _____ ampere.

   (a) 80 ampere   (b) 100 ampere   (c) 125 ampere   (d) none of these
     • Answer: (b) 100 ampere

Step 1: ➥ Convert 125% to a decimal: 1.25

Step 2: ➥ Multiply the load rating by the multiplier: 80 ampere × 1.25 = 100 ampere

❏ Limit to 80 Percent

The maximum continuous load on an overcurrent protection device is limited to 80 percent of the device rating. If the device is rated 50 ampere, what is the maximum continuous load?

   (a) 80 ampere   (b) 125 ampere   (c) 50 ampere   (d) 40 ampere
     • Answer: (d) 40 ampere

Step 1: ➥ Convert 80% to a decimal: 0.8

Step 2: ➥ Multiply the overcurrent protection device rating by the multiplier: 50 ampere × 0.8 = 40 ampere

1–5 PARENTHESES

Whenever numbers are in *(parentheses)*, we must complete the mathematical function within the parentheses before proceeding with the rest of the problem.

❏ Parenthesis

What is the voltage drop of two No. 14 conductors carrying 16 ampere for a distance of 100 feet? Use the following example for the answer:

$$VD = \frac{2 \times K \times I \times D}{CM}, = \frac{(2 \text{ wires} \times 12.9 \text{ ohm} \times 16 \text{ ampere} \times 100 \text{ feet})}{4,110 \text{ circular mils}}$$

   (a) 3 volts   (b) 3.6 volts   (c) 10.04 volts   (d) none of these
     • Answer: (c) 10.04 volts

Step 1: ➥ Calculate the value of the parentheses first: (2 wires × 12. 9 ohm × 16 ampere × 100 feet) = 41,280.

Step 2: ➥ Divide the top number by the bottom number: 41,280/4,110 = 10.4 volts dropped.

1–6 PERCENTAGES

A *percentage (%)* is a ratio of two numbers. When changing a percent to a decimal or whole number, simply move the decimal point two places to the left (Figure 1–2).

❏ Percentage

   32.5% = 0.325   100% = 1.00   125% = 1.25   300% = 3.00

1–7 PERCENT INCREASE

Increasing a number by a specific percentage is accomplished by:

Step 1: ➥ Converting the percentage to a decimal.

Step 2: ➥ Determining the multiplier: Add one to the decimal value from Step 1.

Step 3: ➥ Multiply the number by the multiplier from Step 2.

❏ **Percent Increase**

Increase the whole number 45 by 35%.

(a) 61   (b) 74   (c) 83   (d) 104

• Answer: (a) 61

Step 1: ➥ Convert 35% to 0.35.

Step 2: ➥ Multiplier: Add one to the decimal value from Step 1: 1 + 0.35 = 1.35.

Step 3: ➥ Multiply the number 45 by the multiplier: 45 × 1.35 = 60.75.

Step 4: ➥ Round 60.75 up to 61.

1–8 PERCENTAGE RECIPROCALS

A reciprocal is a whole number converted into a fraction with the number *one* as the numerator (top number). This fraction is then converted to a decimal.

Step 1: ➥ Convert the number to a decimal.

Step 2: ➥ Divide the number into one.

❏ **Reciprocal**

What is the reciprocal of 80 percent?

(a) 0.80 percent   (b) 100 percent   (c) 125 percent   (d) none of these

• Answer: (c) 1.25 or 125 percent

Step 1: ➥ Convert 80% to a decimal: 80% = 0.8.

Step 2: ➥ Divide 0.80 into one: $^1/_{0.80}$ = 1.25, which is the same as 125%.

❏ **Reciprocal**

A continuous load requires an overcurrent protection device sized no smaller than 125 percent of the load [*Section 210–20(a)*]. What is the maximum continuous load permitted on a 100-ampere overcurrent protection device?

(a) 100 ampere   (b) 125 ampere   (c) 80 ampere   (d) none of these

• Answer: (c) 80 ampere

Step 1: ➥ Convert 125% to a decimal: 125% = 1.25

Step 2: ➥ Divide 1.25 into one: $^1/_{1.25}$ = 0.8 or 80%.

> If the overcurrent device is sized no less than 125 percent of the load, the load is limited to 80 percent of the overcurrent protection device rating (reciprocal). Therefore, the maximum load is limited to:
> 100 ampere × 0.8 = ampere.

1–9 ROUNDING

Numbers below 5 are rounded down, while numbers 5 and above are rounded up.

Note: Rounding to three significant figures should be sufficient for most calculations, such as:

0.1245 = 0.125
1.674   = 1.67
21.94   = 21.9
367.28 = 367

Rounding for Exams

You should always round your answer to the same magnitude as the answers. Do not choose "none of these" in an exam until you have checked all the answers. If after rounding your answer to the exam format there is no answer, then you should choose "none of these."

❑ **Rounding**

The sum of 12, 17, 28, and 40 is equal to _____?

(a) 80   (b) 90   (c) 100   (d) none of these

   • Answer: (c) 100

The answer is actually 97, but there is no 97 as a choice. Do not choose "none of these" in an exam until you have checked how the choices are rounded off. The choices in this case are all rounded off to the nearest ten.

1–10 SQUARING

Squaring a number is multiplying a number by itself, such as: $23^2 = 23 \times 23 = 529$.

❑ **Squaring**

What is the power consumed, in watts, of a No. 12 conductor that is 200 feet long and has a resistance of 0.4 ohm? The current flowing in the circuit is 16 ampere. Formula: $I^2 \times R$

(a) 50 watts   (b) 150 watts   (c) 100 watts   (d) 200 watts

   • Answer: (c) 100 watts

Step 1: ➛ $P = I^2 \times R$, I = 16 ampere, R = 0.4 ohm.

Step 2: ➛ P = 16 ampere $^2 \times$ 0.4 ohm.

Step 3: ➛ P = 102.4 watts; answers are rounded to 50s.

❑ **Squaring**

What is the area, in square inches, of a 1-inch trade size raceway whose actual internal diameter is 1.049 inches? Formula: Area = $\pi \times r^2$, π = 3.14, r = radius ($^1/_2$ the diameter).

(a) 1 square inch   (b) 0.86 square inch   (c) 0.34 square inch   (d) 0.5 square inch

   • Answer: (b) 0.86 square inch

Step 1: ➛ Raceway area = $\pi \times r^2$, = 3.14 $\times$ ($^1/_2 \times$ 1.049) ^{2}, = 3.14 $\times$ 0.5245^2.

Step 2: ➛ Raceway area = 3.14 $\times$ (0.5245 $\times$ 0.5245), = 3.14 $\times$ 0.2751 = 0.86 square inch.

1–11 SQUARE ROOT

The *square root* of a number is the opposite of squaring a number. For all practical purposes you must use a calculator with a square root key to determine the square root of a number. For an electrician's exam, the only square root number you need to know is the square root of $\sqrt{3}$ is 1.732. To multiply, divide, add, or subtract a number by a square root value, determine the square root value first, then perform the math function.

Step 1: ➛ Enter the number in the calculator.

Step 2: ➛ Press the $\sqrt{\ }$ key of the calculator.

❑ **Square Root Sample**

What is the $\sqrt{3}$?

(a) 1.55   (b) 1.73   (c) 1.96   (d) none of these

   • Answer: (b) 1.73

Step 1: ➛ Enter the number 3 in a calculator.

Step 2: ➛ Press the $\sqrt{\ }$ key = 1.732.

❑ **Square Root**

36,000 watts/(208 volts $\times \sqrt{3}$) is equal to _____ ampere?

(a) 120   (b) 208   (c) 360   (d) 100

   • Answer: (d) 100

Step 1: ➛ Determine the $\sqrt{3}$ = 1.732.

Step 2: ➛ Multiply 208 volts $\times$ 1.732 = 360 volts.

Step 3: ➛ 36,000 watts/360 volts = 100 ampere.

❏ **Square Root**

The phase voltage is equal to $\dfrac{208 \text{ volts}}{\sqrt{3}}$ _____?

(a) 120 volts   (b) 208 volts   (c) 360 volts   (d) none of these

• Answer: (a) 120 volts

Step 1: ➥ Determine the $\sqrt{3} = 1.732$.

Step 2: ➥ Divide 208 volts by 1.732 = 120 volts.

1–12 TRANSPOSING FORMULAS

Transposing is an algebraic function used to rearrange formulas. Example: the formula $I = {}^P/_E$ can be transposed to $E = P/I$ or $P = E \times I$ (Figure 1–3):

❏ **Transpose**

Transpose the formula $CM = \dfrac{(2 \times K \times I \times D)}{VD}$

to find the voltage drop of the circuit (Figure 1–4).
(a) VD = CM
(b) VD = $(2 \times K \times I \times D)$
(c) VD = $(2 \times K \times I\ D)/CM$
(d) none of these

• Answer: (c)

$$VD = \dfrac{(2 \times K \times I \times D)}{CM}$$

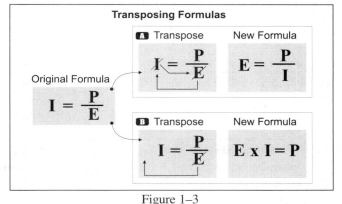

Figure 1–3
Transposing Formulas

Note: Most people have no idea of how to transpose formulas. I only add this to the book to refresh the memory of those who do understand. If you don't understand, it's okay; don't worry about it.

PART B – BASIC ELECTRICAL FORMULAS

1–13 ELECTRICAL CIRCUIT

An electric circuit consists of power source, conductors, and load. For current to travel in the circuit, there must be a complete path from one terminal of the power supply through the conductors and the load and back to the other terminal of the power supply (Figure 1–5).

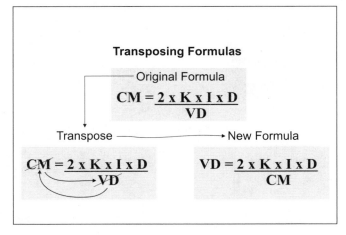

Figure 1–4
Transposing Formulas

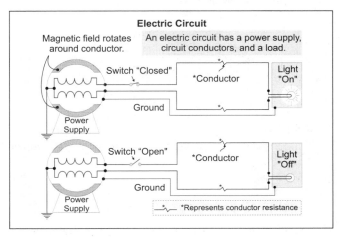

Figure 1–5
Electric Circuit

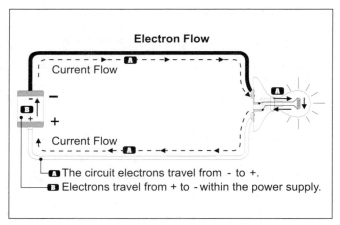

Figure 1–6
Electron Flow

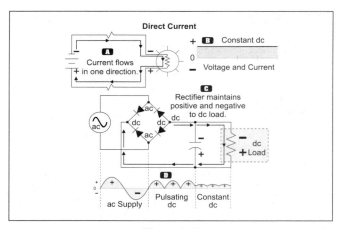

Figure 1–7
Direct Current

1–14 ELECTRON FLOW

Inside a direct current power source (such as a battery) the electrons travel from the positive terminal to the negative terminal; however, outside of the power source, electrons travel from the negative terminal to the positive terminal (Figure 1–6).

1–15 POWER SOURCE

In any completed circuit, it takes a force to push the electrons through the power source, conductor, and load. The two types of electric current are *direct current* and *alternating current*.

Direct Current

The polarity from direct current power sources never changes. That is, the current always flows out of the negative terminal of the power source in the same direction. When the power supply is a *battery,* the polarity and the voltage magnitude remain the same (Figure 1–7).

Alternating Current

Alternating current power sources produce a voltage and current that has a constant change in polarity and magnitude at a constant frequency. Alternating current flow is produced by a *generator* or an *alternator* (Figure 1–8).

1–16 CONDUCTANCE AND RESISTANCE

Conductance

Conductance is the property of metal that permits current to flow. The best conductors, in order of their conductivity, are: silver, copper, gold, and aluminum. Although silver is a better conductor of electricity than copper, copper is used most widely because it is less expensive (Figure 1–9, Part A).

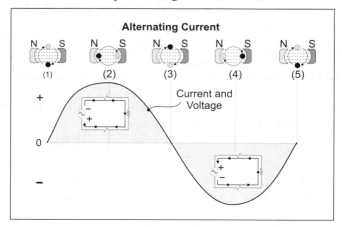

Figure 1–8
Alternating Current

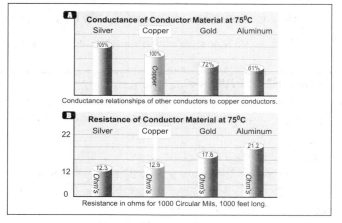

Figure 1–9
Conductance and Resistance

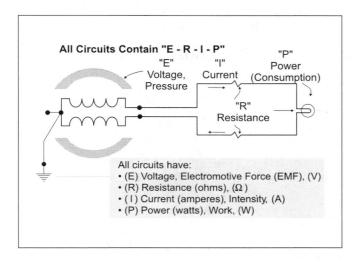

Figure 1–10
Electrical Circuit Values

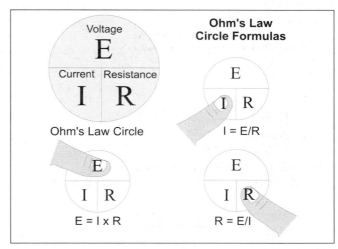

Figure 1–11
Ohm's Law Circle Formulas

Resistance

Resistance is the opposite of conductance. It is the property that opposes the flow of electric current. The resistance of a conductor is measured in ohms according to a standard length of 1,000 feet (Figure 1–9, Part B). This value is listed in the *National Electrical Code®, Chapter 9, Table 8* for direct current circuits and in *Chapter 9, Table 9* for alternating current circuits.

1–17 ELECTRICAL CIRCUIT VALUES

In an electrical circuit there are four circuit values that must be understood. They are *voltage, resistance, current,* and *power* (Figure 1–10).

Voltage

Electron pressure is called *electromotive force (E or EMF),* measured by the unit *volt (V),* and abbreviated by the letters *E, EMF* or *V.* Voltage is also a term used to described the difference of potential between any two points.

Resistance

The friction opposition to the flow of electrons is called *resistance (R),* and the unit of measurement is the *ohm.* Every component of an electric circuit contains resistance, including the power supply.

Current

Free electrons moving in the same direction in a conductor produce an electrical *current* sometimes called *intensity (I).* The rate at which electrons move is measured by the unit called *ampere (A).*

Power

The rate of work that can be produced by the movement of electrons is called *power (P),* and the unit is the *watt (W).*

Note: A 100 watt lamp consumes 100 watts of power per hour.

1–18 OHM's LAW, I = E/R

Ohm's Law, $I = E/R$, demonstrates the relationship between current, voltage, and resistance in a direct current, or in an alternating current circuit that supplies only resistive loads (Figure 1–11). Other derived formulas include:

I = E/R   **E = I × R**   **R = E/I**

Ohm's Law states that:

Current is directly proportional to voltage. If the voltage is increased by a given percentage, current increases by that same percentage. If the voltage is decreased by a given percentage, current decreases by the same percentage (Figure 1–12, Part A).

Current is inversely proportional to resistance. An increase in resistance results in a decrease in current. A decrease in resistance results in an increase in current (Figure 1–12, Part B).

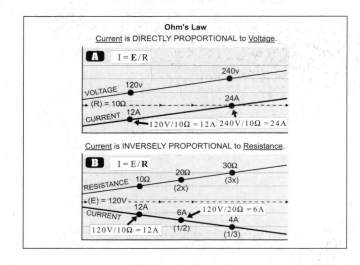

Figure 1–12
Part A – Current Proportional to Voltage
Part B – Current Inversely Proportional to Resistance

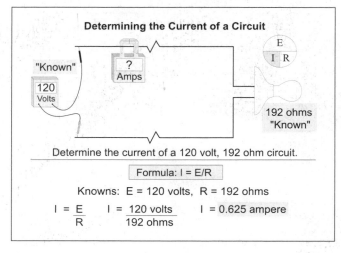

Figure 1–13
Determining the Current
of a Circuit

Opposition to Current Flow

In a direct current (dc) circuit, the physical resistance of the conductor opposes the flow of electrons. In an alternating current (ac) circuit, three factors oppose current flow. They are *conductor resistance, inductive reactance*, and *capacitive reactance*. The opposition to current flow, due to a combination of resistance and reactance, is called *impedance,* measured in ohms, and abbreviated with the letter Z. Impedance will be covered later, so for now assume that all circuits have very little or no reactance.

❏ Ampere

A 120 volt power source supplies a lamp with a resistance of 192 ohm. What is the current flow of the circuit (Figure 1–13)?

(a) 0.6 ampere (b) 0.5 ampere (c) 2.5 ampere (d) 1.3 ampere

• Answer: (a) 0.6 ampere

Step 1: ➤ *What is the question?* What is the current, (I)?

Step 2: ➤ *What do you know?*
E = 120 volts, R = 192 ohm.

Step 3: ➤ The formula is: I = E/R

Step 4: ➤ I = 120 volts/192 ohm = 0.625 ampere.

❏ Voltage

What is the voltage drop of two No. 12 conductors that supply a 16 ampere load located 50 feet from the power supply? The total resistance of both conductors is 0.2 ohm (Figure 1–14).

(a) 16 volts (b) 32 volts
(c) 1.6 volts (d) 3.2 volts

• Answer: (d) 3.2 volts

Step 1: ➤ *What is the question?* What is voltage drop, (E)?

Step 2: ➤ *What do you know about the conductors?*
I = 16 ampere, R = 0.2 ohm.

Step 3: ➤ The formula is: E = I × R.

Step 4: ➤ E = 16 ampere × 0.2 ohm = 3.2 volts.

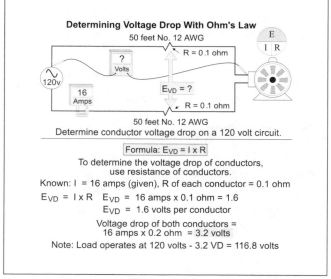

Figure 1–14
Voltage Example

❑ **Resistance**

What is the resistance of the circuit conductors when the conductor voltage drop is 3 volts and the current flowing through the conductors is 100 ampere (Figure 1–15)?

    (a) 0.03 ohm   (b) 0.2 ohm   (c) 3 ohm   (d) 30 ohm

     • Answer: (a) 0.03 ohm

Step 1: ➤ *What is the question?* What is the resistance, (R)?

Step 2: ➤ *What do you know about the conductors?* E = 3 volts dropped, I = 100 ampere.

Step 3: ➤ The formula is: R = E/I.

Step 4: ➤ R = 3 volts/100 ampere, = 0.03 ohm.

1–19 PIE CIRCLE FORMULA

The PIE circle formula shows the relationships between power, current, and voltage (Figure 1–16).

    P = E × I   **I = P/E**   **E = P/I**

❑ **Power**

What is the power loss, in watts, for two conductors that carry 12 ampere and have a voltage drop of 3.6 volts (Figure 1–17)?

    (a) 4.3 watts   (b) 43 watts   (c) 432 watts   (d) none of these

     • Answer: (b) 43 watts

Step 1: ➤ *What is the question?* It is: What is the power, (P)?

Step 2: ➤ *What do you know?* E = 3.6 volts dropped, I = 12 ampere.

Step 3: ➤ The formula is: P = E × I.

Step 4: ➤ The answer is: P = 3.6 volts × 12 ampere = 43.2 watts per hour.

❑ **Current**

What is the current flow, in ampere, in the circuit conductors that supply a 7.5 kW heat strip rated 240 volts when connected to a 240-volt power supply (Figure 1–18)?

    (a) 25 ampere   (b) 31 ampere   (c) 39 ampere   (d) none of these

     • Answer: (b) 31 ampere

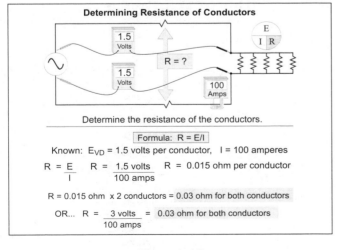

Figure 1–15
Resistance Example

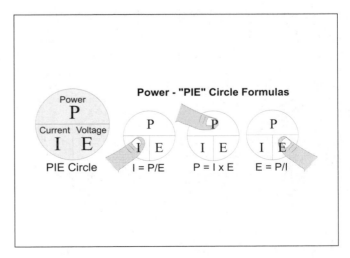

Figure 1–16
Pie Circle Formulas

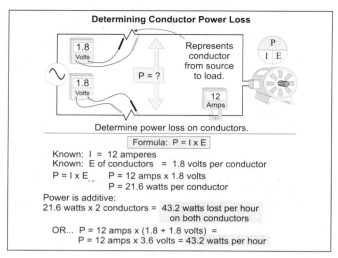

Figure 1–17
Determining Conductor Power Loss

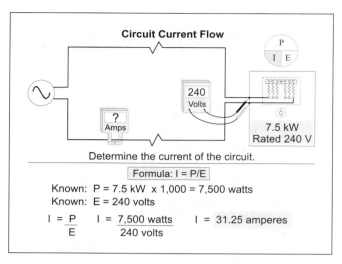

Figure 1–18
Circuit Current Flow

Step 1: ➤ *What is the question? What is the current* (I)?

Step 2: ➤ *What do you know?* P = 7,500 watts, E = 240 volts

Step 3: ➤ The formula is: I = P/E.

Step 4: ➤ I = 7,500 watts/240 volts = 31.25 ampere

1–20 FORMULA WHEEL

The formula wheel combines Ohm's Law and the PIE formulas. The formula wheel is divided up into four sections with three formulas in each section (Figure 1–19).

❏ **Resistance**

What is the resistance of a 75-watt light bulb rated 120 volts (Figure 1–20)?

(a) 100 ohm   (b) 192 ohm   (c) 225 ohm   (d) 417 ohm

• Answer: (b) 192 ohm

R = E^2/P (Formula 1 of 12), E = 120-volt rating, P = 75-watt rating, R = 120 volts^2/75 watts = 192 ohm

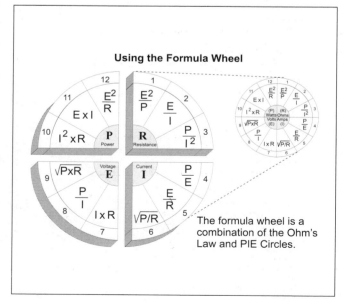

Figure 1–19
Using the Formula Wheel

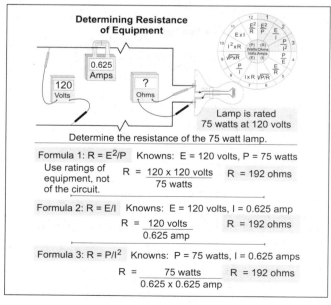

Figure 1–20
Determining Resistance of Equipment

❏ **Current**

What is the current flow of a 10 kW heat strip connected to a 230-volt (single-phase) power supply (Figure 1–21)?

(a) 13 ampere   (b) 26 ampere

(c) 43 ampere   (d) 52 ampere

• Answer: (c) 43 ampere

$I = P/E$ (Formula 4 of 12)

$P = 10,000$ watts, $E = 230$ volts

$I = 10,000$ watts/230 volts = 43 ampere

Note: Always assume single-phase, unless three-phase is specified in the question.

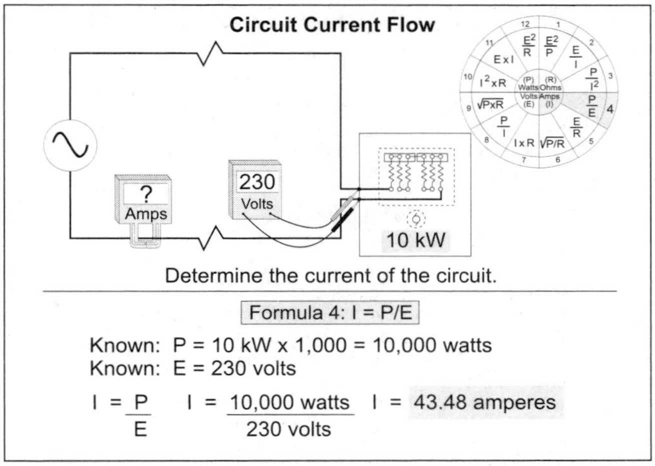

❏ **Voltage**

What is the voltage drop of 200 feet of No. 12 conductor that carries 16 ampere (Figure 1–22)? The resistance of No. 12 copper conductor is 2 ohm per 1,000 feet.

(a) 1.6 volts dropped   (b) 2.9 volts dropped

• Answer: (d) 6.4 volts dropped

$E = I \times R$ (Formula 7 of 12)

$I = 16$ ampere, $R = 2$ ohm/1,000 = 0.002 ohm per foot $\times$ 200 feet = 0.4 ohm

$E = 16$ ampere $\times$ 0.4 ohm = 6.4 volts dropped

(c) 3.2 volts dropped   (d) 6.4 volts dropped

Circuit Current Flow

Determine the current of the circuit.

Formula 4: I = P/E

Known: P = 10 kW x 1,000 = 10,000 watts
Known: E = 230 volts

$I = \dfrac{P}{E}$ $I = \dfrac{10,000 \text{ watts}}{230 \text{ volts}}$ $I = 43.48$ amperes

Figure 1–21
Circuit Current Flow

❏ **Power**

The total resistance of two No. 12 copper conductors, 75 feet long, is 0.3 ohm (0.15 ohm for each conductor). The current of the circuit is 16 ampere. What is the power loss of the conductors in watts per hour (Figure 1–23)?

(a) 19 watts   (b) 77 watts   (c) 172.8 watts   (d) none of these

• Answer: (b) 77 watts

$P = I^2 R$ (Formula 10 of 12)

$I = 16$ ampere, $R = 0.3$ ohm

$P = 16$ ampere$^2 \times 0.3$ ohm = 76.8 watts per hour

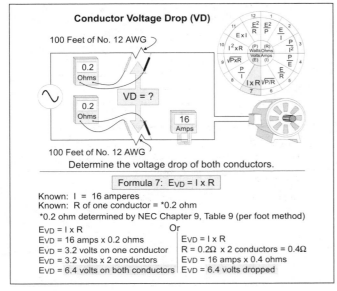

Figure 1–22
Conductor Voltage Drop (VD)

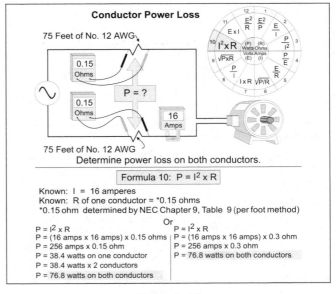

Figure 1–23
Conductor Power Loss

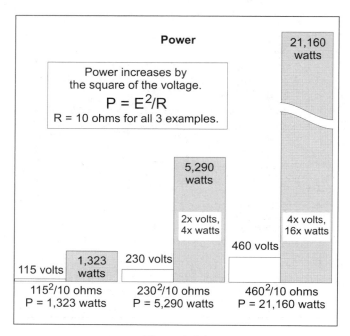

Figure 1–24
Power Changes with the Square of the Voltage

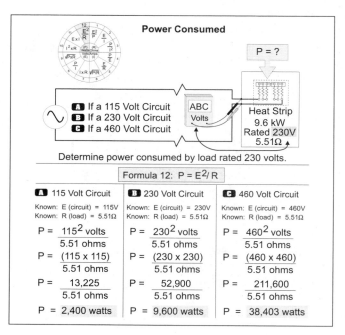

Figure 1–25
Power Changes with the Square of the Voltage

1–21 POWER CHANGES WITH THE SQUARE OF THE VOLTAGE

The power consumed by a resistor is affected by the voltage applied. Power is proportional to the square of the voltage and directly proportional to the resistance (Figure 1–24).

$$P = E^2/R$$

❑ **Power Changes with the Square of the Voltage**

What is the power consumed of a 9.6 kW heat strip rated 230 volts connected to 115-, 230-, and 460-volt power supplies? Note: The resistance of the heat strip is 5.51 ohm (Figure 1–25).

Step 1: ➥ *What is the question?* What is the power consumed (P)?

Step 2: ➥ *What do you know about the heat strip?*
E = 115 volts, 230 volts, and 460 volts, R = 5.51 ohm

Step 3: ➥ The formula to determine power is $P = E^2/R$.
P at 115 volts = 115 volts^2/5.51 ohm, P = 2,400 watts
P at 230 volts = 230 volts^2/5.51 ohm, P = 9,600 watts (2 times volts = 4 times power)
P at 460 volts = 460 volts^2/5.51 ohm, P = 38,403 watts (4 times volts = 16 times power)

1–22 ELECTRIC METERS

Basic electrical meters use a helically wound coil of conductor called a *solenoid* to produce a strong electromagnetic field to attract an iron bar inside the coil. The iron bar that moves inside the coil is called an *armature*. When a meter has positive (+) and negative (–) shown for the meter leads, the meter is said to be *polarized*. The negative (–) lead must be connected to the negative terminal of the power source, and the positive (+) lead must be connected to the positive terminal.

Ammeter

An ammeter is a meter that has a *helically* (spirally) wound coil, and it uses the circuit energy to measure direct current. As current flows through the meter's coil, the coil's electromagnetic field draws in the iron bar (armature). The greater the current flow through the meter's coil, the greater the electromagnetic field and the further the armature is drawn into the coil. Ammeters are connected in series with the circuit and are used to measure only direct current (Figure 1–26).

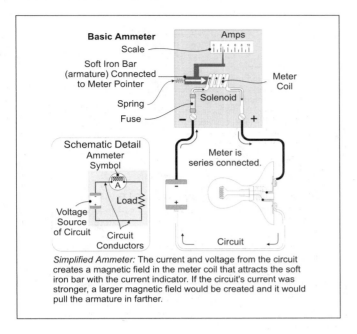

Figure 1–26
Basic Ammeter

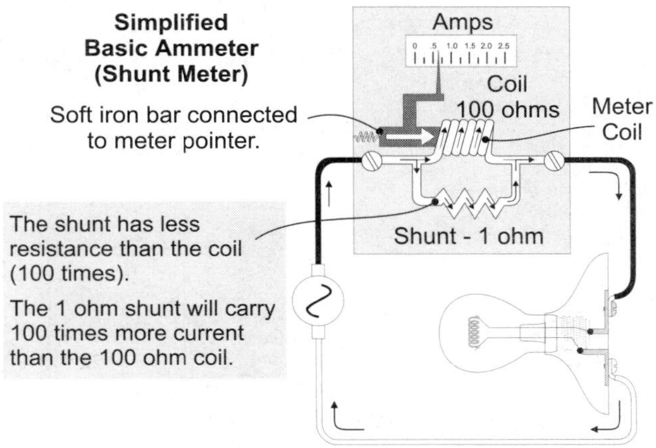

Figure 1–27
Simplified Basic Ammeter (Shunt Meter)

Ammeters are connected in *series* with the power supply and the load. If the ammeter is accidentally connected in *parallel* to the power supply, the current flow through the meter will be extremely high. The excessive high current through the meter will destroy the meter due to excessive heat.

If the ammeter is not connected to the proper *polarity* when measuring direct current, the meter's needle will quickly move in the reverse direction and possibly damage the meter's calibration.

Ammeters that measure currents larger than 10 milliampere often contain a device called a *shunt* which is placed in parallel with the meter coil. This permits the current flow to divide between the meter's coil and the shunt bar. The current through the meter coil depends on the resistance of the shunt bar.

❏ **Shunt Ammeter Current**

What is the current flow through the meter if the shunt bar is 1 ohm and the coil is 100 ohm (Figure 1–27)?

(a) the same as the shunt
(b) 1/10 the shunt amperage
(c) 1/100 the shunt amperage
(d) 1/1,000 the shunt amperage

• Answer: (c) 1/100 the shunt amperage

Since the shunt is 100 times less resistant than the meter's coil, the shunt bar will carry 100 times more current than the meter's coil, or the meter's coil will carry 1/100 the shunt amperage.

Clamp-on Ammeter

Clamp-on ammeters are used to measure alternating current. They are connected *perpendicularly* (at a 90° angle) around the conductor without breaking the circuit. A clamp-on ammeter indirectly utilizes the circuit energy by induction of the electromagnetic field. The clamp-on ammeter is actually a transformer (sometimes called a *current transformer*). The primary winding is the phase conductor, and the secondary winding is the meter's coil (Figure 1–28).

The electromagnetic field around the phase conductor expands and collapses, which causes electrons to flow in the meter's circuit. As current flows through the meter's coil, the electromagnetic field of the meter draws in the armature. Since the phase conductor serves as the primary (one turn), the current to be measured must be high enough to produce an electromagnetic field that is strong enough to cause the meter to operate.

Ohmmeter and Megohmeter

Ohmmeters are used to measure the resistance of a circuit or component and can be used to locate open circuits or shorts. An ohmmeter has an armature, a coil, and its own power supply, generally a battery. Ohmmeters are always connected to de-

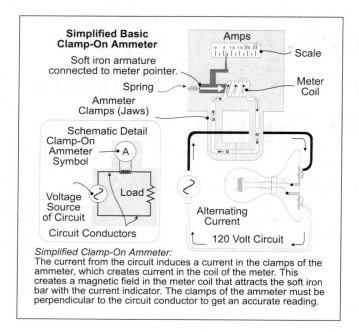

Simplified Basic Clamp-On Ammeter

Soft iron armature connected to meter pointer.

Spring

Ammeter Clamps (Jaws)

Schematic Detail
Clamp-On Ammeter Symbol

Voltage Source of Circuit

Circuit Conductors

Load

Amps

Scale

Meter Coil

Alternating Current

120 Volt Circuit

Simplified Clamp-On Ammeter:
The current from the circuit induces a current in the clamps of the ammeter, which creates current in the coil of the meter. This creates a magnetic field in the meter coil that attracts the soft iron bar with the current indicator. The clamps of the ammeter must be perpendicular to the circuit conductor to get an accurate reading.

Figure 1–28
Clamp-on Ammeters

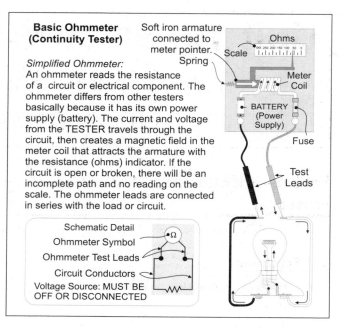

Basic Ohmmeter (Continuity Tester)

Soft iron armature connected to meter pointer.

Simplified Ohmmeter:
An ohmmeter reads the resistance of a circuit or electrical component. The ohmmeter differs from other testers basically because it has its own power supply (battery). The current and voltage from the TESTER travels through the circuit, then creates a magnetic field in the meter coil that attracts the armature with the resistance (ohms) indicator. If the circuit is open or broken, there will be an incomplete path and no reading on the scale. The ohmmeter leads are connected in series with the load or circuit.

Scale

Spring

Ohms

Meter Coil

BATTERY (Power Supply)

Fuse

Test Leads

Schematic Detail

Ohmmeter Symbol

Ohmmeter Test Leads

Circuit Conductors

Voltage Source: MUST BE OFF OR DISCONNECTED

Figure 1–29
Ohm Meter

energized circuits, and polarity need not be observed. When an ohmmeter is used, current flows through the meter's coil, causing an electromagnetic field around the coil and drawing in the armature. The greater the current flow, as a result of lower resistance (I = E/R), the greater the magnetic field and the further the armature is drawn into the coil. A short circuit is indicated by a reading of zero, and an open circuit is indicated by infinity (∞) (Figure 1–29).

A *megohmmeter* or *megohmer* is an instrument designed to measure very high resistances such as those found in cable insulation between motor or transformer windings.

Voltmeter

Voltmeters are used to measure both dc and ac voltage. A voltmeter contains a *resistor* in series with the coil and utilizes the circuit energy for its operation. The purpose of the resistor in the meter is to reduce the current flow through the meter. As current flows through the meter's helically wound coil, the combined electromagnetic field of the coil draws in the iron bar. The greater the circuit voltage, the greater the current flow through the meter's coil (I = $^E/_R$). The greater the current flow through the meter's coil, the greater the electromagnetic field and the further the armature is drawn into the coil (Figure 1–30).

Polarity must be observed when connecting voltmeters to direct current circuits. If the meter is not connected to the proper polarity, the meter's needle will quickly move in the reverse direction and damage the meter's calibration. Polarity is not required when connecting voltmeters to alternating current circuits.

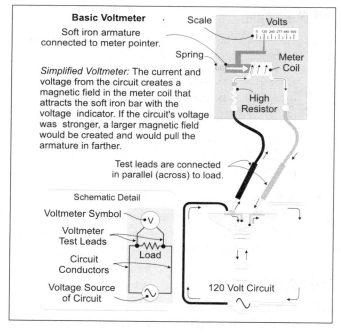

Basic Voltmeter

Soft iron armature connected to meter pointer.

Simplified Voltmeter: The current and voltage from the circuit creates a magnetic field in the meter coil that attracts the soft iron bar with the voltage indicator. If the circuit's voltage was stronger, a larger magnetic field would be created and would pull the armature in farther.

Scale

Spring

Volts

Meter Coil

High Resistor

Test leads are connected in parallel (across) to load.

Schematic Detail

Voltmeter Symbol

Voltmeter Test Leads

Circuit Conductors

Voltage Source of Circuit

Load

120 Volt Circuit

Figure 1–30
Volt Meter

Unit 1 – Electrician's Math and Basic Electrical Formulas Questions

Part A – Electrician's Math (• Indicates that 75% or less of exam takers get the question correct)

Note: Always assume copper conductors for all answers, unless aluminum is specified in the question.

1–1 Fractions

1. The decimal equivalent for the fraction $^1/_2$ is _____ .
   (a) 0.5 (b) 5 (c) 2 (d) 0.2

2. The decimal equivalent for the fraction $^4/_{18}$ is _____ .
   (a) 4.5 (b) 1.5 (c) 2.5 (d) 0.2

1–2 Kilo

3. What is the kW of a 75-watt load?
   (a) 75 kW (b) 7.5 kW (c) 0.75 kW (d) 0.075 kW

1–3 Knowing Your Answer

4. • The output of a transformer is 100 VA. The transformer efficiency is 90 percent. What is the transformer input power? Note: Because efficiency is always less than 100%, the input is always greater than the output.
   (a) 90 watts (b) 110 watts (c) 100 watts (d) 125 watts

1–4 Multiplier

5. The method of increasing a number by multiplying it by another number is call the _____ .
   (a) percentage (b) decimal (c) fraction (d) multiplier

6. An overcurrent protection device (breaker or fuse) is required to be sized no less than 115 percent of the load. If the load is 20 ampere, the overcurrent protection device would have to be sized at no less than _____ .
   (a) 20 ampere (b) 23 ampere (c) 17 ampere (d) 30 ampere

7. The maximum continuous load on an overcurrent protection device is limited to 80 percent of the device rating. If the device is rated 90 ampere, the maximum continuous load is _____ ampere.
   (a) 72 (b) 90 (c) 110 (d) 125

8. A 50 ampere rated wire is required to be adjusted for temperature. If the correct multiplier is 0.80, which of the following statements is/are correct?
   (a) the answer will be less than 50 ampere (b) 80 percent of the ampacity (50) can be used
   (c) the formula is 50 ampere $\times$ 0.8 (d) all the above

1–5 Parentheses

9. What is the distance of two No. 14 conductors carrying 16 ampere with a voltage drop of 10 volts?
   Formula: D = (4,100 circular mils $\times$ 10 volts dropped)/(2 wires $\times$ 12.9 ampere $\times$ 16 ohm)
   (a) 50 feet (b) 75 feet (c) 100 feet (d) 150 feet

10. What is the current in ampere of a three-phase, 18 kW, 208-volt load? Formula: $I = \dfrac{W}{E \times \sqrt{3}}$

    (a) 25 ampere   (b) 50 ampere

    (c) 100 ampere   (d) 150 ampere

1–6 Percentages

11. When changing a percent value to a decimal or whole number, simply move the decimal point two places to the _____ .

    (a) right   (b) left   (c) depends   (d) none of these

12. The decimal equivalent for 75 percent is _____ .

    (a) 0.075   (b) 0.75   (c) 7.5   (d) 75

13. The decimal equivalent for 225 percent is_____ .

    (a) 225   (b) 22.5   (c) 2.25   (d) 0.225

14. The decimal equivalent for 300 percent is _____ .

    (a) 0.03   (b) 0.3   (c) 3   (d) 30.0

1–7 Percent Increase

15. The feeder demand load for an 8 kW load, increased by 20 percent, is _____ kVA.

    (a) 8   (b) 9.6   (c) 6.4   (d) 10

1–8 Percentage Reciprocals

16. What is the reciprocal of 125 percent?

    (a) 0.8   (b) 100 percent   (c) 125 percent   (d) none of these

17. A continuous load requires an overcurrent protection device sized no smaller than 125 percent of the load. What is the maximum continuous load permitted on a 100-ampere overcurrent protection device?

    (a) 100 ampere   (b) 150 ampere   (c) 80 ampere   (d) 110 ampere

1–9 Rounding

18. • The sum of 5, 7, 8, and 9 is approximately _____ .

    (a) 20   (b) 25   (c) 30   (d) 35

1–10 Squaring

19. What is the power consumed in watts of a No. 12 conductor that is 100 feet long and has a resistance of (R) 0.2 ohm, the current (I) in the circuit is 16 ampere? Formula: Power = I^2R.

    (a) 75 watts   (b) 50 watts   (c) 100 watts   (d) 200 watts

20. • What is the area in square inches of a 2-inch raceway?

    Formula: Area = πr^2 π, π = 3.14, r = radius ($^1/_2$ the diameter)

    (a) 1 square inch   (b) 2 square inches   (c) 3 square inches   (d) 4 square inches

21. The numeric equivalent of 4^2 is _____ .

    (a) 2   (b) 8   (c) 16   (d) 32

22. The numeric equivalent of 12^2 is _____ .

    (a) 3.46   (b) 24   (c) 144   (d) 1,728

1–11 Square Root

23. What is the square root of 1,000 ($\sqrt{1000}$)?

    (a) 3   (b) 32   (c) 100   (d) 500

24. The square root of $\sqrt{3}$ is _____ .

    (a) 1.732   (b) 9   (c) 729   (d) 1.5

1–12 Transposing Formulas

25. • Transform the formula $CM = \dfrac{(2 \times K \times I \times D)}{VD}$ to find the distance of the circuit.

    (a) $D = VD \times CM$

    (b) $D = \dfrac{(2 \times K \times I \times C \times M)}{VD}$

    (c) $D = (2 \times K \times I) \times CM$

    (d) $D = \dfrac{CM \times VD}{2 \times K \times I}$

26. If $I = {}^{P}/_{E}$, which of the following statements contain the correct transposed formula?
    (a) $P = {}^{E}/_{I}$ (b) $P = {}^{I}/_{E}$ (c) $P = I \times E$ (d) $P = I^{2}E$

Part B – Basic Electrical Formulas

1–13 Electrical Circuit

27. An electric circuit consists of the _____ .
    (a) power source   (b) conductors   (c) load   (d) all of these

1–14 Electron Flow

28. Inside the power source, electrons travel from the positive terminal to the negative terminal.
    (a) True   (b) False

1–15 Power Source

29. The polarity of a(n) _____ current power source(s) never change. One terminal is always negative, and the other is always positive. _____ current flows out of the negative terminal of the power source at the same polarity.
    (a) Static   (b) Direct   (c) Alternating   (d) all of the above

30. _____ current power sources produce a voltage that has a constant, equal change in polarity and magnitude in both directions.
    (a) Static   (b) Direct   (c) Alternating   (d) all of the above

1–16 Conductance and Resistance

31. Conductance is the property of metal that permits current to flow. The best conductors, in order of conductivity, are:

    _____ .

    (a) gold, silver, copper, aluminum   (b) copper, gold, copper, aluminum
    (c) gold, copper, silver, aluminum   (d) silver, copper, gold, aluminum

1–17 Electrical Circuit Values

32. The _____ is the pressure required to force one ampere of electrons through a one ohm resistor.
    (a) ohm   (b) watt   (c) volt   (d) ampere

33. All conductors have resistance that opposes the flow of electrons. Some materials have more resistance than others. _____ has the lowest resistance, and _____ is more resistant than copper.
    (a) Silver, gold   (b) Gold, aluminum   (c) Gold, silver   (d) None of these

34. Resistance is represented by the letter R, and it is expressed in _____ .
    (a) volts   (b) impedance   (c) capacitance   (d) ohms

35. The opposition to the flow of current can be thought of as restricting the flow of electrons in the circuit. Every component of an electric circuit contains resistance, except the power supplies such as the generator or transformer.
    (a) True   (b) False

36. In electrical systems, the volume of electrons that moves through a conductor is called the _____ of the circuit.
    (a) intensity   (b) voltage   (c) power   (d) resistance

37.   The rate of work that can be produced by the movement of electrons is called _____ .
       (a) voltage   (b) current   (c) power   (d) none of these

1–18 Ohm's Law, I = E/R

38.   The Ohm's Law formula demonstrates that current is _____ proportional to the voltage and _____ proportional to the resistance.
       (a) indirectly, inversely (b) inversely, directly   (c) inversely, indirectly   (d) directly, inversely

39.   In an alternating current circuit, which factors oppose current flow?
       (a) resistance   (b) capacitance reactance   (c) induction reactance   (d) all of these

40.   The opposition to current flow due in an alternating current circuit is called _____ and is often represented by the letter Z.
       (a) resistance   (b) capacitance   (c) induction   (d) impedance

41.   • What is the voltage drop of two No. 12 conductors supplying a 16-ampere load located 100 feet from the power supply? Formula: $E_{VD} = I \times R$, I = 16 ampere, R = 0.4 ohms (200 feet of No. 12 copper wire)
       (a) 6.4 volts   (b) 12.8 volts   (c) 1.6 volts   (d) 3.2 volts

42.   What is the resistance of the circuit conductors when the conductor voltage drop is 7.2 volts and the current flow is 50 ampere?
       (a) 0.14 ohm   (b) 0.3 ohm   (c) 3 ohm   (d) 14 ohm

1–19 Pie Circle Formula

43.   What is the power loss in watts for a conductor that carries 24 ampere and has a voltage drop of 7.2 volts?
       (a) 173 watts   (b) 350 watts   (c) 700 watts   (d) 2,400 watts

44.   What is the current flow of a 10 kW heat strip rated 240 volts, single-phase?
       (a) 35 ampere   (b) 38 ampere   (c) 42 ampere   (d) 60 ampere

1–20 Formula Wheel

45.   • The formulas listed in the formula wheel apply to _____ .
       (a) circuits only
       (b) circuits with unity power factor
       (c) a and b
       (d) none of these

46.   When working any formula, the key to getting the correct answer is following these four simple steps:
       Step 1: ➥   Know what the question is asking.
       Step 2: ➥   Determine the knowns of the circuit or resistor.
       Step 3: ➥   Select the formula.
       Step 4: ➥   Work out the formula calculation.
       (a) True   (b) False

47.   The total resistance of two No. 12 copper conductors 150 feet long is 0.6 ohm and the current of the circuit is 16 ampere. What is the power loss of the conductors in watts per hour?
       (a) 50 watts per hour   (b) 150 watts per hour   (c) 300 watts per hour   (d) 600 watts per hour

48.   • What is the conductor power loss in watts for a 120-volt circuit that has a 3 percent voltage drop and carries a current flow of 12 ampere? The load operates 24 hours per day, 365 days each year.
       (a) 43 watts   (b) 86 watts   (c) 172 watts   (d) 722 watts

49.   • What does it cost per year (24 hours per day, 365 day per year at 8.6 cents per kW) for the power loss of a conductor? The No. 12 copper conductor resistance is 0.3 ohm and the current flow is 12 ampere.
       (a) $33.55   (b) $13.10   (c) $130.50   (d) $140.21

1–21 Power Changes with the Square of the Voltage

50.　• What is the power consumed of a 10 kW heat strip rated 230 volts connected to a 115-volt circuit?
　　(a) 10 kW　　　　(b) 2.5 kW　　　　(c) 5 kW　　　　(d) 20 kW

1–22 Electric Meters

51.　A(n) _____ is connected in series with the load.
　　(a) watt-hour meter　　(b) voltmeter　　　(c) power meter　　　(d) ammeter

52.　• Ammeters are used to measure _____, are connected in series with the circuit, and are said to shunt the circuit.
　　(a) current　　　　(b) power　　　　(c) voltage　　　　(d) all of these

53.　Clamp-on ammeters have one coil connected _____ around the circuit conductor.
　　(a) series　　　　(b) parallel　　　　(c) series-parallel　　　(d) at right angles (perpendicular)

54.　An ohmmeter has _____ connected in series with the resistor. As current flows through the meter coil, the magnetic field around the coil draws in the soft iron bar. The greater the current flow through the circuit, the greater the magnetic field and the further the armature is drawn into the coil.
　　(a) a coil and resistor　　(b) a coil and power supply　　(c) two coils　　　(d) none of these

☆ Challenge Questions

1–2 Kilo

55.　• kVA is equal to _____ .
　　(a) 100 VA　　　　(b) 1,000 volts　　　(c) 1,000 watts　　　(d) 1,000 VA

1–16 Conductance and Resistance

56.　• _____ is not an insulator.
　　(a) Bakelite　　　　(b) Oil　　　　(c) Air　　　　(d) Salt water

1–17 Electrical Circuit Values

57.　• _____ is not the force that moves electrons.
　　(a) EMF　　　　(b) Voltage　　　　(c) Potential　　　　(d) Current

58.　Conductor resistance varies with _____ .
　　(a) material　　　　(b) voltage　　　　(c) current　　　　(d) power

1–18 Ohm's Law, I = $\frac{E}{R}$

59.　• If the contact resistance of a connection increases and the current remains the same, the voltage drop across the connection _____ .
　　(a) will increase　　(b) will decrease　　(c) will remain the same　　(d) cannot be determined

60.　• To double the current of a circuit when the voltage remains constant, the resistance (R) must be _____ .
　　(a) doubled　　　　(b) reduced by half　　(c) increased　　　(d) none of these

61.　• An ohmmeter is being used to test a relay coil. The equipment instructions indicate that the resistance of the coil should be between 30 and 33 ohm. The ohmmeter indicates that the actual resistance is less than 22 ohm. This reading would most likely indicate_____ .
　　(a) the coil is okay　　(b) an open coil　　(c) a shorted coil　　(d) a meter problem

1–20 Formula Wheel

62.　• To calculate the power consumed by a resistive appliance, one needs to know _____ .
　　(a) voltage and current　(b) current and resistance　(c) voltage and resistance　(d) any of these

63.   • The number of watts of heat given off by a resistor is expressed by the formula $I^2 \times R$. If 10 volts is applied to a 5 ohm resistor, then _____ of heat will be given off.
    (a) 500 watts   (b) 250 watts   (c) 50 watts   (d) 20 watts

64.   • Power loss in a circuit because of heat can be determined by the formula _____ .
    (a) $P = R \times I$ (b) $P = I \times R$ (c) $P = I^2 \times R$ (d) none of these

65.   • If current remains the same and resistance increases, the circuit will consume _____ power.
    (a) more   (b) less

66.   When a lamp that is rated 500 watts at 115 volts is connected to a 120-volt power supply, the current of the circuit will be _____ . Tip: Does power remain the same when voltage is changed?
    (a) 3.8 ampere   (b) 4.5 ampere   (c) 2.7 ampere   (d) 5.5 ampere

1–21 Power Changes with the Square of the Voltage

67.   A 120-volt rated toaster will produce _____ heat when supplied by 115 volts.
    (a) more   (b) less   (c) the same   (d) none of these

68.   • When a resistive load is operated at a voltage 10 percent higher than the nameplate rating of the appliance, the appliance will _____ .
    (a) have a longer life   (b) draw a lower current   (c) use more power   (d) none of these

69.   • A 1,500-watt heater rated 230 volts is connected to a 208-volt supply. The power consumed for this load is _____ watts. Tip: When the voltage is reduced, will the power be greater or less?
    (a) 1,625   (b) 1,750   (c) 1,850   (d) 1,225

70.   • The total resistance of a circuit is 12 ohm; the load is 10 ohm, and the wire 2 ohm. If the current of the circuit is 3 ampere, then the power consumed by the circuit conductors is _____ .
    (a) 28 watts   (b) 18 watts   (c) 90 watts   (d) 75 watts

1–22 Electric Meters

71.   • The best instrument for detecting an electric current is a(n) _____ .
    (a) ohmmeter   (b) voltmeter   (c) ammeter   (d) wattmeter

72.   The polarity of a circuit being tested must be observed when connecting an ohmmeter to _____ .
    (a) an alternating current circuit
    (b) a direct current circuit
    (c) any circuit
    (d) polarity does not matter because the circuit is not energized

73.   • When the test leads of an ohmmeter are shorted together, the meter will read _____ on the scale.
    (a) zero ohm   (b) 1,000   (c) infinity   (d) all of these

74.   • A short circuit is indicated by a reading of _____ when tested with an ohmmeter.
    (a) zero   (b) ohm   (c) infinity   (d) R

75.   Voltmeters are used to measure _____ .
    (a) voltages to ground   (b) voltage differences   (c) ac voltages only   (d) dc voltages only

76.   Voltmeters must be connected in _____ with the circuit component being tested.
    (a) series   (b) parallel   (c) series-parallel   (d) multiwire

77.   To measure the voltage across a load, you would connect a(n) _____ .
    (a) voltmeter across the load   (b) ammeter across the load
    (c) voltmeter in series with the load   (d) ammeter in series with the load

78.   A voltmeter is connected in _____ to the load.
    (a) series   (b) parallel   (c) series-parallel   (d) none of these

79.   • In the course of normal operation, the least effective instrument in indicating that a generator may overheat because it is overloaded is a(n) _____ .
(a) ammeter   (b) voltmeter   (c) wattmeter   (d) none of these

80.   • A dc voltmeter (not a digital meter) can be used to measure _____ .
(a) power   (b) frequency   (c) polarity   (d) power factor

81.   • Polarity must be observed when connecting an analog voltmeter to _____ current circuit.
(a) an alternating   (b) direct   (c) any   (d) polarity does not matter

82.   The minimum number of wattmeters necessary to measure the power in the load of a balanced 3-phase, 4-wire system is _____ .
(a) 1   (b) 2   (c) 3   (d) 4

NEC® Questions from Section 90-1 through Sections 230-51(c)

The following *National Electrical Code®* questions are in consecutive order. Questions with • indicate that 75 percent or less of exam takers get the question correct.

Article 90 – Introduction

Article 90 is the introduction of the *NEC.®*

83.   The *National Electrical Code®* is _____ .
(a) intended to be a design manual
(b) meant to be used as an instruction guide for untrained persons
(c) for the practical safeguarding of persons and property
(d) published by the Bureau of Standards

84.   Compliance with the provisions of the *Code* will result in _____ .
(a) good electrical service   (b) an efficient electrical system
(c) an electrical system free from hazard   (d) all of these

85.   • The purpose of this *Code* is the practical safeguarding of persons and property from hazards arising from the use of electricity. This *Code* contains provisions considered necessary for safety regardless of _____ .
(a) efficient use
(b) convenience
(c) good service or future expansion
(d) all of these

86.   • The *Code* applies to the installation of _____ .
(a) electrical conductors and equipment within or on public and private buildings
(b) outside conductors and equipment on the premises.
(c) optical fiber cable.
(d) all these

87.   The *Code* does not cover installations in ships, watercraft, railway rolling stock, aircraft, or automotive vehicles.
(a) True   (b) False

88.   Installations of communications equipment under the exclusive control of communications utilities located outdoors or in building spaces used exclusively for such installations _____ covered by the *Code*.
(a) are   (b) are sometimes   (c) are not   (d) might be

89.   Chapters 1 through 4 of the *NEC®* apply _____ .
(a) generally to all electrical installations   (b) to special installations and conditions
(c) to special equipment and material   (d) all of these

90. The authority having jurisdiction of enforcement of the *Code* has the responsibility _____ .
    (a) for making interpretations of the rules of the *Code*
    (b) for deciding upon the approval of equipment and materials
    (c) for waving specific requirements in the *Code* and allowing alternate methods and material if safety is maintained
    (d) all of these

91. Explanatory material, such as references to other standards, references to related Sections of the *Code*, or information related to a *Code* rule is included in the *Code* book in the form of a Fine Print Note (FPN).
    (a) True   (b) False

92. Equipment listed by a qualified electrical testing laboratory is not required to have the _____ wiring to be reinspected at the time of installation.
    (a) external   (b) associated   (c) internal   (d) all of these

NEC® Chapter 1 – General Requirements

As the name implies, this Chapter contains the general installation requirements. Article 90 and Chapter 1 provide a good understanding of the *Code* book structure, terminology, and general assumptions, which make the other *NEC®* Chapters much easier to use and understand.

Article 100 – Definition

Article 100 explains many of the terms used that apply to the *NEC®*. It is very important for you to understand the meanings of these terms.

93. Capable of being removed or exposed without damaging the building structure or finish, or not permanently closed in by the structure or finish of the building defines _____ .
    (a) accessible (equipment)
    (b) accessible (wiring methods)
    (c) accessible, readily
    (d) all of these

94. • A synthetic nonflammable insulating medium which, when decomposed by electric arcs, produces predominantly nonflammable gaseous mixtures is known as _____ .
    (a) oil   (b) geritol   (c) askarel   (d) phenol

95. • The connection between the grounded (neutral) conductor and the equipment grounding conductor at the service is accomplished with the use of a _____ jumper.
    (a) main bonding
    (b) bonding
    (c) equipment bonding
    (d) circuit bonding

96. A branch circuit that supplies only one utilization equipment is a(n) _____ branch circuit.
    (a) individual   (b) general-purpose   (c) isolated   (d) special purpose

97. A circuit breaker is a device designed to _____ a circuit by nonautomatic means and to open the circuit automatically on a predetermined overcurrent without damage to itself when properly applied within its rating.
    (a) blow   (b) disconnect   (c) connect   (d) open and close

98. • A conductor encased within material of composition or thickness that is not recognized by the *Code* is called a _____ conductor.
    (a) noninsulating   (b) bare   (c) covered   (d) protected

99. A load is considered to be continuous if it is expected to continue for _____ hour(s) or more.
    (a) $\frac{1}{2}$ (b) 1   (c) 2   (d) 3

100. Which of the following does the *Code* recognize as a device?
(a) switch
(b) switch and light bulb
(c) lock nut and switch
(d) lock nut and bushing

101. So constructed or protected that dust will not interfere with its successful operation is called _____ .
(a) dusttight   (b) dustproof   (c) dust rated   (d) all of these

102. Varying duty is defined as _____ .
(a) intermittent operation in which the load conditions are regularly recurrent
(b) operation at a substantially constant load for an indefinite length of time
(c) operation for alternate intervals of load and rest; or load, no load, and rest
(d) operation at loads and for intervals of time, both of which may be subject to wide variations

103. • The case or housing of apparatus or the fence surrounding an installation to prevent accidental contact of persons to energized parts is called _____ .
(a) guarded   (b) covered   (c) protection   (d) an enclosure

104. When the term exposed is used by the *Code* it refers to _____ .
(a) capable of being inadvertently touched or approached nearer than a safe distance by a person
(b) parts that are not suitably guarded, isolated, or insulated
(c) wiring on, or attached to, the surface or behind panels being designed to allow access
(d) all of these

105. Connected to earth or to some conducting body that serves in place of the earth is called _____ .
(a) grounding   (b) bonded   (c) grounded   (d) all of the above

106. A device intended for the protection of personnel that functions to de-energize a circuit within an established period of time when a current to ground exceeds some predetermined value less than that required to operate the overcurrent protection device of the supply circuit is a(n) _____ .
(a) dual-element fuse
(b) inverse-time breaker
(c) ground-fault circuit-interrupter
(d) safety switch

107. Recognized as suitable for the specific purpose, function, use, environment, and application is the definition of _____ .
(a) labeled   (b) identified   (c) listed   (d) approved

108. Equipment or materials to which a symbol or other identifying mark acceptable to the authority having jurisdiction has been attached is known as _____ .
(a) listed   (b) labeled   (c) approved   (d) rated

109. The environment of a wiring method under the eave of a house with a roof open porch would be considered a _____ location.
(a) dry   (b) damp   (c) wet   (d) moist

110. A circuit in which any arc or thermal effect produced, under intended operating conditions of the equipment or due to opening, shorting, or grounding of field wiring, is not capable, under specified test conditions, of igniting the flammable gas, vapor, or dust-air mixture is a _____ circuit.
(a) nonconductive   (b) branch   (c) nonincendive   (d) closed

111. • An overload may be caused by a short-circuit or ground-fault.
(a) True   (b) False

112. Something constructed, protected, or treated so as to prevent rain from interfering with the successful operation of the apparatus under specified test conditions is defined as _____ .
(a) raintight   (b) waterproof   (c) weathertight   (d) rainproof

113.   A single receptacle is a _____ contact device with no other contact device on the same yoke.
    (a) dual   (b) single   (c) multiple   (d) live

114.   A _____ system is where a premise wiring is derived from a transformer and that has no direct electrical connection, including a solidly connected grounded circuit conductor, to supply conductors originating in another system.
    (a) separately derived   (b) classified   (c) direct   (d) emergency

115.   Overhead service conductors from the last pole or other aerial support to and including the splices if any are called _____ conductors.
    (a) service-entrance   (b) service drop   (c) service   (d) overhead service

116.   • Service conductors between the street main and the first point of connection to the service entrance run underground are known as the _____ .
    (a) utility service   (b) service lateral   (c) service drop   (d) main service conductors

117.   The total components and subsystems that, in combination, convert solar energy into electrical energy is called a _____ system.
    (a) solar   (b) solar voltaic   (c) separately derived source   (d) solar photovoltaic

118.   A form of general-use switch so constructed that it can be installed in flush device boxes or on outlet box covers or otherwise used in conjunction with wiring systems recognized by the *Code* is a _____ switch.
    (a) transfer   (b) motor-circuit   (c) general-use snap   (d) bypass isolation

119.   Utilization equipment is equipment that utilizes electricity for _____ .
    (a) chemicals   (b) heating   (c) lighting   (d) any of these

120.   Something constructed so that moisture will not enter the enclosure under specific test conditions is called _____ .
    (a) watertight   (b) moisture proof   (c) waterproof   (d) rainproof

Article 110 – General Requirements

121.   Equipment is required to be installed and used according to its _____ instructions.
    (a) listed or published   (b) labeled or design   (c) listed and labeled   (d) any of the above

122.   All wiring shall be installed so that the completed system will be free from _____ .
    (a) short circuits   (b) grounds   (c) a and b   (d) none of these

123.   Electrical equipment shall have sufficient short-circuit _____ rating to permit the circuit protection device to clear a fault without extensive damage to the electrical components of the circuit.
    (a) withstand   (b) current   (c) overload   (d) all of these

124.   Unless identified for use in the operating environment, no conductors or equipment shall be _____ having a deteriorating effect on the conductors or equipment.
    (a) located in damp or wet locations
    (b) exposed to fumes, vapors, and gases
    (c) exposed to liquids and excessive temperatures
    (d) all of these

125.   The *NEC*® requires that electrical work be installed _____ .
    (a) in a neat and workmanlike manner
    (b) under the supervision of a qualified person
    (c) completed before being inspected
    (d) all of these

126.   Electrical equipment that depends on the _____ principles for cooling of exposed surfaces shall be installed so that airflow over such surfaces is not prevented by walls or by adjacent installed equipment.
    (a) Peter
    (b) natural circulation of air and convection
    (c) artificial cooling and circulation
    (d) air-conditioning

127. Connection by means of wire binding screws or studs and nuts having upturned lugs or the equivalent shall be permitted for _____ or smaller conductors.

(a) No. 10   (b) No. 8   (c) No. 6   (d) none of these

128. • What size THHN conductor is required for a 50-ampere circuit if the equipment terminals are listed for 75°C conductor sizing? Tip: Table 310-16 lists conductor ampacities.

(a) No. 10   (b) No. 8   (c) No. 6   (d) all of these

129. Separately installed pressure connectors shall be used with conductors at the _____ not exceeding the ampacity at the listed and identified temperature rating of the connector.

(a) voltages   (b) temperatures   (c) listings   (d) ampacities

130. Sufficient access and _____ shall be provided and maintained about all electrical equipment to permit ready and safe operation and maintenance of such equipment.

(a) ventilation   (b) cleanliness   (c) circulation   (d) working space

131. The minimum working clearance on a circuit 120 volts to ground, with exposed live parts on one side and no live or grounded parts on the other side of the working space is _____ feet.

(a) 1   (b) 3   (c) 4   (d) 6

132. • The required working clearance for access to live parts operating at 300 volts, nominal, to ground where there are exposed live parts on one side and grounded parts on the other side, is _____ feet according to Table 110-26(a).

(a) 3   (b) $3^{1}/_{2}$ (c) 4   (d) $4^{1}/_{2}$

133. Equipment such as raceways, cables, wireways, cabinets, panels, etc. can be located above or below other electrical equipment where the associated equipment does not extend more than _____ inches from the front of the electrical equipment.

(a) 3   (b) 6   (c) 12   (d) 30

134. Where the required work space is doubled, _____ entrance(s) to the working space is/are required.

(a) 1   (b) 2   (c) 3   (d) 4

135. The minimum headroom of working spaces about service equipment, switchboards, panelboards, or motor control centers shall be 6´6˝, except service equipment or panelboards in existing dwelling units that do not exceed 200 amperes.

(a) True   (b) False

136. The dedicated space above a panelboard extends from the floor to the structural ceiling. A suspended ceiling is considered the structural ceiling.

(a) True   (b) False

137. When live parts of electrical equipment are guarded by suitable permanent, substantial partitions, or screens, any openings in such partitions or screens shall be so sized and located that persons are not likely to come into accidental contact with the live parts or to bring _____ into contact with them.

(a) dust   (b) conducting objects   (c) wires   (d) contaminating parts

138. Electrical installations over 600 volts located in _____ , where access is controlled by lock and key or other approved means, shall be considered to be accessible to qualified persons only.

(a) a room or closet

(b) a vault

(c) an area surrounded by a wall, screen, or fence

(d) any of these

139. • On circuits over 600 volts, nominal, where energized live parts are exposed, the minimum clear work space shall not be less than _____ high for over 600 volts.

(a) 3´   (b) 5´   (c) 6´3˝   (d) 6´6˝

140. _____ shall be provided to give safe access to the working space around equipment over 600 volts installed on platforms, balconies, mezzanine floors, or in attic or roof rooms or spaces.
     (a) Ladders
     (b) Platforms or ladders
     (c) Permanent ladders or stairways
     (d) Openings

141. • Switches or other equipment operating at 600 volts, nominal, or less and serving only equipment within a high-voltage vault, room, or enclosure shall be permitted to be installed in the _____ enclosure, room, or vault if accessible to qualified persons only.
     (a) restricted   (b) medium-voltage   (c) sealed   (d) high-voltage

142. The lighting outlets provided for illumination about electrical equipment over 600 volts shall be so arranged that persons changing lamps or making repairs on the _____ will not be endangered by live parts or other equipment.
     (a) lighting system   (b) electrical system   (c) electrical equipment   (d) panelboard

Chapter 2 – Wiring and Protection

Chapter 2 of the *NEC®* is a general rules chapter as applied to wiring and protection of conductors and equipment. The rules in Chapter 2 apply everywhere in the *NEC®* except as modified in Chapters 5, 6, or 7. Along with Chapter 3, it can be considered the heart of the *Code*. Many of the everyday applications of the *NEC®* are found in this chapter.

Article 200 – Use and Identification of Grounded Conductor

Article 200 covers requirements for identification of terminals, grounded neutral conductors in premises wiring systems, and identification of grounded conductors.

The general rule is: All premises wiring systems shall have an identified grounded neutral conductor. A grounded conductor is usually the system neutral or white wire. Be careful not to confuse white grounded wires (neutral) with green grounding wires.

143. Premises wiring shall not be electrically connected to a supply system unless the supply system contains, for any grounded conductor of the interior system, a corresponding conductor that is grounded.
     (a) True   (b) False

144. Distinctive marking at the terminals during the process of installation shall identify the grounded conductors of _____ metal-sheathed cable.
     (a) armored   (b) mineral-insulated   (c) copper   (d) aluminum

145. • Where a cable containing an insulated conductor with a white or natural gray outer finish is used for 3-way or 4-way switch loops, the white or natural gray conductor can be used for the supply to the switch. Reidentification of the white or natural gray conductor is not required.
     (a) True   (b) False

146. No _____ shall be attached to any terminal or lead so as to reverse designated polarity.
     (a) grounded conductor   (b) grounding conductor
     (c) ungrounded conductor   (d) grounding connector

Article 210 – Branch Circuit

Article 210 covers the requirements for branch circuits. A branch circuit is the conductors between the final overcurrent protection device and the outlet or utilization equipment. Article 210 applies generally to most branch circuits, but not to all!

147. Where more than one nominal voltage system exists in a building, each ungrounded system conductor shall be identified by phase and system. The means of identification shall be permanently posted at each branch-circuit panelboard.
     (a) True   (b) False

148. In dwelling units, the voltage between conductors shall not exceed 120 volts, nominal, between conductors that supply the terminals of _____ .
(a) lighting fixtures
(b) cord- and plug-connected loads of less than 1,440 VA, nominal
(c) cord- and plug-connected loads of more than 1,440 VA, nominal
(d) a and b

149. • When replacing an ungrounded receptacle in a bedroom of a dwelling unit, if a grounding means does not exist in the receptacle enclosure, you must use a _____ .
(a) nongrounding receptacle   (b) grounding receptacle
(c) GFCI-type receptacle   (d) a or c

150. Ground-fault circuit-interrupter (GFCI) protection for personnel is required for all 125-volt, single-phase, 15- and 20-ampere receptacles installed in a dwelling-unit _____ .
(a) attic   (b) garage   (c) laundry   (d) all of these

151. GFCI protection of personnel is required for fixed electric snow-melting or de-icing equipment receptacles that are not readily accessible and are supplied by a dedicated branch circuit.
(a) True   (b) False

152. Receptacle outlets shall not be installed in a face-up position in the work surfaces or countertops of a wet bar sink.
(a) True   (b) False

153. Where the load is computed on a volt-amperes-per-square-foot basis, the load shall be evenly proportioned among multioutlet branch circuits within the _____ .
(a) premises   (b) branch circuits   (c) panelboard(s)   (d) dwelling unit

154. Effective January 1, 2002, all branch circuits that supply 125-volt, 15- and 20-ampere receptacles in dwelling-unit bedrooms shall be _____ protected.
(a) AFCI   (b) GFCI   (c) a and b   (d) none of these

155. The neutral conductor of a 3-wire branch circuit supplying a household electric range shall be permitted to be smaller than the ungrounded conductors where the maximum demand of an $8^3/_4$ kW range has been computed according to Column A of Table 220-19. However, the neutral ampacity shall not be less than _____ percent of the branch circuit rating and not be smaller than No. _____ .
(a) 50, 6   (b) 70, 6   (c) 50, 10   (d) 70, 10

156. Where connected to a branch circuit supplying _____ or more receptacles or outlets, a receptacle shall not supply a total cord- and plug-connected load in excess of the maximum specified in Table 210-21(b)(2).
(a) two   (b) three   (c) four   (d) five

157. • It shall be permitted to base the _____ rating of a range receptacle on a single range demand load specified in Table 220-19.
(a) circuit   (b) voltage   (c) ampere   (d) load

158. Multioutlet circuits rated 15 or 20 amperes can supply fixed appliances (utilization equipment fastened in place) as long as the fixed appliances do not exceed _____ percent of the circuit rating.
(a) 125   (b) 100   (c) 75   (d) 50

159. No point along the floor line in any wall space may be more than _____ feet from an outlet.
(a) 12   (b) 10   (c) 8   (d) 6

160. Receptacle outlets shall, insofar as practicable, be spaced equal distances apart in a dwelling unit. Receptacle outlets in floors shall not be counted as part of the required number of receptacle outlets unless they are located within _____ of the wall.
(a) 6 inches   (b) 12 inches   (c) 18 inches   (d) close to the wall

161. Two 20-ampere small appliance circuits can supply more than one kitchen of a dwelling.
(a) True   (b) False

162.   One receptacle outlet shall be installed at each island or peninsular countertop with a long dimension of _____ inches or greater and a short dimension of _____ inches or greater.

(a) 12, 24   (b) 24, 12   (c) 24, 48   (d) 48, 24

163.   Kitchen and dinning room countertop receptacle outlets in dwelling units must be installed above the countertop surface and not more than ___ inches above the countertop.

(a) 12   (b) 18   (c) 24   (d) none of these

164.   A one-family or two-family dwelling unit requires a minimum of _____ GFCI receptacle(s) located outdoors.

(a) zero   (b) one   (c) two   (d) three

165.   If a portion of the dwelling-unit basement is finished, then a GFCI-protected 125-volt 15- or 20-ampere, receptacle outlet must be installed in the unfinished area of the basement.

(a) True   (b) False

166.   At least one receptacle outlet shall be installed directly above a show window for each _____ .

(a) 12 square feet   (b) 12 linear feet   (c) 10 linear feet   (d) 15 horizontal feet

167.   A lighting outlet shall be installed at the exterior side of each outdoor entrance or exit that has grade level access.

(a) True   (b) False

168.   Illumination on the exterior side of outdoor entrances or exits that have grade level access can be controlled by _____ .

(a) home automation devices
(b) motion sensors
(c) photocells
(d) any of these

169.   For other than dwelling units, a wall-switched lighting outlet is required near equipment requiring servicing in attics or underfloor spaces, and the switch must be located at the point of entrance to the attic or underfloor space.

(a) True   (b) False

Article 215 – Feeders

Article 215 covers feeders, which are the conductors between the service point to the service equipment.

170.   Where installed in a metal raceway, all feeder conductors using a common neutral shall be _____ .

(a) insulated for 600 volts
(b) enclosed within the same raceway
(c) shielded
(d) none of these

171.   • Ground-fault protection is required for the feeder disconnect if _____ .

(a) the feeder is rated 1,000 amperes or more
(b) it is a solidly grounded wye system
(c) the system voltage is 480Y/277 volts
(d) all of these

Article 220 – Branch Circuit and Feeder Calculations

Article 220 covers the requirements feeders and service load calculations for residential, commercial, and industrial occupancies.

172.   • The 3 volt-ampere per square foot for dwelling unit general lighting includes all 15 and 20 ampere general use receptacles. The floor area shall _____ open porches, garages, or unused or unfinished spaces, not adaptable for future use.

(a) include   (b) not include

173.   • When determining the load for recessed lighting fixtures for branch circuits, the load shall be based on the _____ .
(a) wattage rating of the fixture
(b) VA of the equipment and lamps
(c) wattage rating of the lamps
(d) none of these

174.   • The demand factors of Table 220-11 shall apply to the computed load of feeders to areas in hospitals, hotels, and motels where the entire lighting is likely to be used at one time, as in operating rooms, ballrooms, or dining rooms.
(a) True   (b) False

175.   • Receptacle loads for non-dwelling units computed at not more than 180 VA per outlet in accordance with Section 220-3(c)(9) shall be permitted to be _____ .
(a) added to the lighting loads and made subject to demand factors of Table 220-11
(b) made subject to the demand factors of Table 220-13
(c) made subject to the lighting demand loads of Table 220-3(b)
(d) a or b

176.   Using standard load calculations, the feeder demand factor for five household clothes dryers is _____ percent.
(a) 70   (b) 80   (c) 50   (d) 100

177.   The demand factors of Table 220-20 apply to space heating, ventilating, and air-conditioning equipment.
(a) True   (b) False

178.   There shall be no reduction in the size of the neutral conductor on _____-type lighting loads.
(a) dwelling unit   (b) hospital   (c) nonlinear   (d) motel

179.   Under the optional method for calculating a single-family dwelling, general loads beyond the initial 10 kW are to be assessed at a _____ percent demand factor.
(a) 40   (b) 50   (c) 60   (d) 75

180.   • The connected load to which the demand factors of Table 220-32 apply shall include the _____ rating of all appliances that are fastened in place, permanently connected or located to be on a specific circuit: ranges, wall-mounted ovens, counter-mounted cooking units, clothes dryers, water heaters, and space heaters.
(a) calculated
(b) nameplate
(c) circuit
(d) overcurrent protection

181.   • Service-entrance or feeder conductors whose demand load is determined by the optional calculation as permitted in Section 220-36 shall not be permitted to have the neutral load determined by Section 220-22.
(a) True   (b) False

Article 225 – Outside Branch Circuits and Feeders

This Article contains the requirements for proper installation of outside branch circuits and feeders

182.   Conductors installed between buildings, structures, or poles as well as wiring and equipment located on or attached to the outside of buildings, structures, or poles must comply with Article 225.
(a) True   (b) False

183.   On circuits of 600 volts or less, overhead spans up to 50 feet in length shall have conductors not smaller than _____ .
(a) No. 14   (b) No. 12   (c) No. 6   (d) No. 10

184.   Open conductors installed outside shall be separated from open conductors of other circuits by not less than _____ inches.
(a) 4   (b) 6   (c) 8   (d) 10

185. Conductors on poles shall have a separation of not less than 1 foot where not placed on racks or brackets. Power conductors rated 300 volts or less installed above communication conductors supported on poles shall provide a horizontal climbing space not less than _____ inches.
   (a) 12   (b) 24   (c) 6   (d) none of these

186. Where the voltage between conductors does not exceed 300 volts, and the roof has a slope of not less than 4 inches in 12 inches, a reduction in clearance to _____ feet shall be permitted.
   (a) 1   (b) 2   (c) 3   (d) 4

187. Overhead conductors to a building shall maintain a vertical clearance of _____ feet above platforms, projections or surfaces from which they might be reached. This vertical clearance shall extend 3 feet measured horizontally from the platforms, projections, or surfaces from which they might be reached.
   (a) 6   (b) 8   (c) 10   (d) 12

188. Raceways on exterior surfaces of buildings shall be _____ and arranged to drain.
   (a) raintight   (b) rainproof   (c) watertight   (d) waterproof

189. Which of the following cannot be attached to vegetation?
   (a) overhead conductor spans
   (b) overhead conductor spans for temporary wiring
   (c) lighting fixtures
   (d) all of these

Article 230 – Services

   Article 230 covers service conductors and equipment for control and protection of services and their installation requirements.

190. A single building or other structure sufficiently large to require two or more services is permitted by _____ .
   (a) architects
   (b) special permission
   (c) written authorization
   (d) master electricians

191. _____ shall not be installed beneath openings through which material may be moved, such as openings in farm and commercial buildings, and shall not be installed where they will obstruct entrance to these building openings.
   (a) Overcurrent protection devices   (b) Overhead service conductors
   (c) Grounding conductors   (d) Wiring systems

192. The minimum size service drop conductor permitted by the *Code* is No. _____ copper and No. _____ aluminum.
   I. 6 cu   II. 8 cu   III. 6 al   IV. 8 al
   (a) II and III   (b) I and IV   (c) I and III   (d) II and IV

193. 120/240-volt service conductors terminate at a through-the-roof raceway, and less than 6 feet of the conductors pass over the roof overhang. The minimum clearance above the roof for these service conductors is _____ .
   (a) 12 inches   (b) 18 inches   (c) 2 feet   (d) 5 feet

194. The minimum clearance for service drops, not exceeding 600 volts, over commercial areas subject to truck traffic is _____ feet.
   (a) 10   (b) 12   (c) 15   (d) 18

195. Where raceway-type service masts are used, all raceway fittings shall be _____ for use with service masts.
   (a) identified   (b) approved   (c) heavy-duty   (d) listed

196. Underground copper service conductors shall not be smaller than No. _____ .
   (a) 3   (b) 4   (c) 6   (d) 8

197. Where two to six service disconnecting means in separate enclosures are grouped at one location and supply separate loads from one service drop or lateral, _____ set(s) of service-entrance conductors shall be permitted to supply each or several such service equipment enclosures.
    (a) one   (b) two   (c) three   (d) four

198. Wiring methods permitted for service conductors include _____ .
    (a) mineral-insulated cable   (b) electrical metallic tubing
    (c) liquidtight nonmetallic conduit   (d) all of these

199. Service conductors or cables shall be protected against physical damage.
    (a) True   (b) False

200. Where individual open conductors are exposed to _____, the conductors shall be mounted on insulators or on insulating supports attached to racks, brackets, or other approved means.
    (a) a corrosive environment
    (b) the weather
    (c) the general public
    (d) any inspector

Unit 2

Electrical Circuits

PART A – SERIES CIRCUITS

INTRODUCTION TO SERIES CIRCUITS

A series circuit is a circuit in which the current leaves the voltage source and flows through every electrical device with the same intensity before it returns to the voltage source. If any part of a series circuit is opened, the current stops flowing in the entire circuit (Figure 2–1).

For most practical purposes, series (closed-loop) circuits are not used for building wiring, but they are important for the operation of many control and signal circuits (Figure 2–2). Motor control circuit stop switches are generally wired in series with the starter's coil and the line conductors. Dual-rated motors, such as 460/230 volts, have their windings connected in series when supplied by the higher voltage, and in parallel when supplied by the lower voltage.

2–1 UNDERSTANDING SERIES CALCULATIONS

It is important to understand the relationship between resistance, current, voltage, and power of series circuits, (Figure 2–3).

Calculating Resistance Total

In a series circuit, the total resistance of the circuit is equal to the sum of all the series resistors' resistance, according to the formula:

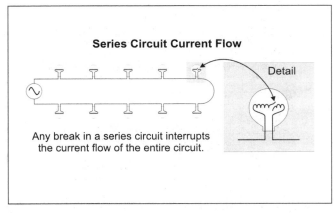

Series Circuit Current Flow

Any break in a series circuit interrupts
the current flow of the entire circuit.

Detail

Figure 2–1
Series Circuit Current Flow

$$R_T = R_1 + R_2 + R_3 + R_4$$

❏ Total Resistance

What is the total resistance of the loads (Figure 2–4)?

(a) 2.5 ohm (b) 5.5 ohm

(c) 7.5 ohm (d) 10 ohm

 • Answer: (c) 7.5 ohm

R_1 – Power Supply	0.05 ohm
R_2 – Conductor No. 1	0.15 ohm
R_3 – Appliance	7.15 ohm
R_4 – Conductor No. 2	0.15 ohm
Total Resistance:	7.50 ohm

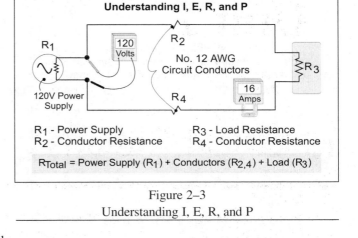

Figure 2–3

Understanding I, E, R, and P

Calculating Voltage Drop

The result of current flowing through a resistor is a reduction in voltage across the resistor, which is called voltage drop. In a closed loop circuit, the sum of the voltage drops of all the loads is equal to the voltage source (Figure 2–5). This is known as Kirchoff's First Law of Voltage: The voltage drop (VD) of each resistor can be determined by the formula:

$$E_{VD} = I \times R$$

I = Current of the circuit

R = Resistance of the resistor

❏ Voltage Drop

What is the voltage drop across each resistor (Figure 2–6) using the formula VD = I × R?

 • Answer: VD = I × R

VD = I × R

VD_{R1} – Power Supply	16 ampere × 0.05 ohm	=	0.8 volts dropped
VD_{R2} – Conductor No. 1	16 ampere × 0.15 ohm	=	2.4 volts dropped
VD_{R3} – Appliance	16 ampere × 7.15 ohm	=	114.4 volts dropped

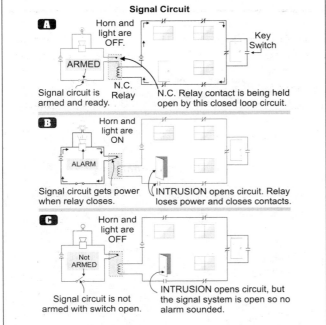

Figure 2–2

Signal Circuit

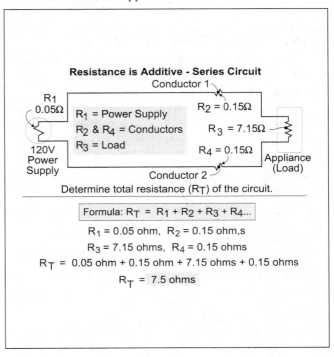

Figure 2–4

Resistance Is Additive

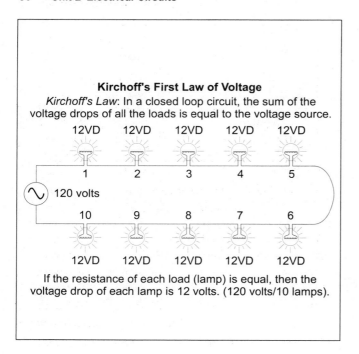

Figure 2–5
Kirchoff's First Law of Voltage

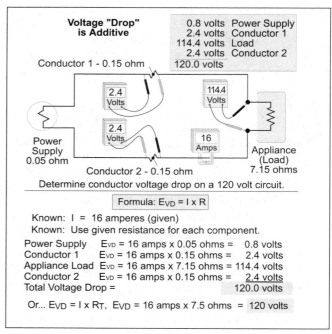

Figure 2–6
Voltage Drop Is Additive

VD$_{R4}$ – Conductor No. 2	16 ampere × 0.15 ohm	=	2.4 volts dropped
Total Voltage Drop	16 ampere × 7.50 ohm	=	120.0 volts dropped

Note: Due to rounding, the sum of the voltage drops may be slightly different than the voltage source.

Voltage of Series Connected Power Supplies

When *power supplies* are connected in series, the voltage of each power supply will add together (sum), providing that all the polarities are connected properly.

❏ Series Connected Power Supplies

What is the total voltage output of four 1.5-volt batteries connected in series (Figure 2–7)?

(a) 1.5 volts   (b) 3.0 volts

(c) 4.5 volts   (d) 6.0 volts

• Answer: (d) 6 volts

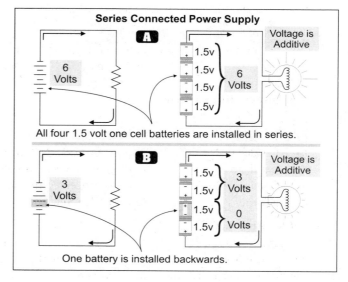

Figure 2–7
Series Connected Power Supply

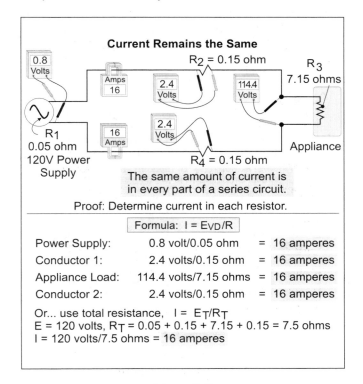

Current Remains the Same

R₂ = 0.15 ohm
R₃ 7.15 ohms
0.8 Volts
Amps 16
2.4 Volts
114.4 Volts
16 Amps
2.4 Volts
R₁ 0.05 ohm 120V Power Supply
Appliance
R₄ = 0.15 ohm

The same amount of current is in every part of a series circuit.

Proof: Determine current in each resistor.

Formula: I = E$_{VD}$/R

Power Supply:	0.8 volt/0.05 ohm	= 16 amperes
Conductor 1:	2.4 volts/0.15 ohm	= 16 amperes
Appliance Load:	114.4 volts/7.15 ohms	= 16 amperes
Conductor 2:	2.4 volts/0.15 ohm	= 16 amperes

Or... use total resistance, I = E$_T$/R$_T$
E = 120 volts, R$_T$ = 0.05 + 0.15 + 7.15 + 0.15 = 7.5 ohms
I = 120 volts/7.5 ohms = 16 amperes

Figure 2–8
Current Remains the Same

Series Circuit - Power

Conductor 1
R₁ 0.05 ohm
Power (watts) is additive.
R₂ = 0.15 ohm 38.4 watts
R₃ = 7.15 ohms
12.8 watts
R₄ = 0.15 ohm 38.4 watts
120V Power Supply
16 Amps
1,830 watts Appliance (Load)
Conductor 2

Determine the total power consumed by the circuit.

Formula: P = I² x R

Known: I = 16 amperes
Known: R of each load as given

Power Supply:	16² amps x 0.05 ohm	= 12.8 watts
Conductor 1:	16² amps x 0.15 ohms	= 38.4 watts
Appliance Load:	16² amps x 7.15 ohms	= 1,830.4 watts
Conductor 2:	16² amps x 0.15 ohm	= 38.4 watts
	Total Watts	= 1,920.0 watts

Or... P = I² x R$_T$ = 16² amps x 7.5 ohms = 1,920 watts

Figure 2–9
Series Circuit – Power Is Additive

Current of Resistor

In a series circuit, the current throughout the circuit is constant and does not change. The current through each resistor of the circuit can be determined by the formula:

$I = E/R$

E = Voltage drop of the load or circuit

R = Resistance of the load or circuit

Note: If the resistance is not given, you can determine the resistance of a resistor (if you know the nameplate voltage and power rating of the load) by the formula: $R = E^2/P$, E = Nameplate voltage rating (squared), P = Nameplate power rating

❏ Current of Circuit

What is the current flow through the series circuit (Figure 2–8)? Note: Current remains the same.

(a) 4 ampere   (b) 8 ampere
(c) 12 ampere   (d) 16 ampere

   • Answer: (d) 16 ampere

$I = E/R$

I$_{R1}$ – Power Supply	0.8 volt drop/0.05 ohm	= 16 ampere
I$_{R2}$ – Conductor No. 1	2.4 volts dropped /0.15 ohm	= 16 ampere
I$_{R3}$ – Appliance	114.4 volts dropped/7.15 ohm	= 16 ampere
I$_{R4}$ – Conductor No. 2	2.4 volts dropped/0.15 ohm	= 16 ampere
Circuit Current	120 volts/7.5 ohm	= 16 ampere

Power of Resistor or Circuit

The power consumed in a series circuit is equal to the sum of the power of all of the resistors in the series circuit. The resistor with the highest resistance will consume the most power, and the resistor with the smallest resistance will consume the least power. You can calculate the power consumed (watts) of each resistor or of the circuit by the formula:

$P = I^2R$, I^2 = Current of the circuit (squared), R = Resistance of circuit or resistor

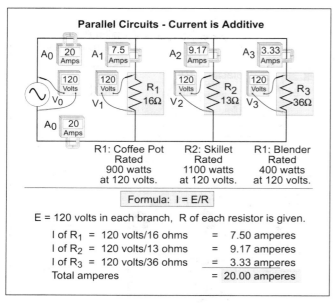

Figure 2–15
Parallel Circuits – Current Is Additive

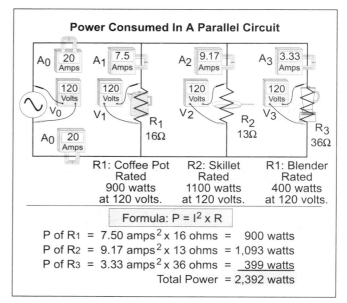

Figure 2–16
Power Consumed in a Parallel Circuit

Current through Each Branch

The current from the power supply is equal to the sum of the branch circuit currents. The current in each branch depends on the branch voltage and branch resistance and can be calculated by the formula:

$$I = \frac{E}{R}$$

E = Voltage of Branch

R = Resistance of Branch

❏ Current through Each Branch

What is the current of each appliance (Figure 2–15)?

• Answer: $I = \frac{E}{R}$

$I = \frac{E}{R}$

I_{R1} – Coffee Pot	120 volts/16 ohm	=	7.50 ampere
I_{R2} – Skillet	120 volts/13 ohm	=	9.17 ampere
I_{R3} – Blender	120 volts/36 ohm	=	3.33 ampere
Total Current (7.5 ampere + 9.17 ampere + 3.33 ampere)			20.00 ampere

Power Consumed of Each Branch

The total power consumed of any circuit is equal to the sum of the branch powers. Each branch power depends on the branch current and resistance. The power can be found by the formula:

$$P = I^2R$$

I = Current of Each Branch

R = Resistance of Each Branch

❏ Power of Each Branch

What is the power consumed of each appliance (Figure 2–16)?

• Answer: $P = I^2R$

$P = I^2R$

I_{R1} – Coffee Pot	7.5 ampere^2 ×16 ohm	=	900 watts
I_{R2} – Skillet	9.17 ampere^2 × 13 ohm	=	1,093 watts
I_{R3} – Blender	3.33 ampere^2 × 36 ohm	=	399 watts
Total Power (900 watts + 1,093 watts + 399 watts)			2,392 watts

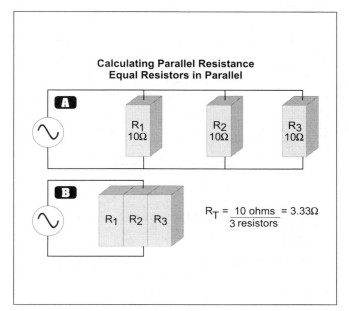

Figure 2–17
Calculating Parallel Resistance–Equal Resistors in Parallel

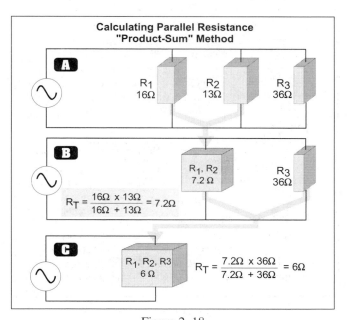

Figure 2–18
Calculating Parallel Resistance "Product–Sum" Method

2–5 PARALLEL CIRCUIT RESISTANCE CALCULATIONS

In a parallel circuit, the total circuit resistance is always less than the smallest resistor and can be determined by one of three methods:

Equal Resistor Method

The equal resistor method can be used when all the resistors of the parallel circuit have the same resistance. Simply divide the resistance of one resistor by the number of resistors in parallel.

$R_T = R/N$

R = Resistance of One Resistor, N = Number of Resistors

❏ Equal Resistors Method

The total resistance of three 10 ohm resistors is _____ (Figure 2–17).

(a) 10 ohm   (b) 20 ohm   (c) 30 ohm   (d) none of these

• Answer: (d) none of these

$$R_T = \frac{\text{Resistance of One Resistor}}{\text{Number of Resistors}}, = \frac{10 \text{ ohms}}{3 \text{ resistors}}, = 3.33 \text{ ohm}$$

The product over the sum method

The product over the sum method can be used to calculate the resistance of two resistors.

$$R_T = \frac{R_1 \times R_2 \text{ (Product)}}{R_1 + R_2 \text{ (Sum)}}$$

The term *product* means the answer of numbers that are multiplied together. The term *sum* is the answer to numbers that are added together. The product of the sum method can be used for more than two resistors, but only two can be calculated at a time.

❏ Product of the Sum Method

What is the total resistance of a 16 ohm coffee pot and a 13 ohm skillet connected in parallel (Figure 2–18)?

(a) 16 ohm   (b) 13.09 ohm   (c) 29.09 ohm   (d) 7.2 ohm

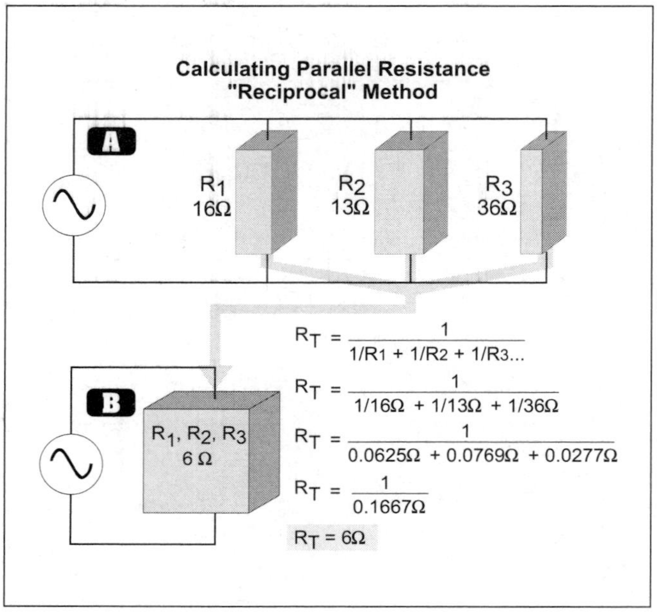

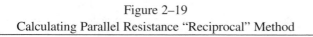

Calculating Parallel Resistance "Reciprocal" Method

A

R_1 16Ω R_2 13Ω R_3 36Ω

$R_T = \dfrac{1}{1/R_1 + 1/R_2 + 1/R_3...}$

$R_T = \dfrac{1}{1/16Ω + 1/13Ω + 1/36Ω}$

B

R_1, R_2, R_3
6 Ω

$R_T = \dfrac{1}{0.0625Ω + 0.0769Ω + 0.0277Ω}$

$R_T = \dfrac{1}{0.1667Ω}$

$R_T = 6Ω$

Figure 2–19
Calculating Parallel Resistance "Reciprocal" Method

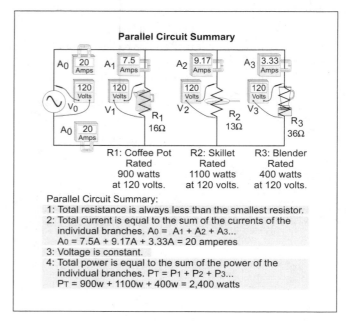

Parallel Circuit Summary

| A_0 20 Amps | A_1 7.5 Amps | A_2 9.17 Amps | A_3 3.33 Amps |

V_0 V_1 R_1 16Ω V_2 R_2 13Ω V_3 R_3 36Ω

120 Volts (each)

A_0 20 Amps

| R1: Coffee Pot Rated 900 watts at 120 volts. | R2: Skillet Rated 1100 watts at 120 volts. | R3: Blender Rated 400 watts at 120 volts. |

Parallel Circuit Summary:
1: Total resistance is always less than the smallest resistor.
2: Total current is equal to the sum of the currents of the individual branches. $A_0 = A_1 + A_2 + A_3...$
   $A_0 = 7.5A + 9.17A + 3.33A = 20$ amperes
3: Voltage is constant.
4: Total power is equal to the sum of the power of the individual branches. $P_T = P_1 + P_2 + P_3...$
   $P_T = 900w + 1100w + 400w = 2,400$ watts

Figure 2–20
Parallel Circuit Summary

• Answer: (d) 7.2 ohm

The total resistance of a parallel circuit is always less than the smallest resistor (13 ohm).

$$R_T = \frac{R_1 \times R_2}{R_1 + R_2} = \frac{16 \text{ ohms} \times 13 \text{ ohms}}{16 \text{ ohms} + 13 \text{ ohms}} = 7.2 \text{ ohm}$$

Reciprocal Method

The advantage of the reciprocal method is that this formula can be used for an unlimited number of parallel resistors.

$$R_T = \frac{1}{^1/R_1 + ^1/R_2 + ^1/R_3 \ldots}$$

❏ **Reciprocal Method**

What is the resistance total of a 16 ohm, 13 ohm, and 36 ohm resistor connected in parallel (Figure 2–19)?

(a) 13 ohm   (b) 16 ohm   (c) 36 ohm   (d) 6 ohm

• Answer: (d) 6 ohm

$$R_T = \frac{1}{^1/_{16} \text{ ohm} + ^1/_{13} \text{ ohm} + ^1/_{36} \text{ ohm}} \quad R_T = \frac{1}{0.0625 \text{ ohm} + 0.0769 \text{ ohm} + 0.0278 \text{ ohm}}$$

$$R_T = \frac{1}{0.1672 \text{ ohm}} \quad R_T = 6 \text{ ohm}$$

2–6 PARALLEL CIRCUIT SUMMARY

Note 1:   The total resistance of a parallel circuit is always less than the smallest resistor (Figure 2–20).

Note 2:   Current total of a parallel circuit is equal to the sum of the currents of the individual branches.

Note 3:   Voltage is constant.

Note 4:   Power total is equal to the sum of the power in all the individual branches.

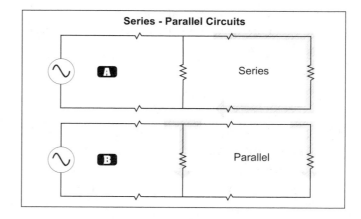

Series - Parallel Circuits

A   Series

B   Parallel

Figure 2–21
Series – Parallel Circuits

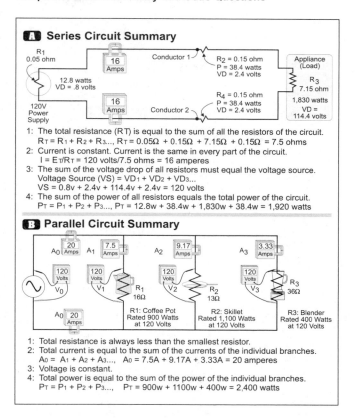

A Series Circuit Summary

1: The total resistance (RT) is equal to the sum of all the resistors of the circuit.
$R_T = R_1 + R_2 + R_3...$, $R_T = 0.05\Omega + 0.15\Omega + 7.15\Omega + 0.15\Omega = 7.5$ ohms
2: Current is constant. Current is the same in every part of the circuit.
$I = E_T/R_T = 120$ volts/7.5 ohms = 16 amperes
3: The sum of the voltage drop of all resistors must equal the voltage source.
Voltage Source (VS) = $VD_1 + VD_2 + VD_3...$
VS = 0.8v + 2.4v + 114.4v + 2.4v = 120 volts
4: The sum of the power of all resistors equals the total power of the circuit.
$P_T = P_1 + P_2 + P_3...$, $P_T = 12.8w + 38.4w + 1,830w + 38.4w = 1,920$ watts

B Parallel Circuit Summary

1: Total resistance is always less than the smallest resistor.
2: Total current is equal to the sum of the currents of the individual branches.
$A_0 = A_1 + A_2 + A_3...$, $A_0 = 7.5A + 9.17A + 3.33A = 20$ amperes
3: Voltage is constant.
4: Total power is equal to the sum of the power of the individual branches.
$P_T = P_1 + P_2 + P_3...$, $P_T = 900w + 1100w + 400w = 2,400$ watts

Figure 2–22
Series Circuit Summary

Series - Parallel Resistance

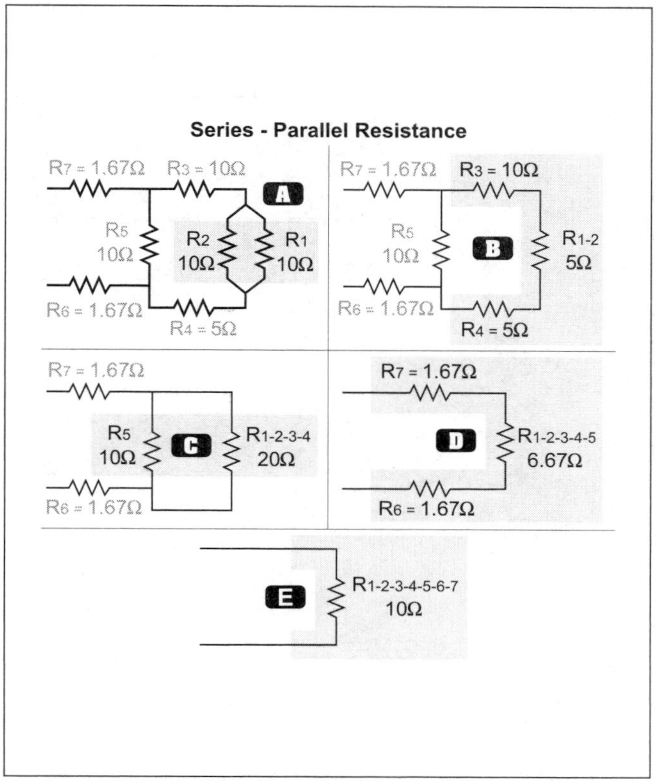

Figure 2–23
Series-Parallel Resistance

PART C – SERIES-PARALLEL CIRCUITS

INTRODUCTION TO SERIES-PARALLEL CIRCUITS

A series-parallel circuit is a circuit that contains some resistors in series and some resistors in parallel to each other (Figure 2–21).

2–7 REVIEW OF SERIES AND PARALLEL CIRCUITS

For a better understanding of series-parallel circuits, let's review the rules for series and parallel circuits. That portion of the circuit that contains resistors in series must comply with the rules of series circuits, and that portion of the circuit that is connected in parallel must comply with the rules of parallel circuits (Figure 2–22).

Series Circuit Rules:

Note 1:   Resistance is additive.

Note 2:   Current is constant.

Note 3:   Voltage is additive.

Note 4:   Power is additive.

Parallel Circuit Rules:

Note 1:   Resistance is less than the smallest resistor.

Note 2:   Current is additive.

Note 3:   Voltage is constant.

Note 4:   Power is additive.

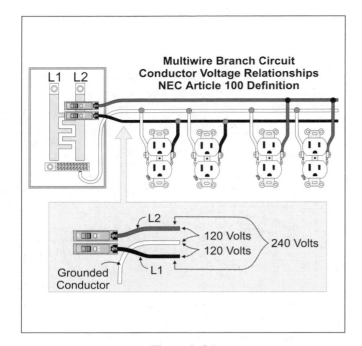

Figure 2–24
Multiwire Branch Circuit

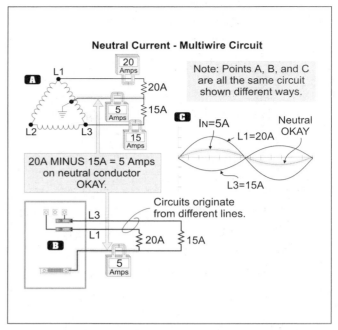

Figure 2–25
Neutral Current in Multiwire Branch Circuit

2–8 SERIES-PARALLEL CIRCUIT RESISTANCE CALCULATIONS

When determining the resistance total of a series-parallel circuit, it is best to redraw the circuit so you can see the series components and the parallel branches. Determine the resistance of the series components first or the parallel components depending on the circuit, then determine the resistance total of all the branches. Keep breaking the circuit down until you have determined the total effective resistance of the circuit as one resistor.

❑ **Series-Parallel Circuit Resistance**

What is the resistance total of the circuit (Figure 2–23, Part A)?

(a) 2 ohms   (b) 5 ohms   (c) 7 ohms   (d) 10 ohms

• Answer: (d) 10 ohms

Step 1: ➠ Determine the equal parallel resistors: R_1 (10 ohm) and R_2 (10 ohm) (Figure 2–23, Part A):
$R_T = {}^R/_N$, R = Resistance of one resistor, N = Number of resistors
R_T = 10 ohm/2 resistors, R_T = 5 ohm

Step 2: ➠ Redraw the circuit (Figure 2–23, Part B)
Determine the series resistance of $R_{1,2}$ (5 ohm), plus R_3 (10 ohm), plus R_4 (5 ohm).
Series resistance total is equal to $R_{1,2} + R_3 + R_4$
R_T = 5 ohm + 10 ohm + 5 ohm, R_T = 20 ohm

Step 3: ➠ Redraw the circuit (Figure 2–23, Part C)
Determine the parallel resistance of R_5, plus the resistance of $R_{1,2,3,4}$. Remember the resistance total of a parallel circuit is always less than the smallest parallel branch (10 ohm). Since we have only two parallel branches, the resistance can be determined by the product of the sum method.
$$R_T = \frac{R_5 \times R_{1,2,3,4}\ (Product)}{R_1 + R_2\ (Sum)} = \frac{(10\ ohms \times 20\ ohms)}{(10\ ohms + 20\ ohms)} = 6.67\ ohm$$

Step 4: ➠ Redraw the circuit (Figure 2–23, Part D)
Determine the series resistance total of R_7, plus R_6, plus $R_{1,2,3,4,5}$
Series resistance total is equal to $R_7 + R_6 + R_{1,2,3,4,5}$
R_T = 1.67 ohm + 1.67 ohm + 6.67 ohm, R_T = 10 ohm

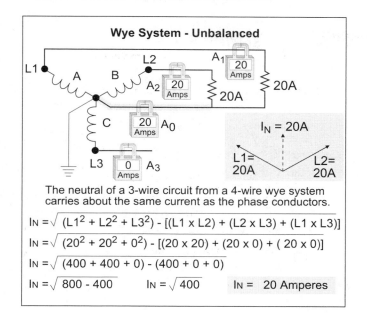

The neutral of a 3-wire circuit from a 4-wire wye system carries about the same current as the phase conductors.

$$I_N = \sqrt{(L1^2 + L2^2 + L3^2) - [(L1 \times L2) + (L2 \times L3) + (L1 \times L3)]}$$

$$I_N = \sqrt{(20^2 + 20^2 + 0^2) - [(20 \times 20) + (20 \times 0) + (20 \times 0)]}$$

$$I_N = \sqrt{(400 + 400 + 0) - (400 + 0 + 0)}$$

$$I_N = \sqrt{800 - 400} \qquad I_N = \sqrt{400} \qquad I_N = 20 \text{ Amperes}$$

Figure 2–26
Neutral Current – Unbalanced Wye Circuit

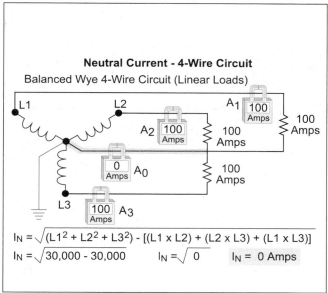

$$I_N = \sqrt{(L1^2 + L2^2 + L3^2) - [(L1 \times L2) + (L2 \times L3) + (L1 \times L3)]}$$

$$I_N = \sqrt{30,000 - 30,000} \qquad I_N = \sqrt{0} \qquad I_N = 0 \text{ Amps}$$

Figure 2–27
Neutral Current – Balanced Wye 4-Wire Circuit

PART D – MULTIWIRE BRANCH CIRCUITS

INTRODUCTION TO MULTIWIRE BRANCH CIRCUITS

A multiwire branch circuit consists of two or more ungrounded conductors (hot or phase conductors) having a potential difference between them and an equal difference of potential between each hot wire and grounded (neutral) conductor (Figure 2–24).

Note: The *National Electrical Code*® contains specific requirements on multiwire branch circuits. See Article 100 definition of multiwire branch circuit, Section 210-4 branch circuit requirements, and Section 300-13(b) requirements for pigtailing neutral conductors.

2–9 NEUTRAL CURRENT CALCULATIONS

When current flows in the *neutral* conductor of a multiwire branch circuit, the current is called *unbalanced current*. This current can be determined according to the following:

120/240-volt, 3-Wire Circuits

In a 120/240-volt, 3-wire circuit consisting of two hot wires and a grounded (neutral) conductor, the third wire will carry no current when the circuit is balanced. However, the neutral (or grounded) conductor will carry unbalanced current when the circuit is not balanced. The neutral current can be calculated as:

$$I_{Neutral} = N_{Line\ 1} - N_{Line\ 2}$$

$N_{Line\ 1}$ = Line 1 Neutral Current

$N_{Line\ 2}$ = Line 2 Neutral Current

❏ 120/240-volt, 3-Wire Neutral Current

What is the neutral current if current is: Line 1 = 20 ampere and Line 2 = 15 ampere (Figure 2–25)?

(a) 0 ampere   (b) 5 ampere   (c) 10 ampere   (d) 35 ampere

• Answer: (b) 5 ampere

$I_{Neutral} = N_{Line\ 1} - N_{Line\ 2}$

$I_{Neutral}$ = 20 ampere less 15 ampere

$I_{Neutral}$ = 5 ampere

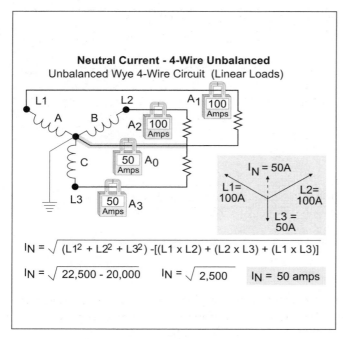

Figure 2–28
Neutral Current – 4-Wire Wye – Unbalanced

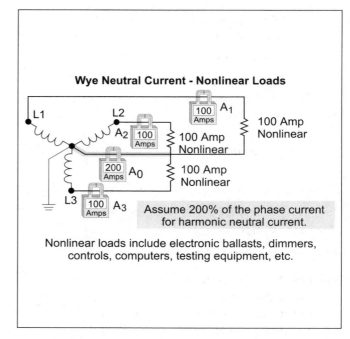

Figure 2–29
Neutral Current – Nonlinear Loads

Wye 3-Wire Neutral Current

A 3-wire, 208Y/120- or 480Y/277-volt circuit consisting of two phases and a neutral always carries unbalanced current. The current on the grounded (neutral) conductor is determined by the formula:

$$I_{Neutral} = \sqrt{(N_{Line\ 1}^2 + N_{Line\ 2}^2) - (N_{Line\ 1} \times N_{Line\ 1})}$$

❏ Three-wire Wye Circuit Neutral Current

What is the neutral current for a 20 ampere, 3-wire circuit (two hots and a neutral)? Power is supplied from a 208Y/120-volt feeder (Figure 2–26).

(a) 40 ampere   (b) 20 ampere   (c) 60 ampere   (d) 0 ampere

• Answer: (b) 20 ampere

$$I_{Neutral} = \sqrt{(N_{Line\ 1}^2 + N_{Line\ 2}^2) - (N_{Line\ 1} \times N_{Line\ 1})}$$

$$I_{Neutral} = \sqrt{(20\ ampere^2 + 20\ ampere^2) - (20\ ampere \times 29\ ampere)} = \sqrt{400}$$

$$I_{Neutral} = 20\ amperes$$

❏ Wye 4-wire Circuit Neutral Current

A 4-wire wye, 120/208Y- or 277/480Y-volt circuit will carry no current when the circuit is balanced, but will carry unbalanced current when the circuit is not balanced.

$$I_{Neutral} = \sqrt{(N_{Line\ 1}^2 = N_{Line\ 2}^2 = N_{Line\ 3}^2) - (N_{Line\ 1} \times N_{Line\ 2}) + (N_{Line\ 2} \times N_{Line\ 3}) + (N_{Line\ 1} \times N_{Line\ 3})}$$

$N_{Line\ 1}$ = Neutral Current Line 1

$N_{Line\ 2}$ = Neutral Current Line 1

$N_{Line\ 3}$ = Neutral Current Line 1

❏ Balanced Circuits

What is the neutral current for a 4-wire, 208Y/120-volt feeder where $L_1 = 100$ ampere, $L_2 = 100$ ampere, and $L_3 = 100$ ampere (Figure 2–27)?

(a) 50 ampere   (b) 100 ampere   (c) 125 ampere   (d) 0 ampere

• Answer: (d) 0 ampere

$$I_{Neutral} = \sqrt{(N_{Line\ 1}^2 \times N_{Line\ 2}^2 + N_{Line\ 3}^2) - (N_{Line\ 1} \times N_{Line\ 2}) + (N_{Line\ 2} \times N_{Line\ 3}) + (N_{Line\ 1} \times N_{Line\ 3})}$$

$$I_{Neutral} = \sqrt{(100\ Amps^2 + 100\ Amps^2 + 100\ Amps^2) - [(100A \times 100A) + (100A \times 100A) + (100A \times 100A)]} = \sqrt{0}$$

$I_{Neutral} = 0$ amperes

❏ Unbalanced Circuits

What is the neutral current for a 4-wire, 208Y/120-volt feeder where $L_1 = 100$ ampere, $L_2 = 100$ ampere, and $L_3 = 50$ ampere (Figure 2–28)?

(a) 50 ampere   (b) 100 ampere   (c) 125 ampere   (d) 0 ampere

   • Answer: (a) 50 ampere

$$I_{Neutral} = \sqrt{(N_{Line\ 1}^2 \times N_{Line\ 2}^2 + N_{Line\ 3}^2) - (N_{Line\ 1} \times N_{Line\ 2}) + (N_{Line\ 2} \times N_{Line\ 3}) + (N_{Line\ 1} \times N_{Line\ 3})}$$

$$I_{Neutral} = \sqrt{(100A^2 + 100A^2 + 50A^2) - [(100A \times 100A) + (100A \times 50A) + (100A \times 50A)]}$$

$$= \sqrt{2,500}$$

$I_{Neutral} = 50$ amperes

Nonlinear Load Neutral Current

The neutral conductor of a balanced wye, 4-wire circuit will carry current when supplying power to nonlinear loads, such as computers, copy machines, laser printers, fluorescent lighting, etc. The current can be as much as two times the phase current, depending on the harmonic content of the loads.

❏ Nonlinear Loads

What is the neutral current for a 4-wire, 208Y/120-volt feeder supplying power to a nonlinear load, where $L_1 = 100$ ampere, $L_2 = 100$ ampere, and $L_3 = 100$ ampere? Assume that the harmonic content results in neutral current equal to 200 percent of the phase current (Figure 2–29).

(a) 80 ampere   (b) 100 ampere   (c) 125 ampere   (d) 200 ampere

   • Answer: (d) 200 ampere

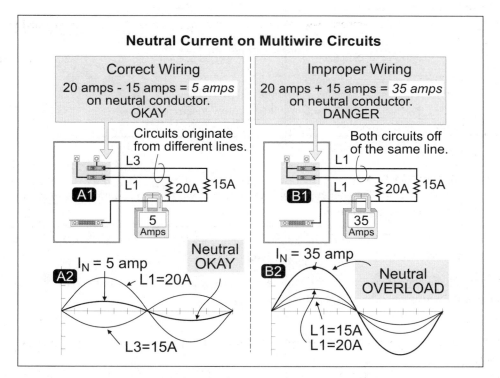

Figure 2–30
Neutral Current on Multiwire Circuits

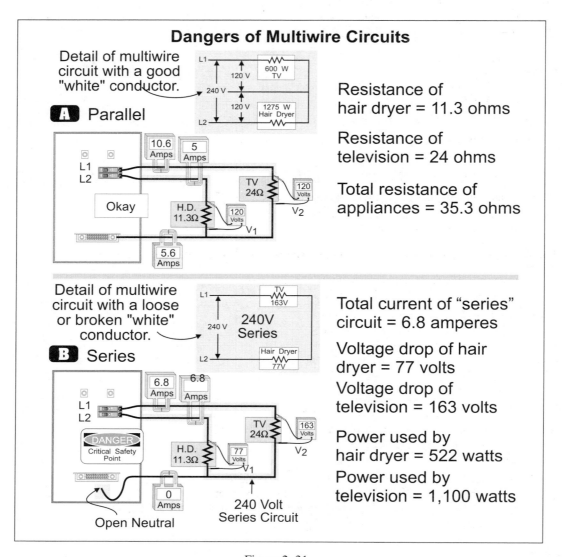

Figure 2–31
Neutral Current on Multiwire Circuits

2–10 DANGERS OF MULTIWIRE BRANCH CIRCUIT

Improper wiring, or mishandling of multiwire branch circuits, can cause excessive neutral current (overload) or destruction of electrical equipment because of overvoltage if the neutral conductor is opened.

Overloading the Neutral

If the ungrounded conductors (hot wires) of a multiwire branch circuit are connected to different phases, the current on the grounded (neutral) conductor will cancel. If the ungrounded conductors are not connected to different phases, the current from each phase will add on the neutral conductor. This can result in an overload of the neutral conductor (Figure 2–30).

Note: Overloading of the neutral conductor will cause the insulation to look discolored due to excessive heat. Now you know why the neutral wires sometimes look burned.

Overvoltage

If the neutral conductor of a multiwire branch circuit is opened, the multiwire branch circuit changes from a parallel circuit into a series circuit. Instead of two 120-volt circuits, there is now one 240-volt circuit, which can result in fires and the destruction of the electric equipment because of overvoltage (Figure 2–31).

To determine the operating voltage of each load in an open multiwire branch circuit, use the following steps:

Step 1: ➡ Determine the resistance of each appliance, $R = E^2/P$.
E = Appliance voltage nameplate rating
P = Appliance power nameplate rating

Step 2: ➡ Determine the circuit resistance, $R_T = R_1 + R_2$.

Step 3: ➡ Determine the current of the circuit, $I = E_S/R_T$.
E_S = Voltage Source
R_T = Resistance total, from Step 2

Step 4: ➡ Determine the voltage for each appliance, $E = I_T \times R$.
I_T = Current of the circuit
R = Resistance of each resistor

Step 5: ➡ Determine the power consumed of each appliance, $P = E^2/R$.
E = Voltage the appliance operates at (squared)
R = Resistance of the appliance

❏ **At what voltage does each of the loads operate if the neutral is opened (Figure 2–31)?**

Step 1: ➡ Determine the resistance of each appliance, $R = E^2/P$.
Hair dryer rated 1,275 watts at 120 volts.
$R = 120$ volts^2/1,275 watts = 11.3 ohms
Television rated 600 watts at 120 volts.
$R = 120$ volts^2/600 watts = 24 ohm

Step 2: ➡ Determine the circuit resistance, $R_T = R_1 + R_2$.
$R_T = 11.3$ ohm + 24 ohm = 35.3 ohm

Step 3: ➡ Determine the current of the circuit, $I = E_S/R_T$.
E_S = Voltage Source, R_T = Resistance Total
$I = \dfrac{240 \text{ volts}}{35.3 \text{ ohms}} = 6.8$ ampere

Step 4: ➡ Determine the voltage for each appliance, $E = I_T \times R$.
I_T = Current of the circuit, R = Resistance of each resistor
Hair dryer: 6.8 ampere $\times$ 11.3 ohm = 76.84 volts
Television: 6.8 ampere $\times$ 24 ohm = 163.2 volts

Step 5: ➡ Determine the power consumed of each appliance, $P = E^2/R$.
E^2 = Voltage the appliance operates at (squared)
R = Resistance of the appliance
Hair Dryer: $P = 76.8$ volts^2/11.3 ohm = 522 watts
Television: $P = 163.2$ volt^2/24 ohm = 1,110 watts

Note: The 600 watt, 120-volt rated TV operates at 163 volts and consumes 1,110 watts. You can kiss this TV good-bye. Because of the dangers associated with multiwire branch circuits, do not use them for sensitive or expensive equipment such as computers, stereos, etc.

Unit 2 – Electrical Circuits Summary Questions

Part A – Series Circuits

Introduction to Series Circuits

1. A closed-loop circuit is a circuit in which a specific amount of current leaves the voltage source and flows through every electrical device in a single path before it returns to the voltage source.
   (a) True   (b) False

2. Closed-loop circuits are typically used for signal and control circuits.
   (a) True   (b) False

2–1 Understanding Series Calculations

3. Resistance opposes the flow of electrons. In a series circuit, the total circuit resistance is equal to the sum of all the series resistors.
   (a) True   (b) False

4. The opposition to current flow results in _____ .
   (a) current   (b) voltage   (c) voltage drop   (d) none of these

5. In a series circuit, the current is _____ through the transformer, the conductors, and the appliance.
   (a) proportional   (b) distributed   (c) additive   (d) constant

6. • When power supplies are connected in series, the voltage remains the same, provided that all the polarities are connected properly.
   (a) True   (b) False

7. The power consumed in a series circuit is equal to the power of the largest resistor in the series circuit.
   (a) True   (b) False

Part B – Parallel Circuits

Introduction to Parallel Circuits

8. A _____ circuit is a circuit in which current leaves the voltage source, branches through different parts of the circuit in different magnitudes, and then branches back to the voltage source.
   (a) series   (b) parallel   (c) series-parallel   (d) multiwire

2–4 Understanding Parallel Calculations

9. • The power supply provides the pressure needed to move the electrons; however, the _____ oppose(s) the current flow.
   (a) power supply   (b) conductors   (c) appliances   (d) all of these

10. When power supplies are connected in parallel, the amp-hour capacity remains the same.
    (a) True   (b) False

11. The total current of a parallel circuit is equal to the sum of the branch currents. The current in each branch can be calculated by the formula: $I = {}^E/_R$.
    (a) True   (b) False

12. When current flows through a resistor, power is consumed. The power consumed of each branch can be determined by the formula: $P = I^2 \times R$. The total power consumed in a parallel circuit is equal to the largest branch power.
     (a) True   (b) False

2–5 Parallel Circuit Resistance Calculations

13. The basic method(s) of calculating total resistance of a parallel circuit is/are _____ .
     (a) equal resistor method   (b) product of the sum method
     (c) reciprocal method   (d) all of these

14. The total resistance of three 6 ohm resistors in parallel is _____ .
     (a) 6 ohm   (b) 12 ohm   (c) 18 ohm   (d) none of these

15. The circuit resistance of a 600-watt coffee pot and a 1,000-watt skillet is _____ when connected to a 120-volt parallel circuit.
     (a) 24 ohm   (b) 14.4 ohm   (c) 38.4 ohm   (d) 9 ohm

16. The resistance total of a 20 ohm, 20 ohm, and 10 ohm resistor in parallel is _____ .
     (a) 5 ohm   (b) 20 ohm   (c) 30 ohm   (d) 50 ohm

2–6 Parallel Circuit Summary

17. • Which of the following statements is/are true about parallel circuits?
     I. The total resistance of a parallel circuit is less than the smallest resistor of the circuit.
     II. Current total is equal to the sum of the branch currents.
     III. The power of all resistors is equal to the sum of the branch powers.
     (a) I and II   (b) I, II, and III   (c) II and III   (d) I and III

Part C – Series-Parallel and Multiwire Branch Circuits

Introduction to Series-Parallel Circuits

18. A _____ is a circuit that contains some resistors in series and some resistors in parallel to each other.
     (a) parallel circuit   (b) series circuit   (c) series-parallel circuit   (d) none of these

Part D – Multiwire Branch Circuits

Introduction to Multiwire Branch Circuits

19. A multiwire branch circuit has two or more ungrounded conductors having a potential difference between them and an equal difference of potential between each ungrounded conductor and the grounded (neutral) conductor.
     (a) True   (b) False

2–9 Neutral Current Calculations

20. • The current on the grounded (neutral) conductor of a 2-wire circuit will be _____ of the current on the ungrounded conductor.
     (a) 50%   (b) 70%   (c) 80%   (d) 100%

Three-Wire Circuits

21. • A balanced 120/240 volt, single-phase, 3-wire circuit is connected so the ungrounded conductors are from different transformer phases (Line 1 and Line 2). The current on the grounded (neutral) conductor will be _____ of the ungrounded conductor current.
     (a) 0%   (b) 70%   (c) 80%   (d) 100%

22.   A 3-wire, 120/240 volt circuit will carry 10 ampere unbalanced neutral current if:
       Line 1 = 20 ampere and Line 2 = 10 ampere.
       (a) True   (b) False

23.   • What is the neutral current for a 20 ampere, 3-wire, 208Y/120-volt circuit?
       (a) 0 ampere   (b) 10 ampere   (c) 20 ampere   (d) 40 ampere

Four-Wire Circuits

24.   The neutral of a 4-wire, 208Y/120- or 480Y/277-volt circuit will carry the unbalanced current when the circuit is
       balanced.
       (a) True   (b) False

25.   What is the neutral current for a 4-wire, 208Y/120-volt circuit, L1 = 20 ampere, L2 = 20 ampere, L3 = 20 ampere?
       (a) 0 ampere   (b) 10 ampere   (c) 20 ampere   (d) 40 ampere

26.   • The grounded (neutral) conductor of a balanced wye 4-wire circuit will carry no current when supplying power to
       balanced nonlinear loads.
       (a) True   (b) False

27.   • A 3-phase, 4-wire multiwire branch circuit has 20 ampere on each 120-volt phase. The neutral conductor could carry
       as much as _____ .
       (a) 0 amperes   (b) 10 amperes   (c) 15 amperes   (d) 40 amperes

2–10 Dangers of Multiwire Branch Circuit

28.   Improper wiring or mishandling of multiwire branch circuits can cause _____ connected to the circuit.
       (a) overloading of the ungrounded conductors
       (b) overloading of the grounded (neutral) conductors
       (c) destruction of equipment because of overvoltage
       (d) b and c

29.   • Because of the dangers associated with the open neutral (grounded conductor), the continuity of the _____ conductor
       cannot be dependent on the receptacle [*Section 300-13(b)*].
       (a) ungrounded   (b) grounded   (c) a and b   (d) none of these

☆ Challenge Questions

Part A – Series Circuits

30.   • Two resistors, one rated 4 ohm and one rated 8 ohm, are connected in series. If the voltage drop across both resistors is
       12 volts, then the current that would pass through the 4 ohm resistor is _____ .
       (a) 1 ampere   (b) 2 ampere
       (c) 4 ampere   (d) 8 ampere

31.   • A series circuit has four 40 ohm resistors and the
       power supply is 120 volts. The voltage drop of each
       resistor would be _____ .
       (a) one-quarter of the source voltage
       (b) 30 volts
       (c) the same across each resistor
       (d) all of these

32.   • The power consumed in a series circuit is _____ .
       (a) the sum of the power consumed of each load
       (b) determined by the formula $P_T = I^2 \times R_T$
       (c) determined by the formula $P_T = E \times I$
       (d) all of these

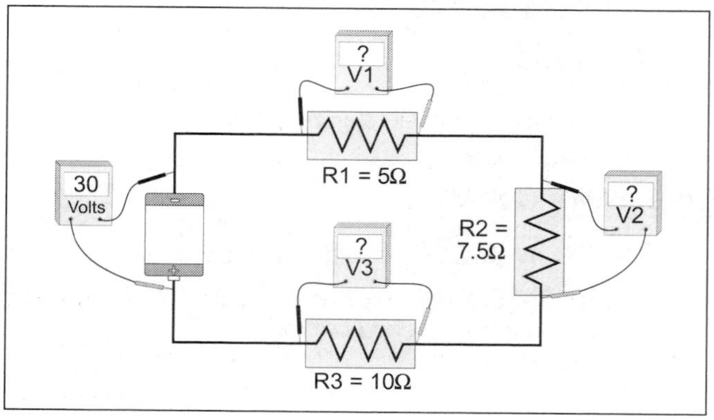

Figure 2–32

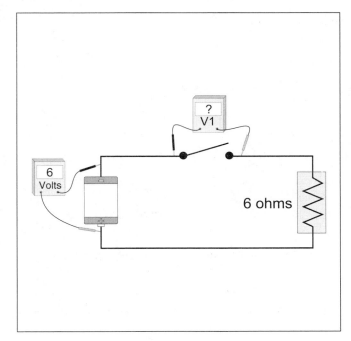

Figure 2–33

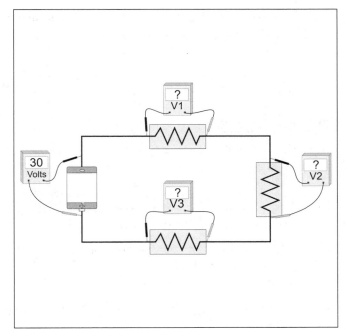

Figure 2–34

33.　• The reading on voltmeter 2 (V_2) is _____ (Figure 2–32).
　　(a) 5 volts　　　　　(b) 7 volts　　　　　(c) 10 volts　　　　　(d) 6 volts

34.　The voltmeter connected across the switch would read _____ (Figure 2–33).
　　(a) 3 volts　　　　　(b) 12 volts　　　　　(c) 6 volts　　　　　(d) 18 volts

Part B – Parallel Circuits

35.　• In general, when multiple light bulbs are wired in a single fixture, they are connected in _____ to each other.
　　(a) series　　　　　(b) series-parallel　　　　　(c) parallel　　　　　(d) order of wattage

36.　• A single-phase, dual-rated, 120/240-volt motor will have its winding connected in _____ when supplied by 120 volts.
　　(a) series　　　　　(b) parallel　　　　　(c) series-parallel　　　　　(d) parallel-series

37.　• The voltmeters shown in Figure 2–34 are connected _____ each of the loads.
　　(a) in series to　　　　　(b) across　　　　　(c) in parallel to　　　　　(d) b and c

38.　• If the supply voltage remains constant, resistors will consume the most power when they are connected _____ .
　　(a) all in series　　　　　　　　　　(b) all in parallel
　　(c) with two parallel pairs in series　　　　　(d) with one pair in parallel and the other two in series

Circuit Resistance

39.　A parallel circuit has three resistors. One resistor is rated 2 ohm, one is 3 ohm, and the other is 7 ohm. The total resistance of the parallel circuit is _____ . Remember the total resistance of any parallel circuit is always less than the smallest resistor.
　　(a) 12 ohm　　　　　(b) 1 ohm　　　　　(c) 42 ohm　　　　　(d) 1.35 ohm

Figure 2–35 applies to the next three questions.

40.　• The total current of the circuit can be measured by ammeter _____ (Figure 2–35).
　　(a) 1　　　　　(b) 2　　　　　(c) 3　　　　　(d) none of these

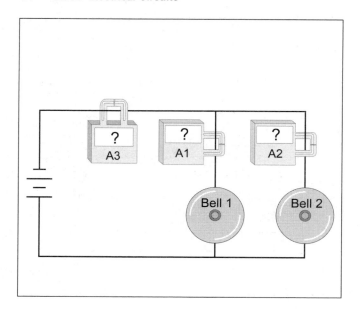

Figure 2–35

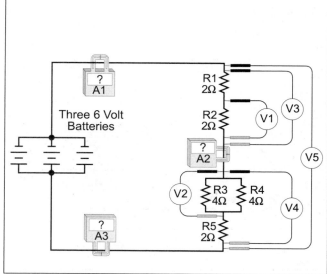

Figure 2–36

41.   • If Bell 2 consumed 12 watts of power when supplied by two 12-volt batteries (connected in series), the resistance of this bell would be _____ (Figure 2–35).

(a) 9.6 ohm   (b) 44 ohm   (c) 576 ohm   (d) 48 ohm

42.   Determine the total circuit resistance of the parallel circuit based on the following facts (Figure 2–35):
1. The current on ammeter 1 reads 0.75 ampere.
2. The voltage of the circuit is 30 volts.
3. Bell 2 has a resistance of 48 ohm.
Tip: Total resistance of a parallel circuit is always less than the smallest resistor.

(a) 22 ohm   (b) 48 ohm   (c) 1920 ohm   (d) 60 ohm

Part C – Series-Parallel Circuits

Figure 2–36 applies to the next three questions:

43.   • The total current of this circuit can be read on _____ (Figure 2–36).

(a) ammeter 1   (b) ammeter 2   (c) ammeter 3   (d) all of these

44.   • The reading of V_2 is _____ (Figure 2–36).

(a) 1.5 volts   (b) 4 volts   (c) 4 volts   (d) 8 volts

45.   • The reading of V_4 is _____ (Figure 2–36).

(a) 1.5 volts   (b) 3 volts   (c) 5 volts   (d) 8 volts

Figure 2–37 applies to the next two questions:

46.   Resistor R_1 has a resistance of 5 ohm, resistors R_2, R_3, and R_4 have a resistance of 15 ohm each. The total resistance of this series-parallel circuit is _____ (Figure 2–37).

(a) 50 ohm   (b) 35 ohm   (c) 25 ohm   (d) 10 ohm

47.   • What is the voltage drop across R_1? R_1 is 5 ohm and total resistance of R_2, R_3, and R_4 is 5 ohm (Figure 2–37).

(a) 60 volts   (b) 33 volts   (c) 40 volts   (d) 120 volts

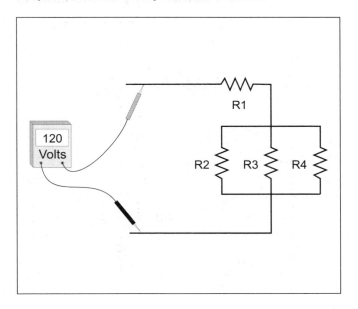

Figure 2–37

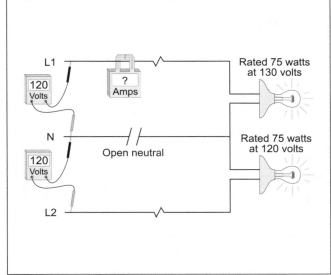

Figure 2–38

Part D – Multiwire Branch Circuits

48. • If the neutral of the circuit in the diagram is opened, the circuit becomes one series circuit of 240 volts. Under this condition, the current of the circuit is _____ . Tip: Determine the total resistance (Figure 2–38).
    (a) 0.67 ampere   (b) 0.58 ampere   (c) 2.25 ampere   (d) 0.25 ampere

NEC® Questions from Section 230-52 through Sections 346-10

The following *National Electrical Code*® questions are in consecutive order. Questions with • indicate that 75 percent or less of exam takers get the question correct.

Article 230 – Services

49. Where individual open conductors enter a building or other structure through tubes, _____ shall be formed on the conductors before they enter the tubes.
    (a) drop loops   (b) knots   (c) drip loops   (d) none of these

50. Service cables shall be formed in a gooseneck and taped and painted or taped with a self-sealing, weather-resistant _____ .
    (a) cover   (b) protected   (c) thermoplastic   (d) none of these

51. Service-drop conductors and service-entrance conductors shall be arranged so that _____ will not enter service raceway or equipment.
    (a) moisture   (b) condensation   (c) water   (d) a and c

52. The service disconnecting means shall be _____ .
    (a) accessible   (b) readily accessible   (c) outdoors   (d) indoors

53. Each service disconnecting means shall be suitable for _____ .
    (a) hazardous locations
    (b) disconnecting the service
    (c) protection of person's
    (d) the prevailing conditions

54. The additional service disconnecting means for fire pumps or for emergency, legally required standby, or optional standby services permitted by Section 230-2 shall be installed sufficiently remote from the one to six service disconnecting means for normal service to minimize the possibility of _____ interruption of supply.
    (a) accidental   (b) intermittent   (c) simultaneous   (d) prolonged

55. Where the service disconnecting means is a power-operated circuit breaker, it shall be able to be opened by hand in the event of a _____ .
    (a) ground-fault   (b) short-circuit   (c) power surge   (d) power supply failure

56. For installations consisting of not more than two 2-wire branch circuits, the service disconnecting means shall have a rating of not less than _____ amperes.
    (a) 15   (b) 20   (c) 25   (d) 30

57. Where the service disconnecting means consists of more than one switch or circuit breaker, the combined ratings of all the switches or circuit breakers used _____ than the rating required by Section 230–79.
    (a) can be more than   (b) shall not be less   (c) must be more than   (d) none of these

58. In a multiple-occupancy building, each occupant shall have access to his or her own _____ .
    (a) overcurrent protection devices   (b) receptacles
    (c) service-entrance assembly   (d) panelboard

59. • Where necessary to prevent tampering, an automatic overcurrent protection device protecting service conductors supplying only a specific load, such as a water heater, shall be permitted to be _____ where located so as to be accessible.
    (a) locked   (b) sealed   (c) a or b   (d) none of these

60. • The maximum setting of the ground-fault protection in a service disconnecting means shall be _____ amperes.
    (a) 800   (b) 1,000   (c) 1,200   (d) 2,000

61. Ground-fault protection that functions to open the service disconnecting means _____ protect(s) service conductors or the service disconnecting means.
    (a) will   (b) will not   (c) adequately   (d) totally

62. For services exceeding 600 volts, nominal, cable tray systems shall be permitted to support cables identified as _____ conductors.
    (a) service-entrance   (b) overcurrent protection   (c) two-wire circuit   (d) three-wire circuit

Article 240 – Overcurrent Protection

Article 240 covers overcurrent protection is the first line of defense against the potential of electricity to cause damage to life and property. Proper sizing and application of overcurrent protection devices is critical for every system.

63. In general, conductors other than flexible cords and fixture wires shall be protected against overcurrent in accordance with their _____ as specified in Section 310-15.
    (a) rating   (b) markings   (c) ampacities   (d) listings

64. • A feeder to a panelboard is run with No. 6 THHN conductor. The maximum size overcurrent protection for these conductors are _____ amperes. Note: Terminal rating of 60°C.
    (a) 60   (b) 40   (c) 90   (d) 100

65. Which of the following are not standard size fuses or inverse time circuit breakers?
    (a) 45   (b) 70   (c) 75   (d) 80

66. Which of the following statements about supplementary overcurrent protection is correct?
    (a) Shall not be used in lighting fixtures.
    (b) May be used as a substitute for a branch circuit overcurrent protection device.
    (c) May be used to protect internal circuits of equipment.
    (d) Shall be readily accessible.

67. Ground-fault protection of equipment shall be provided in accordance with the provisions of Section 230-95 for solidly grounded wye electrical systems of more than 150 volts to ground, but not exceeding 600 volts phase-to-phase for each building or structure main disconnecting means rated _____ amperes or more.
    (a) 1,000   (b) 1,500   (c) 2,000   (d) 2,500

68. • Circuit breakers shall _____ all ungrounded conductors of the circuit.
    (a) open   (b) close   (c) protect   (d) inhibit

69. No tap conductor shall supply another conductor.
    (a) True   (b) False

70. Conductors can be tapped from the transformer secondary if the total lengths of the tap conductors are no longer than _____ feet. The tap conductors must have an ampacity not less than the ampacity of the overcurrent protection and must terminate in a single circuit breaker, or set of fuses having a rating not greater than the secondary conductor ampacity.
    (a) 10   (b) 25   (c) 100   (d) no limit

71. Overcurrent protection devices shall be _____ .
    (a) accessible (as applied to wiring methods)
    (b) accessible (as applied to equipment)
    (c) readily accessible
    (d) inaccessible to unauthorized personnel

72. Enclosures for overcurrent protection devices must be mounted in a _____ position.
    (a) vertical
    (b) horizontal
    (c) vertical or horizontal
    (d) there are no requirements

73. Plug fuses with Edison bases have a maximum rating of _____ amperes.
    (a) 20   (b) 30   (c) 40   (d) 50

74. • Which of the following statements about Type S fuses are not true?
    (a) Adapters shall fit Edison-base fuse holders.
    (b) Adapters are designed to be easily removed.
    (c) Type S fuses shall be classified as not over 125 volts and 30 amperes.
    (d) a and c

75. Dimensions of Type S fuses, fuseholders, and adapters shall be standardized to permit interchange ability regardless of the _____ .
    (a) model   (b) manufacturer   (c) amperage   (d) voltage

76. Fuses are required to be marked with _____ .
    (a) ampere and voltage rating
    (b) interrupting rating where other than 10,000 amperes
    (c) the name or trademark of the manufacturer
    (d) all of these

77. A _____ shall be of such design that any alteration of its trip point (calibration) or the time required for its operation will require dismantling of the device or breaking of a seal for other than intended adjustments.
    (a) Type S fuse   (b) Edison-base fuse   (c) circuit breaker   (d) fuseholder

78. Circuit breakers rated at _____ amperes or less and _____ volts or less shall have the ampere rating molded, stamped, etched, or similarly marked into their handles or escutcheon areas.
    (a) 100, 600   (b) 600, 100   (c) 1,000, 6,000   (d) 6,000, 1,000

79. A circuit breaker with a _____ voltage rating, e.g., 240 volt or 480 volt, can be used where the nominal voltage between any two conductors does not exceed the circuit breaker's voltage rating.
    (a) straight   (b) slash   (c) high   (d) low

Article 250 – Grounding and Bonding

Grounding is one of the largest, most important, and least understood articles in the *NEC®*. As specified in Section 90-1(a), safety is the key element and purpose of the *NEC®*. Proper grounding and bonding is essential for maximum protection of life and property. If overcurrent protection is considered the first line of defense, grounding could be considered the last line of defense.

80. The path to ground from circuits, equipment, and metal conductor enclosures shall _____ .
    (a) be permanent and continuous
    (b) have the capacity to conduct safely any fault current likely to be imposed on it
    (c) have sufficiently low impedance to limit the voltage to ground and to facilitate the operation of the circuit protective devices
    (d) all of these

81. Grounding and bonding conductors shall not be connected by _____ .
    (a) pressure connections
    (b) solder
    (c) lugs
    (d) approved clamps

82. • When grounding service-supplied alternating-current systems, the grounding electrode conductor shall be connected (bonded) to the grounded service conductor (neutral) at _____ .
    (a) the load end of the service drop   (b) the meter equipment
    (c) the service disconnect   (d) any of these

83. When service entrance conductors exceed 1,100 kcmil for copper, the required grounded conductor for the service must be sized not less than _____ percent of the area of the largest phase conductor.
    (a) 15   (b) 19   (c) 12   (d) 25

84. Main and equipment bonding jumpers shall be a _____ .
    (a) screw   (b) wire   (c) bus   (d) any of these

85. Where livestock is housed, that portion of the equipment grounding conductor run underground to the remote building disconnecting means shall be insulated or covered _____ .
    (a) aluminum   (b) copper   (c) copper-clad aluminum   (d) none of these

86. • Because of the increasing use of nonmetallic repairs to the interior metal water pipes, interior metal water pipe located more than _____ feet from the point of entrance to the building shall not be used to serve as a conductor for the purpose of the bonding of the different building electrodes.
    (a) 2   (b) 4   (c) 5   (d) 6

87. The metal frame of the building that is effectively grounded is considered part of the grounding electrode system.
    (a) True   (b) False

88. The smallest diameter pipe or conduit permitted for a made electrode shall be _____ inch.
    (a) $\frac{1}{2}$ (b) $\frac{3}{4}$ (c) 1   (d) none of these

89. The upper end of the made electrode shall be _____ ground level unless the aboveground end and the grounding electrode conductor attachment are protected against physical damage.
    (a) above the   (b) flush with   (c) below the   (d) b or c

90. • Where the resistance to ground of a single made electrode exceeds 25 ohms, _____ .
(a) additional electrodes must be added until the resistance to ground is less than 25 ohms
(b) one additional electrode must be added
(c) no additional electrodes are required
(d) the electrode can be omitted

91. Two or more electrodes that are effectively bonded together shall be considered as a single electrode in this sense.
(a) True   (b) False

92. When aluminum grounding electrode conductors are used outdoors to connect to the grounding electrode, the aluminum shall not be installed within _____ inches of the earth.
(a) 6   (b) 12   (c) 15   (d) 18

93. • The largest size grounding electrode conductor required for any service is a _____ copper.
(a) No. 6   (b) No. 1/0   (c) No. 3/0   (d) 250 kcmil

94. Grounding electrode conductor connections to a concrete-encased or buried grounding electrode are required to be readily accessible.
(a) True   (b) False

95. The grounding conductor connection to the grounding electrode shall be made by _____ .
(a) listed lugs   (b) exothermic welding
(c) listed pressure connectors   (d) any of these

96. Metal enclosures used to provide support or protection of _____ from physical damage shall not be required to be grounded.
(a) conductors   (b) feeders   (c) cables   (d) none of these

97. Service equipment, service raceways, and service conductor enclosures shall be bonded by _____ .
(a) service metal enclosures, raceways, and cables can be bonded to the grounded (neutral) service conductor.
(b) threaded raceways into enclosures, couplings, hubs, conduit bodies, etc.
(c) bonding-type locknuts can be used on raceway or connector terminations where concentric or eccentric knockouts are not encountered.
(d) all of these

98. When bonding enclosures, metal raceways, frames, fittings, and other metal noncurrent-carrying parts, any nonconductive paint, enamel, or similar coating shall be removed at _____ .
(a) contact surfaces   (b) threads   (c) contact points   (d) all of these

99. Where oversized, concentric, or eccentric knockouts are not encountered or where a box with concentric or eccentric knockouts is listed for the purpose, then the raceway or cable can terminate with _____ .
(a) two locknuts   (b) fitting with one locknut   (c) listed fitting   (d) all of these

100. A service is supplied by three metal raceways. Each raceway contains 600 kcmil phase conductors. Determine the size of the service bonding jumper for each raceway.
(a) No. 1/0   (b) No. 2/0   (c) 225 kcmil   (d) 500 kcmil

101. The equipment bonding jumper can be installed on the outside of a raceway providing the length of the run is not more than _____ inches and the bonding jumper is routed with the raceway.
(a) 12   (b) 24   (c) 36   (d) 72

102. The lightning protection system grounding electrode _____ be bonded to the building grounding electrode system.
(a) shall   (b) shall not   (c) can   (d) none of these

103. An electrically operated pipe organ shall have both the generator and motor frame grounded or _____ .
(a) the generator and motor shall be effectively insulated from the ground
(b) the generator and motor shall be effectively insulated from the ground and from each other
(c) the generator shall be effectively insulated from the ground and from the motor driving it
(d) both shall have double insulation

104. Flexible metal conduit shall be permitted for grounding if the length in any ground return path does not exceed 6 feet and an overcurrent protection device that does not exceed _____ amperes protects the circuit conductors.

(a) 15   (b) 20   (c) 30   (d) 60

105. Equipment grounding conductors shall be the same size as the circuit conductors for _____ ampere circuits.

(a) 15   (b) 20   (c) 30   (d) all of these

106. • What size equipment grounding conductor is required for a nonmetallic raceway that contains the following three circuits? No. 12 circuit protected by a 20 ampere device, No. 10 circuit protected by a 30 ampere device, and No. 8 circuit protected by a 40 ampere device.

(a) No. 10   (b) No. 6   (c) No. 8   (d) No. 12

107. The terminal for the connection of the equipment grounding conductor shall be identified by a green-colored, _____ .

(a) not readily removable terminal screw with a hexagonal head

(b) hexagonal, not readily removable terminal nut

(c) pressure wire connector

(d) all of these

108. A grounded circuit conductor shall not be used for grounding noncurrent-carrying metal parts of equipment on the load side of _____ .

(a) the service disconnecting means

(b) separately derived system disconnecting means

(c) overcurrent protection devices for separately derived system not having a main disconnecting means

(d) all of these

109. An equipment bonding jumper for a grounding-type receptacle must be installed between the receptacle and a flush mounted outlet box, even where there is direct metal-to-metal contact between the metal yoke of the receptacle and the metal box.

(a) True   (b) False

110. The *NEC*® permits metal device boxes installed flush with the surface to meet the requirement for grounding receptacle contacts without having to resort to a separate equipment bonding jumper for the receptacle.

(a) True   (b) False

111. • Where one or more equipment grounding conductors enter a box, all equipment grounding conductors shall be spliced together (in the enclosure). This does not apply to insulated equipment grounding conductors for isolated ground receptacles for electronic equipment.

(a) True   (b) False

112. Secondary circuits of current and potential instrument transformers shall be grounded where the primary windings are connected to circuits of _____ volts or more to ground and, where on switchboards, shall be grounded irrespective of voltage.

(a) 300   (b) 600   (c) 1,000   (d) 150

Article 280 – Surge Arresters

113. A surge arrester is a protective device for limiting surge voltages by _____ or bypassing surge current.

(a) decreasing   (b) discharging   (c) limiting   (d) derating

114. The conductor between a lightning arrester and the line and surge arrester and the grounding connection shall not be smaller than (installations operating at 1,000 volts or more) _____ copper.

(a) No. 4   (b) No. 6   (c) No. 8   (d) No. 2

Chapter 3 – Wiring Methods and Material

NEC® Chapter 3 general rules for wiring methods and material is generally related to wiring raceways, cables, junction boxes, cabinets, etc. The rules in this chapter apply everywhere in the *NEC®* except as modified (such as in Chapters 5, 6, and 7). Along with Chapter 2, Chapter 3 can be considered the heart of the *Code*.

Article 300 – Wiring Methods

115. All conductors of a circuit, including the grounded and equipment grounding conductors must be contained within the same _____ .
   (a) raceway   (b) cable   (c) trench   (d) all of these

116. Where nonmetallic-sheathed cable is installed through punched or factory-made holes in metal studs or metal framing members, _____ must be installed before the cable is installed.
   (a) grommets   (b) bushings   (c) plates   (d) a or b

117. Wiring methods installed behind panels that allow access, such as the space above a dropped ceiling, are required to be _____ according to their applicable articles.
   (a) supported   (b) painted   (c) in a metal raceway   (d) all of these

118. • When installing raceways underground, rigid nonmetallic conduit and other approved raceways must have a minimum of _____ inches of cover.
   (a) 6   (b) 12   (c) 18   (d) 22

119. The _____ is defined as the area between the top of direct burial cable and the finished grade.
   (a) notch   (b) cover   (c) gap   (d) none of these

120. Direct buried conductors or cables shall be _____, providing those splices are made with approved methods and identified materials.
   (a) permitted to be spliced without a splice box
   (b) spliced only in an approved junction box
   (c) spliced only in a weatherproof splice box
   (d) none of these

121. • Where installing direct buried cables, a _____ shall be used at the end of a conduit that terminates underground.
   (a) splice kit   (b) terminal fitting   (c) bushing   (d) b or c

122. Which of the following metal parts shall be protected from corrosion inside and out?
   (a) ferrous raceways   (b) metal elbows   (c) boxes   (d) all of these

123. Circulation of warm and cold air in _____ must be prevented.
   (a) raceways   (b) boxes   (c) cabinets   (d) panels

124. The independent support wires shall be distinguishable from fire rated suspended ceiling framing support wires by _____ .
   (a) color   (b) tagging   (c) other effective means   (d) all of these

125. In multiwire circuits, the continuity of the _____ conductor shall not be dependent upon the device connections.
   (a) ungrounded   (b) grounded   (c) grounding   (d) a and b

126. Fittings shall be used only with the specific wiring methods for which they are designed and listed.
   (a) True   (b) False

127. The number of conductors permitted in a raceway shall be limited to _____ .
   (a) permit heat to dissipate
   (b) prevent damage to insulation during installation
   (c) prevent damage to insulation during removal of conductors
   (d) all of these

128. Metal raceways shall not be _____ by welding unless the raceway is specifically designed to be or otherwise specifically permitted to be in the *Code*.
   (a) supported   (b) terminated   (c) connected   (d) all of these

129. Conductors in metal raceways and enclosures shall be so arranged as to avoid heating the surrounding metal by alternating current induction. To accomplish this, the _____ conductor(s) shall be grouped together.
   (a) phase   (b) neutral   (c) equipment grounding   (d) all of these

130. No wiring of any type shall be installed in ducts used to transport _____ .
   (a) dust   (b) flammable vapors   (c) loose stock   (d) all of these

131. Equipment and devices shall be permitted within ducts or plenum chambers used to transport environmental air only if necessary for their direct action upon, or sensing of, the _____ .
   (a) contained air   (b) air quality   (c) air temperature   (d) none of these

132. • Wiring methods permitted in the drop ceiling area used for environmental air include _____ .
   (a) electrical metallic tubing   (b) flexible metal conduit of any length
   (c) armored cable (Type AC) of any length   (d) all of these

Article 305 – Temporary Wiring

Article 305 contains specific rules for temporary wiring methods.

133. There is no time limit for temporary electrical power and lighting except that it must be removed upon completion for which of the following?
   (a) construction or remodeling   (b) maintenance or repair
   (c) demolition of buildings   (d) all of these

134. The following applies to the temporary wiring of branch circuits
   (a) No open wiring conductors shall be laid on the floor.
   (b) All circuits shall originate in an approved panelboard.
   (c) Overcurrent devices in accordance with Article 240 shall protect all conductors.
   (d) all of these

135. • All receptacles for temporary branch circuits are required to be electrically connected to the _____ conductor.
   (a) grounded   (b) grounding
   (c) equipment grounding   (d) grounding electrode

136. At construction sites, boxes are not required for temporary wiring splices of _____ .
   (a) multiconductor cords   (b) multiconductor cables
   (c) a or b   (d) none of these

137. All 125-volt, single-phase, 15, 20, and 30 ampere receptacle outlets used by personnel for temporary power shall have ground-fault circuit-interrupter protection for personnel. GFCI protection can be by a _____ or other devices such as GFCI adapters incorporating listed GFCI protection.
   (a) circuit breaker   (b) receptacle   (c) cord sets   (d) all of these

138. Ground-fault protection for personnel is required for receptacles rated more than 30 amperes or other than 125 volts. The ground-fault protection shall be supplied by:
   (a) GFCI protection device
   (b) Assured Equipment Grounding Conductor Program (AEGCP)
   (c) AFCI protection device
   (d) a or b

Article 310 – Conductors for General Wiring

Article 310 – Conductors for General Wiring, along with Article 300, are the two general-use articles in Chapter 3 that most electricians use on a daily basis. Article 310 covers the general requirements for conductors and their type designations, insulation, markings, mechanical strengths, ampacity ratings, and uses. These requirements do not apply to conductors that form an integral part of equipment, such as motors and appliances.

139. Conductors smaller than No. 1/0 can be connected in parallel to supply control power, provided _____ .
    (a) they are all contained within the same raceway or cable
    (b) each parallel conductor has an ampacity sufficient to carry the entire load
    (c) the circuit overcurrent protection device rating does not exceed the ampacity of any individual parallel conductor
    (d) all of these

140. When conductors are run in parallel, the currents need to be evenly divided between the individual parallel conductors so that each conductor is evenly heated. This is accomplished by insuring that each of the conductors within a parallel set has the same_____ and all conductors terminate in the same manner.
    (a) length   (b) material   (c) cross-sectional area   (d) all of these

141. Solid dielectric insulated conductors operated above 2,000 volts in permanent installations shall have _____ insulation and shall be shielded.
    (a) ozone-resistant   (b) asbestos   (c) high-temperature   (d) perfluoro-alkoxy

142. The _____ rating of a conductor is the maximum temperature at any location along its length that the conductor can withstand over a prolonged period of time, without serious degradation.
    (a) ambient   (b) temperature   (c) maximum withstand   (d) short-circuit

143. Conductors with thermoplastic and fibrous outer braid Type TBS insulation are used for wiring _____ .
    (a) switchboards only   (b) in a dry location   (c) in a wet location   (d) lighting fixtures

144. Lettering on conductor insulation indicates intended condition of use. THWN is rated _____ .
    (a) 75°C   (b) for wet locations   (c) a and b   (d) not enough information

145. The ampacity of a conductor can be different along the length of the conductor. The higher ampacity is permitted to be used for the lower ampacity, if the lower ampacity is no more than _____ feet or no more than _____ percent of the length of the higher ampacity.
    (a) 10, 20   (b) 20, 10   (c) 10, 10   (d) 15, 15

146. • When bare grounding conductors are allowed, their ampacities are limited to _____ .
    (a) 60°C
    (b) 75°C
    (d) 90°C
    (d) that permitted for the insulated conductors of the same size

147. When determining the number of conductors that are considered current-carrying, a grounding conductor is _____ .
    (a) counted as one conductor
    (b) considered to be a current-carrying conductor
    (c) considered to be a noncurrent-carrying conductor and is not counted
    (d) counted as one conductor for each ground wire in the raceway

148. The ampacity of a single insulated No. 1/0 THHN copper conductor in free air is _____ amperes.
    (a) 260   (b) 300   (c) 185   (d) 215

149. For applications where underground circuits must be buried deeper than shown in the underground ampacity Tables 310-60 through 310-84, the following ampacity derating factor shall be permitted to be used: _____ percent per increased foot of depth for all values of Rho.
    (a) 3   (b) 6   (c) 9   (d) 12

Article 318 – Cable Trays

150. A cable tray is a unit or assembly of units or Sections, and associated fittings, that forms a _____ structural system used to securely fasten or support cables and raceways.
    (a) rigid   (b) flexible   (c) movable   (d) secure

151. Cable trays shall _____ .
    (a) include fittings for changes in direction and elevation
    (b) have side rails or equivalent structural members
    (c) be made of corrosion-resistant material
    (d) all of these

152. _____ can be supported from cable trays.
    (a) Raceway   (b) Cables   (c) Outlet boxes   (d) All of these

153. • Aluminum cable trays shall not be used as an equipment grounding conductor for circuits with ground-fault protection above _____ amperes.
    (a) 2,000   (b) 300   (c) 500   (d) 1,200

154. A box shall not be required where cables or conductors from cable trays are installed in bushed conduit and tubing used for support or for protection against _____ .
    (a) abuse   (b) unauthorized access   (c) physical damage   (d) tampering

155. For the ampacity of cables, rated 2,000 volts or less, in cable trays, the derating factors of the Notes to Ampacity Tables of 0 to 2,000 Volts shall apply only to multiconductor cables with more than _____ current-carrying conductors.
    (a) one   (b) two   (c) three   (d) four

156. Where nails are used to mount knobs for the support of open wiring on insulators, they shall not be smaller than _____ penny.
    (a) six   (b) eight   (c) ten   (d) none of these

157. A conductor used for open wiring, where sleeved, shall be carried through a _____ .
    (a) separate sleeve or tube
    (b) weatherproof tube
    (c) tube of absorbent material
    (d) grounded metallic tube

Article 320 – Open Wiring on Insulators

158. • Open wiring on insulators within _____ feet from the floor shall be considered exposed to physical damage.
    (a) 4   (b) 2   (c) 7   (d) none of these

159. _____ shall be permitted to be installed in messenger supported wiring.
    (a) Multiconductor service entrance cable
    (b) Type MI cable
    (c) Multiconductor underground feeder cable
    (d) all of these

Article 324 – Concealed Knob-and-Tube Wiring

160. Concealed knob-and-tube wiring is permitted to be used only for extensions of existing installations and elsewhere only by special permission as in the hollow spaces of _____ .
    (a) walls   (b) ceilings   (c) a and b   (d) none of these

161. Where concealed solid knobs are used, conductors shall be securely tied thereto by _____ equivalent to that of the conductor.
    (a) wires having insulation   (b) wires having an AWG
    (c) nonconductive material   (d) none of these

Article 326 — Medium Voltage Cable

162. • Type MV is defined as a single or multiconductor solid dielectric insulated cable rated _____ volts or higher.
    (a) 601   (b) 1,001   (c) 2,001   (d) 6,001

Article 328 – Flat Conductor Cable (Type FCC)

163. FCC systems shall be permitted both for general-purpose and appliance branch circuits; they shall not be permitted for individual branch circuits.
  (a) True   (b) False

164. Flat conductor cable cannot be installed in _____ .
  (a) residential units   (b) schools   (c) hospitals   (d) any of these

165. Floor-mounted-type flat conductor cable and fittings shall be covered with carpet squares no larger than _____ .
  (a) 36 square inches
  (b) 36 inches square
  (c) 30 square inches
  (d) 24 inches square

166. Metal shields for flat conductor cable must be electrically continuous to the _____ .
  (a) floor
  (b) cable
  (c) equipment grounding conductor
  (d) none of these

167. Type FCC cable shall be clearly and durably marked _____ .
  (a) on the top side at intervals not exceeding 30 inches
  (b) on both sides at intervals not exceeding 24 inches
  (c) with conductor material, maximum temperature, and ampacity
  (d) b and c

Article 330 – Mineral-Insulated Cable (Type MI)

168. Type MI cable shall be securely supported at intervals not exceeding _____ feet.
  (a) 3   (b) $3^1/_2$ (c) 5   (d) 6

169. The radius of the inner edge of any bend in Type MI cable shall not be less than five times the external diameter of the metallic sheath for cable not more than _____ inch in external diameter.
  (a) $^1/_2$ (b) $^3/_4$ (c) $^5/_8$ (d) $1^1/_2$

170. Type MI cable conductors shall be of _____ with a cross-sectional area corresponding to standard AWG sizes.
  (a) solid copper
  (b) solid or stranded copper
  (c) stranded copper
  (d) solid copper or aluminum

171. The outer sheath of Type MI cable is made up of _____ .
  (a) aluminum   (b) steel alloy   (c) copper   (d) b or c

Article 331 – Electrical Nonmetallic Tubing

172. Where electrical nonmetallic tubing is installed concealed in walls, floors, and ceilings of buildings exceeding three floors above grade, a barrier must be provided having a minimum _____ minute finish rating as listed for fire rated assemblies.
  (a) 5   (b) 10   (c) 15   (d) 30

173. The _____ of conductors used in prewired manufactured assemblies shall be identified by means of a printed tag or label attached to each end of the manufactured assembly.
  (a) type   (b) size   (c) quantity   (d) all of these

174. • Electrical nonmetallic tubing is not permitted in places of assembly, unless it is encased in at least _____ of concrete.
  (a) 1 inch   (b) 2 inches   (c) 3 inches   (d) 4 inches

175.   Electrical nonmetallic tubing must be secured in place every _____ inches.
       (a) 12   (b) 18   (c) 24   (d) 36

Article 333 – Armored Cable (Type AC)

176.   Armored cable is limited or not permitted _____ .
       (a) in commercial garages
       (b) where subject to physical damage
       (c) in theaters
       (d) all of these

177.   The radius of the curve of the inner edge of any bend shall not be less than _____ for AC cable.
       (a) five times the largest conductor within the cable
       (b) three times the diameter of the cable
       (c) five times the diameter of the cable
       (d) six times the outside diameter of the conductors

178.   Where AC cable is run across the top of a floor joist in an attic without permanent ladders or stairs, substantial guard strips within _____ feet of the scuttle hole shall protect the cable.
       (a) 7   (b) 6   (c) 5   (d) 3

Article 334 – Metal-Clad Cable (Type MC)

179.   MC cables can be installed _____ .
       (a) direct burial
       (b) in concrete
       (c) in cinder fill
       (d) none of these

180.   The minimum size conductor permitted for Type MC cable is No. _____ .
       (a) 18 copper   (b) 14 copper   (c) 12 copper   (c) none of these

Article 336 – Nonmetallic-Sheathed Cable

181.   • Type nonmetallic sheathed cables shall be permitted to be used in _____ .
       (a) one-family dwellings
       (b) multifamily dwellings
       (c) other structures
       (d) all of these

182.   NM cable shall not be used _____ .
       (a) in commercial buildings
       (b) in the air void of masonry block not subject to excessive moisture
       (c) for exposed work
       (d) embedded in poured cement, concrete, or aggregate

183.   Nonmetallic-sheathed cable shall be secured in place within _____ inches of every cabinet, box, or fitting.
       (a) 6   (b) 10   (c) 12   (d) 18

184.   Nonmetallic-sheathed cable installed within accessible ceilings for the connections to lighting fixtures and equipment do not need to be secured within 12 inches from the lighting fixture or equipment, where the free length does not exceed ___ feet.
       (a) $4^1/_2$ (c) $2^1/_2$ (c) $3^1/_2$ (d) any of these

185.   • The difference in the overall covering between NM cable and NMC cable is that it is _____ .
   (c) corrosion-resistant
   (b) flame-retardant
   (c) fungus-resistant
   (d) moisture-resistant

Article 338 – Service-Entrance Cable (Types SE and USE)

186.   • Type USE or SE cable must have a minimum of _____ conductors (including the uninsulated one) in order for one of the conductors to be uninsulated.
   (a) one   (b) two   (c) three   (d) four

Article 339 – Underground Cable (Type UF)

187.   • The maximum size underground feeder cable is No. _____ copper.
   (a) 14   (b) 10   (c) 1/0   (d) 4/0

188.   Type UF cable shall not be used in _____ .
   (a) motion picture studios
   (b) storage battery rooms
   (c) hoistways
   (d) all of these

Article 340 – Power and Control Tray Cable (Type TC)

189.   Type TC cable shall be permitted to be used _____ .
   (a) for power and lighting circuits
   (b) in cable trays in hazardous locations
   (c) in Class I control circuits
   (d) all of these

Article 342 – Nonmetallic Extensions

190.   Nonmetallic surface extensions are permitted in _____ .
   (a) buildings not over three stories
   (b) buildings over four stories
   (c) all buildings
   (d) none of these

191.   Each run of nonmetallic extensions shall terminate in a fitting that covers the _____ .
   (a) device
   (b) box
   (c) end of the extension
   (d) end of the assembly

192.   Aerial cable suspended over work benches, not accessible to pedestrians, shall have a clearance of not less than _____ feet above the floor.
   (a) 12   (b) 15   (c) 14   (d) 8

Article 343 – Nonmetallic Underground Conduit with Conductors

193.   Nonmetallic underground conduit with conductors shall not be used _____ .
   (a) in exposed locations
   (b) inside buildings
   (c) in hazardous (classified) locations
   (d) all of these

Article 345 – Intermediate Metal Conduit (IMC)

194.   Where practical, contact of dissimilar metals shall be avoided anywhere in a raceway system to prevent _____ .
        (a) corrosion   (b) galvanic action   (c) shorts   (d) none of these

195.   • The cross sectional area of 1-inch IMC is approximately _____ square inch(es).
        (a) 1.22   (b) 0.62   (c) 0.96   (d) 2.13

196.   Running threads of intermediate metal conduit shall not be used on conduit for connection at couplings.
        (a) True   (b) False

197.   One inch intermediate metal conduit must be supported every _____ feet.
        (a) 8   (b) 10   (c) 12   (d) 14

Article 346 – Rigid Metal Conduit

198.   • Aluminum fittings and enclosures shall be permitted to be used with _____ conduit.
        (a) steel rigid metal
        (b) aluminum rigid metal
        (c) rigid nonmetallic
        (d) a and b

199.   Materials such as straps, bolts, etc., associated with the installation of rigid metal conduit in a wet location are required
        to be _____ .
        (a) weatherproof   (b) weathertight   (c) corrosion resistant   (d) none of these

200.   • The minimum radiuses of a field bend on 1 1/4 inch rigid metallic conduit is _____ inches.
        (a) 7   (b) 8   (c) 14   (d) 10

Unit 3

Understanding Alternating Current

OBJECTIVES

After reading this unit, the student should be able to briefly explain the following concepts:

Part A – Alternating Current Fundamentals	Part B and C – Induction and Capacitance	Part D – Power Factor and Efficiency
Alternating current	Charge, discharging and testing of capacitors	Apparent power (volt-amperes)
Armature turning frequency	Conductor impedance	Efficiency
Current flow	Conductor shape	Power factor
Generator	Induced voltage and applied current	True power (watts)
Magnetic cores	Magnetic cores	
Phase differences in degrees	Use of capacitors	
Phase – in and out		
Values of alternating current		
Waveform		

After reading this unit, the student should be able to briefly explain the following terms:

Part A – Alternating Current Fundamentals		
Ampere-turns	RMS	Inductive reactance
Armature speed	Root-mean-square	Phase relationship
Coil	Self-inductance	Self-inductance
Conductor cross-sectional area	Skin effect	Skin effect
Effective (RMS)	Waveform	**Part D – Power Factor and Efficiency**
Effective to peak	**Part B and C – Induction and Capacitance**	Apparent power
Electromagnetic field	Back-EMF	Efficiency
Frequency	Capacitance	Input watts
Impedance	Capacitive reactance	Output watts
Induced voltage	Capacitor	Power factor
Magnetic field	Coil	True power
Magnetic flux lines	Counterelectromotive force	Volts-amperes
Peak	Eddy currents	Wattmeter
Peak to effective	Farads	
Phase relationship	Frequency	
	Henrys	
	Impedance	
	Induced voltage	
	Induction	

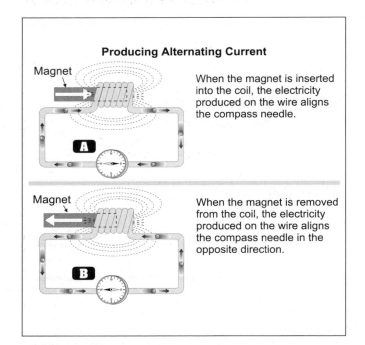

Figure 3–1
Producing Alternating Current

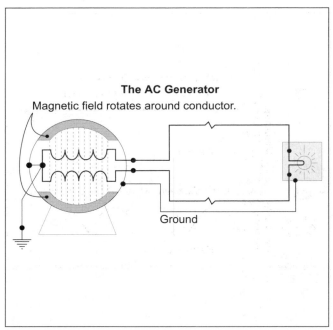

Figure 3–2
The AC Generator

PART A – ALTERNATING CURRENT FUNDAMENTALS

3–1 CURRENT FLOW

For current to flow in a circuit, the circuit must be a closed loop and the power supply must push the electrons through the completed circuit. The transfer of electrical energy can be accomplished by electrons flowing in one constant direction (direct current) or by electrons flowing in one direction and then reversing in the other direction (alternating current).

3–2 ALTERNATING CURRENT

Alternating current is generally used instead of direct current for electrical systems and building wiring. This is because alternating current can easily have voltage variations (transformers). In addition, alternating current can be transmitted inexpensively at high voltage over long distances resulting in reduced voltage drop of the power distribution system as well as smaller power distribution wire and equipment. Alternating current can be used for certain applications for which direct current is not suitable.

Alternating current is produced when electrons in a conductor are forced to move because there is a moving magnetic field. The lines of force from the magnetic field cause the electrons in the wire to flow in a specific direction. When the lines of force of the magnetic field move in the opposite direction, the electrons in the wire are forced to flow in the opposite direction (Figure 3–1). Electrons will flow only when there is relative motion between the conductors and the magnetic field.

3–3 ALTERNATING CURRENT GENERATOR

An alternating current generator consists of many loops of wire that rotate between the flux lines of a magnetic field. Each conductor loop travels through the magnetic lines of force in opposite directions, causing the electrons within the conductor to move in a specific direction (Figure 3–2). The force on the electrons caused by the magnetic flux lines is called voltage, or electromotive force (EMF), and is abbreviated as V or E.

The magnitude of the electromotive force is dependent on the number of turns (wraps) of wire, the strength of the magnetic field, and the speed at which the coil rotates. The rotating conductor loop mounted on a shaft is called a rotor or armature. Slip, or collector rings, and carbon brushes are used to connect the output voltage from the generator to an external circuit.

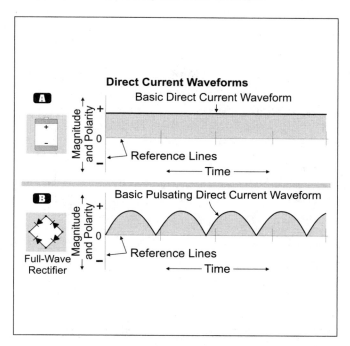

Figure 3–3
Direct Current Waveforms

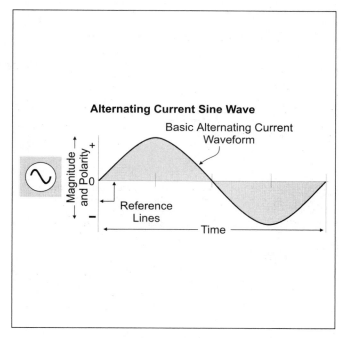

Figure 3–4
Alternating Current Sine Wave

3–4 WAVEFORM

A waveform is a pictorial view of the shape and magnitude of the level and direction of current or voltage over a period of time. The waveform represents the magnitude and direction of the current or voltage over time in relationship with the generators armature.

Direct Current Waveform

The polarity of the voltage of direct current is positive, and the current flows through the circuit in the same direction at all times. In general, direct current voltage and current remain the same magnitude, particularly when supplied from batteries or rectifiers (Figure 3–3).

Alternating Current Waveform

The waveform for alternating current displays the level and direction of the current and voltage for every instant of time for one full revolution of the armature. When the waveform of an alternating current circuit is symmetrical, with positive polarity above and negative below the zero reference level, the waveform is called a sine wave (Figure 3–4).

3–5 ARMATURE TURNING FREQUENCY

Frequency is a term used to indicate the number of times a generators armature turns one full revolution (360°) in one second. This is expressed as hertz, or cycles per second, and is abbreviated as Hz or cycles per second (cps). In the United States, frequency is not a problem for most electrical work because it remains constant at 60 hertz (Figure 3–5).

Armature Speed

A current or voltage that is produced when the generators armature makes one cycle (360°) in 1/60th of a second is called 60 Hz. This is because the armature travels at

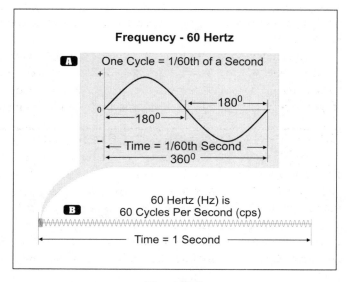

Figure 3–5
Frequency – 60 Hertz

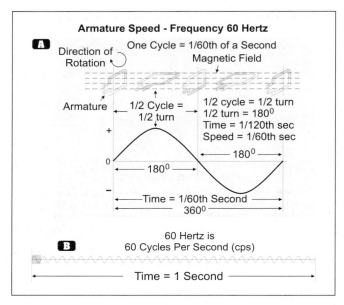

Figure 3–6
Armature Speed

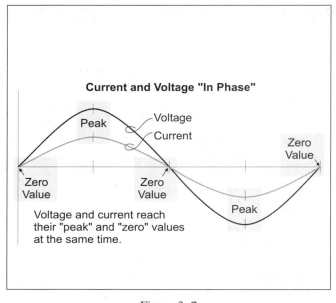

Figure 3–7
Current and Voltage "In Phase"

a speed of one cycle every 1/60th of a second. A 60 Hz circuit takes 1/60th of a second for the armature to complete one full cycle (360°), and it will take 1/120th of a second to complete $^1/_2$ cycle (180°) (Figure 3–6).

3–6 PHASE – IN AND OUT

Phase is a term used to indicate the time relationship between two waveforms, such as voltage to current, or voltage to voltage. *In phase* means that two waveforms are in step with each other, that is, they cross the horizontal zero axis at the same time (Figure 3–7). The terms lead and lag are used to describe the relative position, in degrees, between two waveforms. The waveform that is ahead *leads* and the waveform behind *lags* (Figure 3–8).

3–7 PHASE DIFFERENCES IN DEGREES

Phase differences are expressed in *degrees,* with one full revolution of the armature equal to 360°. Single-phase, 120/240 volt power is out of phase by 180°, and three-phase power is out of phase by 120° (Figure 3–9).

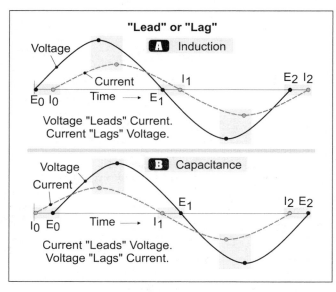

Figure 3–8
Lead or Lag

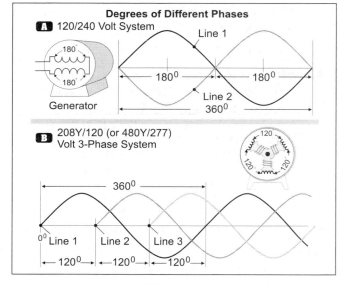

Figure 3–9
Degrees of Different Phases

3–8 VALUES OF ALTERNATING CURRENT

There are many different types of values in alternating current, the most important are peak (maximum) and effective. In alternating current systems, effective is the value that will cause the same amount of heat to be produced, as in a direct current circuit containing only resistance. Another term for effective is RMS, which is root-mean-square. In the field, whenever you measure voltage, you are always measuring effective voltage. Newer types of clamp-on ammeter have the ability to measure peak as well as effective (RMS) current (Figure 3–10).

Note: The values of direct current generally remains constant and are equal to the effective values of alternating current.

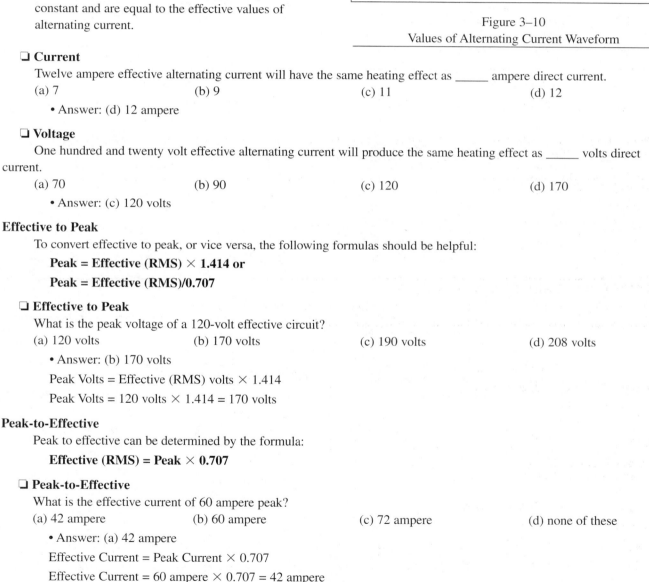

Values of Alternating Current Waveform

Peak Value = Effective (RMS) x 1.414
Effective (RMS) Value = Peak x .707
Average Value = Peak x .637

0 (Zero) Reference Line

Instantaneous Value
Zero Value
Average Value
Effective (RMS) Value
Peak Value
Instantaneous Value

The effective value of voltage is also called True RMS. This is what most voltmeters read. The values shown in this figure apply to voltage waveforms as well as current waveforms.

Figure 3–10
Values of Alternating Current Waveform

❏ **Current**

Twelve ampere effective alternating current will have the same heating effect as _____ ampere direct current.

(a) 7   (b) 9   (c) 11   (d) 12

  • Answer: (d) 12 ampere

❏ **Voltage**

One hundred and twenty volt effective alternating current will produce the same heating effect as _____ volts direct current.

(a) 70   (b) 90   (c) 120   (d) 170

  • Answer: (c) 120 volts

Effective to Peak

To convert effective to peak, or vice versa, the following formulas should be helpful:

Peak = Effective (RMS) × 1.414 or

Peak = Effective (RMS)/0.707

❏ **Effective to Peak**

What is the peak voltage of a 120-volt effective circuit?

(a) 120 volts   (b) 170 volts   (c) 190 volts   (d) 208 volts

  • Answer: (b) 170 volts

  Peak Volts = Effective (RMS) volts × 1.414

  Peak Volts = 120 volts × 1.414 = 170 volts

Peak-to-Effective

Peak to effective can be determined by the formula:

Effective (RMS) = Peak × 0.707

❏ **Peak-to-Effective**

What is the effective current of 60 ampere peak?

(a) 42 ampere   (b) 60 ampere   (c) 72 ampere   (d) none of these

  • Answer: (a) 42 ampere

  Effective Current = Peak Current × 0.707

  Effective Current = 60 ampere × 0.707 = 42 ampere

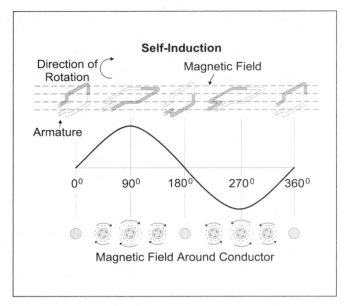

Figure 3–11
Self-Induction

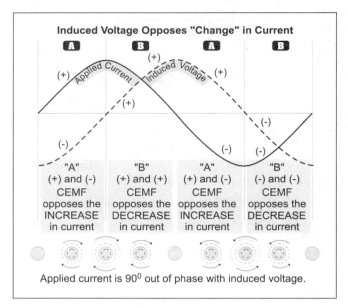

Figure 3–12
Induced Voltage

PART B – INDUCTION

INDUCTION INTRODUCTION

When a magnetic field moves through a conductor, it causes the electrons in the conductor to move. The movement of electrons caused by electromagnetism is called *induction*. In an alternating current circuit, the current movement of the electrons increases and decreases with the rotation of the *armature* through the magnetic field. As the current flowing through the conductor increases or decreases, it causes an expanding and collapsing electromagnetic field within the conductor (Figure 3–11). This varying electromagnetic field within the conductor causes the electrons in the conductor to move at 90° to the flowing electrons within the conductor. This movement of electrons because of the conductor's electromagnetic field is called self-induction. Self-induction (induced voltage) is commonly called *counterelectromotive-force (CEMF)*, or back-EMF.

3–9 INDUCED VOLTAGE AND APPLIED CURRENT

The induced voltage in the conductor because of the conductor's counterelectromotive-force (CEMF) is always 90° out of phase with the applied current (Figure 3–12). When the current in the conductor increases, the induced voltage tries to prevent the current from increasing by storing some of the electrical energy in the magnetic field around the conductor. When the current in the conductor decreases, the induced voltage attempts to keep the current from decreasing by releasing the energy from the magnetic field back into the conductor. The opposition to any change in current because of induced voltage is called inductive reactance, which is measured in ohms, abbreviated as X_L, and known as *Lenz's Law*. Inductive reactance changes proportionally with frequency and can be found by the formula:

$X_L = 2\pi fL$

$\pi = 3.14$

f = frequency

L = Inductance in henrys

❏ **Inductive Reactance**

What is the inductive reactance of a conductor in a transformer that has an inductance of 0.001 henrys? Calculate at 60 Hz, 180 Hz, and 300 Hz (Figure 3–13).

• Answers:

$X_L = 2\pi fL$

X_L at 60 Hz, $= 2 \times 3.14 \times 60\ \text{Hz} \times 0.001\ \text{henrys} = 0.377\ \text{ohm}$

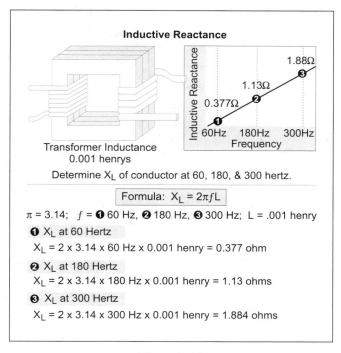

Inductive Reactance

Transformer Inductance
0.001 henrys

Determine X_L of conductor at 60, 180, & 300 hertz.

Formula: $X_L = 2\pi f L$

$\pi = 3.14$; $f =$ ❶ 60 Hz, ❷ 180 Hz, ❸ 300 Hz; L = .001 henry

❶ X_L at 60 Hertz

$X_L = 2$ x 3.14 x 60 Hz x 0.001 henry = 0.377 ohm

❷ X_L at 180 Hertz

$X_L = 2$ x 3.14 x 180 Hz x 0.001 henry = 1.13 ohms

❸ X_L at 300 Hertz

$X_L = 2$ x 3.14 x 300 Hz x 0.001 henry = 1.884 ohms

Figure 3–13
Inductive Reactance

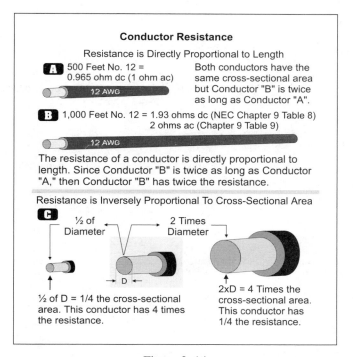

Conductor Resistance

Resistance is Directly Proportional to Length

A 500 Feet No. 12 = 0.965 ohm dc (1 ohm ac) | Both conductors have the same cross-sectional area but Conductor "B" is twice as long as Conductor "A".

B 1,000 Feet No. 12 = 1.93 ohms dc (NEC Chapter 9 Table 8) 2 ohms ac (Chapter 9 Table 9)

The resistance of a conductor is directly proportional to length. Since Conductor "B" is twice as long as Conductor "A," then Conductor "B" has twice the resistance.

Resistance is Inversely Proportional To Cross-Sectional Area

C ½ of Diameter / 2 Times Diameter

½ of D = 1/4 the cross-sectional area. This conductor has 4 times the resistance.

2xD = 4 Times the cross-sectional area. This conductor has 1/4 the resistance.

Figure 3–14
Conductor Resistance

X_L at 180 Hz, = $2 \times 3.14 \times 180$ Hz $\times 0.001$ henrys = 1.13 ohm

X_L at 300 Hz, = $2 \times 3.14 \times 300$ Hz $\times 0.001$ henrys = 1.884 ohm

3–10 CONDUCTOR IMPEDANCE

The resistance of a conductor is a physical property of the conductors that oppose the flow of electrons. This resistance is directly proportional to the conductor length and is inversely proportional to the conductor cross-sectional area. This means that the longer the wire, the greater the conductor resistance; the smaller the cross-sectional area of the wire, the greater the conductor resistance (Figure 3–14).

The opposition to alternating current flow due to inductive reactance and conductor resistance is called impedance (Z). Impedance is only present in alternating current circuits and is measured in ohms. This opposition to alternating current (Z) is greater than the resistance (R) of a direct current circuit because of *counterelectromotive-force, eddy currents*, and *skin effect*.

Eddy Currents and Skin Effect

Eddy currents are small independent currents that are produced as a result of the expanding and collapsing magnetic field of alternating current flow (Figure 3–15). The expanding and collapsing magnetic field of an alternating current circuit also induces a counterelectromotive-force in the conductors that repel the flowing electrons towards the surface of the conductor (Figure 3–16). The counterelectromotive-force causes the circuit current to flow near the surface of the conductor rather than at the conductor's center. The flow of electrons near the surface is known as skin effect. Eddy currents and skin effect decrease the effective conductor cross-sectional area for current flow, which results in an increase in the conductor's impedance which is measured in ohms.

3–11 INDUCTION AND CONDUCTOR SHAPE

The shape into which the conductor is configured affects the amount of *self-induced voltage* in the conductor. If a conductor is coiled into adjacent loops *(winding)*, the electromagnetic field *(flux lines)* of each conductor loop adds together to create a stronger overall magnetic field (Figure 3–17). The amount of self-induced voltage created within a conductor is directly proportional to the current flow, the length of the conductor, and the frequency at which the magnetic fields cut through the conductors (Figure 3–18).

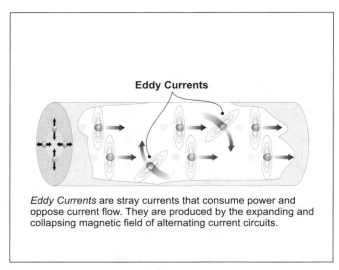

Figure 3–15
Eddy Currents

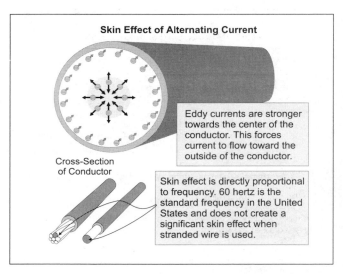

Figure 3–16
Skin Effect of Alternating Current

3–12 INDUCTION AND MAGNETIC CORES

Self-induction in a *coil* of conductors is affected by the winding current magnitude, the core material, the number of conductor loops *(turns)*, and the spacing of the winding (Figure 3–19). An *iron core* within the conductor winding permits an easy path for the electromagnetic flux lines which produce a greater counterelectromotive-force than air core windings. In addition, conductor coils are measured in *ampere-turns*, that is, the current in ampere times the number of conductor loops or turns.

Ampere-Turns = Coil Ampere × Number of Coil Turns

❏ **Ampere-Turns**

A magnetic coil in a transformer has 1,000 turns and carries 5 ampere. What is the ampere-turns rating of this transformer (Figure 3–20)?

(a) 5,000 ampere-turns   (b) 7,500 ampere-turns   (c) 10,000 ampere-turns   (d) none of these

• Answer: (a) 5,000 ampere-turns

Ampere-Turns = Ampere × Number of Coil Turns

Ampere-Turns = 5 ampere × 1,000 turns = 5,000

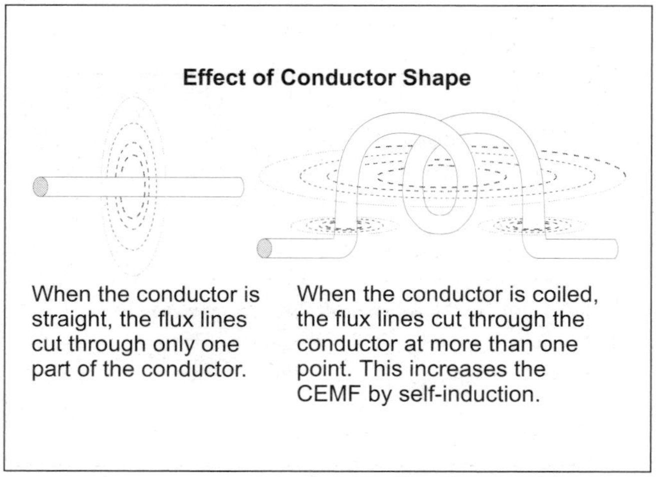

Figure 3–17
Effect of Conductor Shape

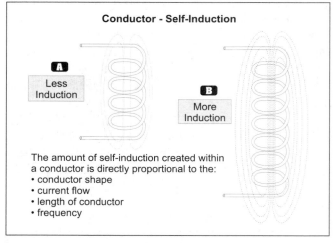

Figure 3–18
Conductor – Self-Induction

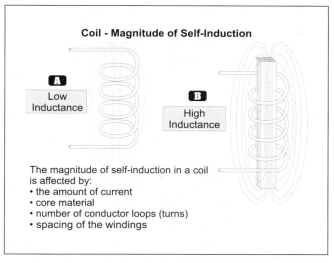

Figure 3–19
Coil – Magnitude of Self-Induction

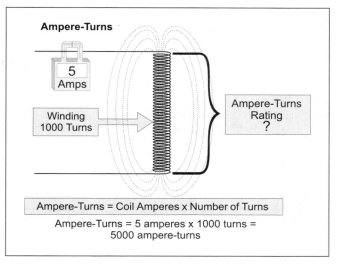

Figure 3–20
Ampere Turns

PART C – CAPACITANCE

CAPACITANCE INTRODUCTION

Capacitance is the property of an electric circuit that enables the storage of electric energy by means of an electrostatic field, much as a spring stores mechanical energy (Figure 3–21). Capacitance exists whenever an insulating material separates two objects that have a difference of potential. A capacitor is two metal foils separated by insulating material (wax paper) rolled together. Devices that intentionally introduce capacitance into circuits are called capacitors or condensers.

3–13 CHARGE, TESTING, AND DISCHARGING

When a capacitor has a potential difference between the plates, it is said to be *charged*. One plate has an excess of free electrons (–), and the other plate has a lack of electrons (+). Because of the insulation *(dielectric)* between the capacitor foils, electrons cannot flow from one plate to the other. However, there are electric lines of force between the plates, and this force is called the *electric field* (Figure 3–22).

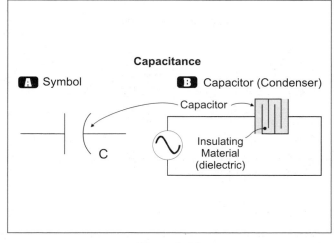

Figure 3–21
Capacitance

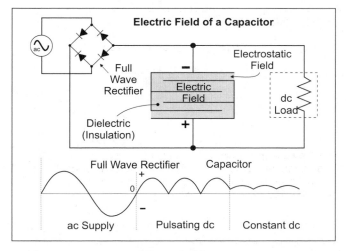

Figure 3–22
Electric Field of a Capacitor

Factors that determine capacitance are: the surface area of the plates, the distance between the plates, and the dielectric insulating material between the plates. Capacitance is measured in *farads,* but the opposition offered to the flow of current by a capacitor is called *capacitive reactance*, which is measured in *ohms*. Capacitive reactance is abbreviated X_C, is inversely proportional to frequency and can be calculated by the equation:

$$X_C = \frac{1}{2\pi fC}$$

$\pi = 3.14$, f = frequency, C = Capacitance measure in farads

❏ Capacitive Reactance

What is the capacitive reactance of a 0.000001 farad capacitor supplied by 60 Hz, 180 Hz, and 300 Hz (Figure 3–23)?

Answer: $X_C = \frac{1}{2\pi fC}$.

$$X_C \text{ at } 60 \text{ Hz} = \frac{1}{2 \times 3.14 \times 60 \text{ Hz} \times 0.000001 \text{ farads}} = 2{,}654 \text{ ohm}$$

$$X_C \text{ at } 180 \text{ Hz} = \frac{1}{2 \times 3.14 \times 180 \text{ Hz} \times 0.000001 \text{ farads}} = 885 \text{ ohm}$$

$$X_C \text{ at } 300 \text{ Hz} = \frac{1}{2 \times 3.14 \times 300 \text{ Hz} \times 0.000001 \text{ farads}} = 530.46 \text{ ohm}$$

Capacitor Short and Discharge

The more the capacitor is charged, the stronger the electric field. If the capacitor is overcharged, the electrons from the negative plate could be pulled through the dielectric insulation to the positive plate, resulting in a short of the capacitor. To discharge a capacitor, all that is required is a conducting path between the capacitor plates.

Testing a Capacitor

If a capacitor is connected in series with a direct current power supply and a test lamp, the lamp will be continuously illuminated if the capacitor is shorted, but will remain dark if the capacitor is good (Figure 3–24).

3–14 USE OF CAPACITORS

Capacitors (condensers) are used to prevent arcing across the *contacts* of low ampere switches and direct current power supplies for electronic loads such as a computer (Figure 3–25). In addition, capacitors cause the current to lead the voltage by

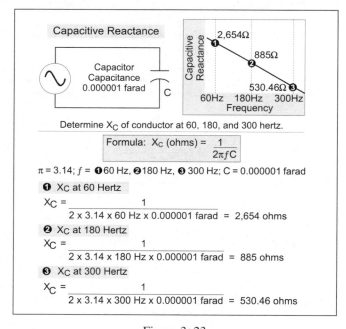

Figure 3–23
Capacitive Reactance

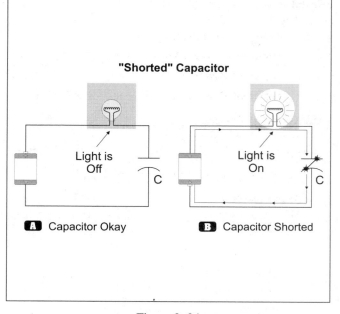

Figure 3–24
Shorted Capacitor

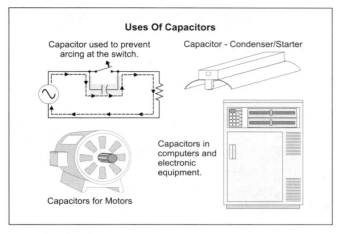

Figure 3–25
Uses of Capacitors

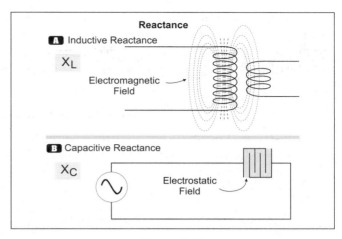

Figure 3–26
Reactance

as much as 90° and can be used to correct poor power factor due to inductive reactive loads such as motors. Poor power factor, due to induction, causes current to lag voltage by as much as 90°. Correction of the power factor to *unity* (100%) is known as *resonance* and occurs when inductive reactance is equal to capacitive reactance in a series circuit $X_L = X_C$.

PART D – POWER FACTOR AND EFFICIENCY

POWER FACTOR INTRODUCTION

Alternating current circuits develop inductive and capacitive reactance which causes some power to be stored temporarily in the electromagnetic field of the inductor or in the electrostatic field of the capacitor (Figure 3–26). Because of the temporary storage of energy, the voltage and current of inductive and capacitive circuits are not in phase. This out of phase relationship between voltage and current is called *power factor* (Figure 3–27). Power factor affects the calculation of power in alternating current circuits, but not in direct current circuits.

3–15 APPARENT POWER (VOLT-AMPERE)

In alternating current circuits, the value from multiplying volts times ampere equals *apparent power* and is commonly referred to as *volt-ampere*, VA, or *kilovolt-amperes* kVA. Apparent power (VA) is equal to or greater than *true power* (watts) depending on the *power factor*. When sizing circuits or equipment, always size the circuit equipment according to the apparent power (volt-ampere), not the true power (watt).

Apparent Power = Volts $\times$ Ampere (single-phase)

Apparent Power = Volts $\times$ Ampere $\times \sqrt{3}$ (three-phase)

Apparent Power = True Power/Power Factor

❏ **Apparent Power**

What is the apparent power of a 16 ampere load operating at 120 volts (Figure 3–28)?

(a) 2,400 VA   (b) 1,920 VA   (c) 1,632 VA   (d) none of these

• Answer: (b) 1,920 VA

❏ **Apparent Power**

What is the apparent power of a 250-watt fixture that has a power factor of 80 percent (Figure 3–29)?

(a) 200 VA   (b) 250 VA   (c) 313 VA   (d) 375 VA

• Answer: (c) 313 VA

Apparent Power = True Power/Power Factor

Apparent Power = 250 watts/0.8

Apparent Power = 313 volt-ampere

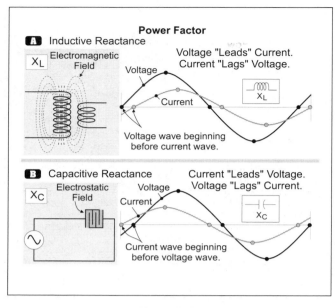

Figure 3–27
Power Factor

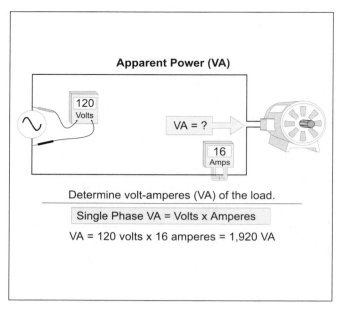

Figure 3–28
Apparent Power (VA)

3–16 POWER FACTOR

Power factor is the ratio of active power (watts) to apparent power (volt-ampere) expressed as a percent that does not exceed 100 percent. Power factor is a form of measurement of how far the voltage and current are out of phase with each other. In an alternating current circuit that supplies power to resistive loads such as incandescent lighting, heating elements, etc., the circuit voltage and current will be *in phase*. The term in phase means that the voltage and current reach their zero and peak values at the same time, resulting in a power factor of 100 percent or *unity*. Unity can occur if the circuit only supplies resistive loads or capacitive reactance (X_C) is equal to inductive reactance (X_L).

The formulas for determining power factor are:

Power Factor = True Power/Apparent Power

Power Factor = Watts/Volt-Ampere

Power Factor = Resistance (R)/Impedance (Z)

The formulas for current:

I = Watts/(Volts Line to Line × Power Factor) – Single-Phase

I = Watts/(Volts Line to Line × 1.732 × Power Factor) – Three-Phase

❏ **Power Factor**

What is the power factor of a fluorescent lighting fixture that produces 160 watts of light and has a ballast rated for 200 volt-ampere (Figure 3–30)?

(a) 80 percent   (b) 90 percent   (c) 100 percent   (d) none of these

    • Answer: (a) 80 percent

$$\text{Power Factor} = \frac{\text{True Power}}{\text{VA}} = \frac{160 \text{ watts}}{200 \text{ volt-ampere}} = 0.80 \text{ or } 80\%$$

❏ **Current**

What is the current flow for three 5 kW, 230-volt, single-phase loads that have a power factor of 90 percent?

(a) 72 ampere   (b) 90 ampere   (c) 100 ampere   (d) none of these

    • Answer: (a) 72 ampere

    I = Watts/(Volts × Power Factor) = 15,000 Watts/(230 volts × 0.9) = 72 ampere

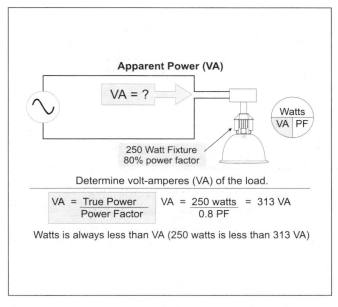

Figure 3–29
Apparent Power (VA)

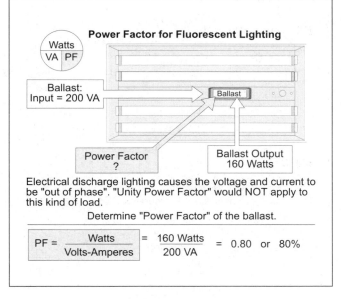

Figure 3–30
Power Factor for Fluorescent Lighting

3–17 TRUE POWER (WATTS)

ALTERNATING CURRENT

True power is the energy consumed for work, expressed by the unit called the watt. Power is measured by a wattmeter which is connected series-parallel [series (measures ampere) and parallel (measures volts)] to the circuit conductors. The true power of a circuit that contains inductive or capacitance reactance can be calculated by the use of one of the following formulas:

True Power (alternating current) = Volts $\times$ Ampere $\times$ Power Factor (single-phase)

True Power (alternating current) = Volts $\times$ Ampere $\times$ Power Factor $\times \sqrt{3}$ (three-phase)

Note: True power for a direct current circuit is calculated as Volts $\times$ Ampere.

❏ **True Power**

What is the true power of a 16-ampere load operating at 120 volts with a power factor of 85 percent (Figure 3–31)?

(a) 2,400 watts   (b) 1,920 watts   (c) 1,632 watts   (d) none of these

• Answer: (c) 1,632 watts

True Power = Volt-ampere $\times$ Power Factor

True Power = (120 volts $\times$ 16 ampere) $\times$ 0.85 = 1,632 watts per hour

Note: True Power is always equal to or less than apparent power.

DIRECT CURRENT

In direct current or alternating current circuits at unity power factor:

True Power = Volts $\times$ Ampere

❏ **True Power**

What is true power of a 30 ampere, 240-volt (single-phase) resistive load that has unity power factor (Figure 3–32)?

(a) 4,300 watts   (b) 7,200 watts   (c) 7,200 VA   (d) none of these

• Answer: (b) 7,200 watts

True Power = Volts $\times$ Ampere

True Power = 240 volts $\times$ 30 ampere = 7,200 watts

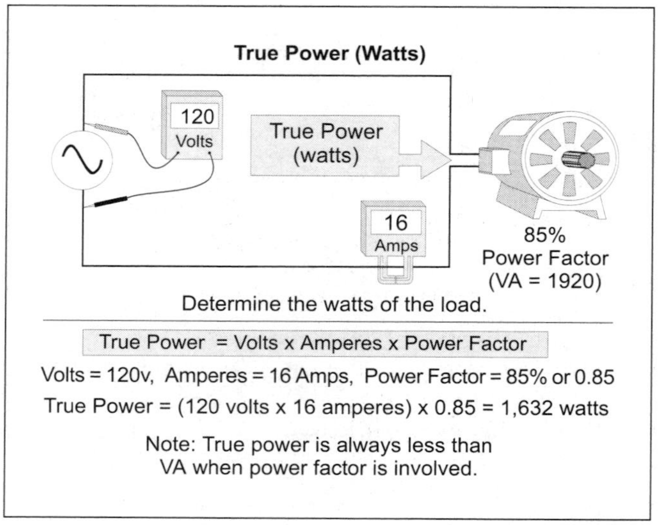

True Power (Watts)

True Power
(watts)

Determine the watts of the load.

120
Volts

16
Amps

85%
Power Factor
(VA = 1920)

True Power = Volts x Amperes x Power Factor

Volts = 120v, Amperes = 16 Amps, Power Factor = 85% or 0.85

True Power = (120 volts x 16 amperes) x 0.85 = 1,632 watts

Note: True power is always less than
VA when power factor is involved.

Figure 3–31
True Power (Watts)

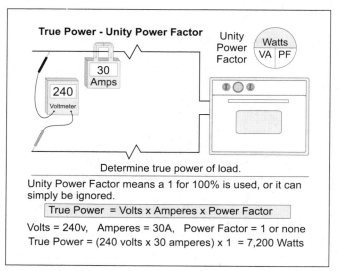

True Power - Unity Power Factor

Unity Power Factor

Watts
VA | PF

30 Amps

240
Voltmeter

Determine true power of load.

Unity Power Factor means a 1 for 100% is used, or it can
simply be ignored.

True Power = Volts x Amperes x Power Factor

Volts = 240v, Amperes = 30A, Power Factor = 1 or none
True Power = (240 volts x 30 amperes) x 1 = 7,200 Watts

Figure 3–32
True Power – Unity Power Factor

3–18 EFFICIENCY

Efficiency has nothing to do with power factor. Efficiency is the ratio of output power to input power, where power factor is the ratio of true power (watts) to apparent power (volt-ampere). Energy that is not used for its intended purpose is called power loss. Power losses can be caused by conductor resistance, friction, mechanical loading, etc. Power losses of equipment are expressed by the term efficiency, which is the ratio of the input power to the output power (Figure 3–33). The formulas for efficiency are:

Efficiency = Output watts/Input watts

Input watts = Output watts/Efficiency

Output watts = Input watts × Efficiency

❑ **Efficiency**

If the input of a load is 800 watts and the output is 640 watts, what is the efficiency of the equipment (Figure 3–34)?

(a) 60 percent   (b) 70 percent   (c) 80 percent   (d) 100 percent

   • Answer: (c) 80 percent

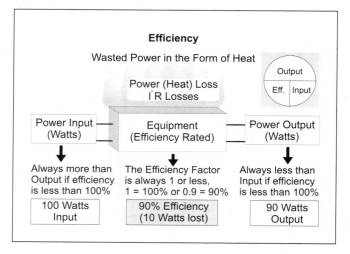

Figure 3–33
Efficiency

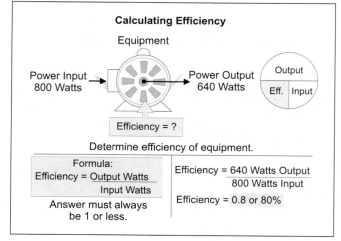

Figure 3–34
Calculating Efficiency

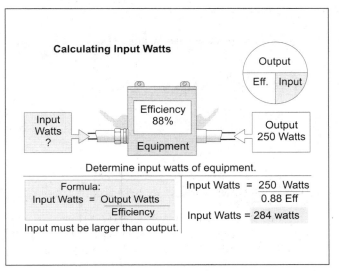

Figure 3–35
Calculating Input Watts

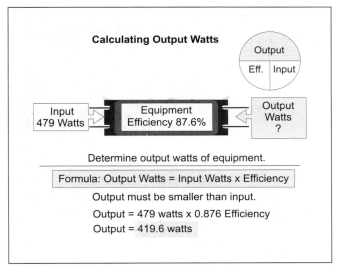

Figure 3–36
Calculating Output Watts

$$\text{Efficiency} = \frac{\text{Output}}{\text{Input}}$$

$$\text{Efficiency} = \frac{640 \text{ watts}}{800 \text{ watts}} = 0.8 \text{ or } 80\%$$

Note: Efficiency is always less than 100 percent.

❏ **Input**

If the output of a load is 250 watts and the equipment is 88 percent efficient, what are the input watts (Figure 3–35)?

(a) 200 watts   (b) 250 watts   (c) 285 watts   (d) 325 watts

• Answer: (c) 285 watts

$$\text{Input} = \frac{\text{Output}}{\text{Efficiency}}$$

$$\text{Input} = \frac{250 \text{ watts}}{0.88 \text{ efficiency}}$$

Input = 284 watts

Note: Input is always greater than the output.

❏ **Output**

If a load is 87.6 percent efficient, for every 479 watts of input there will be _____ watts of output (Figure 3–36).

(a) 440   (b) 480   (c) 390   (d) 420

• Answer: (d) 420 watts

Output = Input × Efficiency

Output = 479 × .876

Output = 419.6 watts

Note: Output is always less than input.

Unit 3 – Understanding Alternating Current Summary Questions

Part A – Alternating Current Fundamentals

3–1 Current Flow

1. The effects of electron movement are the same regardless of the direction of the current flow.
    (a) True   (b) False

3–2 Alternating Current

2. Alternating current is primarily used because it can be transmitted inexpensively and because it can be used for certain applications for which direct current is not suitable.
    (a) True   (b) False

3. Faraday's experiments revealed that when a magnetic field moves through a coil of wire, the lines of force of the magnetic field make the electrons in the wire flow in a specific direction. When the magnetic field moves in the opposite direction, electrons in the wire flow continue to flow in the same direction.
    (a) True   (b) False

3–3 Alternating Current Generator

4. A simple alternating current generator consists of a rotating loop of wire between the lines of force of opposite poles of a magnet. The magnitude of the voltage is dependent on the _____ .
    (a) number of turns of wire   (b) strength of the magnetic field
    (c) speed at which the coil rotates   (d) all of these

3–4 Waveform

5. • A waveform is used to display the level and direction of current and voltage. The waveform for _____ circuits displays the level and direction of the current and voltage for every instant of time for one full revolution of the armature.
    (a) direct current   (b) alternating current   (c) a and b   (d) none of these

6. For alternating current circuits, the waveform is called the _____ .
    (a) frequency   (b) cycle   (c) degree   (d) none of these

3–5 Armature Turning Frequency

7. The number of times the armature turns in one second is called the frequency. Frequency is expressed as _____ or cycles per second.
    (a) degrees   (b) sign wave   (c) phase   (d) hertz

3–6 Phase – In and Out

8. When two waveforms are in step with each other, they are said to be in phase. In a purely resistive alternating current circuit, the current and voltage are in phase. This means that they both reach their zero and _____ values at the same time.
    (a) peak   (b) effective   (c) average   (d) none of these

3–7 Phase Differences in Degrees

9. Phase differences are expressed in _____ .
    (a) sign   (b) phase   (c) hertz   (d) degrees

10. The terms _____ and _____ are used to describe the relative positions in time of two waveforms (voltage or current).
    (a) hertz, phase   (b) frequency, phase   (c) sine, degrees   (d) lead, lag

3–8 Values of Alternating Current

11. • _____ is the value of the voltage or current at any one particular moment of time.
    (a) Peak   (b) Root-mean-square   (c) Effective   (d) Instantaneous

12. _____ is the maximum value that alternating current or voltage reaches.
    (a) Peak   (b) Root-mean-square   (c) Instantaneous   (d) None of these

13. • "Effective" is the alternating current voltage or current value that produces the same amount of heat in a resistor that would be produced by the same amount of direct current voltage or current. _____ is the same as effective.
    (a) Peak   (b) Root-mean-square   (c) Instantaneous   (d) None of these

Part B – Induction

Induction Introduction

14. The movement of electrons because of electromagnetism is called _____ .
    (a) flux lines   (b) voltage   (c) induction   (d) magnetic field

15. The induction of voltage in a conductor because of expanding and collapsing magnetic fields is known as _____ .
    (a) flux lines   (b) power   (c) self-induced voltage   (d) magnetic field

16. Change in current produces a magnetic field through the conductor which produces an induced voltage that always opposes the change in current. The induced voltage that opposes the change in current is called _____ .
    (a) CEMF
    (b) counterelectromotive-force
    (c) back-EMF
    (d) all of these

3–9 Induced Voltage and Applied Current

17. When the conductor current increases, the direction of the induced voltage (CEMF) in the conductor is opposite to the direction of the conductor current and tries to prevent the current from _____ .
    (a) decreasing   (b) increasing

18. When the current in the conductor decreases, the direction of the induced voltage in the conductor attempts to keep the current from decreasing by releasing the energy from the magnetic field back into the conductor.
    (a) True   (b) False

3–10 Conductor Impedance

19. In direct current circuits, the only property that affects current and voltage flow is _____ , which is a physical property of the conductors that oppose current flow.
    (a) voltage   (b) CEMF   (c) back-EMF   (d) none of these

20. Conductor resistance is directly proportional to the conductor length and inversely proportional to conductor cross-sectional area.
    (a) True   (b) False

21.   Alternating currents produce a CEMF that is set up inside the conductor which increases the effective resistance of the conductor because of eddy currents and skin effect.
(a) True   (b) False

22.   The opposition to current flow in a conductor because of resistance and induction is called _____ .
(a) resistance   (b) capacitance   (c) induction   (d) impedance

23.   • Eddy currents are small independent currents that are produced as a result of direct current. Eddy currents flow erratically through a conductor, consume power, and increase the effective conductor resistance by opposing the current flow.
(a) True   (b) False

24.   The expanding and collapsing magnetic field of the conductors caused by the current flow in a conductor induces a voltage in the conductors which repels the flowing electrons towards the surface of the conductor. This has the effect of decreasing the effective conductor cross-sectional area, which causes an increase in the conductor impedance. The flow of electrons near the surface is known as _____ .
(a) eddy currents   (b) induced voltage   (c) impedance   (d) skin effect

3–11 Induction and Conductor Shape

25.   The amount of the self-induced voltage created within the conductor is directly proportional to the current flow, the length of the conductor, and frequency at which the magnetic fields cut through the conductors.
(a) True   (b) False

3–12 Induction and Magnetic Cores

26.   • Self-inductance (CEMF) in a coil is effected by the _____ .
(a) winding length and shape   (b) core material
(c) frequency   (d) all of these

Part C – Capacitance

Capacitance Introduction

27.   _____ is a property of an electric circuit that enables it to store electric energy by means of an electrostatic field and to release this energy at a latter time.
(a) Capacitance   (b) Induction   (c) Self-induction   (d) None of these

3–13 Charge, Testing, and Discharging

28.   When a capacitor has a potential difference between the plates, it is said to be _____ . One plate has an excess of free electrons, and the other plate has fewer electrons.
(a) induced   (b) charged   (c) discharged   (d) shorted

29.   If the capacitor is overcharged, the electrons from the negative plate may be pulled through the insulation to the positive plate. The capacitor is said to have _____ .
(a) charged   (b) discharged   (c) induced   (d) shorted

30.   To discharge a capacitor, all that is required is a _____ path between the capacitor plates.
(a) conducting   (b) insulating   (c) isolating   (d) none of these

3–14 Uses of Capacitors

31.   What helps prevent arcing across the contacts of electric switches?
(a) springs   (b) condenser   (c) inductor   (d) resistor

32. The current in a purely capacitive circuit _____ .
     (a) leads the applied voltage by 90°   (b) lags the applied voltage by 90°
     (c) leads the applied voltage by 180°   (d) lags the applied voltage by 180°

33. Circuits containing inductive or capacitive reactance temporarily store power in the electromagnetic field of induction and the electrostatic field of capacitors.
     (a) True   (b) False

Part D – Power Factor and Efficiency

3–15 Apparent Power (Volt-Ampere)

34. If you measure voltage and current in an inductive or capacitive circuit and then multiply them together, you would obtain the circuit's _____ .
     (a) true power   (b) power factor   (c) apparent power   (d) power loss

35. Apparent power is equal to or greater than true power depending on the power factor.
     (a) True   (b) False

36. When sizing circuits or equipment, always size the circuit components and transformers according to the apparent power, not the true power.
     (a) True   (b) False

3–16 Power Factor

37. Power factor is a measurement of how far the voltage and current are out of phase with each other. Power factor is the ratio between true power (resistive load) to apparent power (reactive load). Power factor can be expressed by the formula: _____ .
     (a) $PF = {}^P/_E$ (b) $PF = {}^R/_Z$ (c) $PF = I^2R$ (d) $PF = {}^Z/_R$

38. • When current and voltage are in phase, the power factor is _____ .
     (a) 100 percent   (b) unity   (c) 90°   (d) a and b

39. In an alternating current circuit that supplies power to resistive loads such as incandescent lighting, heating elements, etc., the circuit voltage and current are said to be _____, resulting in a power factor of unity.
     (a) out of phase   (b) leading by 90°
     (c) 90° out of phase   (d) none of these

3–17 True Power (Watts)

40. True power is the energy consumed for work expressed by the term watts. To determine the true power of a circuit that contains inductive or capacitance reactance, we must multiply the volts times the current times the _____ .
     (a) efficiency   (b) sine wave   (c) power factor   (d) none of these

41. In direct current circuits, the voltage and current are constant and the true power is simply volts times ampere.
     (a) True   (b) False

42. What does it cost per year (9 cents per kWH) for ten 150-watt recessed fixtures to operate if they are on 6 hours per day?
     (a) $153   (b) $296   (c) $235   (d) $180

Power Factor

43. A 2 × 4 recessed fixture contains four 34-watt lamps, and the ballast is rated 1.5 ampere at 120 volts. What is the power factor of the ballast assuming 100 percent efficiency?
     (a) 55%   (b) 65%   (c) 70%   (d) 75%

44.   • What is the apparent power of a 20-ampere load operating at 120 volts with a power factor of 85 percent?
       (a) 2,400 watts   (b) 1,920 watts   (c) 1,632 watts   (d) 2,400 VA

45.   • What is the true power of a 20-ampere load operating at 120 volts with a power factor of 85 percent?
       (a) 2,400 watts   (b) 1,920 watts   (c) 1,632 watts   (d) none of these

46.   • Since power factor cannot be greater than 100 percent, true power is equal to or less than the apparent power. Because of power factor, the VA of the load is greater than the watts, which results in less loads per circuit, more circuits, and larger kVA transformers.
       (a) True   (b) False

47.   • What is the true power of a 10-ampere circuit operating at 120 volts with unity (100%) power factor?
       (a) 1,200 VA   (b) 2,400 VA   (c) 1,200 watts   (d) 2,400 watts

48.   What size transformer is required for a 125 ampere, 240-volt, single-phase load?
       (a) 3 kVA   (b) 30 kVA   (c) 12.5 kVA   (d) 15 kVA

49.   What size transformer is required for a 30 kW-load that has a power factor of 85 percent?
       (a) 12.5 kVA   (b) 35 kVA   (c) 7.5 kVA   (d) 15 kVA

50.   • How many 20 ampere, 120-volt circuits are required for forty-two, 300-watt recessed fixtures (noncontinuous load)? Note: Only so many fixtures are permitted on a circuit.
       (a) 3 circuits   (b) 4 circuits   (c) 5 circuits   (d) 6 circuits

51.   • How many 20 ampere, 120-volt circuits are required for forty-two, 300-watt recessed fixtures (noncontinuous load) with a power factor of 85 percent?
       (a) 5 circuits   (b) 6 circuits   (c) 7 circuits   (d) 8 circuits

3–18 Efficiency

52.   Efficiency is the ratio of the output to input power.
       (a) True   (b) False

53.   If the output is 1,320 watts and the input is 1,800 watts, what is the efficiency of the equipment?
       (a) 62 percent   (b) 73 percent   (c) 0 percent   (d) 100 percent

54.   If the output is 160 watts and the equipment is 88 percent efficient, what is the input ampere at 120 volts?
       (a) 0.75 ampere   (b) 1.500 ampere   (c) 2.275 ampere   (d) 3.250 ampere

55.   A transformer that is 97 percent efficient produces _____ watts output for every 1 kW input.
       (a) 970 watts   (b) 1,000 watts   (c) 1,030 watts   (d) 1,300 watts

☆ Challenge Questions

Part A – Alternating Current Fundamentals

3–2 Alternating Current

56.   The primary reason(s) for high voltage transmission lines is (are) _____ .
       (a) reduced voltage drop   (b) smaller wire
       (c) smaller equipment   (d) all of these

57.   One of the advantages of a higher-voltage system as compared to a lower-voltage system (for the same wattage loads) is _____ .
       (a) reduced voltage drop   (b) reduced power use
       (c) large currents   (d) lower electrical pressure

58. The advantage of alternating current over direct current is that alternating current provides for _____ .
     (a) better speed control
     (b) ease of voltage variation
     (c) lower resistance at high currents
     (d) none of these

3–4 Waveform

59. A waveform represents _____ .
     (a) the magnitude and direction of current or voltage
     (b) how current or voltage can vary with time
     (c) how output voltage can vary with the generator armature
     (d) all of these

3–5 Armature Turning Frequency

60. Frequency of an alternating current waveform is the number of times the current or voltage goes through 360° in _____ .
     (a) $1/10th$ second   (b) 5 seconds   (c) 1 second   (d) 60 seconds

61. How much time does it take for 60 Hz alternating current to travel through 180°?
     (a) $1/120$ second   (b) $1/40$ second   (c) $1/180$ second   (d) none of these

3–8 Values of Alternating Current

62. • The heating effects of 10 ampere alternating current as compared with 10 ampere direct current is _____ .
     (a) the same   (b) less   (c) greater   (d) none of these

63. If the maximum value of an alternating current system is 50 ampere, the RMS value would be approximately _____ .
     (a) 25 ampere   (b) 30 ampere   (c) 35 ampere   (d) 40 ampere

64. • The maximum value of 120-volt direct current is equal to the maximum value of an equivalent alternating current.
     (a) True   (b) False

65. • You are getting a 120 volt reading on your voltmeter. This is an indication of the _____ value of the voltage source.
     (a) average   (b) peak   (c) effective   (d) instantaneous

Part B – Induction

3–9 Induced Voltage and Applied Current

66. _____ Law states that a change in current produces a counterelectromotive-force whose direction is such that it opposes the change in current.
     (a) Kirchoff's Second   (b) Kirchoff's First   (c) Lenz's   (d) Hertz

67. Inductive reactance is abbreviated as _____ .
     (a) I^2R (b) L_X (c) X_L (d) Z

68. Inductive reactance changes proportionally with frequency.
     (a) True   (b) False

69. Inductive reactance is measured in _____ .
     (a) farads   (b) watts   (c) ohms   (d) coulombs

70. • If the frequency is constant, the inductive reactance of a circuit will _____ .
     (a) remain constant regardless of the current and voltage changes
     (b) vary directly with the voltage
     (c) vary directly with the current
     (d) not affect the impedance

3–10 Conductor Impedance

71. The total opposition to current flow in an alternating current circuit is expressed in ohms and is called _____ .
    (a) impedance   (b) conductance   (c) reluctance   (d) none of these

72. Impedance is present in _____ type circuit(s).
    (a) resistance   (b) direct current   (c) alternating current   (d) none of these

73. Conductor resistance to alternating current flow is _____ the resistance to direct current.
    (a) higher than   (b) lower than   (c) the same as   (d) none of these

Part C – Capacitance

3–13 Charge, Testing, and Discharging

74. If a test lamp is placed in series with a capacitor with a direct current voltage source and the lamp is continuously illuminated, it is an indication that the capacitor is _____ .
    (a) fully charged   (b) shorted   (c) fully discharged   (d) open-circuit

75. The insulating material between the surface plates of a capacitor is called the _____ .
    (a) inhibitor   (b) electrolyte   (c) dielectric   (d) regulator

76. Capacitors are measured in _____ .
    (a) watts   (b) volts   (c) farads   (d) henrys

77. Capacitive reactance is measured in _____ .
    (a) ohms   (b) volts   (c) watts   (d) henrys

3–14 Uses of Capacitors

78. • In a circuit that has only capacitive reactance (X_C), the voltage and current are said to be out of phase to each other because the voltage _____ .
    (a) leads the current by 90°
    (b) lags the current by 90°
    (c) leads the current
    (d) none of these

79. Resonance occurs when _____ .
    (a) only resistance occurs in the system
    (b) the power factor is equal to 0
    (c) $X_L = X_C$ are in a series circuit
    (d) when R = Z

Part D – Power Factor and Efficiency

3–15 Apparent Power (Volt-Ampere)

80. If you multiply the voltage times the current in an inductive or capacitive circuit, the answer you obtain will be the _____ of the circuit.
    (a) watts   (b) true power   (c) apparent power   (d) all of these

81. The apparent power of a 19.2 ampere, 120-volt load is _____ .
    (a) 2,304 kVA   (b) 2.3 kVA   (c) 2.3 VA   (d) 230 kVA

3–16 Power Factor

82.   Power factor in an alternating current circuit will be unity (100%), if the circuit contains only _____ .
      (a) induction motors   (b) transformers   (c) reactance coils   (d) resistive loads

83.   • Three 8 kW electric discharge lighting bank circuits, which have a 92 percent power factor, are connected to a 230-volt, 3-phase source. The current flow of these lights is _____ .
      (a) 37 ampere   (b) 50 ampere   (c) 65 ampere   (d) 75 ampere

3–17 True Power (Watts)

84.   • A wattmeter is connected in _____ in the circuit.
      (a) series   (b) parallel   (c) series-parallel   (d) none of these

85.   • True power is always voltage times current for _____ .
      (a) all alternating current circuits
      (b) direct current circuits
      (c) alternating current circuits at unity power factor
      (d) b and c

86.   Power consumed in either a single-phase alternating current or direct current system is always equal to _____ .
      (a) $E \times I$ (b) $E \times R$ (c) $I^2 \times R$ (d) $E/(I \times R)$

87.   • The power consumed on a 76 ampere, 208-volt, 3-phase circuit that has a power factor of 89 percent is _____ .
      (a) 27,379 watts   (b) 35,808 watts   (c) 24,367 watts   (d) 12,456 watts

88.   • The true power of a single-phase, 2.1 kVA load with a power factor of 91 percent is _____ .
      (a) 2.1 kW   (b) 1.91 kW   (c) 1.75 kW   (d) 1,911 kW

3–18 Efficiency

89.   • Motor efficiency can be determined by which of the following formulas?
      (a) hp $\times$ 746
      (b) hp $\times$ 746/VA Input
      (c) hp $\times$ 746/Watts Input
      (d) hp $\times$ 746/kVA Input

90.   The efficiency ratio of a 4,000 VA transformer with a secondary VA of 3,600 VA is _____ .
      (a) 80%   (b) 70%   (c) 90%   (d) 110%

NEC® Questions from Section 346-12 through Sections 427-2

The following *National Electrical Code*® questions are in consecutive order. Questions with • indicate that 75 percent or less of exam takers get the question correct.

Article 346 – Rigid Metal Conduit

91.   Supporting rigid metal conduit by a bored or punched hole in a _____ meets the support requirements of the *Code*.
      (a) wall   (b) truss   (c) rafter   (d) framing member

92.   Rigid metal conduit shall be _____ every 10 feet as required by Section 110-21.
      (a) stamped
      (b) clearly and durably identified
      (c) marked
      (d) none of these

Article 347 – Rigid Nonmetallic Conduit

93. Rigid, nonmetallic conduit and fittings can be used in portions of dairies, laundries, canneries, or other wet locations and in locations where walls are frequently washed; however, the entire conduit system including boxes and _____ shall be installed to prevent water from entering the conduit.

    (a) couplings   (b) fittings   (c) supports   (d) all of these

94. Rigid, nonmetallic conduit shall not be used:
    (a) in hazardous (classified) locations.
    (b) for the support of lighting fixtures or other equipment.
    (c) where subject to physical damage unless identified for such use.
    (d) all of these

95. Rigid, nonmetallic conduit shall be securely fastened within _____ inches of each box.
    (a) 6   (b) 24   (c) 12   (d) 36

96. • Rigid, nonmetallic conduit requires bushings or adapters to protect conductors No. _____ and larger from abrasion.
    (a) 8   (b) 6   (c) 3   (d) 4

Article 348 – Electrical Metallic Tubing (EMT)

97. Couplings and connectors used with electrical metallic tubing shall be made up _____ .
    (a) of metal   (b) in accordance with industry standards
    (c) tight   (d) none of these

98. The _____ for use with electrical metallic tubing shall have a circular cross section.
    (a) tubing   (b) bends   (c) elbows   (d) all of these

Article 349 – Flexible Metallic Tubing

99. Where 3/8-inch flexible metallic tubing has a fixed bend for installation purposes and not flexed for service, the minimum radius of the bend shall not be less than _____ inches.
    (a) 8   (b) $12^1/_2$ (c) $3^1/_2$ (d) 4

Article 350 – Flexible Metal Conduit

100. Flexible metal conduit can be installed exposed or concealed where not subject to physical damage.
    (a) True   (b) False

101. The largest size conductor permitted in $^3/_8$-inch flexible metal conduit is No. _____ .
    (a) 12   (b) 16   (c) 14   (d) 10

102. In a concealed flexible metal conduit installation, _____ connectors shall not be used.
    (a) straight   (b) angle   (c) grounding type   (d) none of these

Article 351 – Liquidtight Flexible Conduit

103. The maximum number of No. 14 THHN permitted in 3/8 inch liquidtight flexible metal conduit with inside fittings is _____ .
    (a) 3   (b) 7   (c) 5   (d) 6

104. • $^1/_2$-inch listed liquidtight flexible metal conduit with fittings listed for grounding can be used as a grounding path when _____ .
    (a) a 20 ampere overcurrent protection device protects circuit conductors, total length of the liquidtight is 6 feet or less
    (b) a 60 ampere overcurrent protection device protects conductors and the total length of the liquidtight is 6 feet or less
    (c) a and b
    (d) none of these

105. The *Code* permits liquidtight flexible nonmetallic conduit to be installed exposed or concealed in lengths over _____ feet if it is essential for required flexibility.
(a) 2   (b) 3   (c) 6   (d) 10

Article 352 – Surface Raceways

106. The derating factors of Section 310-15(b)(2)(a) Notes to Ampacity Tables of 0 to 2,000 Volts, shall not apply to conductors installed in surface metal raceways where _____ .
(a) the cross-sectional area exceeds 4 square inches
(b) the current-carrying conductors do not exceed thirty
(c) the total cross-sectional area of all conductors does not exceed 20 percent of the interior cross-sectional area of the raceway
(d) all of these

107. The conductors, including splices and taps, in a metal surface raceway shall not fill the raceway to more than _____ percent of its cross-sectional area at that point.
(a) 75   (b) 40   (c) 38   (d) 53

108. • The use of surface nonmetallic raceways shall be permitted _____ .
(a) in dry locations   (b) where concealed   (c) in hoistways   (d) all of these

Article 354 – Underfloor Raceways

109. Loop wiring in an underfloor raceway _____ to be a splice or tap.
(a) is considered   (b) shall not be considered   (c) is permitted   (d) none of these

110. Inserts set in fiber underfloor raceways after the floor is laid shall be _____ into the raceway.
(a) taped   (b) glued   (c) screwed   (d) mechanically secured

Article 356 – Cellular Metal Floor Raceways

111. • Loop wiring shall _____ in a cellular metal raceway.
(a) not be permitted   (b) not be considered a splice or tap
(c) be considered a splice or tap when used   (d) be permitted on conductor sizes No. 10 or less

112. A transverse metal raceway for electrical conductors, furnishing access to predetermined cells of precast cellular concrete floors, which permits installation of electrical conductors from a distribution center to the floor cells is usually known as a _____ .
(a) cell   (b) header   (c) open-bottom raceway   (d) none of these

Article 358 – Cellular Concrete Floor Raceways

113. A header that attaches to a floor duct shall be installed _____ to the cell.
(a) parallel   (b) straight   (c) right angle   (d) none of these

114. In cellular concrete floor raceways, a grounding conductor shall connect the insert receptacle to a _____ .
(a) negative ground connection provided in the raceway
(b) negative ground connection provided on the header
(c) positive ground connection provided on the header
(d) grounded terminal located within the insert

Article 362 – Metal Wireways and Nonmetallic Wireways

115. • Wireways shall be permitted for _____ .
(a) exposed work   (b) concealed work
(c) wet locations if of raintight construction   (d) a and c

116.   Splices and taps shall be permitted within a wireway provided they are accessible. The conductor, including splices and taps, shall not fill the wireway to more than _____ percent of its area at that point.
(a) 25   (b) 80   (c) 125   (d) 75

117.   Extensions from wireways are not permitted.
(a) True   (b) False

Article 363 – Flat Cable Assemblies (Type FC)

118.   Flat cable assemblies shall not be installed _____ .
(a) where subject to corrosive vapors unless suitable for the application
(b) in hoistways
(c) in any hazardous (classified) location
(d) all of these

119.   Tap devices used in FC assemblies shall be rated at not less than _____ amperes, or more than 300 volts, and they shall be color-coded in accordance with the requirements of Section 363-20.
(a) 20   (b) 15   (c) 30   (d) 40

Article 364 – Busways

120.   Busways shall not be installed _____ .
(a) where subject to severe physical damage
(b) outdoors or in wet or damp locations unless identified for such use
(c) in hoistways
(d) all of these

121.   • It shall be permissible to extend busways vertically through dry floors if totally enclosed (unventilated) where passing through and for a minimum distance of _____ feet above the floor to provide adequate protection from physical damage.
(a) 6   (b) $6^1/_2$ (c) 8   (d) 10

122.   Where bus enclosures terminate at machines cooled by flammable gas, _____ shall be provided to prevent accumulation of flammable within the bus enclosures.
(a) seal-off bushings   (b) baffles   (c) a and b   (d) none of these

Article 365 – Cablebus

123.   Cablebus shall not be permitted for _____ .
(a) a service   (b) branch circuits   (c) exposed work   (d) concealed work

124.   The individual conductors in a cablebus shall be supported at intervals not greater than _____ feet for vertical runs.
(a) $^1/_2$ (b) 1   (c) $1^1/_2$ (d) 2

Article 370 – Outlet, Device, Pull, and Junction Boxes

Article 370 is another article that has numerous applications. It contains information about the number of conductors permitted in boxes as well as how to size junction boxes. It also contains the requirements for supports and mounting equipment as well as many other requirements.

125.   Round boxes shall not be used with any wiring method connector where a locknut or bushing is connected to the side of the box.
(a) True   (b) False

126.   Boxes, conduit bodies, and fittings installed in wet locations need not be listed for use in wet locations.
(a) True   (b) False

127. When counting the number of conductors in a box, a conductor running through the box is counted as _____ conductor(s).
   (a) one   (b) two   (c) zero   (d) none of these

128. • A reduction of _____ conductors shall be made for outlet box fill for a hickey and two clamps in a box.
   (a) 1   (b) 2   (c) 3   (d) zero

129. • Where nonmetallic-sheathed cable is used with nonmetallic boxes no larger than $2^1/_4 \times 4$ inches, the cable is not required to be secured to the box if the cable is fastened within _____ inches.
   (a) 6   (b) 8   (c) 10   (d) 12

130. Surface extensions from recessed boxes can be made by mechanically attaching an extension ring over the recessed box and attaching the extension wiring method to the extension ring.
   (a) True   (b) False

131. In nonstructural mounting of boxes and fittings, it shall be permissible to make a _____ installation in existing covered surfaces where adequate support is provided by clamps, anchors, or fittings.
   (a) temporary   (b) workmanlike   (c) permanent   (d) flush

132. Enclosures that are not over _____ cubic inches in size and have threaded entries, and do not contain devices or support lighting fixtures shall be considered to be adequately supported where two or more conduits are threaded wrench-tight into the enclosure.
   (a) 50   (b) 75   (c) 100   (d) 125

133. • The minimum size box that does not enclose a flush device shall not be less than _____ inch(es) deep.
   (a) $^{15}/_{16}$ (b) $1^5/_8$ (c) 1   (d) $1^1/_2$

134. Wall mounted fixture weighing not more than 6 pounds and not exceeding _____ inches in any dimensions can be supported to a device box with two No. 6 screws.
   (a) 4   (b) 8   (c) 12   (d) 16

135. Pull boxes or junction boxes that have any dimension over _____ feet shall have all conductors cabled or racked up in an approved manner.
   (a) 3   (b) 6   (c) 9   (d) 12

136. A standard sheet metal outlet box shall be made from steel not less than _____ .
   (a) 0.0625 inch   (b) 0.0757 inch   (c) 15 MSG   (d) 16 MSG

137. A straight pull shall not be less than _____ for a junction box containing conductors over 600 volts.
   (a) 18 times the diameter of the largest raceway
   (b) 48 times the diameter of the largest raceway
   (c) 48 times the outside diameter of the largest shielded conductor
   (d) 36 times the largest conductor

138. • For angle or U-pulls, the distance between the conductor entry (for systems over 600 volts, nominal) and the opposite wall of the box shall not be less than _____ times the outside diameter, over sheath, of the largest cable or conductor.
   (a) 6   (b) 12   (c) 24   (d) 36

Article 373 – Cabinets, Cutout Boxes, and Meter Enclosures

139. Unused openings in cabinets and cutout boxes must be closed with a protective fitting _____ that of the wall of the enclosure.
   (a) equivalent to   (b) the same as   (c) larger than   (d) not required

140. For a steel cabinet or cutout box, the metal shall not be less than _____ inch uncoated.
   (a) 0.53   (b) 0.035   (c) 0.053   (d) 1.35

163.   Lampholders installed over highly combustible material shall be of the _____ type.
(a) industrial   (b) switched   (c) unswitched   (d) residential

164.   • Incandescent lighting fixtures that have open lamps and pendant type lighting fixtures, can be installed in clothes closets where proper clearance is maintained from combustible products.
(a) True   (b) False

165.   Coves for lighting fixtures shall have adequate space and shall be so located that the lamps and equipment can be properly installed and _____ .
(a) maintained   (b) protected from physical damage
(c) tested   (d) inspected

166.   The hand hole of metal fixture poles can be omitted for metal poles _____ feet or less above finish grade. This is only permitted if the pole is provided with a hinged base and the grounding terminal is accessible within the hinged base.
(a) 8   (b) 18   (c) 20   (d) none of these

167.   Lighting fixtures shall be supported independently of the outlet box where the weight exceeds _____ pounds.
(a) 60   (b) 50   (c) 40   (d) 30

168.   Exposed conductive parts of lighting fixtures shall be _____ .
(a) grounded   (b) painted   (c) bonded   (d) a and b

169.   Fixture wires used for lighting fixtures shall not be smaller than No. _____ .
(a) 22   (b) 18   (c) 16   (d) 14

170.   Wiring on fixture chains and other movable parts shall be _____ .
(a) rated for 110ºC   (b) stranded   (c) hard usage rated   (d) none of these

171.   A listed lighting fixture or a listed fixture assembly shall be permitted to be cord-connected if located _____ the outlet box and the cord is continuously visible for its entire length outside the fixture and is not subject to strain or physical damage.
(a) within   (b) directly below   (c) directly above   (d) adjacent to

172.   All lighting fixtures requiring ballasts or transformers shall be plainly marked with their electrical _____ and the manufacturer's name, trademark, or other suitable means of identification.
(a) voltage   (b) rating   (c) amperage   (d) none of these

173.   Lampholders with Edison base screwshells are designed for lampholders and screw-in receptacle adapters.
(a) True   (b) False

174.   Switched lampholders shall be of such construction that the switching mechanism interrupts the electrical connection to the _____ .
(a) lampholder   (b) center contact   (c) branch circuit   (d) all of these

175.   Receptacle faceplate covers made of insulating material shall be noncombustible and not less than _____ inch in thickness.
(a) 0.10   (b) 0.04   (c) 0.01   (d) 0.22

176.   Receptacles, cord connectors, and attachment plugs shall be constructed so that the receptacle or cord connectors will not accept an attachment plug with a different _____ or current rating than that for which the device is intended.
(a) voltage   (b) amperage   (c) heat   (d) all of these

177.   Which of the following statements is true for receptacle covers in a wet location?

I.   The receptacle cover shall be "listed as weatherproof while the attachment plug is inserted" for stationary or fixed loads that are intended to have an attachment plug inserted into the receptacle.

II.   The receptacle cover shall be "listed as weatherproof while the attachment plug is not inserted" for portable loads such as appliances and power tools.
(a) I only   (b) II only   (c) I and II   (d) none of these

178. • A receptacle outlet installed outdoors shall be located so that _____ is not likely to touch the outlet cover or plate.
   (a) a person   (b) water accumulation   (c) metal   (d) none of these

179. Recessed incandescent lighting fixtures shall have _____ protection and shall so be identified as thermally protected.
   (a) physical   (b) corrosion   (c) thermal   (d) all of these

180. The *NEC®* permits lighting fixture tap conductors to be a minimum of ___ inches long.
   (a) 12   (b) 13   (c) 14   (d) 18

181. Ballasts for fluorescent or electric discharge lighting installed indoors must have _____ protection.
   (a) short circuit   (b) overcurrent   (c) integral thermal   (d) none of these

182. An autotransformer, which is used to raise voltage to more than 300 volts, as part of a ballast for supplying lighting units, shall be supplied by a _____ system.
   (a) system   (b) grounded   (c) listed   (d) identified

183. Lighting track is a manufactured assembly and its length may not be altered by the addition or subtraction of sections of track.
   (a) True   (b) False

184. Lighting track shall not be installed _____ .
   (a) where subject to physical damage
   (b) in wet or damp locations
   (c) a and b
   (d) none of these

Article 422 – Appliances

Article 422 contains specific requirements for common appliances. Most branch circuit calculations are contained in Article 220, but Article 422 should be consulted for appliances.

185. Branch circuit conductors to individual appliances shall not be sized _____ than required by the appliance markings or instructions.
   (a) larger   (b) smaller

186. The rating or setting of an overcurrent protection device for a 13.3 ampere single non-motor operated appliance should not exceed _____ amperes.
   (a) 15   (b) 35   (c) 25   (d) 45

187. Infrared lamps for industrial heating appliances shall have overcurrent protection not exceeding _____ amperes.
   (a) 30   (b) 40   (c) 50   (d) 60

188. The cord for a dishwasher and trash compactor shall not be longer than _____ feet measured from the back of the appliance.
   (a) 2   (b) 4   (c) 6   (d) 8

189. The maximum allowable rating of a permanently connected appliance where the branch circuit overcurrent protection device is used as the appliance disconnecting means is _____ horsepower.
   (a) 1/8   (b) $^1/_4$ (c) $^1/_2$ (d) $^3/_4$

190. • Electric heaters of the cord- and plug-connected immersion type shall be so constructed and installed that current-carrying parts are effectively _____ from electrical contact with the substance in which they are immersed.
   (a) isolated   (b) protected   (c) insulated   (d) all of these

191. Each electric appliance shall be provided with a _____ giving the identifying name and the rating in volts and amperes, or in volts and watts.
   (a) plaque   (b) nameplate   (c) sticker   (d) directory

Article 424 – Fixed Electric Space Heating Equipment

192. Fixed electric space heating equipment requiring supply conductors with insulation rated over _____°C shall be clearly marked.
      (a) 75   (b) 60   (c) 90   (d) all of these

193. • Electric heating appliances employing resistance-type heating elements rated more than _____ amperes shall have the heating elements subdivided.
      (a) 60   (b) 50   (c) 48   (d) 35

194. On space heating cables, blue leads indicate a cable rated _____ volts.
      (a) 120   (b) 240   (c) 208   (d) 277

195. Electric space heating cables shall not be installed over cabinets whose clearance from the ceiling is less than the minimum _____ dimension of the cabinet to the nearest cabinet edge that is open to the room or area.
      (a) horizontal   (b) vertical   (c) overall   (d) depth

196. When installing duct heaters, sufficient clearance shall be maintained to permit replacement and adjustment of controls and heating elements.
      (a) True   (b) False

Article 426 – Outdoor Electric De-icing and Snow-Melting Equipment

197. For electrode-type boilers, each boiler shall be designed so that in normal operation there is no change in state of the heat transfer medium, and shall be equipped with a temperature sensitive _____ .
      (a) protective device   (b) limiting means   (c) shut-off device   (d) all of these

198. Resistance heating elements of de-icing _____ shall not be installed where they bridge expansion joints unless adequately protected from expansion and contraction.
      (a) heating cables   (b) units   (c) panels   (d) all of these

199. An impedance heating system that is operating at a _____ greater than 30, but not more than 80, shall be grounded at designated point(s).
      (a) voltage   (b) amperage   (c) wattage   (d) temperature

Article 427 – Electric Heating Equipment For Pipelines and Vessels

200. Types of pipeline resistive heaters are heating _____ .
      (a) blankets   (b) tape   (c) barrels   (d) a and b

Unit 4

Motors and Transformers

OBJECTIVES

After reading this unit, the student should be able to briefly explain the following concepts:

Part A – Motors	Nameplate ampere	Transformer primary vs. secondary
Alternating current motors	Reversing DC motors	Transformer current
Dual voltage motors	Reversing AC motors	Transformer kVA rating
Horsepower/watts	Volt-ampere calculations	Transformer power losses
Motor speed control	**Part B – Transformers**	Transformer turns ratio

After reading this unit, the student should be able to briefly explain the following terms:

Part A – Motors	Watts rating	Excitation current
Armature winding	**Part B – Transformers**	Flux leakage loss
Commutator	Auto transformers	Hysteresis losses
Dual voltage	Circuit impedance	Kilo volt-ampere
Field winding	Conductor losses	Line current
Horsepower ratings	Core losses	Ratio
Magnetic field	Counter-electromotive force	Self-excited
Motor full load current	Current transformers (CT)	Step-down transformers
Nameplate ampere	Eddy current losses	Step-up transformers
Torque		

PART A – MOTORS

MOTOR INTRODUCTION

The *electric motor* operates on the principle of the attracting and repelling forces of magnetic fields. One magnetic field (permanent or electromagnetic) is *stationary* and the other magnetic field (called the *armature* or *rotor*) rotates between the poles of the stationary magnet (Figure 4–1). The turning or repelling forces between the magnetic fields are called *torque.* Torque is dependent on the strength of the stationary magnetic field, the strength of magnetic field of the armature, and the physical construction of the motor.

The repelling force of like magnetic polarities, and the attraction force of unlike polarities, causes the armature to rotate in the electric motor. For a direct current motor, a device called a *commutator* is placed on the end of the conductor loop or armature. The purpose of the commutator is to maintain the proper polarity of the loop, so as to keep the armature or rotor turning.

The *armature winding* of a motor carries a starting of current when voltage is first applied to the motor windings. As the armature turns, it cuts the lines of force of the *field winding* resulting in an increase in counterelectromotive-force, (CEMF) (Figure 4–2). The increased CEMF results in an increase in inductive reactance, which increases the circuit impedance. The increased circuit impedance causes a decrease in the motor armature running current (Figure 4–3).

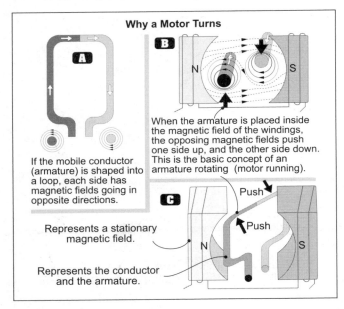

Figure 4–1
Why a Motor Turns

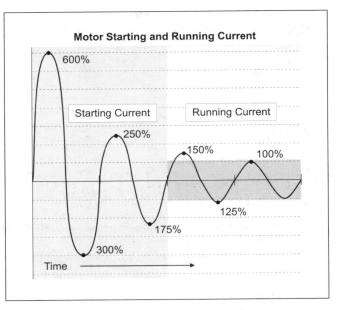

Figure 4–2
Motor Starting and Running Current

4–1 MOTOR SPEED CONTROL

One of the advantages of *direct current motor* over the alternating current motor is the motor's ability to maintain a constant speed. But series direct current motors are susceptible to run away (increase rotational speed) when not connected to a load.

If the speed of a direct current motor is increased, the armature winding is cut by the magnetic field at an increasing rate, resulting in an increase of the armatures CEMF. The increased CEMF in the armature acts to cut down on the armature current, resulting in the motor slowing down. Placing a load on a direct current motor causes the motor to slow down, which reduces the rate at which the armature winding is cut by the magnetic field flux lines. A reduction of the armature flux lines results in a decrease in the armature's CEMF and an increase in motor speed.

4–2 REVERSING A DIRECT CURRENT MOTOR

To reverse a direct current motor, you must reverse either the *magnetic field* of the field winding or the magnetic field of the armature. This is accomplished by reversing either the field or armature current flow (Figure 4–4). Because most direct current motors have the field and armature winding connected to the same direct current power supply, reversing the polarity of the power supply changes both the field and armature simultaneously. To reverse the rotation of a direct current motor, you must reverse either the field or armature leads, but not both.

4–3 ALTERNATING CURRENT MOTORS

Fractional horsepower motors that can operate on either alternating or direct current are called universal motors. A motor that will not operate on direct current is called an induction motor.

The induction motor is the purest form of an alternating current motor, with no physical connection between its rotating member *(rotor),* and stationary member *(stator).* Two common types of alternating current induction motors are synchronous and wound rotor motors.

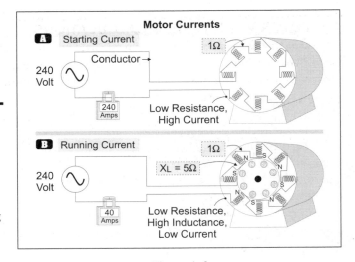

Figure 4–3
Motor Currents

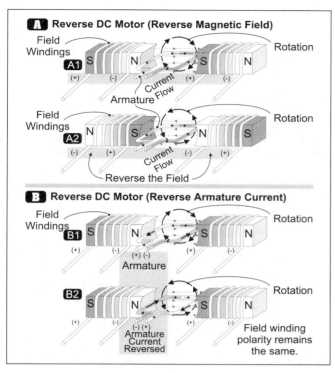

Figure 4–4
Reverse DC Motor

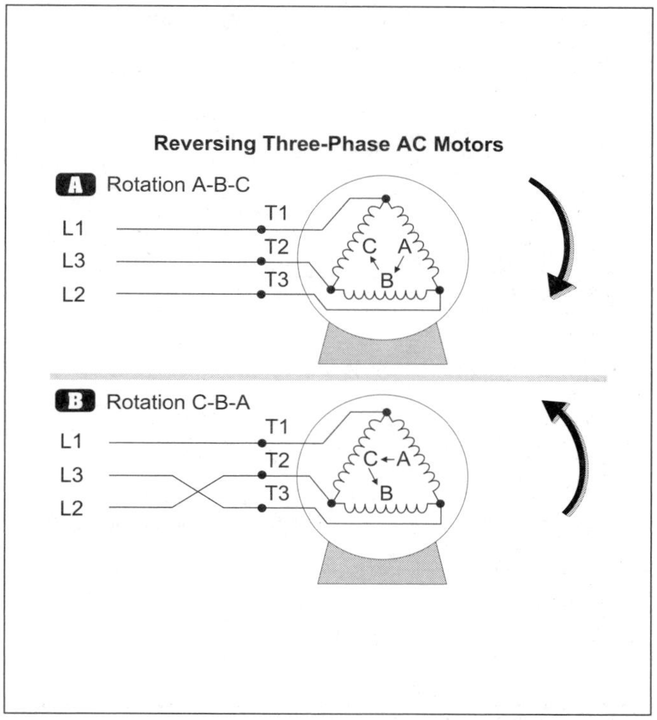

Figure 4–5
Reversing Three-Phase AC Motors

Synchronous Motors

The synchronous motor's rotor is locked in step with the rotating stator field. Synchronous motors maintain their speed with a high degree of accuracy and are used for electric clocks and other timing devices.

Wound Rotor Motors

Because of their high starting torque requirements, wound rotor motors are used only in special applications. They only operate on three-phase alternating current.

4–4 REVERSING ALTERNATING CURRENT MOTORS

Three-phase alternating current motors can be reversed by reversing any two of the three line conductors that supply the motor (Figure 4–5).

4–5 MOTOR VOLT-AMPERE CALCULATIONS

Dual voltage motors are made with two field windings, each rated for the lower voltage marked on the nameplate of the motor. When the motor is wired for the lower voltage, the field windings are connected in parallel; when wired for the higher voltage, the motor windings are connected in series (Figure 4–6).

Motor Input VA

Regardless of the voltage connection, the *power* consumed by a motor is the same at either voltage. To determine the *motor input* apparent power (VA), use the following formulas:

Motor VA (single-phase) = Volts $\times$ Ampere

Motor VA (three-phase) = Volts $\times$ Ampere $\times$ $\sqrt{3}$

❏ Motor VA Single-Phase

What is the motor VA of a single-phase 115/230-volt, 1 horsepower motor that has a current rating of 16 ampere at 115 volts and 8 ampere at 230 volts (Figure 4–7)?

(a) 1,450 VA   (b) 1,600 VA   (c) 1,840 VA   (d) 1,920 VA

• Answer: (c) 1,840 VA

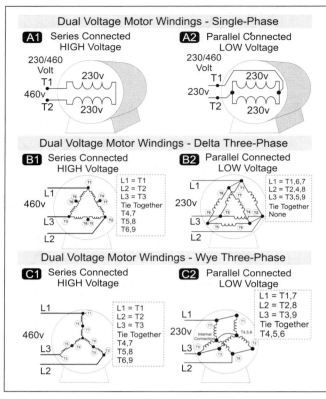

Figure 4–6
Dual Voltage Motor Windings – Single-Phase

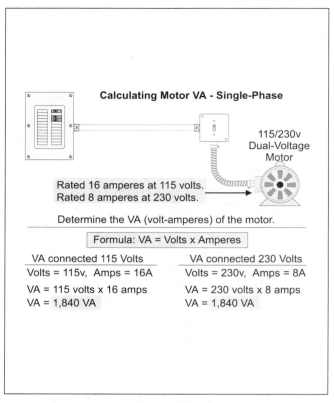

Figure 4–7
Calculating Motor VA – Single-Phase

Motor VA = Volts × Ampere

Volts = 115 or 230, Ampere = 16 or 8 ampere

Motor VA = 115 volts × 16 ampere, or = 230 volts × 8 ampere = 1,840 VA

❑ **Motor VA Three-Phase**

What is the motor VA of a 3-phase, 230/460 volt, 30 horsepower motor that has a current rating of 40 ampere at 460 volts (Figure 4–8)?

(a) 41,450 VA   (b) 31,600 VA   (c) 21,840 VA   (d) 31,869 VA

   • Answer: (d) 31,869 VA

Motor VA = Volts × Ampere × $\sqrt{3}$

Volts = 460 volts, Ampere = 40 ampere, $\sqrt{3}$ = 1.732

Motor VA = 460 volts × 40 ampere × 1.732 = 31,869 VA

4–6 MOTOR HORSEPOWER/WATTS

The mechanical work (output) of a motor is rated in horsepower and can be converted to electrical energy as 746 watts per horsepower.

Horsepower = Output Watts/746 Watts

Motor Output Watts = Horsepower × 746

❑ **Motor Horsepower**

What size horsepower motor is required to produce 15 kW output (Figure 4–9)?

(a) 5 horsepower   (b) 10 horsepower   (c) 20 horsepower   (d) 30 horsepower

   • Answer: (c) 20 horsepower

$$\text{Horsepower} = \frac{\text{Output Watts}}{746} = \frac{15,000 \text{ watts}}{746 \text{ watts}} = 20 \text{ horsepower}$$

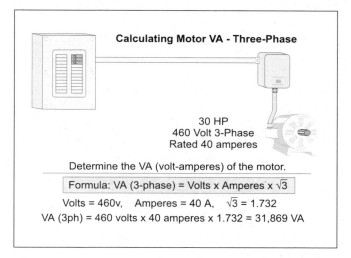

Figure 4–8
Calculating Motor VA – Three-Phase

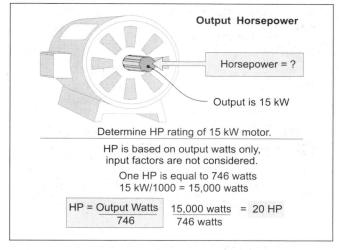

Figure 4–9
Output Horsepower

❏ Motor Output Watt

What is the output watt rating of a 10 horsepower motor (Figure 4–10)?

(a) 3 kW (b) 2.2 kVA (c) 3.2 kW (d) 7.5 kW

• Answer: (d) 7.5 kW

Output Watts = 10 horsepower × 746 Watts = 7,460 watts

$$\text{Output watts} = \frac{7,460 \text{ watts}}{1,000} = 7.46 \text{ kW}$$

4–7 MOTOR NAMEPLATE AMPERE

The motor nameplate indicates the motor operating voltage and current. The actual current drawn by the motor depends on how much the motor is loaded. It is important not to overload a motor above its rated horsepower because the current of the motor will increase to a point that the motor winding will be destroyed from excess heat. The nameplate current can be calculated by the following formulas:

$$\text{Motor Nameplate (single-phase) Ampere} = \frac{\text{Motor Horsepower} \times 746 \text{ Watts}}{(\text{Volts} \times \text{Efficiency} \times \text{Power Factor})}$$

$$\text{Motor Nameplate (three-phase) Ampere} = \frac{\text{Motor Horsepower} \times 746 \text{ Watts}}{(\text{Volts} \times \sqrt{3} \times \text{Efficiency} \times \text{Power Factor})}$$

❏ Nameplate Ampere – Single-Phase

What is the nameplate ampere for a 7.5 horsepower motor rated 240 volts, single-phase? The efficiency is 87 percent and the power factor is 93 percent (Figure 4–11).

(a) 19 ampere (b) 24 ampere (c) 19 ampere (d) 29 ampere

• Answer: (d) 29 ampere

$$\text{Nameplate ampere} = \frac{\text{Horsepower} \times 746 \text{ Watts}}{(\text{Volts} \times \text{Efficiency} \times \text{Power Factor})}$$

$$\text{Nameplate ampere} = \frac{7.5 \text{ Horsepower} \times 746 \text{ watts}}{(240 \text{ volts} \times 0.87 \text{ Efficiency} \times 0.93 \text{ Power Factor})} = 28.81 \text{ ampere}$$

❏ Nameplate Ampere – Three-Phase

What is the nameplate ampere of a 40 horsepower motor rated 208 volts, 3-phase? The efficiency is 80 percent and the power factor is 90 percent (Figure 4–12).

(a) 85 ampere (b) 95 ampere (c) 105 ampere (d) 115 ampere

• Answer: (d) 115 ampere

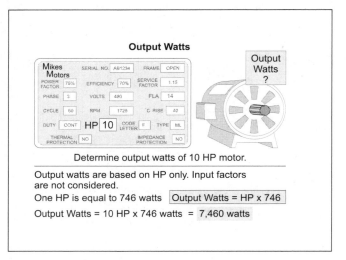

Figure 4–10
Output Watts

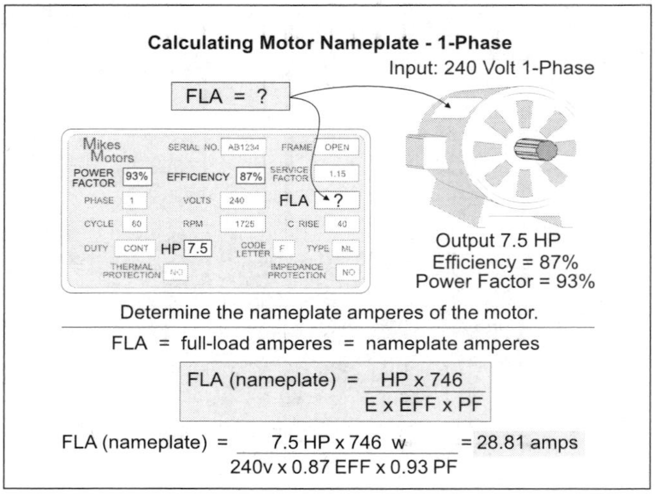

Calculating Motor Nameplate - 1-Phase

Figure 4–11
Calculating Motor Nameplate

$$\text{Nameplate ampere} = \frac{\text{Horsepower} \times 746 \text{ Watts}}{(\text{Volts} \times \sqrt{3} \times \text{Efficiency} \times \text{Power Factor})}$$

$$\text{Nameplate} = \frac{40 \text{ Horsepower} \times 746 \text{ Watts}}{(208 \text{ volts} \times 1.732 \times 0.8 \text{ Efficiency} \times 0.9 \text{ Power Factor})} = 29{,}840/259.4 = 115 \text{ ampere}$$

PART B – TRANSFORMER BASICS

TRANSFORMER INTRODUCTION

A *transformer* is a stationary device used to raise or lower voltage. Transformers have the ability to transfer electrical energy from one circuit to another by *mutual induction* between two conductor coils. Mutual induction occurs between two conductor coils *(windings)* when electromagnetic lines of force within one winding induces a voltage into a second winding (Figure 4–13).

Current transformers use the circuit conductors as the primary winding and step the current down for metering. Often, the *ratio* of the current transformer is 1,000 to 1. This means that if there are 400 ampere on the phase conductors, the current transformer steps the current down to 0.4 ampere for the meter.

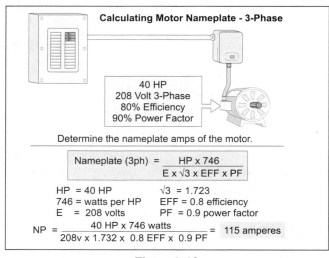

Figure 4–12
Calculating Motor Nameplate – 3-Phase

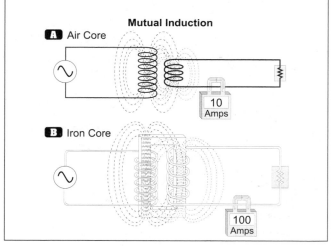

Figure 4–13
Mutual Induction

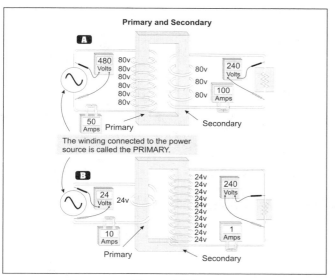

Figure 4–14
Transformer Primary and Secondary

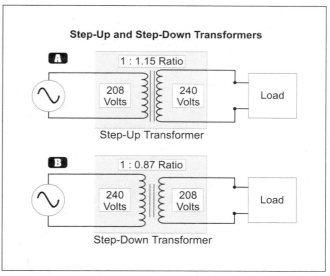

Figure 4–15
Step-Up and Step-Down Trransformers

The *magnetic coupling (magnetomotive force, MMF)* between the primary and secondary winding can be increased by increasing the winding *ampere-turns*. Ampere-turns can be increased by increasing the number of coils and/or the current through each coil. When the current in the core of a transformer is raised to a point where there is high *flux density*, additional increases in current will produce few additional flux lines. The transformer iron core is said to be *saturated*.

4–8 TRANSFORMER PRIMARY AND SECONDARY

The transformer *winding* connected to the source is called the *primary winding*. The transformer winding connected to the load is called the *secondary winding*.

4–9 TRANSFORMER SECONDARY AND PRIMARY VOLTAGE

Voltage induced in the secondary winding of a transformer is equal to the sum of the voltages induced in each *loop* of the secondary winding. The voltage induced in the secondary of a transformer depends on the number of secondary conductor turns cut by the primary magnetic flux lines. The greater the number of secondary conductor loops, the greater the secondary voltage (Figure 4–14).

Step-Up and Step-Down Transformers

The secondary winding of a *step-down transformer* has fewer turns than the primary winding, resulting in a lower secondary voltage. The secondary winding of a *step-up transformer* has more turns than the primary winding, resulting in a higher secondary voltage (Figure 4–15).

4–10 AUTOTRANSFORMERS

Autotransformers are transformers that use a common winding for both the primary and the secondary. The disadvantage of an auto transformer is the lack of isolation between the primary and secondary conductors, but they are often used because they are less expensive (Figure 4–16).

4–11 TRANSFORMER POWER LOSSES

When current flows through the winding of a transformer, power is dissipated in the form of heat. This loss is referred to as conductor I^2R loss. In addition, losses include flux leakage, core loss from eddy currents, and hysteresis heating losses (Figure 4–17).

Conductor Resistance Loss

Transformer windings are made of many turns of wire. The resistance of the conductors is directly proportional to the length of the conductors and inversely proportional to the cross-sectional area of the conductor. The more turns there are, the longer the conductor is, and the greater the conductor resistance. Conductor losses can be determined by the formula: $P = I^2R$.

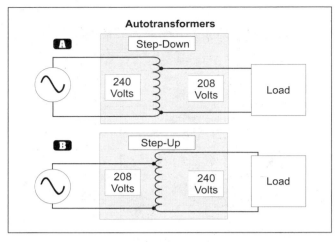

Figure 4–16
Autotransformers

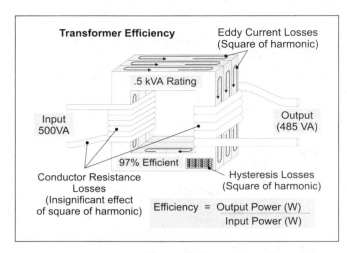

Figure 4–17
Transformer Efficiency

Flux Leakage Loss

The *flux leakage loss* represents the electromagnetic flux lines between the primary and secondary winding that are not used to convert electrical energy from the primary to the secondary. They represent wasted energy.

Core Losses

Iron is the only metal used for transformer cores because it offers low resistance to magnetic flux lines *(low reluctance)*. Iron cores permit more flux lines between the primary and secondary winding, thereby increasing the magnetic coupling between the primary and secondary windings. However, alternating currents produce electromagnetic fields within the windings that induce a circulating current in the iron core. These circulating currents *(eddy currents)* flow within the iron core producing power losses that cannot be transferred to the secondary winding. Transformer iron cores are laminated to have a small cross-sectional area to reduce the eddy currents and their associated losses.

Hysteresis Losses

Each time the primary magnetic field expands and collapses, the transformer's iron core molecules realign themselves to the changing polarity of the electromagnetic field. The energy required to realign the iron core molecules to the changing electromagnetic field is called hysteresis losses.

Heating by the Square of the Frequency

Hysteresis and eddy current losses are affected by the square of the alternating current frequency. For this reason, care must be taken when iron core transformers are used in applications involving high frequencies and nonlinear loads. If the transformer operates at the third harmonic (180 Hz), the losses will be the square of the multiple of the fundamental frequency. The third harmonic frequency (180 Hz) is three times the fundamental frequency (60 Hz).

❏ Transformer Losses by Square of Frequency

What is the effect on a 60 Hz rated transformer of third harmonic loads (180 Hz)?

(a) heating increases three times   (b) heating increases six times

(c) heating increases nine times   (d) no significant heating

   • Answer: (c) heating increases nine times (3^2)

4–12 TRANSFORMER TURNS RATIO

The relationship of the primary winding voltage to the secondary winding voltage is the same as the relation between the number of primary turns as compared to the number of secondary turns. This relationship is called turns ratio or voltage ratio.

❏ Delta Winding Turns Ratio

What is the turns ratio of a delta-delta transformer? The primary winding is 480 volts, and the secondary winding voltage is 240 volts (Figure 4–18)?

(a) 4:1   (b) 1:4   (c) 2:1   (d) 1:2

   • Answer: (c) 2:1

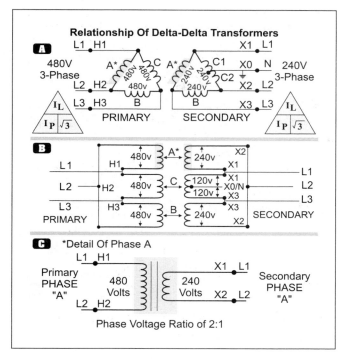

Figure 4–18
Relationships of Delta-Delta Transformers

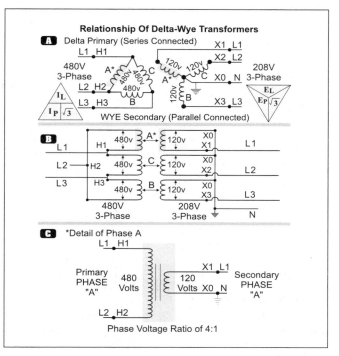

Figure 4–19
Relationships of Delta-Wye Transformers

The primary winding voltage is 480 and the secondary winding voltage is 240.

This results in a ratio of 480:240 or 2:1.

❏ **Wye Winding Ratio**

What is the turns ratio of a delta-wye transformer? The primary winding is 480 volts, and the secondary winding is 120 volts (Figure 4–19).

(a) 4:1   (b) 1:4   (c) 2.3:1   (d) 1:2.3

• Answer: (a) 4:1

The primary winding voltage is 480, and the secondary winding voltage is 120.

This results in a ratio of 480:120 or 4:1.

4–13 TRANSFORMER kVA RATING

Transformers are rated in kilovolt-ampere, abbreviated as kVA.

4–14 TRANSFORMER CURRENT

Whenever the number of primary turns is greater than the number of secondary turns, the secondary voltage is less than the primary voltage. This results in secondary current being greater than the primary current because the power remains the same, but the voltage changes. Since the secondary voltage of most transformers is less than the primary voltage, the secondary conductors carry more current than the primary (Figure 4–20). Primary and secondary line current can be calculated by:

Current (single-phase) = Volt-Ampere/Volts

Current (three-phase) = Volt-Ampere/(Volts × 1.732)

❏ **Transformer Current – Single-Phase**

What is the primary and secondary line current for a single-phase 25 kVA transformer, rated 480 volts primary and 240 volts secondary (Figure 4–21)?

(a) 52/104 ampere   (b) 104/52 ampere   (c) 104/208 ampere   (d) 208/104 ampere

• Answer: (a) 52/104 ampere

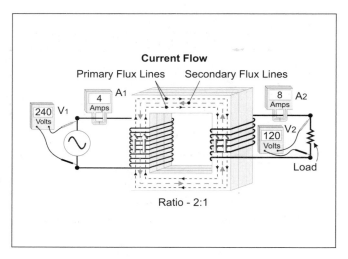

Figure 4–20
Transformer Current Flow

Primary current = VA/E

$$\text{Primary current} = \frac{25{,}000 \text{ VA}}{480 \text{ volts}}$$

Primary current = 52 ampere

Secondary current = VA/E

$$\text{Secondary current} = \frac{25{,}000 \text{ VA}}{240 \text{ volts}}$$

Secondary current = 104 ampere

❏ **Transformer Current – Three-Phase**

What is the primary and secondary line current for a 3-phase 37.5 kVA transformer, rated 480 volts primary and 208 volts secondary (Figure 4–22)?

(a) 45/104 ampere   (b) 104/40 ampere
(c) 208/140 ampere   (d) 140/120 ampere

• Answer: (a) 45/104 ampere

$$\text{Primary current} = \frac{\text{VA}}{\text{E} \times \sqrt{3}}$$

$$\text{Primary current} = \frac{37{,}500 \text{ VA}}{(480 \text{ volts} \times 1.732)}$$

Primary current = 45 ampere

$$\text{Secondary current} = \frac{\text{VA}}{\text{E} \times \sqrt{3}}$$

$$\text{Secondary current} = \frac{37{,}500 \text{ VA}}{(208 \text{ volts} \times 1.732)}$$

Secondary current = 104 ampere

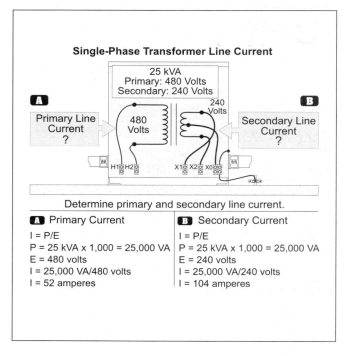

Figure 4–21
Transformer Line Current

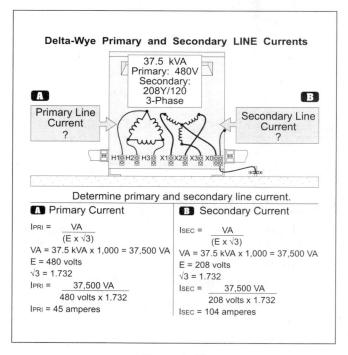

Figure 4–22
Delta-Wye Transformers Primary and
Secondary Line Currents

Unit 4 – Motors and Transformers Summary Questions

Part A – Motors

Motor Introduction

1. When voltage is applied to the motor's armature, short-circuit current will flow and the armature will begin to turn. As the armature starts turning, it cuts the lines of force of the field winding, resulting in induced CEMF in the armature conductor, which reduces the short-circuit current.
   (a) True   (b) False

2. An electric motor works because of the effects a _____ has against a wire carrying an electric current.
   (a) magnetic field   (b) commutator   (c) voltage source   (d) none of these

3. The rotating part of a direct current motor or generator is called the _____ .
   (a) shaft   (b) rotor   (c) capacitor   (d) field

4. For a direct current motor, a device called a _____ is placed on the end of the conductor loop (armature). The polarity of the loop is maintained with the proper magnetic field to keep the opposing magnetic fields pushing each other in the same direction. This in turn keeps the loop turning.
   (a) coil   (b) resistor   (c) commutator   (d) none of these

4–1 Motor Speed Control

5. • One of the great advantages of the direct current motor is the motor's ability to maintain a constant speed. If the speed of a direct current motor is increased, the armature will cut through the field winding magnetic flux at an increasing rate resulting in a lower CEMF that acts to cut down on the increased armature current, which slows the motor back down.
   (a) True   (b) False

6. • Placing a load on a direct current motor causes the motor to slow down, which increases the rate at which the field flux lines are cut by the armature. As a result, the armature CEMF increases, resulting in an increase in the applied armature voltage and current. The increase in current results in an increase in motor speed.
   (a) True   (b) False

4–2 Reversing a Direct Current Motor

7. To reverse a direct current motor, you must reverse the direction of the _____ .
   (a) field current   (b) armature current   (c) a or b   (d) a and b

4–3 Alternating Current Motors

8. In a(n) _____ motor, the rotor is actually locked in step with the rotating stator field and is dragged along at the synchronous speed of the rotating magnetic field. _____ motors maintain their speed with a high degree of accuracy and are used for electric clocks and other timing devices.
   (a) alternating current   (b) universal   (c) wound rotor   (d) synchronous

9. _____ rotor motors are used only as special applications because of their high starting torque requirements and only operate on 3-phase alternating current power.
   (a) Alternating current   (b) Universal   (c) Wound rotor   (d) Synchronous

10. _____ motors are fractional horsepower motors that operate equally well on alternating current and direct current and are used for vacuum cleaners, electric drills, mixers, and light household appliances.
    (a) Alternating current   (b) Universal   (c) Wound rotor   (d) Synchronous

4–4 Reversing Alternating Current Motors

11. • Three-phase alternating current motors can be reversed by changing the wiring from the line wiring from ABC phase configuration to_____ .
    (a) BCA   (b) CAB   (c) CBA   (d) ABC

4–5 Motor Volt-Ampere Calculations

12. Dual-voltage 277/480 volt motors are made with two field windings, each rated at 277 volts. The field windings are connected in parallel for _____ volt operation and in series for _____ volt operation.
    (a) 277, 480   (b) 480, 277   (c) 277, 277   (d) 480, 480

4–6 Motor Horsepower/Watts

13. What size motor is required to produce 30 kW output?
    (a) 20 horsepower   (b) 30 horsepower   (c) 40 horsepower   (d) 50 horsepower

14. What are the output watts of a 15 horsepower motor?
    (a) 11 kW   (b) 15 kVA   (c) 22 kW   (d) 31 kW

15. What are the output watts of a 5 horsepower motor, 3-phase 480 volt, efficiency 75 percent and power factor 70 percent?
    (a) 3.75 kW   (b) 7.5 kVA   (c) 7.5 kW   (d) 10 kW

4–7 Motor Nameplate Ampere

16. • For practical purposes you will not need to calculate motor nameplate current; but you should understand how it is calculated.
    (a) True   (b) False

17. What are the nameplate ampere for a 5 horsepower motor, 240 volt, single-phase, efficiency at 90 percent and power factor of 80 percent?
    (a) 19.3 ampere   (b) 21.6 ampere   (c) 28.2 ampere   (d) 31.1 ampere

18. What are the nameplate ampere of a 20 horsepower 208 volt, 3-phase motor, power factor 90 percent, and efficiency of 80 percent?
    (a) 50 ampere   (b) 58 ampere   (c) 65 ampere   (d) 80 ampere

Part B – Transformer Basics

Transformer Introduction

19. A _____ is a stationary device used to raise or lower the voltage and has the ability to transfer electrical energy from one circuit to another, with no physical connection between the two.
    (a) capacitor   (b) motor   (c) relay   (d) transformer

20. Transformers operate on the principle of _____ .
    (a) magnetoelectricity   (b) triboelectric effect   (c) thermocouple   (d) mutual induction

4–8 Primary and Secondary Transformer

21. The transformer winding that is connected to the source is called the _____ winding and the transformer winding that is connected to the load is called the _____ . Transformers are reversible; that is, either winding can be used as the primary or secondary.
    (a) secondary, primary   (b) primary, secondary   (c) depends on the wiring   (d) none of these

22. Voltage induced in the secondary winding of a transformer is dependent on the number of secondary turns as compared to the number of primary turns.
    (a) True   (b) False

23. The secondary winding of a step-down transformer has _____ turns than the primary, resulting in a _____ secondary voltage as compared to the primary.
    (a) less, higher   (b) more, lower   (c) less, lower   (d) more, higher

24. The secondary winding of a step-up transformer has _____ turns than the primary, resulting in a _____ secondary voltage as compared to the primary.
    (a) less, higher   (b) more, lower   (c) less, lower   (d) more, higher

4–10 Autotransformers

25. Autotransformers use the same winding for both the primary and secondary. The disadvantage of an autotransformer is the lack of _____ between the primary and secondary conductors.
    (a) power   (b) voltage   (c) isolation   (d) grounding

4–11 Transformer Power Losses

26. The most common causes of power losses for transformers are _____ .
    (a) conductor resistance (b) eddy currents   (c) hysteresis   (d) all of these

27. The leakage of the electromagnetic flux lines between the primary and secondary winding represents wasted energy.
    (a) True   (b) False

28. • The expanding and collapsing electromagnetic field of the transformer also induces a voltage in the transformer core. The induced voltage causes _____ to flow within the core, which removes energy from the transformer winding and represents wasted power.
    (a) eddy currents   (b) flux   (c) inductive   (d) hysteresis

29. Eddy currents can be reduced by dividing the core into many flat sections or laminations. Because the laminations have a _____ cross-sectional area, the resistance offered to the eddy currents is greatly increased.
    (a) round   (b) porous   (c) large   (d) small

30. As current flows through the transformer, the iron core is temporarily magnetized by the electromagnetic field created by the alternating current. Each time the primary magnetic field expands and collapses, the core molecules realign themselves to the changing polarity of the electromagnetic field. The energy required to realign the core molecules to the changing electromagnetic field is called the _____ loss of the core.
    (a) eddy current   (b) flux   (c) inductive   (d) hysteresis

4–12 Transformer Turns Ratio

31. The relationship of the primary winding voltage to the secondary winding voltage is the same as the relation between the number of conductor turns on the primary as compared to the secondary. This relationship is called _____ .
    (a) ratio   (b) efficiency   (c) power factor   (d) none of these

32. • The primary phase voltage is 240 and the secondary phase is 480. This results in a ratio of _____ .
    (a) 1:2   (b) 2:1   (c) 4:1   (d) 1:4

4–13 Transformer kVA Rating

33. Transformers are rated in _____ .
    (a) VA   (b) kW   (c) watts   (d) kVA

4–14 Transformer Current

34.   • The current flow in the secondary transformer winding creates an electromagnetic field that opposes the primary electromagnetic field resulting in less primary CEMF. The primary current automatically increases in direct proportion to the secondary current.
(a) True   (b) False

35.   • The transformer winding with the _____ number of turns will have the lower current and the winding with the _____ number of turns will have the higher current.
(a) lesser, greater   (b) most, most   (c) least, least   (d) greater, lesser

☆ Challenge Questions

Part A – Motors

4–1 Motor Speed Control

36.   A(n) _____ type of electric motor tends to run away if it is not always connected to its load.
(a) direct current series   (b) direct current shunt
(c) alternating current induction   (d) alternating current synchronous

37.   • A(n) _____ motor has a wide speed range.
(a) alternating current   (b) direct current   (c) synchronous   (d) induction

4–2 Reversing a Direct Current Motor

38.   • If the two line (supply) leads of a direct current series motor are reversed, the motor will _____ .
(a) not run   (b) run backwards   (c) run the same as before   (d) become a generator

39.   • To reverse a direct current series motor, one may simply reverse the supply (power) leads.
(a) True   (b) False

4–3 Alternating Current Motors

40.   The _____ induction motor is used only in special applications and is always operated on 3-phase alternating current power.
(a) compound   (b) synchronous   (c) split phase   (d) wound rotor

41.   • The rotating part of a direct current motor or generator is called the _____ .
(a) shaft   (b) rotor   (c) armature   (d) b or c

4–5 Motor Volt-Ampere Calculations

42.   The input volt-ampere of a 5 horsepower (15.2 ampere), 230 volt, 3-phase motor are closest to _____ .
(a) 7,500 VA   (b) 6,100 VA   (c) 5,300 VA   (d) 4,600 VA

43.   The input volt-ampere of a 1 horsepower (16 ampere), 115 volt, single-phase motor is _____ .
(a) 2,960 VA   (b) 1,840 VA   (c) 3,190 VA   (d) 1,650 VA

Part B – Transformer Basics

4–11 Transformer Power Losses

44.   • When the current in the core of a transformer has risen to a point where high flux density has been reached and additional increases in current produce few additional flux lines, the metal core is said to be _____ .
(a) maximum   (b) saturated   (c) full   (d) none of these

45.   • Magnetomotive force (MMF) can be increased by increasing the _____ .
       (a) number of ampere-turns   (b) current in the coils
       (c) number of coils   (d) all of these

4–12 Transformer Turns Ratio

46.   • The secondary current of a transformer that has a turns ratio of 2:1 is _____ the primary current.
       (a) higher than   (b) lower than   (c) the same as   (d) none of these

4–13 Transformer kVA Rating

47.   The primary kVA rating for a transformer that is 100 percent efficient, with a secondary of 12 volts (E) and with secondary current (I) of 5 ampere is _____ .
       (a) 600 kVA   (b) 30 kVA   (c) 6 kVA   (d) 0.06 kVA

4–14 Transformer Current

48.   • The primary of a transformer has 100 turns and the secondary has 10 turns. What is the primary current of the transformer if the secondary current (I) is 5 ampere?
       (a) 25 ampere   (b) 10 ampere   (c) 5 ampere   (d) 0.5 ampere

49.   • Which winding of a current transformer will carry more current?
       Note: A clamp-on ammeter is a current transformer, the meter acts as the secondary.
       (a) primary   (b) secondary   (c) interwinding   (d) tertiary

50.   A transformer primary winding has 900 turns and the secondary winding has 90 turns, which winding of the transformer has a larger conductor?
       (a) primary   (b) secondary   (c) interwinding   (d) none of these

51.   The primary of a transformer is 480 volts and the secondary is 240 volts, which winding of this transformer has a larger conductor?
       (a) tertiary   (b) secondary   (c) primary   (d) windings are equal

52.   If the transformer voltage turns ratio is 5:1 and the secondary has 10 ampere-turns, the primary current would be _____ if the secondary current was 10 ampere.
       (a) 25 ampere   (b) 10 ampere   (c) 2 ampere   (d) cannot be determined

53.   The transformer primary is 240 volts, the secondary is 12 volts, and the load is two 100-watt lamps. The transformer is 92 percent efficient. The secondary current of this transformer is _____ .
       (a) 1 ampere   (b) 17 ampere   (c) 28 ampere   (d) none of these

54.   The transformer primary is 240 volts, the secondary is 120 volts, and the load is 1,500 watt. The transformer is 92 percent efficient. The primary current of this transformer is _____ ampere.
       (a) 6.8   (b) 8.6   (c) 9.9   (d) 7.8

55.   • The primary kVA of this transformer is _____
       (Figure 4–23)?
       (a) 72.5 kVA   (b) 42 kVA
       (c) 30 kVA   (d) 21 kVA

56.   • The primary current of this transformer is _____
       (Figure 4–23)?
       (a) 90 ampere   (b) 50 ampere
       (c) 25 ampere   (d) 12 ampere

57.   • The secondary voltage of this transformer is _____
       (Figure 4–24)?
       (a) 6 volts   (b) 12 volts
       (c) 24 volts   (d) 30 volts

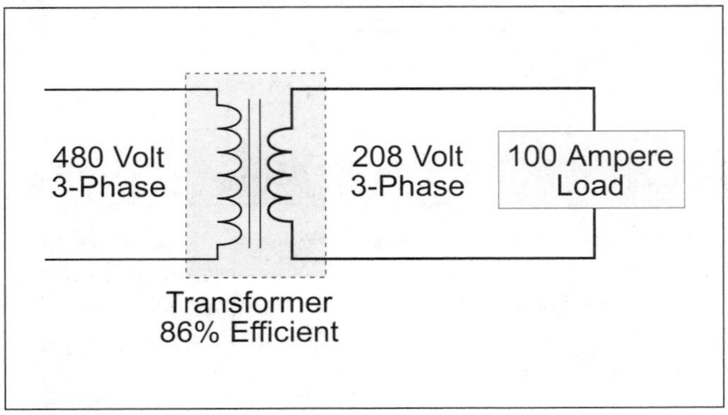

480 Volt
3-Phase

208 Volt
3-Phase

100 Ampere
Load

Transformer
86% Efficient

Figure 4–23

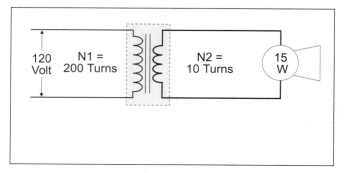

Figure 4–24

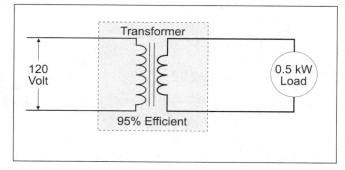

Figure 4–25

58.   The primary current of the transformer is _____ (Figure 4–25)?
   (a) 0.416 ampere   (b) 4.38 ampere   (c) 3.56 ampere   (d) 41.6 ampere

59.   The primary power for the transformer is _____ (Figure 4–25)?
   (a) 526 VA   (b) 400 VA   (c) 475 VA   (d) 550 VA

Transformer Miscellaneous

60.   The transformer primary is 240 volts, the secondary is 12 volts, and the load is two 100-watt lamps. The transformer is 92 percent efficient. The secondary VA of this transformer _____ .
   (a) is 200 VA   (b) is the same as the primary VA
   (c) cannot be calculated   (d) none of these

61.   • The secondary of a transformer is 24 volts with a load of 5 ampere. The primary is 120 volts and the transformer is 100 percent efficient. The secondary VA of this transformer _____ .
   (a) is 120 VA   (b) is the same as the primary VA
   (c) cannot be calculated   (d) a and b

62.   The transformer primary is 240 volts, the secondary is 12 volts, and the load is two 100-watt lamps. The transformer is 92 percent efficient. The primary VA of this transformer _____ .
   (a) is 185 VA   (b) is 217 VA   (c) is 0.217 VA   (d) cannot be calculated

63.   The output of a generator is 20 kW. The input kVA is _____ if the efficiency rating is 65 percent.
   (a) 11 kVA   (b) 20 kVA   (c) 31 kVA   (d) 33 kVA

NEC® Questions from Section 427-22 through Sections 830-10(i)

   The following *National Electrical Code*® questions are in consecutive order. Questions with • indicate that 75 percent or less of exam takers get the question correct.

Article 427 – Electric Heating Equipment for Pipelines and Vessels

64.   Ground-fault protection of equipment shall be provided for electric heat tracing and heating panels.
   (a) True   (b) False

Article 430 – Motors, Motor Circuits, and Controllers

   Article 430 is a very important exam preparation chapter. Diagram 430-1 (FPN) is very helpful for determining which Part of Article 430 you need to use. Many people using the *NEC*® have a difficult time sizing conductors and overcurrent protection for motors because they try to apply general rules. Article 430 has many exceptions to the general rules, and sizing conductors and overcurrent protection should be done using Article 430 requirements.

65.   A motor for usual use shall be marked with a time rating of _____ .
       (a) continuous   (b) 30 or 60 minutes   (c) 5 or 15 minutes   (d) all of these

66.   • Where motors are provided with a terminal housing, the housing shall be of _____ and be of substantial construction.
      (a) steel   (b) iron   (c) metal   (d) copper

67.   Open motors having commutators shall be located or protected so that sparks cannot reach adjacent combustible material, but this shall not prohibit the installation of these motors _____ .
      (a) on wooden floors   (b) over combustible fiber
      (c) under combustible material   (d) none of these

68.   • Conductors that supply two or more motors shall have an ampacity _____ .
      (a) not less than the total sum of the full-load current rating plus 125 percent of the highest motor in the group
      (b) equal to the sum of the full-load current rating of all the motors plus 125 percent of the highest motor in the group
      (c) equal to the sum of the full-load current rating of all the motors plus 25 percent of the highest motor in the group
      (d) not less than 125 percent of the sum of all the motors in the group

69.   Motor overload protection is not required where _____ .
      (a) conductors are oversized by 125 percent   (b) conductors are part of a limited-energy circuit
      (c) it might introduce additional hazards   (d) short-circuit protection is provided

70.   Each continuous duty motor _____ horsepower or less, which is not permanently installed, not automatically started, and is within sight of the controller, shall be permitted to be protected against overload by the branch circuit protective device.
      (a) 1   (b) 2   (c) 3   (d) 4

71.   • Where fuses are used for motor overload protection, a fuse shall be inserted in each ungrounded conductor and also in the grounded conductor if the supply system is _____ with one conductor grounded.
      (a) 2-wire, 3-phase direct current   (b) 2-wire, 3-phase alternating current
      (c) 3-wire, 3-phase direct current   (d) 3-wire, 3-phase alternating current

72.   A motor _____ device that can restart a motor automatically after overload tripping shall not be installed if automatic restarting of the motor can result in injury to persons.
      (a) short-circuit   (b) ground-fault   (c) overcurrent   (d) overload

73.   Where all the conditions of the *Code* are met, several motors not over 1 horsepower are permitted on a single branch circuit providing that the full-load rating of each motor does not exceed _____ amperes.
      (a) 6   (b) 10   (c) 15   (d) 20

74.   Overcurrent protection for motor control circuits shall not exceed 400 percent if the conductor does not extend beyond the motor control equipment enclosure.
      (a) True   (b) False

75.   Motor control circuits shall be so arranged that they will be disconnected from all sources of supply when the disconnecting means is in the open position. Where separate devices are used for the motor and control circuit, they shall be located immediately adjacent one to each other.
      (a) True   (b) False

76.   The controller shall have a horsepower rating at the application voltage not _____ than the horsepower rating of the motor.
      (a) lower   (b) higher   (c) equal to   (d) none of these

77.   • A _____ shall be located in sight from the motor location and the driven machinery location.
      (a) controller   (b) protection device   (c) disconnecting means   (d) all of these

78.   The disconnect for the controller and motor must open all ungrounded conductors for a 3-phase motor.
      (a) True   (b) False

79.   The motor disconnecting means shall be a _____ .
      (a) circuit breaker   (b) motor-circuit switch rated in horsepower
      (c) molded case switch   (d) any of these

80. The disconnecting means for a 50 horsepower 3-phase, 460 volt induction motor (rated 65 amperes) shall have an ampere rating of not less than _____ amperes.
    (a) 126   (b) 75   (c) 91   (d) 63

81. The frames of stationary motors (over 600 volts) shall be grounded _____ .
    (a) where supplied by metal-enclosed wiring   (b) where in a wet location and not isolated or guarded
    (c) if in a hazardous (classified) location   (d) all of these

Article 440 – Air-Conditioning and Refrigerating Equipment

82. The disconnecting means for air-conditioning and refrigeration equipment must be _____ from the air-conditioning or refrigerating equipment.
    (a) readily accessible   (b) within sight   (c) a or b   (d) a and b

83. A hermetic motor-compressor controller shall have a _____ current rating not less than the nameplate rated-load current or branch circuit selection current, whichever is greater.
    (a) continuous-duty full-load   (b) locked-rotor
    (c) a or b   (d) a and b

84. When supplying a room air-conditioner, rated 120 volts, nominal, the length of flexible supply cord shall not exceed _____ feet.
    (a) 4   (b) 6   (c) 8   (d) 10

Article 445 – Generators

85. For phase converters, the ampacity of the single-phase supply conductors shall not be less than _____ percent of the phase converter nameplate single-phase input full-load amperes.
    (a) 75   (b) 100   (c) 125   (d) 150

86. The generator shall be equipped with a disconnecting means to disconnect power for the generator, its protective devices and all control apparatus.
    (a) True   (b) False

Article 450 – Transformers and Transformer Vault

For most electricians, transformers are perhaps the most intimidating of all subjects. As with all equipment, they must have overcurrent protection and they must be grounded.

87. If the primary overcurrent protection device is sized at 250 percent of the primary current, what size secondary overcurrent protection device is required if the secondary current is 42 amperes?
    (a) 40 amperes   (b) 70 amperes   (c) 60 amperes   (d) 90 amperes

88. All transformers and transformer vaults shall be readily accessible to qualified personnel for inspection and maintenance, or:
    (a) Dry-type transformers 600 volts, nominal, or less, located in the open on walls, columns, or structures, shall not be required to be readily accessible.
    (b) Dry-type transformers rated not more than 50 kVA and not over 600 volts can be installed above a hollow space.
    (c) a or b
    (d) none of these

89. • Transformers with nonflammable dielectric fluid rated over 35,000 volts installed in a vault shall be furnished with a _____ when installed indoors.
    (a) fluid confinement area
    (b) pressure relief vent
    (c) means for absorbing or venting any gasses generated by arcing
    (d) all of these

90. Each doorway leading into a vault from the building interior shall be provided with a tight fitting door having a minimum fire rating of _____ hours.
    (a) 2   (b) 4   (c) 5   (d) 3

Article 460 – Capacitors

91. Capacitors shall be _____ so that persons cannot come into accidental contact or bring conducting materials into accidental contact with exposed energized parts, terminals, or busses associated with them.
    (a) enclosed   (b) located   (c) guarded   (d) any of these

92. A capacitor operating at over 600 volts, nominal, shall be provided with means to reduce the residual voltage to 50 volts or less within _____ after disconnecting it from the source of supply.
    (a) 15 seconds   (b) 45 seconds   (c) 1 minute   (d) 5 minutes

Article 470 – Resistors and Reactors

93. Resistors and reactors (over 600 volts, nominal) shall be isolated by _____ to protect personnel from accidental contact with energized parts.
    (a) an enclosure   (b) elevation   (c) a or b   (d) a and b

Article 480 – Storage Batteries

94. Provisions shall be made for sufficient diffusion and ventilation of the gases from the storage battery to prevent the accumulation of an _____ mixture.
    (a) flammable   (b) explosive   (c) toxic   (d) all of these

Chapter 5 Special Occupancies

Articles 500 through 517 have conditions that can be considered as hazardous in nature. Articles 500 through 504 cover the requirements for electrical equipment and wiring for all voltages in locations where fire or explosion hazards may exist due to flammable gasses or vapors, flammable liquids, combustible dust, or ignitable fibers or flyings. Articles 511 through 517 are specific types of hazardous occupancies, such as gas stations and anesthesia areas in health care facilities.

Article 500 – Hazardous (Classified) Locations breaks down and categorizes hazardous area requirements and describes and defines the requirements for electrical installations in these locations.

Article 500 – Hazardous (Classified) Locations

95. Hazardous locations are classified depending on the properties of the _____, which may be present and the likelihood that a flammable or combustible concentration or quantity is present.
    (a) flammable vapors   (b) flammable gases or liquids
    (c) combustible dusts or fibers   (d) all of these

96. Electrical equipment installed in hazardous (classified) locations must be constructed for the class, division, and group. Group "C" atmosphere contains _____ .
    (a) hydrogen   (b) ethylene   (c) propylene oxide   (d) all of these

97. A Group "E" atmosphere contains combustible metal dusts.
    (a) True   (b) False

98. For Class _____ locations, Groups "E, F, and G", the classification involves the tightness of the joints of assembly and shaft openings to prevent entrance of dust in the dust-ignition-proof enclosure. The blanketing effect of layers of dust on the equipment may cause overheating, electrical conductivity of the dust, and the ignition temperature of the dust.
    (a) I   (b) II   (c) III   (d) all of these

99. Equipment installed in hazardous locations must be approved and shall be marked to show the _____ .
     (a) class   (b) group
     (c) operating temperature reference to 40°C ambient   (d) all of these

100. A Class I, Division 2 location usually includes locations where volatile flammable liquids or flammable gases or vapors are used, but which, in the judgment of the authority having jurisdiction, would become hazardous only in case of an accident or of some unusual operating condition.
     (a) True   (b) False

101. Class II locations are those that are hazardous because of the presence of _____ .
     (a) combustible dust   (b) easily ignitable fibers or flyings
     (c) flammable gases or vapors   (d) flammable liquids or gases

102. Locations in which combustible fibers are stored are designated as _____ .
     (a) Class II, Division 2 (b) Class III, Division 1   (c) Class III, Division 2   (d) non-hazardous

Article 501 – Class I Hazardous Locations

103. Which of the following wiring methods are permitted in a Class I, Division 1 location?
     (a) threaded rigid metal conduit   (b) threaded IMC
     (c) MI cable   (d) all of these

104. Boxes, enclosures, fittings, and joints are not required to be explosion-proof in a Class I, Division 2 location. However, if arcs or sparks (such as from make-or-break contacts) can result from equipment being utilized, that equipment must be installed in an explosion-proof enclosure meeting the requirements for Class I, Division 1 locations.
     (a) True   (b) False

105. Sealing compound is employed with Type MI cable terminal fittings in Class I locations for the purpose of _____ .
     (a) preventing the passage of gas or vapor   (b) excluding moisture and other fluids
     (c) limiting a possible explosion   (d) preventing the escape of powder

106. A seal fitting is required for each conduit run passing from a Class I, Division 2 location into an unclassified location located no more than _____ feet from the boundary.
     (a) 3   (b) 6   (c) 10   (d) 20

107. The minimum thickness of sealing compound in Class I, Division 1 and 2 locations shall not be less than the trade size of the conduit or sealing fitting and in no case less than _____ inches.
     (a) $1/8$ (b) $1/4$ (c) $3/8$ (d) $5/8$

108. Cables not in a raceway that contain shielded cables and/or twisted pair cables in a Class I, Division 2 location shall not require the removal of the shielding material or separation of the twisted pairs, provided the termination is by an approved means to minimize the entrance of _____ and prevent propagation of flame into the cable core.
     (a) gas   (b) vapors   (c) dust   (d) a and b

109. Fused or unfused disconnect and isolating switches for transformers or capacitor banks that are not intended to interrupt current in the normal performance of the function for which they are installed shall be permitted to be installed in general purpose enclosures.
     (a) True   (b) False

Article 502 – Class II Hazardous Locations

110. In a Class II location, where electrically-conducting dust is present, flexible connections can be made with _____ .
     (a) flexible metal conduit   (b) Type AC armored cable
     (c) hard-usage cord   (d) liquidtight flexible metal conduit with approved fittings

Article 503 – Class III Hazardous Locations

111. A fixture in a Class III location that may be exposed to physical damage shall be protected by a _____ guard.
(a) substantial   (b) metal   (c) suitable   (d) bronze

Article 511 – Commercial Garages and Storage

112. Parking garages used for parking or storage and where no repair work is done except for exchange of parts and routine maintenance requiring no use of electrical equipment, open flame, welding, or the use of volatile flammable liquids are not classified as a hazardous location.
(a) True   (b) False

113. • In a commercial garage, the pit shall be classified as a _____ location, unless provisions are made for six air changes per hour.
(a) Class I, Division 2   (b) Class II, Division 2   (c) Class II, Division 1   (d) Class I, Division 1

114. Equipment less than _____ feet above the floor level that may produce arcs, sparks, or particles of hot metal shall be of the totally enclosed-type or so constructed as to prevent escape of sparks or hot metal particles.
(a) 6   (b) 10   (c) 12   (d) 18

Article 513 – Aircraft Hangars

115. Stock rooms and similar areas adjacent to aircraft hangars but effectively isolated and adequately ventilated shall be designated as _____ locations.
(a) Class I, Division 2   (b) Class II, Division 1
(c) Class II, Division 2   (d) non-hazardous

Article 514 – Gasoline Dispensing Stations

116. Each circuit leading to or through a dispensing pump shall be provided with a switch or other acceptable means to disconnect _____ from the source of supply all conductors of the circuit, including the grounded neutral, if any.
(a) automatically   (b) simultaneously   (c) manually   (d) individually

117. The emergency controls for attended self-service stations must be located no more than _____ feet from the gasoline dispensers.
(a) 20   (b) 50   (c) 75   (d) 100

118. • Underground gasoline dispenser wiring shall be installed in _____ .
(a) threaded rigid metal conduit
(b) threaded intermediate metal conduit
(c) rigid nonmetallic conduit when buried not less than two feet of earth
(d) any of these

Article 516 – Spray Application, Dipping, and Coating Processes

119. • Locations where flammable paints are dried but in which the ventilating equipment is interlocked with the electrical equipment may be designated as a _____ location.
(a) Class I, Division 2   (b) Unclassified   (c) Class II, Division 2   (d) Class II, Division 1

Article 517 – Health Care Facilities

120. The patient care area is any portion of a health care facility where patients are intended to be _____ .
(a) examined   (b) treated   (c) moved   (d) a or b

121. Each general care area patient bed location shall be provided with _____ receptacle(s).
(a) one single or one duplex   (b) six single or three duplex
(c) two single or one duplex   (d) four single or two duplex

122.   A(n) _____ system must supply major electric equipment necessary for patient care and basic hospital operation.
       (a) emergency   (b) equipment   (c) life safety   (d) none of these

123.   In a location where flammable anesthetics are employed, the area shall be classified as Class I, Division 1, which shall extend _____ .
       (a) upward to the structural ceiling   (b) upward to a level 8 feet above the floor
       (c) upward to a level 5 feet above the floor   (d) 10 feet in all directions

124.   Health care equipment frequently in contact with bodies of persons shall not exceed _____ volts.
       (a) 50   (b) 115   (c) 230   (d) 10

Article 518 – Places of Assembly

Article 518 applies to buildings or portions of buildings designed or intended for the assembly of 100 or more persons. Section 518-2 lists (but does not limit) several types of occupancies where this article applies. Several of the Articles following 518 are for specific places of assembly.

125.   A place of assembly is a building, portion of a building, or structure intended for the assembly of _____ or more persons.
       (a) 50   (b) 100   (c) 150   (d) 200

126.   • A motel conference room is designed for the assembly of more than 100 persons. The fixed wiring methods require _____ .
       (a) rigid nonmetallic conduit   (b) Type MC cable
       (c) nonmetallic sheathed cable   (d) Type AC cable

Article 520 – Theaters, Audience Areas of Motion Picture and Television Studios, and Similar Locations

127.   All theater fixed stage switchboards that are not completely enclosed, dead front and dead rear, or recessed into a wall, shall be provided with a metal hood extending the full length of the board to protect all equipment from falling objects.
       (a) True   (b) False

128.   The pilot light provided within a portable stage switchboard enclosure shall have overcurrent protection rated or set at not over _____ amperes.
       (a) 10   (b) 15   (c) 20   (d) 30

Article 525 – Carnivals, Circuses, Fairs, and Similar Events

129.   Carnival and circuit overhead wiring outside of tents and concession areas shall maintain a vertical clearance of _____ feet above platforms, projections, or surfaces from which they might be reached, and shall extend 3 feet measured horizontally from surfaces from which they might be reached.
       (a) 3   (b) 6   (c) 8   (d) 10

130.   Wiring for an amusement ride, attraction, tent, or similar structure shall not be supported by any other ride or structure unless specifically designed for the purpose.
       (a) True   (b) False

131.   Electrical wiring in and around water-attractions such as Bumper Boats for carnivals, circuses, and fairs must comply with the requirements of Article 680 – Pools and Fountains.
       (a) True   (b) False

Article 530 – Motion Picture and Television Studios and Similar Locations

132.   A switch for the control of parking lights in a theater may be permitted to be installed inside the projection booth.
       (a) True   (b) False

Article 545 – Manufactured Building

133. The *National Electrical Code*® specifies wiring methods for prefabricated buildings (manufactured buildings).
(a) True   (b) False

Article 547 – Agricultural Buildings

134. Receptacles rated 125 volts _____ ampere, must be GFCI protected if they are located in an agricultural livestock building that has an equipotential plane.
(a) 15   (b) 20   (c) 30   (d) a and b

Article 550 – Mobile Homes and Mobile Home Parks

Article 550 is broken down into three Parts: Part A, General, is mostly definitions that apply to this article; Part B, Mobile Homes, applies to the power cord that links a mobile home to its supply, as well as all wiring methods and materials within and on the mobile home, plus calculations; and Part C, Services and Feeders, applies to the park in general and the individual services on the lots. Miscellaneous buildings such as recreation and meeting rooms, or common laundry facilities fall under the regular requirements of Chapters 1 through 4.

135. Which of the following may be used as a feeder from the service equipment to a mobile home?
(a) a permanently installed circuit
(b) two-50 ampere power supply cords approved for mobile homes
(c) a or b
(d) none of these

136. The receptacle outlet for mobile and manufactured home heat tape used to protect cold water inlet piping must be _____ protected and it shall be connected to an interior branch circuit where all of the receptacles of the circuit are _____ protected.
(a) AFCI   (b) GFCI   (c) a or b   (d) none of these

137. A 16 kW nameplate rating of a free-standing range in a mobile home would use _____ VA as the computed load.
(a) 8,000   (b) 8,800   (c) 9,600   (d) 10,000

138. Outdoor disconnecting means for mobile and manufactured homes shall be installed so the bottom of the enclosure is not less than ___ feet above the finished grade or working platform.
(a) 1   (b) 2   (c) 3   (d) 6

Article 551 – Recreational Vehicles and Parks

139. A 50 ampere receptacle at recreational vehicle sites can be supplied from a single-phase, _____ volt, 3-wire system.
(a) 120/240   (b) 120/208   (c) 277/480   (d) a or b

Article 555 – Marinas and Boatyards

140. Receptacles that provide shore power for boats shall be rated not less than _____ .
(a) 20 ampere duplex receptacle
(b) 20 ampere single receptacle of the locking and grounding type
(c) 30 ampere single receptacle ground-fault
(d) none of these

141. The feeder for six 20 ampere receptacles supplying shore power shall be calculated at _____ percent of the sum of the rating of the receptacles.
(a) 70   (b) 80   (c) 90   (d) 100

142. Service equipment for floating docks or marinas must be located _____ the floating structure.
(a) next to   (b) on   (c) in   (d) any of these

Chapter 6 – Special Equipment

NEC® Chapter 6 covers the requirements for common, special, or specific equipment. Chapter 6 modifies the general rules of Chapters 1 through 4 as necessary for the individual equipment. Most of the articles in this chapter are short and very specialized. For exam purposes, these are the easiest articles to find answers.

Many of the articles in this chapter have other documents listed that are related to the subject of that article. For example, Article 610 – Cranes and Hoists, Section 610-1 Scope, (FPN) lists ANSI B,30, Safety *Code* for Cranes, Derricks, Hoists, Jacks, and Slings. In Article 620 – Elevators, etc., Section 620-1, (FPN) lists ANSI/ASME A17.1-1993, Safety *Code* for Elevators and Escalators. It is your responsibility to know if these other documents are necessary for you to do the job properly.

Remember, the *NEC®* is only one of many NFPA safety standards. Appendix A of the *NEC®* lists other NFPA documents from which requirements have been "extracted" (copied) into the *NEC®* Section 90-3, Paragraph 4 states that these extracts are identified by the superscript letter x and identified in Appendix A.

Several of the articles in Chapter 6 contain calculations for some of the special equipment concerning wire size, feeder loads, etc. Always check these articles for special requirements and calculations.

Article 600 – Electric Signs and Outline Lighting

143.   Circuits that supply signs and outline lighting systems containing incandescent, fluorescent, and high-intensity discharge forms of illumination shall be rated not to exceed _____ amperes.
(a) 20   (b) 30   (c) 40   (d) 50

144.   Signs or outline lighting systems operated by electronic or electro-mechanical controllers located external to the sign or outline lighting system shall be permitted to have a disconnecting means located _____ .
(a) within sight of the controller   (b) in the same enclosure with the controller
(c) a or b   (d) none of these

145.   The bottom of sign and outline lighting enclosures shall not be less than _____ feet above areas accessible to vehicles.
(a) 12   (b) 14   (c) 16   (d) 18

Article 604 – Manufactured Wiring Systems

146.   • Manufactured wiring systems shall be constructed with _____ .
(a) listed AC or MC cable
(b) No. 10 or 12 AWG copper-insulated conductors
(c) conductors suitable for nominal 600 volts
(d) all of these

Article 610 – Cranes and Hoists

147.   All exposed metal parts of cranes, hoists, and accessories shall _____ into a continuous electrical conductor.
(a) be bonded with No. 6 or larger conductors to be made
(b) be metallically joined together
(c) be grounded
(d) not be grounded or made

Article 620 – Elevators and Escalators

148.   The minimum size conductor for operating control and signaling circuits in an elevator is No. _____ .
(a) 20   (b) 16   (c) 14   (d) 12

149.   Duty on escalator motors shall be classed as _____ .
(a) full time   (b) continuous   (c) various   (d) long term

Article 625 – Electric Vehicle Charging System

150. Electric vehicle supply equipment must have a listed system of protection against electric shock of personnel set at no more than 5 milliampere.
    (a) True   (b) False

Article 645 – Information Technology Equipment (data processing rooms)

151. • Liquidtight flexible conduit may be permitted to enclose branch circuit conductors for information technology communication equipment.
    (a) True   (b) False

152. • In data processing rooms, a disconnecting means is required _____ .
    (a) to disconnect the air conditioning to the room
    (b) to disconnect the power to electronic computer/data processing equipment
    (c) to disconnect all power and lighting
    (d) a single disconnecting means is permitted to control the air conditioning to the room and power to electronic computer/data processing equipment

Article 660 – X-Ray Equipment

153. The ampacity requirements for a disconnecting means of x-ray equipment shall be based on _____ percent of the input required for the momentary rating of the equipment if greater than the long-time rating.
    (a) 125   (b) 100   (c) 50   (d) none of these

Article 665 – Induction/Dielectric Heating

154. Dielectric heating equipment auxiliary rectifiers used with filter capacitors in the output for bias supplies, tube keyers, etc., bleeder resistors shall be used even though the dc voltage may not exceed _____ volts.
    (a) 24   (b) 50   (c) 115   (d) 240

Article 668 – Electrolytic Cells

155. An assembly of electrically interconnected electrolytic cells supplied by a source of dc power is called a _____ .
    (a) cell line attachment   (b) cell line
    (c) electrolytic cell bank   (d) battery storage bank

Article 670 – Industrial Machinery

156. An electrically driven or controlled machine with one or more motors, not hand-portable, and used primarily to transport and distribute water for agricultural purposes is called an _____ .
    (a) irrigation machine   (b) electric water distribution system
    (c) center pivot irrigation machine   (d) automatic water distribution system

Article 680 – Pools, Spas, and Fountains

157. • A spa or hot tub is a hydromassage pool, or tub, designed for the immersion of people. They are not generally designed or intended to have their contents drained or discharged after each use.
    (a) True   (b) False

158. A pool transformer is required to _____ .
    (a) be of the isolated winding type
    (b) have a grounded metal barrier between the primary and secondary windings
    (c) be identified for the purpose
    (d) all of these

159. Receptacles must not be within 10 feet of the water's edge for _____ .
    (a) pools   (b) outdoor spas   (c) outdoor hot tubs   (d) all of these

160. Receptacles located within _____ feet of the inside walls of a pool shall be protected by a ground-fault circuit-interrupter.
    (a) 8   (b) 10   (c) 15   (d) 20

161. All outdoor 15 and 20 ampere single-phase pool, spa, and hot tub pump motors in other than dwelling units shall be
    _____ .
    (a) AFCI protected   (b) GFCI protected   (c) approved   (d) listed

162. • An underground rigid, nonmetallic raceway must be not less than _____ feet from the inside wall of the pool or spa, unless space limitations prevent otherwise.
    (a) 8   (b) 10   (c) 5   (d) 25

163. An accessible disconnecting means for pools, spas, and hot tub equipment shall be installed. The disconnect must be accessible, located within sight from pool, spa, or hot tub equipment, and shall be located at least 5 ft (1.52 m) horizontally from the inside walls of the pool, spa, or hot tub.
    (a) accessible   (b) within sight from the equipment
    (c) at least 5 feet from the water   (d) all of these

164. When rigid nonmetallic conduit extends from the pool light forming shell to a suitable junction box, a No. 8 _____ conductor shall be installed in the raceway.
    (a) solid bare   (b) solid insulated   (c) stranded insulated   (d) b or c

165. Junction boxes for pool lighting shall not be located less than _____ feet from the inside wall of a pool unless separated by a fence or wall.
    (a) 3   (b) 4   (c) 6   (d) 8

166. • When bonding pool and spa equipment, a solid No. 8 copper conductor must be run back to the service equipment. This conductor must be unbroken.
    (a) True   (b) False

167. A No. 8 or larger solid copper bonding conductor must be extended or attached to the panelboard or service equipment enclosure to eliminate voltage gradients in the pool area.
    (a) True   (b) False

168. Metal conduit and metal piping within _____ feet of the inside walls of the pool, and that are not separated from the pool by a permanent barrier, are required to be bonded.
    (a) 4   (b) 5   (c) 8   (d) 10

169. Wet-niche, dry-niche, or no-niche lighting fixtures shall be connected to an equipment grounding conductor sized not smaller than No. 12 insulated copper conductor installed with the circuit conductors in _____ .
    (a) liquidtight flexible nonmetallic conduit   (b) flexible metal
    (c) rigid nonmetallic conduit   (d) a or c

170. A pool deck area radiant heat cable shall _____ .
    (a) not be installed within 5 feet horizontally from the inside wall of the pool
    (b) be mounted at least 12 feet vertically above the pool deck
    (c) not be permitted
    (d) none of these

171. The wiring for spas and hot tubs installed outdoors, such as receptacle, switches, lighting locations, grounding, and bonding must comply with the same requirements as permanently installed pools.
    (a) True   (b) False

172. The maximum length of exposed cord in a fountain shall be _____ feet.
    (a) 3   (b) 4   (c) 6   (d) 10

173. For hydromassage bathtubs, the following applies:
     (a) GFCI devices must protect all associated electric components.
     (b) Switches must be located a minimum of 5 feet from the tub.
     (c) Lighting fixtures located within 5 feet, must be a minimum of 7 feet 6 inches over the maximum water level and GFCI protected.
     (d) Wiring methods are limited to rigid metal, intermediate metal, or rigid nonmetallic conduit.

Article 695 – Fire Pumps

174. The voltage at the line terminals of a fire pump motor controller under motor starting conditions (locked-rotor current) shall not drop more than _____ percent below controller rated voltage.
     (a) 5   (b) 10   (c) 15   (d) any of these

Chapter 7 – Special Conditions

NEC® Chapter 7 can be thought of as a special power chapter. It contains the requirements for emergency and standby systems, signaling systems, plus a few other different power systems.

Article 700 – Emergency Systems

Article 700 pertains to systems that are essential for safety to human life. These systems are often comprised of backup power for emergency power and lighting. This article covers the requirements for the installation of emergency systems where required by other codes or laws. It does not specify where emergency systems are to be installed. See NFPA 101, Life Safety *Code*s for where emergency systems are required and locations of emergency and exit lights. The FPN after *NEC®* Section 700-1 lists other standards related to this subject.

175. The emergency transfer switch can supply _____ .
     (a) emergency loads   (b) computer equipment
     (c) UPS type equipment   (d) all of these

176. Wiring from emergency source or emergency source distribution overcurrent protection to emergency loads shall be kept entirely independent of all other wiring and equipment, except in _____ .
     (a) transfer equipment enclosures.
     (b) exit or emergency lighting fixtures supplied from two sources.
     (c) a common junction box, attached to exit or emergency lighting fixtures supplied from two sources.
     (d) all of the above

177. A storage battery supplying emergency lighting and power shall maintain not less than $87^1/_2$ percent of full voltage at total load for a period of at least _____ hour(s).
     (a) 1   (b) $1^1/_2$ (c) 2   (d) $2^1/_2$

178. Unit equipment (battery pack) shall be on the same branch circuit as that serving the normal lighting in the area and connected _____ of any local switches.
     (a) with   (b) ahead   (c) after   (d) none of these

Article 701 – Legally Required Standby Systems

Article 701 covers systems intended to provide electrical power to aid in fire fighting, rescue operations, control of health hazards, and similar operations. See Section 701-2 FPN.

The requirements for legally required standby systems are similar to those for emergency systems. When normal power is lost, legally required systems are required to come on in 60 seconds or less, and their wiring can be mixed with general wiring. Emergency systems are required to come on in 10 seconds or less, and the wiring is completely separated from the general wiring.

179.   A generator set for a required standby system shall _____ .

(a) have means for automatically starting the prime movers

(b) have two hours full-demand fuel supply if an internal combustion engine

(c) not be solely dependent on public utility gas system

(d) all of these

Article 720 – Circuits and Equipment Operating at Less Than 50 Volts

180.   • Conductors for an appliance circuit supplying more than one appliance or appliance receptacle in an installation operating at less than 50 volts shall not be smaller than No. _____ copper or equivalent.

(a) 18   (b) 14   (c) 12   (d) 10

Article 725 – Remote-Control, Signaling, and Power-Limited Circuits

Article 725 covers remote-control, signaling, and power-limited circuits that are not an integral part of a device or appliance. See Section 725-1 (FPN). This Article covers signaling type systems such as burglar alarms and coaxial wiring associated with the interconnection of electronic data processing and computer equipment not within a data processing room as covered in Article 645.

A signaling circuit is any electrical circuit that supplies energy to an appliance or device that gives a visual and/or audible signal. Examples are doorbells, fire or smoke detectors, and fire or burglar alarms.

A remote control circuit is any circuit which has as its load device the operating coil of a magnetic motor starter, a contactor, or relay. These circuits control one or more other circuits. Low voltage relay switching of lighting and power loads can be a remote-control circuit.

Power-limited circuits are circuits used for functions other than signaling or remote-control, but in which the source of the energy supply is limited in its power to 30 volts, 1,000 VA. Low voltage lighting using 12-volt lamps in lighting fixtures fed from 120/12 volt transformers, is a typical power-limited circuit application.

Class 1 systems include all signaling and remote-control systems, which do not have the special current limitations of Class 2, and 3 systems.

Class 2 and 3 systems are those in which the current is limited to certain specified low values by fuses or circuit breakers, and by supply transformers which will deliver only small currents, or by other approved means. All Class 2 and 3 circuits must have a power source with power-limiting characteristics as described in Tables 725-31(a) and (b), in addition to overcurrent protection.

181.   Remote-control circuits to safety-control equipment shall be _____ if the failure of the equipment to operate introduces a direct fire or life hazard.

(a) Class 1   (b) Class 2   (c) Class 3   (d) Class 1, Division 1

182.   • Class I control circuit conductors shall have overcurrent protection _____ .

(a) in accordance with the values specified in Table 310-16 through 310-31 for No. 14 and larger

(b) not exceeding 7 amperes for No. 18, and 10 amperes for No. 16

(c) and derating factors do not apply

(d) all of these

183.   The power source for a Class 2 shall be a _____ .

(a) listed Class 2 transformer   (b) listed Class 2 power supply

(c) dry cell battery, rated 30 volts or less   (d) any of these

184.   • In Class 2 alternating current circuit installations, the maximum voltage for wet contact that is most likely to occur is _____ volts peaks.

(a) 10.5   (b) 21.2   (c) 100   (d) none of these

185. • Doorbell wiring, rated as a Class 2 control circuit in a dwelling unit _____ run in the same raceway or enclosure with light and power conductors.
(a) is permitted with 600 volt insulation to be
(b) shall not be
(c) shall be
(d) is permitted, if the insulation is equal to the highest voltage conductor installed, to be

186. Conductors of two or more Class 2 circuits shall be permitted within the same cable, raceway, or enclosure provided all the conductors are _____ .
(a) insulated for the maximum of any conductor
(b) separated by a permanent partition
(c) all of the same voltage and amperage class
(d) all of these

Article 760 – Fire Alarm Systems

Article 760 covers the installation of wiring and equipment for fire alarm systems operating at 600 volts, nominal, or less. Examples of this type of system could include fire alarm, guard tour, sprinkler water flow, and sprinkler supervisory systems. Another NFPA standard used in close association with this type of system is NFPA 72, Fire Alarm *Code*.

187. Article 760 covers the requirements for the installation of wiring and equipment of fire alarm systems operating at _____ volts or less.
(a) 50   (b) 300   (c) 600   (d) 1,000

188. Fire alarm wiring is permitted to be installed in ducts that contain loose stock or vapors, if installed in rigid metal or intermediate metal conduit. Be sure to read Section 300-22.
(a) True   (b) False

189. Exposed power-limited fire alarm cables on the load side of overcurrent protection, transformers, and current-limiting devices for power-limited circuits, are permitted if _____ .
(a) adequately supported
(b) terminated in approved fittings
(c) the building construction is utilized for physical protection
(d) all of these

190. Equipment shall be durably marked where plainly visible to indicate each circuit that it is _____ .
(a) on a power-limited fire alarm circuit
(b) a power-limited fire alarm circuit
(c) a fire alarm circuit
(d) none of these

191. Power-limited fire alarm circuit cables installed within buildings in an air-handling space shall use Type_____ conductors.
(a) FPL   (b) CL3P   (c) OFNP   (d) FPLP

Article 770 – Optical Fiber Cables and Raceways

Article 770 covers the use of optical fiber (fiberoptic) cables where used in conjunction with electrical conductors for communication, signaling, and control circuits. This is a relatively new technology and you will see this article change and grow. It does not cover optical cables under the control of utility companies. One advantage for using optical fiber cable in lieu of metallic conductors for signaling and communications purposes is that they are not affected by electrical noise.

192. Optical fiber cable installed in a raceway must be of a type permitted in Chapter 3 and the raceway shall be installed in accordance with Chapter 3 requirements.
(a) True   (b) False

Chapter 8 – Communications Systems

NEC® Chapter 8 covers communications systems and is independent of the other *NEC*® chapters except where they are specifically referenced in the following Articles.

Article 800 – Communication Circuits covers telephone, telegraph (except radio), outside wiring for fire and burglar alarms and similar central station systems, and telephone systems not connected to a central station system, but using similar types of equipment, methods of installation, and maintenance.

Article 810 – Radio and Television Equipment covers radio and television receiving equipment and amateur radio transmitting and receiving equipment, but not equipment and antennas used for coupling carrier current to power line conductors.

Article 820 – Community Antenna Television and Radio Distribution Systems covers coaxial cable distribution of radio frequency signals typically employed in community antenna television (CATV) systems.

Article 830 – Network Powered Broadband Communications Systems. This Article provides the necessary requirements for network-powered broadband communications systems that provide voice, audio, video, data, and interactive services through a network interface unit. Two classifications of network-powered broadband communications system circuits have been and both systems involve some risk of electric shock. The intent of Article 830 is that the classification limits, together with wiring methods and mechanical protection, should result in an installation equivalent in safety to others now permitted in the *NEC*®.

Article 800 – Communications Circuits

193.   Equipment intended to be electrically connected to a telecommunications network shall be listed for the purpose.
(a) True   (b) False

194.   The metallic sheath of communications cables entering buildings shall be _____ .
(a) grounded at the point of emergence through an exterior wall
(b) grounded at the point of emergence through a concrete floor slab
(c) interrupted as close to the point of entrance as practicable by an insulating joint
(d) any of these

195.   Plenum communications raceways shall be _____ as plenum optical fiber raceways for use in ducts, plenums, and other space used for environmental air.
(a) marked   (b) identified   (c) approved   (d) listed

196.   Floor penetrations requiring Types CMR communication cable shall contain only cables suitable for riser or plenum use except where the listed cables are _____ .
(a) encased in metal raceways
(b) located in a fireproof shaft with firestops at each floor
(c) a or b
(d) none of these

Article 810 – Radio and Television Equipment

197.   An outdoor antenna of a receiving station with a 75-foot span using a copper-clad steel conductor shall not be less than No. _____ .
(a) 10   (b) 12   (c) 14   (d) 17

198.   Unshielded lead-in antenna conductors of amateur transmitting stations attached to building surfaces shall clear the building surface which is wired over by a distance not less than _____ inches.
(a) 1   (b) 2   (c) 3   (d) 4

Article 820 – Community Antenna Television and Radio Distribution Systems

199. The conductor used to ground the outer cover of a coaxial cable shall be _____ .
   (a) insulated   (b) No. 14 minimum   (c) bare   (d) a and b

Article 830 – Network-Powered Broadband Communications Systems

200. Where practical, a separation of at least _____ feet shall be maintained between open conductors of communication systems on buildings and lightning conductors.
   (a) 6   (b) 8   (c) 10   (d) 12

CHAPTER 2
NEC® Calculations and *Code* Questions

Scope of Chapter 2

Unit 5

Raceway, Outlet Box, and Junction Boxes Calculations

OBJECTIVES

After reading this unit, the student should be able to briefly explain the following concepts:

Part A – Raceway Fill Calculations	**Part B – Outlet Box Calculations**	**Part C – Pull and Junction Box Calculations**
Existing raceway calculation	Conductor equivalents	Depth of box and conduit body sizing
Raceway sizing	Sizing box – conductors all the same	Pull and junction box size calculations
Raceway properties	size	
Understanding *NEC*® Chapter 9	Volume of box	

After reading this unit, the student should be able to briefly explain the following terms:

Part A – Raceway Fill Calculations	Lead-covered conductor	Outlet box
Alternating current conductor	*NEC*® errors	Pigtails
resistance	Nipple size	Plaster rings
Bare conductors	Raceway size	Short radius conduit bodies
Bending radius	Spare space area	Size outlet box
Compact aluminum building wire	**Part B – Outlet Box Calculations**	Strap
Conductor properties	Cable clamps	Volume
Conductor fill	Conductor terminating in the box	Yoke
Conduit bodies	Conductor running through the box	**Part C – Pull and Junction Box**
Cross-sectional area of insulated	Conduit bodies	**Calculations**
conductors	Equipment bonding jumpers	Angle pull calculation
Expansion characteristics of PVC	Extension rings	Distance between raceways
Fixture wires	Fixture hickey	Horizontal dimension
Grounding conductors	Fixture stud	Junction boxes

PART A – RACEWAY FILL CALCULATIONS

5–1 UNDERSTANDING THE NATIONAL ELECTRICAL CODE®, CHAPTER 9

Chapter 9 – Tables

Table 1 – Conductor Percent Fill

The maximum percentage of conductor fill is listed in Table 1 of Chapter 9 and is based on common conditions where the length of the conductor and number of raceway bends are within reasonable limits [FPN under Table 1] (Figure 5–1).

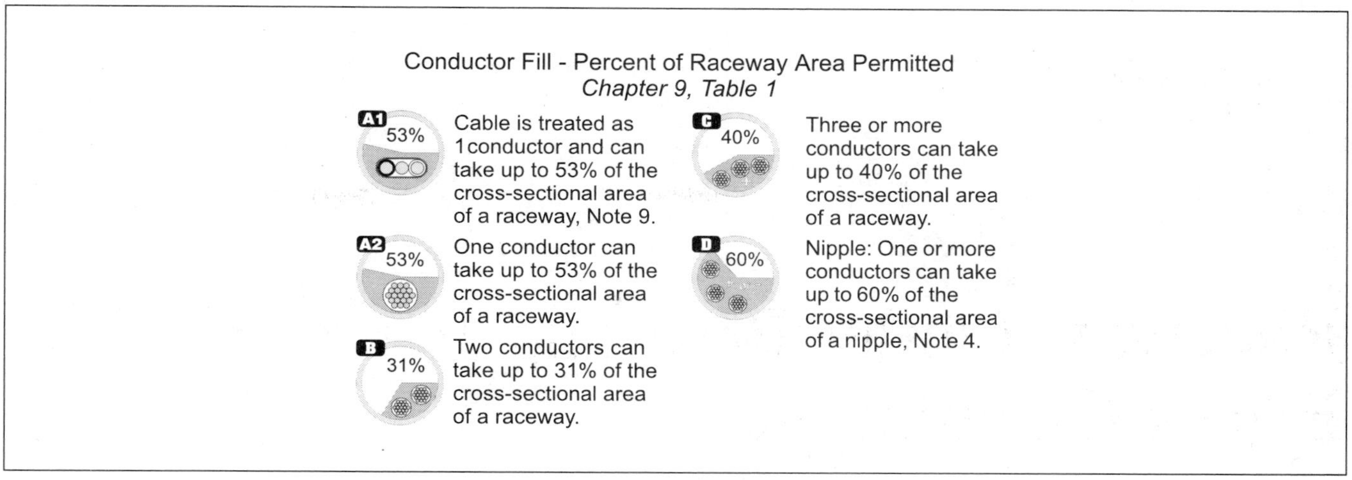

Conductor Fill - Percent of Raceway Area Permitted
Chapter 9, Table 1

A1 53% Cable is treated as 1 conductor and can take up to 53% of the cross-sectional area of a raceway, Note 9.

A2 53% One conductor can take up to 53% of the cross-sectional area of a raceway.

B 31% Two conductors can take up to 31% of the cross-sectional area of a raceway.

C 40% Three or more conductors can take up to 40% of the cross-sectional area of a raceway.

D 60% Nipple: One or more conductors can take up to 60% of the cross-sectional area of a nipple, Note 4.

Figure 5–1
Conductor Fill – Percent of Raceway Area Permitted

Table 1 of Chapter 9, Maximum Percent Conductor Fill	
Number of Conductors	**Percent Fill Permitted**
1 conductor	53% fill
2 conductors	31% fill
3 or more conductors	40% fill
Raceway 24 inches or less	60% fill Chapter 9, Note 4

Table 1, Note 1 – Conductors All the Same Size and Insulation

When all of the conductors are the same size and insulation, the number of conductors permitted in a raceway can be determined simply by looking at the tables located in Appendix C – Conduit and Tubing Fill Tables for Conductors and Fixture Wires of the Same Size.

Tables C1 through C12A are based on maximum percent fill as listed in Table 1 of Chapter 9.

Table C1 – Conductors and fixture wires in electrical metallic tubing
Table C1A – Compact conductors in electrical metallic tubing
Table C2 – Conductors and fixture wires in electrical nonmetallic tubing
Table C2A – Compact conductors in nonelectrical metallic tubing
Table C3 – Conductors and fixture wires in flexible metal conduit
Table C3A – Compact conductors in flexible metal conduit
Table C4 – Conductors and fixture wires in intermediate metal conduit
Table C4A – Compact conductors in intermediate metal conduit
Table C5 – Conductors and fixture wires in liquidtight flexible nonmetallic conduit (gray type)
Table C5A – Compact conductors in liquidtight flexible nonmetallic conduit (gray type)
Table C6 – Conductors and fixture wires in liquidtight flexible nonmetallic conduit (orange type)
Table C6A – Compact conductors in liquidtight flexible nonmetallic conduit (orange type)

Note: The appendix does not have a table for liquidtight flexible nonmetallic conduit of the black type.

Table C7 – Conductors and fixture wires in liquidtight flexible metallic conduit
Table C7A – Compact conductors in liquidtight flexible metal conduit
Table C8 – Conductors and fixture wires in rigid metal conduit

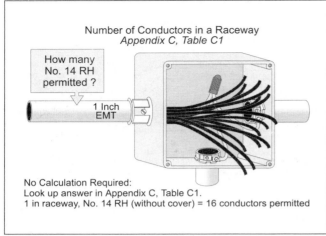

Figure 5–2
Number of Conductors in a Raceway

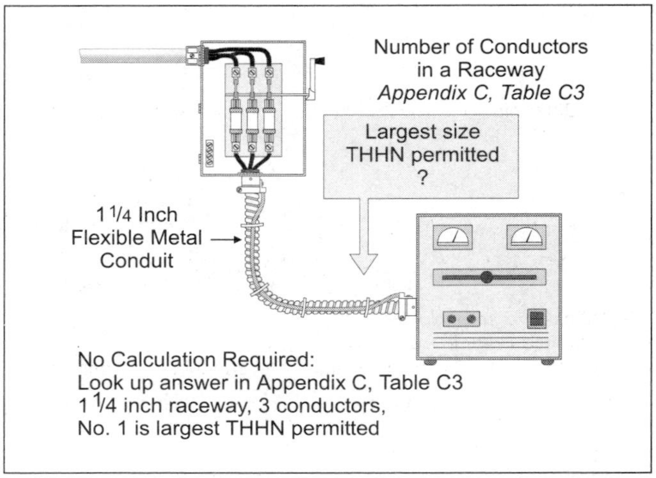

Figure 5–3
Number of Conductors in a Raceway

Table C8A – Compact conductors in rigid metal conduit

Table C9 – Conductors and fixture wires in rigid PVC conduit schedule 80

Table C9A – Compact conductors in rigid PVC conduit schedule 80

Table C10 – Conductors and fixture wires in rigid PVC conduit schedule 40

Table C10A – Compact conductors in rigid PVC conduit schedule 40

Table C11 – Conductors and fixture wires in Type A, rigid PVC conduit

Table C11A – Compact conductors in Type A, PVC conduit

Table C12 – Conductors and fixture wires in Type EB, PVC conduit

Table C12A – Compact conductors in Type EB, PVC conduit

❑ **Appendix C – Table C1**

How many No. 14 RHH conductors (without cover) can be installed in a 1 inch electrical metallic tubing (Figure 5–2)?

(a) 25 conductors   (b) 16 conductors   (c) 13 conductors   (d) 19 conductors

   • Answer: (b) 16 conductors, Appendix C, Table C1

❑ **Appendix C – Table C2A – Compact Conductor**

How many compact No. 6 XHHW conductors can be installed in a 1$^1/_4$ inch nonmetallic tubing?

(a) 10 conductors   (b) 6 conductors   (c) 16 conductors   (d) 13 conductors

   • Answer: (a) 10 conductors, Appendix C, Table C2A

❑ **Appendix C – Table C3**

If 1$^1/_4$ inch flexible metal conduit has three THHN conductors (not compact), what is the largest conductor permitted to be installed (Figure 5–3)?

(a) No. 1   (b) No. 1/0   (c) No. 2/0   (d) No. 3/0

   • Answer: (a) No. 1, Appendix C, Table C3

❑ **Appendix C – Table C4**

How many No. 4/0 RHH conductors (with outer cover) can be installed in 2 inch intermediate metal conduit?

(a) 2 conductors   (b) 1 conductor   (c) 3 conductors   (d) 4 conductors

   • Answer: (c) 3 conductors, Appendix C, Table C4

❑ **Appendix C – Table C7 – Fixture Wire**

How many No. 18 TFFN conductors can be installed in a $^3/_4$ inch liquidtight flexible metallic conduit (Figure 5– 4)?

(a) 40   (b) 26   (c) 30   (d) 39

   • Answer: (d) 39, Appendix C, Table C7

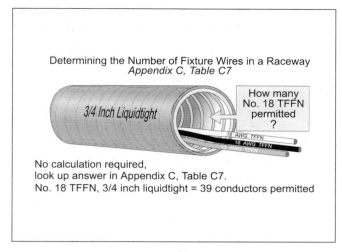

Figure 5–4
Determining the Number of Fixture Wires in a Raceway

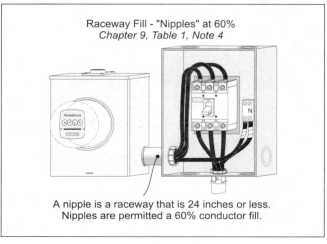

Figure 5–5
Equipment Grounding Conductors

Table 1, Note 3 – Equipment Grounding Conductors

When equipment grounding conductors are installed in a raceway, the actual area of the conductor must be used when calculating raceway fill. Chapter 9, Table 5 can be used to determine the cross-sectional area of insulated conductors, and Chapter 9, Table 8 can be used to determine the cross-sectional area of bare conductors [Note 8 of Table 1, Chapter 9] (Figure 5–5).

Table 1, Note 4 – Nipples, Raceways Not Exceeding 24 Inches

The cross-sectional areas of conduit and tubing can be found in Table 4 of Chapter 9. When a conduit or tubing raceway does not exceed 24 inches in length, it is called a nipple. Nipples are permitted to be filled to 60% of their total cross-sectional area (Figure 5–6).

Table 1, Note 7

When the calculated number of conductors (all of the same size and insulation) results in 0.8 or larger, the next whole number can be used. But be careful: this only applies when the conductors are all the same size and insulation.

Table 1, Note 8

The dimensions for bare conductor are listed in Table 8 of Chapter 9.

Chapter 9, Table 4 – Conduit and Tubing Cross-Sectional Area

Table 4 of Chapter 9 lists the dimensions and cross-sectional area for conduit and tubing. The cross-sectional area of a conduit or tubing is dependent on the raceway type (cross-sectional area of the area) and the maximum percentage fill as listed in Table 1 of Chapter 9.

❏ **Conduit Cross-Sectional Area**

What is the total cross-sectional area of $1\frac{1}{4}$ inch rigid metal conduit (Figure 5–7)?

(a) 1.063 square inches   (b) 1.526 square inches

(c) 1.098 square inches   (d) any of these

    • Answer: (b) 1.526 square inches

    Chapter 9, Table 4

Chapter 9, Table 5 – Dimensions of Insulated Conductors and Fixture Wires

Table 5 of Chapter 9 lists the cross-sectional area of insulated conductors and fixture wires.

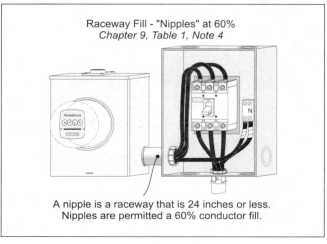

Figure 5–6
Raceway Fill – "Nipples" at 60%

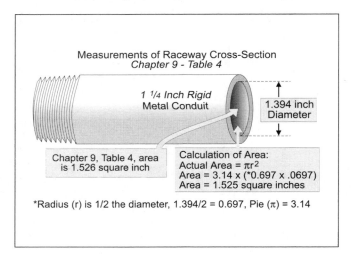

Figure 5–7
Measurements of Raceway Cross-Section

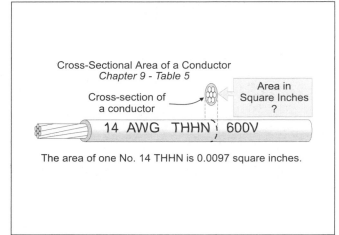

Figure 5–8
Cross-Sectional Area of a Conductor

Table 5 of Chapter 9 – Conductor Cross-Sectional Area							
	RH	RHH/RHW With *Cover*	RHH/RHW With*out* *Cover* or THW	TW	THHN THWN TFN	XHHW	BARE *Stranded* Conductors
Size AWG/kcmil	Approximate Cross-Sectional Area – Square Inches						
Column 1	Column 2	Column 3	Column 4	Column 5	Column 6	Column 7	Chapter 9, Table 8
14	.0293	.0293	.0209	.0139	.0097	.0139	**.004**
12	.0353	.0353	.0260	.0181	.0133	.0181	**.006**
10	.0437	.0437	.0333	.0243	.0211	.0243	**.011**
8	.0835	.0835	.0556	.0437	.0366	.0437	**.017**
6	.1041	.1041	.0726	.0726	.0507	.0590	**.027**
4	.1333	.1333	.0973	.0973	.0824	.0814	**.042**
3	.1521	.1521	.1134	.1134	.0973	.0962	**.053**
2	.1750	.1750	.1333	.1333	.1158	.1146	**.067**
1	.2660	.2660	.1901	.1901	.1562	.1534	**.087**
0	.3039	.3039	.2223	.2223	.1855	.1825	**.109**
00	.3505	.3505	.2624	.2624	.2233	.2190	**.137**
000	.4072	.4072	.3117	.3117	.2679	.2642	**.173**
0000	.4754	.4754	.3718	.3718	.3237	.3197	**.219**

❑ **Table 5 – THHN**

What is the cross-sectional area for one No. 14 THHN conductor (Figure 5–8)?

(a) .0206 square inch   (b) .0172 square inch   (c) .0097 square inch   (d) .0278 square inch

• Answer: (c) .0097 square inch

❏ **Table 5 – RHW** *With Outer Cover*

What is the cross-sectional area for one No. 12 RHW conductor *with outer cover*?

(a) .0206 square inch

(b) .0172 square inch

(c) .0353 square inch

(d) .0278 square inch

   • Answer: (c) .0353 square inch

❏ **Table 5 – RHH** *Without Outer Cover*

What is the cross-sectional area for one No. 10 RHH *without an outer cover*?

(a) .0117 square inch   (b) .0333 square inch

(c) .0252 square inch   (d) .0278 square inch

   • Answer: (b) .0333 square inch

Chapter 9, Table 5A – Compact Aluminum Building Wire Nominal Dimensions and Areas

Tables 5A, Chapter 9 list the cross-sectional area for compact aluminum building wires. We will not use these tables for this unit.

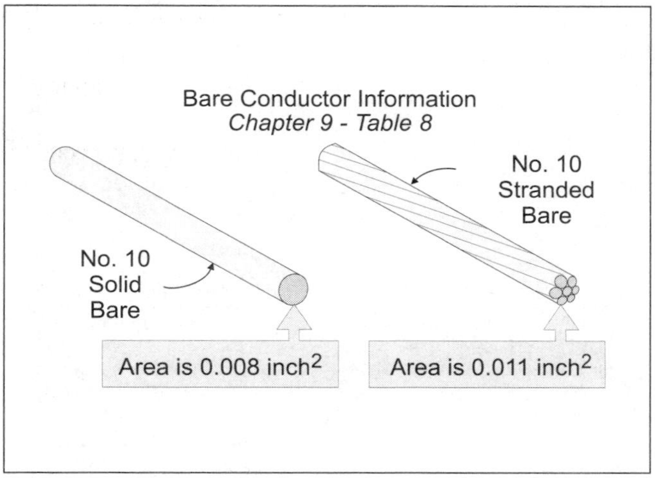

Chapter 9, Table 8 – Conductor Properties

Table 8 contains conductor properties such as: cross-sectional area in circular mils, number of strands per conductor, cross-sectional area in square inches for bare conductors, and conductor's resistance at 75°C for direct current for both copper and aluminum wire.

❏ **Bare Conductor – Cross-Sectional Area**

What is the cross-sectional area in square inches for one No. 10 bare conductor (Figure 5–9)?

(a) .008 solid   (b) .011 stranded   (c) .038   (d) a or b

   • Answer: (d) .008 square inch for solid and .011 square inch for stranded

Chapter 9, Table 9 – AC Resistance for Conductors in Conduit or Tubing

Table 9 contains the alternating current resistance for copper and aluminum conductors.

❏ **Alternating Current Resistance**

What is the alternating current resistance of No. 1/0 copper installed in a metal conduit? Conductor length 1,000 feet.

(a) 0.12 ohm   (b) 0.13 ohm   (c) 0.14 ohm   (d) 0.15 ohm

   • Answer: (a) 0.12 ohm

Bare Conductor Information
Chapter 9 - Table 8

No. 10 Solid Bare

No. 10 Stranded Bare

Area is 0.008 inch²

Area is 0.011 inch²

Figure 5–9
Bare Conductor Information

5–2 RACEWAY AND NIPPLE CALCULATIONS

Appendix C – Tables 1 through 12 cannot be used to determine raceway sizing when conductors of different sizes (or types of insulation) are installed in the same raceway. The following Steps can be used to determine the raceway size and nipple size:

Step 1: ➥ Determine the cross-sectional area (square inches) for each conductor from Table 5 of Chapter 9 for insulated conductors, and Table 8 of Chapter 9 for bare conductors.

Step 2: ➥ Determine the total cross-sectional area for all conductors.

Step 3: ➥ Size the raceway according the percent fill as listed in Table 1 of Chapter 9:
40% for three or more conductors
60% for raceways 24 inches or less in length (nipples).

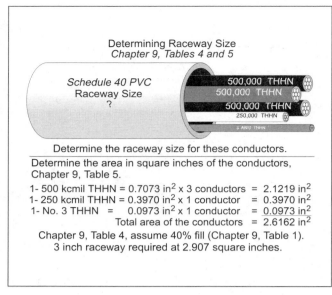

Determining Raceway Size
Chapter 9, Tables 4 and 5

Schedule 40 PVC
Raceway Size
?

500,000 THHN
500,000 THHN
500,000 THHN
250,000 THHN

3 AWG THHN

Determine the raceway size for these conductors.

Determine the area in square inches of the conductors, Chapter 9, Table 5.

1- 500 kcmil THHN = 0.7073 in² x 3 conductors = 2.1219 in²
1- 250 kcmil THHN = 0.3970 in² x 1 conductor = 0.3970 in²
1- No. 3 THHN   =   0.0973 in² x 1 conductor = 0.0973 in²
          Total area of the conductors = 2.6162 in²

Chapter 9, Table 4, assume 40% fill (Chapter 9, Table 1).
3 inch raceway required at 2.907 square inches.

Figure 5–10
Determining Raceway Size

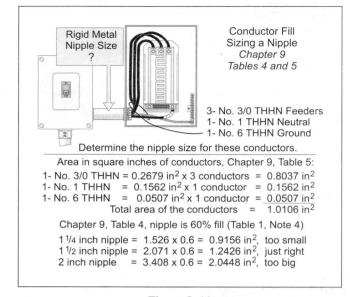

Rigid Metal
Nipple Size
?

Conductor Fill
Sizing a Nipple
*Chapter 9
Tables 4 and 5*

3- No. 3/0 THHN Feeders
1- No. 1 THHN Neutral
1- No. 6 THHN Ground

Determine the nipple size for these conductors.

Area in square inches of conductors, Chapter 9, Table 5:

1- No. 3/0 THHN = 0.2679 in² x 3 conductors = 0.8037 in²
1- No. 1 THHN   = 0.1562 in² x 1 conductor = 0.1562 in²
1- No. 6 THHN   = 0.0507 in² x 1 conductor = 0.0507 in²
        Total area of the conductors   =   1.0106 in²

Chapter 9, Table 4, nipple is 60% fill (Table 1, Note 4)

1 ¼ inch nipple = 1.526 x 0.6 = 0.9156 in², too small
1 ½ inch nipple = 2.071 x 0.6 = 1.2426 in², just right
2 inch nipple   = 3.408 x 0.6 = 2.0448 in², too big

Figure 5–11
Conductor Fill – Sizing a Nipple

❏ Raceway Size

A 400 ampere feeder is installed in schedule 40 rigid nonmetallic conduit. This raceway contains three 500 kcmil THHN conductors, one 250 kcmil THHN conductor, and one No. 3 THHN conductor. What size raceway is required for these conductors (Figure 5–10)?

(a) 2 inch   (b) 2¹/₂ inch   (c) 3 inch   (d) 3¹/₂ inch

   • Answer: (c) 3-inch

Step 1: ➥ Determine cross-sectional area of the conductors, Table 5 of Chapter 9.

   500 kcmil THHN = .7073 square inch × 3 wires = 2.1219 square inch
   250 kcmil THHN = .3970 square inch × 1 wire = 0.3970 square inch
   No. 3 THHN   = .0973 square inch × 1 wire = 0.0973 square inch

Step 2: ➥ Total cross-sectional area of all conductors   = 2.6162 square inch

Step 3: ➥ Size the conduit at 40% fill [Chapter 9, Table 1] using Table 4.
   3 inch schedule 40 PVC has an cross-sectional area of 2.907 square inch for conductors

❏ Nipple Size

What size rigid metal nipple is required for three No. 3/0 THHN conductors, one No. 1 THHN conductor and one No. 6 THHN conductor (Figure 5–11)?

(a) 2 inch   (b) 1 inch   (c) 1¹/₂ inch   (d) none of these

   • Answer: (c) 1¹/₂ inch

Step 1: ➥ Cross-sectional area of the conductors, Table 5 of Chapter 9.

   No. 3/0 THHN   = 0.2679 square inch × 3 wires = 0.8037 square inch
   No. 1 THHN   = 0.1562 square inch × 1 wire = 0.1562 square inch
   No. 6 THHN   = 0.0507 square inch × 1 wire = 0.0507 square inch

Step 2: ➥ Total cross-sectional area of the conductors = 1.0106 square inches.

Step 3: ➥ Size the conduit at 60% fill [Table 1, Note 4 of Chapter 9] using Table 4.

   1¹/₄ inch nipple = 1.5260 × 0.6   = 0.9156 square inch, too small
   1¹/₂ inch nipple = 2.071 × 0.6   = 1.2426 square inches, just right
   2 inch nipple   = 3.408 × 0.6   = 2.0448 square inches, too big

5–3 EXISTING RACEWAY CALCULATIONS

There are times we need to add conductors to an existing raceway. This can be accomplished by using the following steps:

Part 1 – Determine raceway cross-sectional spare space area.

Step 1: ➤ Determine the raceway's cross-sectional area for conductor fill [Table 1 and Table 4 of Chapter 9].

Step 2: ➤ Determine the area of the existing conductors [Table 5 of Chapter 9].

Step 3: ➤ Subtract the cross-sectional area of the existing conductors (Step 2) from the area of permitted conductor fill (Step 1).

Part 2 – To determine the number of conductors permitted in spare space area:

Step 4: ➤ Determine the cross-sectional area of the conductors to be added [Table 5 of Chapter 9 for insulated conductors and Table 8 of Chapter 9 for bare conductors].

Step 5: ➤ Divide the spare space area (Step 3) by the conductors cross-sectional area (Step 4).

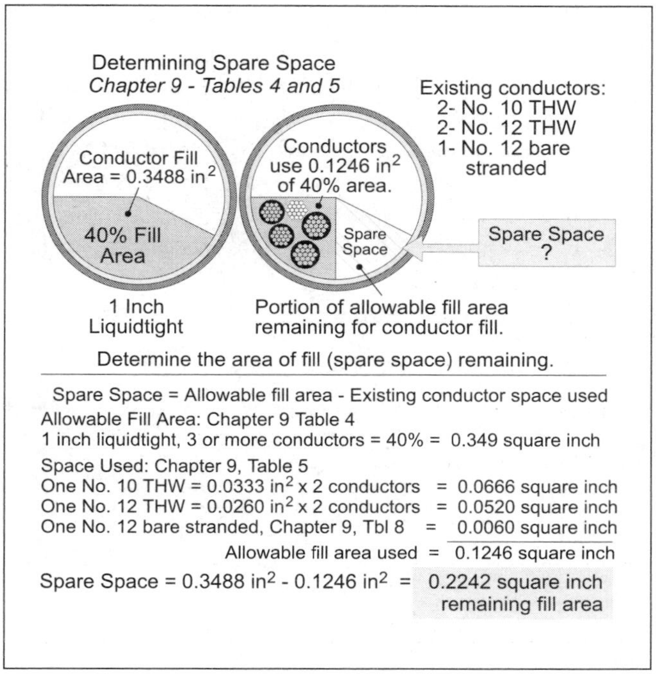

Determining Spare Space
Chapter 9 - Tables 4 and 5

Existing conductors:
2- No. 10 THW
2- No. 12 THW
1- No. 12 bare stranded

Conductor Fill Area = 0.3488 in^2

40% Fill Area

1 Inch Liquidtight

Conductors use 0.1246 in^2 of 40% area.

Spare Space

Spare Space ?

Portion of allowable fill area remaining for conductor fill.

Determine the area of fill (spare space) remaining.

Spare Space = Allowable fill area - Existing conductor space used

Allowable Fill Area: Chapter 9 Table 4
1 inch liquidtight, 3 or more conductors = 40% = 0.349 square inch

Space Used: Chapter 9, Table 5
One No. 10 THW = 0.0333 in^2 x 2 conductors = 0.0666 square inch
One No. 12 THW = 0.0260 in^2 x 2 conductors = 0.0520 square inch
One No. 12 bare stranded, Chapter 9, Tbl 8 = 0.0060 square inch
Allowable fill area used = 0.1246 square inch

Spare Space = 0.3488 in^2 - 0.1246 in^2 = 0.2242 square inch remaining fill area

Figure 5–12
Determining Square Space

❏ **Spare Space Area**

An existing one inch liquidtight flexible metallic conduit contains two No. 12 THW conductors, two No. 10 THW conductors, and one No. 12 bare (stranded). What is the area remaining for additional conductors (Figure 5–12)?

(a) 0.2242 square inch (if the raceway is more than 24 inches long)
(b) 0.3986 square inch (if the raceway is less than 24 inches long)
(c) there is no spare space
(d) a and b
  • Answer: (d) a and b

Step 1: ➤ Conductor cross-sectional area, liquidtight flexible metal conduit [Table 1, Note 4 and Table 4 of Chapter 9].
Raceway   = 0.872 × 0.4   = 0.3488 square inch
Nipple   = 0.872 × 0.6   = 0.5232 square inch

Step 2: ➤ Cross-sectional area of existing conductors.
No. 12 THW = 0.0260 square inch × 2 wires = 0.0520 square inch
No. 10 THW = 0.0333 square inch × 2 wires = 0.0666 square inch
No. 12 bare   = 0.006 square inch × 1 wire = 0.0060 square inch
Total cross-sectional area of existing conductors = 0.1246 square inch

Note: Ground wires must be counted for raceway fill – Table 1, Note 3 of Chapter 9.

Step 3: ➤ Subtract the area of the existing conductors from the permitted area of conductor fill.
Raceway (more than 24 inches long): .3488 square inch – .1246 square inch = .2242 square inch
Nipple (less than 24 inches long):   .5232 square inch – .1246 square inch = .3986 square inch

❏ Conductors in Spare Space Area

An existing 1-inch EMT contains two No. 12 THHN conductors, two No. 10 THHN conductors, and one No. 12 bare (stranded) conductor. How many additional No. 8 THHN conductors can be added to this raceway (Figure 5–13)?

(a) 7 conductors if raceway is more than 24 inches long

(b) 12 conductors if raceway is less than 24 inches long

(c) 15 conductors regardless of the raceway length

(d) a and b

• Answer: (d) a and b

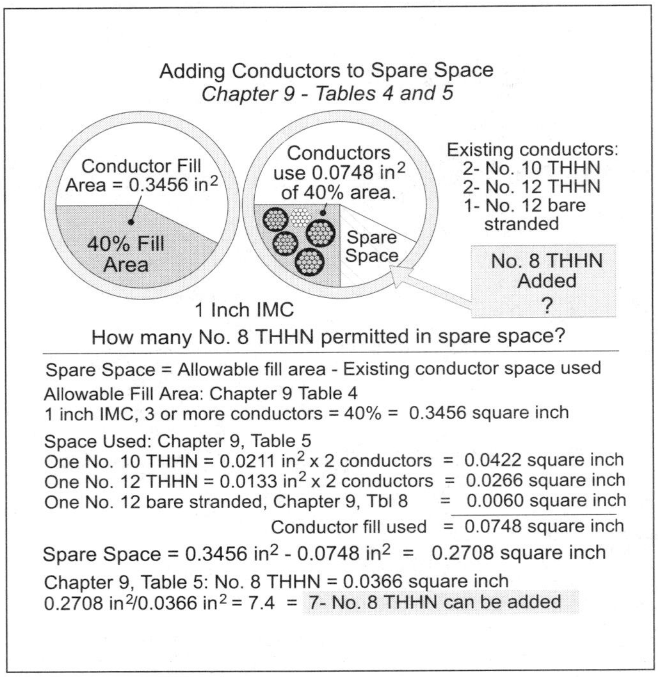

Step 1: ➡ Cross-sectional area permitted for conductor fill   [Table 1, Note 4 and Table 4 of Chapter 9].
Raceway: = .864 × 0.4 = .3456 square inch
Nipple: = .864 × 0.6 = .5184 square inch

Step 2: ➡ Cross-sectional area of existing conductors.
No. 10 THHN
.0211 square inch × 2 = .0422 square inch
No. 12 THHN
.0133 square inch × 2 = .0266 square inch
No. 12 bare
.0060 square inch × 1 = .0060 square inch

Total cross-sectional area of existing conductors = .0748 square inch

Note: Ground wires must be counted for raceway fill see Table 1, Note 3 of Chapter 9.

Step 3: ➡ Subtract the area of the existing conductors from the permitted area of conductor fill.
Raceway more than 24 inches long: .3456 square inch – .0748 square inch = .2708 square inch
Nipple (less than 24 inches long):   .5184 square inch – .0748 square inch = .4436 square inch

Step 4: ➡ Cross-sectional area of the conductors to be installed [Table 5 of Chapter 9].

No. 10 THHN = 0.0211 square inch.

Step 5: ➡ Divide the spare space area (Step 3) by the conductor area.
Raceway: = 0.2708 square inches/.0366 square inch= 7.4 or 7 conductors
Nipple: = 0.4436 square inch/.0366 square inch = 12.1 or 12 conductors.

We must round down to 18 conductors because Note 7 to Table 1 only applies if all of the conductors are the same size and same insulation.

Adding Conductors to Spare Space
Chapter 9 - Tables 4 and 5

Conductor Fill Area = 0.3456 in²
40% Fill Area

Conductors use 0.0748 in² of 40% area.
Spare Space

Existing conductors:
2- No. 10 THHN
2- No. 12 THHN
1- No. 12 bare stranded

No. 8 THHN Added ?

1 Inch IMC
How many No. 8 THHN permitted in spare space?

Spare Space = Allowable fill area - Existing conductor space used
Allowable Fill Area: Chapter 9 Table 4
1 inch IMC, 3 or more conductors = 40% = 0.3456 square inch

Space Used: Chapter 9, Table 5
One No. 10 THHN = 0.0211 in² x 2 conductors = 0.0422 square inch
One No. 12 THHN = 0.0133 in² x 2 conductors = 0.0266 square inch
One No. 12 bare stranded, Chapter 9, Tbl 8   = 0.0060 square inch
                           Conductor fill used = 0.0748 square inch

Spare Space = 0.3456 in² - 0.0748 in² = 0.2708 square inch

Chapter 9, Table 5: No. 8 THHN = 0.0366 square inch
0.2708 in²/0.0366 in² = 7.4 = 7- No. 8 THHN can be added

Figure 5–13
Adding Conductors to Spare Space

5–4 TIPS FOR RACEWAY CALCULATIONS

Tip 1: ➡ Take your time.

Tip 2: ➡ Use a ruler or straight-edge when using tables.

Tip 3: ➡ Watch out for the different types of raceways and conductor insulation, particularly RHH/RHW with or without outer cover.

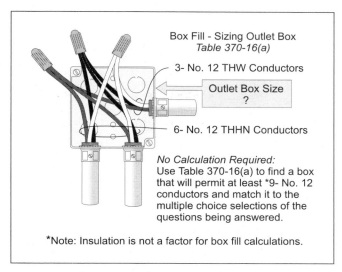

Box Fill - Sizing Outlet Box
Table 370-16(a)

3- No. 12 THW Conductors

Outlet Box Size
?

6- No. 12 THHN Conductors

No Calculation Required:
Use Table 370-16(a) to find a box that will permit at least *9- No. 12 conductors and match it to the multiple choice selections of the questions being answered.

*Note: Insulation is not a factor for box fill calculations.

Figure 5–14
Box Fill – Sizing Outlet Box

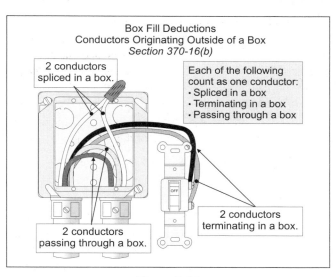

Box Fill Deductions
Conductors Originating Outside of a Box
Section 370-16(b)

2 conductors spliced in a box.

Each of the following count as one conductor:
• Spliced in a box
• Terminating in a box
• Passing through a box

2 conductors terminating in a box.

2 conductors passing through a box.

Figure 5–15
Box Fill Deductions

PART B – OUTLET BOX FILL CALCULATIONS

INTRODUCTION [*SECTION 370-16*]

Boxes shall be of sufficient size to provide free space for all conductors. An outlet box is generally used for the attachment of devices and fixtures and has a specific amount of space (volume) for conductors, devices, and fittings. The volume taken up by conductors, devices, and fittings in a box must not exceed the box fill capacity. The volume of a box is the total volume of its assembled parts, including plaster rings, industrial raised covers, and extension rings. The total volume includes only those fittings that are marked with their volume in cubic inches [*Section 370-16(a)*].

5–5 SIZING BOX – CONDUCTORS ALL THE SAME SIZE [Table 370-16(a)]

When all of the conductors in an outlet box are the same size (insulation doesn't matter), Table 370-16(a) of the *National Electrical Code*® can be used to:

(1) Determine the number of conductors permitted in the outlet box, or

(2) Determine the size outlet box required for the given number of conductors.

Note: Table 370-16(a) applies only if the outlet box contains no switches, receptacles, fixture studs, fixture hickeys, manufactured cable clamps, or grounding conductors (not likely).

❑ **Outlet Box Size**

What size outlet box is required for six No. 12 THHN conductors, and three No. 12 THW conductors (Figure 5–14)?

(a) $4 \times 1^{1}/_{4}$ square   (b) $4 \times 1^{1}/_{2}$ square   (c) $4 \times 1^{1}/_{4}$ round   (d) $4 \times 1^{1}/_{2}$ round

   • Answer: (b) $4 \times 1^{1}/_{2}$ square

Table 370-16(a) permits nine No. 12 conductors, insulation is not a factor.

❑ **Number of Conductors in Outlet Box**

Using Table 370-16(a), how many No. 14 THHN conductors are permitted in a $4 \times 1^{1}/_{2}$ round box?

(a) 7 conductors   (b) 9 conductors   (c) 10 conductors   (d) 11 conductors

   • Answer: (a) 7 conductors

5–6 CONDUCTOR EQUIVALENTS [*SECTION 370-16(b)*]

Table 370-16(a) does not take into consideration the fill requirements of clamps, support fittings, devices, or equipment grounding conductors within the outlet box. In no case can the volume of the box and its assembled sections be less than the fill calculation as listed below:

(1a) Conductor Terminating in the Box. Each conductor that originates outside the box and terminates or is spliced within the box is considered as one conductor (Figure 5–15).

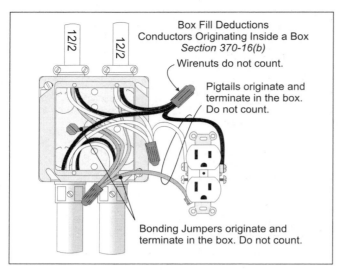

Figure 5–16
Box Fill Conductors Originating Inside a Box

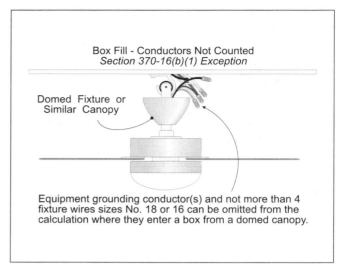

Figure 5–17
Box Fill–Conductors Not Counted

(1b) Conductor Running through the Box. Each conductor that runs through the box is considered as one conductor (Figure 5–15). Conductors, no part of which leaves the box, shall not be counted, this includes equipment bonding jumpers and pigtails (Figure 5–16).

Exception. Fixture wires smaller than No. 14 from a domed fixture or similar canopy are not counted (Figure 5–17).
   (2) Cable Fill. One or more internal cable clamps in the box are considered as one conductor volume in accordance with the volume listed in Table 370-16(b), based on the largest conductor that enters the outlet box (Figure 5–18).

Note: Small fittings such as locknuts and bushings are not counted [*Section 370-16(b)*].
   (3) Support Fittings Fill. One or more fixture studs or hickeys within the box are considered as one conductor volume, based on the largest conductor that enters the outlet box (Figure 5–19).
   (4) Device or Equipment Fill. Each yoke or strap containing one or more devices or equipment is considered as two conductors, based on the largest conductor that terminates on the yoke (Figure 5–20).
   (5) Grounding Conductors. One or more grounding conductors are considered as one conductor volume in accordance with the volume based on the largest grounding conductor that enters the outlet box (Figure 5–21).

Note. Fixture ground wires smaller than No. 14 from a domed fixture or similar canopy are not counted [*Section 370-16(b)(1)* Exception].

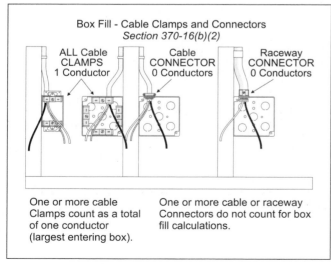

Figure 5–18
Box Fill – Cable Clamps and Connectors

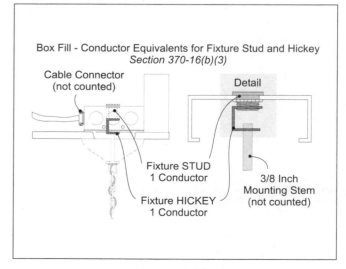

Figure 5–19
Box Fill – Conductor Equivalents for Fixture Stud and Hickey

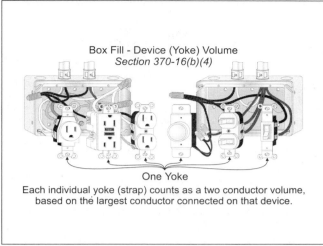

Box Fill - Device (Yoke) Volume
Section 370-16(b)(4)

One Yoke
Each individual yoke (strap) counts as a two conductor volume, based on the largest conductor connected on that device.

Figure 5–20
Box Fill – Device Volume

Note: Grounds are spliced together with one bond wire to the box and one bond wire to the receptacle.

Box Fill - Grounding Conductor
Section 370-16(b)(5)

2- No. 12's
1- No. 14

These 3 ground wires count as One No. 12

14/2 w/G

12/2 w/G

12/2 w/G

Box Fill Based on:
1- No. 12 Ground
4- No. 12 Conductors
5- No. 12's Total

2- No. 14 Conductors
2- No. 14 for Strap
4- No. 14's Total

Figure 5–21
Box Fill – Grounding Conductor

What's Not Counted

Wirenuts, cable connectors, raceway fittings, and conductors that originate and terminate within the outlet box (such as equipment bonding jumpers and pigtails) are not counted for box fill calculations [*Section 370-16(a)*].

❏ Number of Conductors

What is the total number of conductors used for box fill calculations in Figure 5–22?

(a) 5 conductors   (b) 7 conductors   (c) 9 conductors   (d) 11 conductors

• Answer: (d) 11 conductors

Switch –	Five No. 14 conductors, two conductors for device and three conductors terminating
Receptacle –	Four No. 12 conductors, two conductors for the device and two conductors terminating
Ground wire –	One conductor
Cable clamps –	One conductor

5–7 SIZING BOX – DIFFERENT SIZE CONDUCTORS [*SECTION 370-16(b)*]

To determine the size of the outlet box when the conductors are of different sizes (insulation is not a factor), the following steps can be used:

Step 1: ➨ Determine the number and size of conductors equivalents in the box.

Step 2: ➨ Determine the volume of the conductors equivalents from Table 370-16(b).

Step 3: ➨ Size the box by using Table 370-16(a).

Outlet Box Sizing

What size outlet box is required for 14/3 Type NM cable (with ground) that terminates on a switch, with 14/2 NM that terminates on a receptacle, if the box has internal cable clamps factory installed (Figure 5–22)?

(a) 4 × 1¹⁄₄ square   (b) 4 × 1¹⁄₂ square   (c) 4 × 2¹⁄₈ square   (d) any of these

• Answer: (c) 4 × 2¹⁄₈ square

Step 1: ➨ Determine the number and size of conductors.

14/3 NM	3 – No. 14
14/2 NM	2 – No. 14
Cable clamps	1 – No. 14
Switch	2 – No. 14
Receptacles	2 – No. 14
Ground wires	1 – No. 14
Total	11– No. 14

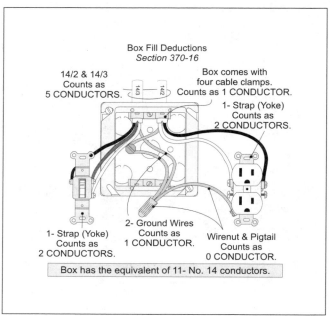

Figure 5–22
Box Fill Deductions

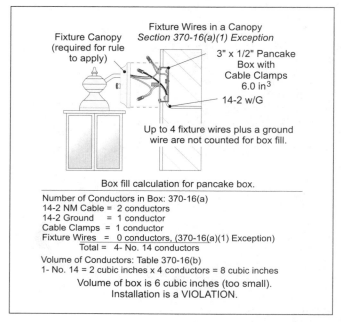

Figure 5–23
Fixture Wires in a Canopy

Step 2: ➡ Determine the volume of the conductors, [Table 370-16(b)].

No. 14 = 2 cubic inches

No. 14 conductors volume = 11 wires × 2 cubic inches = 22 cubic inches

Step 3: ➡ Select the outlet box from Table 370-16(a).

4 × 1¹/₂ square, 21 cubic inches, too small

4 × 2¹/₈ square, 30.3 cubic inches, just right

❏ Domed Fixture Canopy [*Section 370-16(a)*, **Exception**].

A round 3 × ¹/₂ box has a total volume of 6 cubic inches and has factory internal cable clamps. Can this pancake box be used with a lighting fixture that has a domed canopy? The branch circuit wiring is 14/2 nonmetallic sheath cable and the paddle fan has three fixture wires and one ground wire all smaller than No. 14 (Figure 5–23).

(a) Yes   (b) No

• Answer: (b) No

The box is limited to 6 cubic inches, and the conductors total 8 cubic inches [*Section 370-16(a)*].

Step 1: ➡ Determine the number and size of conductors within the box.

14/2 NM	2 – No. 14
Cable clamps	1 – No. 14
Ground wire	1 – No. 14
Total	4 – No. 14 conductors

Step 2: ➡ Determine the volume of the conductors [Table 370–16(b)].

No. 14 = 2 cubic inches

Four No. 14 conductors = 4 wires × 2 cubic inches = 8 cubic inches

❏ Conductors Added to Existing Box

How many No. 14 THHN conductors can be pulled through a 4 × 2¹/₈ square box that has a plaster ring of 3.6 cubic inches? The box already contains two receptacles, five No. 12 THHN conductors, and one No. 12 bare grounding conductor (Figure 5–24).

(a) 4 conductors   (b) 5 conductors   (c) 6 conductors   (d) 7 conductors

• Answer: (b) 5 conductors

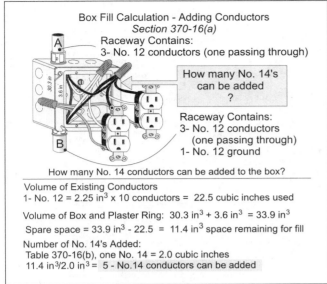

Box Fill Calculation - Adding Conductors
Section 370-16(a)
Raceway Contains:
3- No. 12 conductors (one passing through)

How many No. 14's can be added ?

Raceway Contains:
3- No. 12 conductors (one passing through)
1- No. 12 ground

How many No. 14 conductors can be added to the box?

Volume of Existing Conductors
1- No. 12 = 2.25 in³ x 10 conductors = 22.5 cubic inches used

Volume of Box and Plaster Ring: 30.3 in³ + 3.6 in³ = 33.9 in³

Spare space = 33.9 in³ - 22.5 = 11.4 in³ space remaining for fill

Number of No. 14's Added:
Table 370-16(b), one No. 14 = 2.0 cubic inches
11.4 in³/2.0 in³ = 5 - No.14 conductors can be added

Figure 5–24
Box Fill Calculation – Adding Conductors

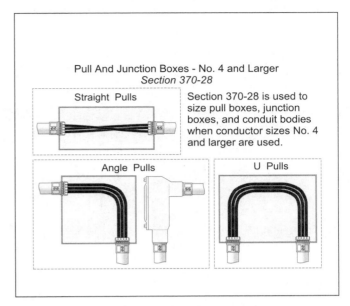

Pull And Junction Boxes - No. 4 and Larger
Section 370-28

Straight Pulls

Section 370-28 is used to size pull boxes, junction boxes, and conduit bodies when conductor sizes No. 4 and larger are used.

Angle Pulls   U Pulls

Figure 5–25
Pull and Junction Boxes

Step 1: ➤ Determine the number and size of the existing conductors.

Two receptacles	4 – No. 12 conductors (2 yokes × 2 conductors)
Five No. 12's	5 – No. 12 conductors
One ground	1 – No. 12 conductor
Total	10 - No. 12 conductors

Step 2: ➤ Determine the volume of the existing conductors [Table 370-16(b)].
No. 12 conductor = 2.25 cubic inches, 10 wires × 2.25 cubic inches = 22.5 cubic inches

Step 3: ➤ Determine the space remaining for the additional No. 14 conductors.
Remaining space = Total space less existing conductors
Total space = 30.3 cubic inches (box) [Table 370-16(a)] + 3.6 cubic inches (ring) = 33.9 cubic inches
Remaining space = 33.9 cubic inches – 22.5 cubic inches (ten No. 12 conductors)
Remaining space = 11.4 cubic inches

Step 4: ➤ Determine the number of No. 14 conductors permitted in the spare space.
Conductors added = Remaining space/added conductors volume
Conductors added = 11.4 cubic inches/2 cubic inches [Table 370-16(b)]
Conductors added = Five No. 14 conductors

PART C – PULL, JUNCTION BOXES, AND CONDUIT BODIES

INTRODUCTION

Pull boxes, junction boxes, and *conduit bodies* must be sized to permit conductors to be installed so that the conductor insulation will not be damaged. For conductors No. 4 and larger, we must size pull boxes, junction boxes, and conduit bodies according to the requirements of Section 370-28 (Figure 5–25).

5–8 PULL AND JUNCTION BOX SIZE CALCULATIONS

Straight Pull Calculation [*Section 370-28(a)(1)*]

A straight pull calculation applies when conductors enter one side of a box and leave through the opposite wall of the box. The minimum distance from where the raceway enters to the opposite wall must not be less than eight times the trade size of the largest raceway (Figure 5–26).

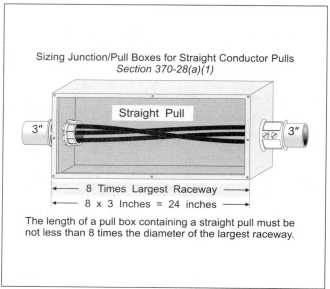

Figure 5–26
Sizing Junction/Pull Boxes for Straight Conductor Pulls

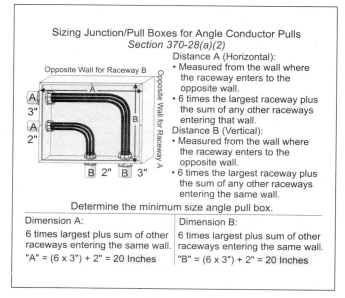

Figure 5–27
Sizing Junction/Pull Boxes for Angle Conductor Pulls

Angle Pull Calculation [*Section 370-28(a)(2)*]

An angle pull calculation applies when conductors enter one wall and leave the enclosure not opposite the wall of the conductor entry. The distance for angle pull calculations from where the raceway enters to the opposite wall must not be less than six times the trade diameter of the largest raceway, plus the sum of the diameters of the remaining raceways on the same wall and *row* (Figure 5–27). When there is more than one row, each row shall be calculated separately, and the row with the largest calculation shall be considered the minimum angle pull dimension.

U-Pull Calculations [*Section 370-28(a)(2)*]

A U-pull calculation applies when the conductors enter and leave from the same wall. The distance from where the raceways enter to the opposite wall must not be less than six times the trade diameter of the largest raceway, plus the sum of the diameters of the remaining raceways on the same wall (Figure 5–28).

Distance between Raceways Containing the Same Conductor Calculation [*Section 370-28(a)(2)*]

After sizing the pull box, the raceways must be installed so that the distance between raceways enclosing the same conductors shall not be less than six times the trade diameter of the largest raceway. This distance is measured from the nearest edge of one raceway to the nearest edge of the other raceway (Figures 5–28 and 5–29).

5–9 DEPTH OF BOX AND CONDUIT BODY SIZING [*SECTION 370-28(a)(2)*, Exception]

When conductors enter an enclosure opposite a *removable cover*, such as the back of a pull box or conduit body, the distance from where the conductors enter to the removable cover shall not be less than the distances listed in Table 373-6(a); one wire per terminal (Figure 5–30).

❏ Depth of Pull Or Junction Box

A 24" × 24" pull box has two 2-inch conduits that enter the back of the box with No. 4/0 conductors. What is the minimum depth of the box?

    (a) 4 inches   (b) 6 inches
    (c) 8 inches   (d) 10 inches
      • Answer: (a) 4 inches, Table 373-6(a)

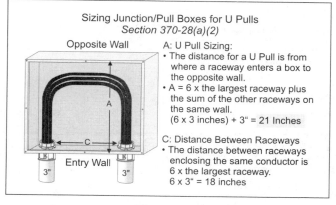

Figure 5–28
Sizing Junction/Pull Boxes for U-Pulls

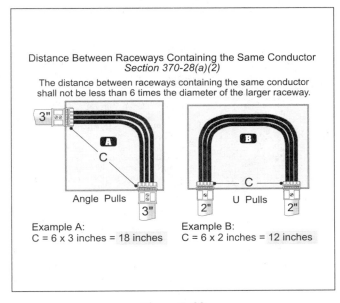

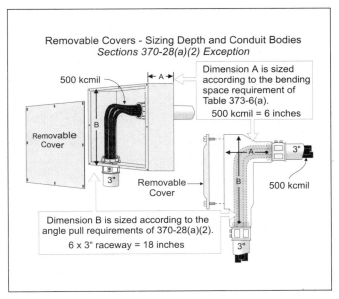

Figure 5–29
Distance between Raceways Containing the
Same Conductor

Figure 5–30
Removable Covers – Sizing Depth and
Conduit Bodies

5–10 JUNCTION AND PULL BOX SIZING TIPS

When sizing pull and junction boxes, the following suggestions should be helpful.

Step 1: ➡ Always draw out the problem.

Step 2: ➡ Calculate the HORIZONTAL distance(s):
- Left to right straight calculation
- Left to right angle or U-pull calculation
- Right to left straight calculation
- Right to left angle or U-pull calculation

Step 3: ➡ Calculate the VERTICAL distance(s):
- Top to bottom straight calculation
- Top to bottom angle or U-pull calculation
- Bottom to top straight calculation
- Bottom to top angle or U-pull calculation

5–11 PULL BOX EXAMPLES

❏ Pull Box Sizing

A junction box contains two 3 inch raceways on the left side and one 3 inch raceway on the right side. The conductors from one of the 3 inch raceways on the left wall are pulled through the 3 inch raceway on the right wall. The other 3 inch raceway's conductors are pulled through a raceway at the bottom of the pull box (Figure 5–31).

↪ Horizontal Dimension

What is the horizontal dimension of this box?

(a) 18 inches   (b) 21 inches   (c) 24 inches   (d) none of these

- Answer: (c) 24 inches, Section 370-28 (Figure 5–31, Part A).

Left wall to the right wall angle pull   = (6 × 3 inches) + 3 inches = 21 inches

Left wall to the right wall straight pull   = 8 × 3 inches = 24 inches

Right wall to left wall angle pull   = No calculation

Right wall to the left wall straight pull   = 8 × 3 inches = 24 inches

⇨ **Vertical Dimension**

What is the vertical dimension of this box?

(a) 18 inches   (b) 21 inches

(c) 24 inches   (d) none of these

• Answer: (a) 18 inches, Section 370-28

(Figure 5–31, Part B).

Top to bottom angle	= No calculation
Top to bottom straight	= No calculation
Bottom to top angle	= 6 × 3 inches = 18 inches
Bottom to top straight	= No calculation

⇨ **Distance between Raceways**

What is the minimum distance between the two 3 inch raceways that contain the same conductors?

(a) 18 inches   (b) 21 inches

(c) 24 inches   (d) none of these

• Answer: (a) 18 inches, 6 × 3 inches,

Section 370-28 (Figure 5–31, Part C)

❏ **Pull Box Sizing**

A pull box contains two 4 inch raceways on the left side and two 2 inch raceways on the top.

⇨ **Horizontal Dimension**

What is the horizontal dimension of the box?

(a) 28 inches   (b) 21 inches   (c) 24 inches   (d) none of these

• Answer: (a) 28 inches, Section 370-28(a)(2)

Left wall to the right wall angle pull	= (6 × 4 inches) + 4 inches = 28 inches
Left wall to the right wall straight pull	= No calculation
Right wall to left wall angle pull	= No calculation
Right wall to the left wall straight pull	= No calculation

⇨ **Vertical Dimension**

What is the vertical dimension of the box?

(a) 18 inches   (b) 21 inches   (c) 24 inches   (d) 14 inches

• Answer: (d) 14 inches, Section 370-28(a)(2)

Top to bottom wall angle pull	= (6 × 2 inches) + 2 inches = 14 inches
Top to bottom wall straight pull	= No calculation
Bottom to top wall angle pull	= No calculation
Bottom to top wall straight pull	= No calculation

⇨ **Distance between Raceways**

What is the minimum distance between the two 4 inch raceways that contain the same conductors?

(a) 18 inches   (b) 21 inches   (c) 24 inches   (d) None of these

• Answer: (c) 24 inches, Section 370-28(a)(2)

6 × 4 inches = 24 inches

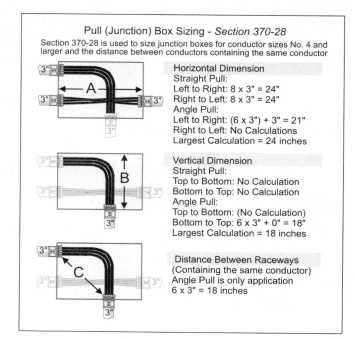

Figure 5–31
Pull (Junction) Box Sizing

Unit 5 – Raceway, Outlet Box, and Junction Boxes Calculations Summary Questions

Part A – Raceway Fill Calculations

5–1 Understanding Chapter 9 Tables

1. When all the conductors are the same size and insulation, the number of conductors permitted in a raceway can be determined simply by looking at the Tables listed in _____ .
    (a) Chapter 9   (b) Appendix B   (c) Appendix C   (d) Appendix D

2. When equipment grounding conductors are installed in a raceway, the actual area of the conductor must be used when calculating raceway fill.
    (a) True   (b) False

3. When a raceway does not exceed 24 inches, the raceway is permitted to be filled to _____ of its cross–section area.
    (a) 53%   (b) 31%   (c) 40%   (d) 60%

4. How many No. 16 TFFN conductors can be installed in a $^3/_8$ inch electrical metallic tubing?
    (a) 40   (b) 26   (c) 30   (d) 29

5. How many No. 6 RHH (without outer cover) can be installed in a 1 inch nonmetallic tubing?
    (a) 25   (b) 16   (c) 13   (d) 7

6. How many No. 1/0 XHHW can be installed in a 2 inch flexible metal conduit?
    (a) 7   (b) 6   (c) 16   (d) 13

7. How many No. 12 RHH (with outer cover) can be installed in a 1 inch IMC raceway?
    (a) 7   (b) 11   (c) 5   (d) 4

8. If we have a 2 inch rigid metal conduit and we want to install three THHN aluminum compact conductors, what is the largest aluminum compact conductor permitted to be installed?
    (a) No. 4/0   (b) 250 kcmil   (c) 300 kcmil   (d) 500 kcmil

9. The actual area of conductor fill is dependent on the raceway size and the number of conductors installed. If there are three or more conductors installed in a raceway the total area of conductor fill is limited to _____ percent.
    (a) 53   (b) 31   (c) 40   (d) 60

10. • What is the area in square inches for a No. 10 THW?
    (a) 0.0333   (b) 0.0172   (c) 0.0252   (d) 0.0278

11. What is the area in square inches for a No. 14 RHW (without cover)?
    (a) 0.0209   (b) 0.0172   (c) 0.0252   (d) 0.0278

12. What is the area in square inches for a No. 10 THHN?
    (a) 0.0117   (b) 0.0172   (c) 0.0252   (d) 0.0211

13. What is the area in square inches for a No. 12 RHH (with outer cover)?
    (a) 0.0117   (b) 0.0353   (c) 0.0252   (d) 0.0327

14. • What is the area in square inches of a No. 8 bare solid?
    (a) 0.013   (b) 0.027   (c) 0.038   (d) 0.045

5–2 Raceway and Nipple Calculations

15. The number of conductors permitted in a raceway is dependent on _____ .
    (a) the area of the raceway
    (b) the percent area fill as listed in Chapter 9, Table 1
    (c) the area of the conductors as listed in Chapter 9, Tables 5 and 8
    (d) all of these

16. A 200 ampere feeder installed in rigid nonmetallic conduit schedule 80, has three No. 3/0 THHN, one No. 2 THHN, and one No. 6 THHN. What size raceway is required?
    (a) 2 inch   (b) $2^1/_2$ inch   (c) 3 inch   (d) $3^1/_2$ inch

17. What size rigid metal nipple is required for three No. 4/0 THHN, one No. 1/0 THHN and No. 4 THHN?
    (a) $1^1/_2$ inch   (b) 2 inch   (c) $2^1/_2$ inch   (d) none of these

5–3 Existing Raceway Calculations

18. • An existing rigid metal nipple contains four No. 10 THHN and one No. 10 (bare stranded) ground wire in a 3/8 inch rigid metal conduit. How many additional No. 10 THHN can be installed?
    (a) 5   (b) 7   (c) 9   (d) 11

Part B – Outlet Box Fill Calculations

5–5 Sizing Box–Conductors All the Same Size [Table 370–16]

19. What size box is required for 6 No. 14 THHN and 3 No. 14 THW?
    (a) $4 \times 1^1/_4$ square   (b) $4 \times 1^1/_2$ round   (c) $4 \times 1^1/_4$ round   (d) none of these

20. How many No. 10 THHN are permitted in a $4 \times 1^1/_2$ square box?
    (a) 8 conductors   (b) 9 conductors   (c) 10 conductors   (d) 11 conductors

5–6 Conductor Equivalents [*Section 370-16*]

21. Table 370-16 does not take into consideration the volume of _____ .
    (a) switches and receptacles   (b) fixture studs and hickeys
    (c) manufactured cable clamps   (d) all of these

22. When determining the number of conductors for box fill calculations, which of the following statements are true?
    (a) A fixture stud or hickey is considered as one conductor for each type, based on the largest conductor that enters the outlet box.
    (b) Internal factory cable clamps are considered as one conductor for one or more cable clamps, based on the largest conductor that enters outlet box.
    (c) The device yoke is considered as two conductors, based on the largest conductor that terminates on the strap (device mounting fitting).
    (d) all of these

23. • When determining the number of conductors for box fill calculations, which of the following statements are true?
    (a) Each conductor that runs through the box without loop (without splice) is considered as one conductor.
    (b) Each conductor that originates outside the box and terminates in the box is considered as one conductor.
    (c) Wirenuts, cable connectors, raceway fittings, and conductors that originate and terminate within the outlet box (equipment bonding jumpers and pigtails) are not counted for box fill calculations.
    (d) all of these

24. It is permitted to omit one equipment grounding conductor and not more than _____ that enter a box from a fixture canopy.
    (a) four fixture wires   (b) four No. 16 fixture wires
    (c) four No. 18 fixture wires   (d) b and c

25. Can a round $4 \times 1/2$ box marked as 8 cubic inches with manufactured cable clamps supplied with 14/2 NM be used with a fixture that has two No. 18 TFN and a canopy cover?
    (a) Yes   (b) No

5–7 Sizing Box–Different Size Conductors [*Section 370-16(b)*]

26. What size outlet box is required for: one 12/2 NM cable that terminates on a switch, one 12/3 NM cable that terminates on a receptacle, and the box has manufactured cable clamps.
    (a) $4 \times 1^{1}/_{4}$ square   (b) $4 \times 1^{1}/_{2}$ square   (c) $4 \times 2^{1}/_{8}$ square   (d) none of these

27. • How many No. 14 THHN conductors can be pulled through a $4 \times 1^{1}/_{2}$ square box with a plaster ring of 3.6 cubic inches? The box contains a two duplex receptacles, five No. 14 THHN and two grounding conductors.
    (a) 1   (b) 2   (c) 3   (d) 4

Part C – Pull, Junction Boxes, and Conduit Bodies

5–8 Pull and Junction Box Size Calculations

28. When conductors No. 4 and larger are installed in boxes and conduit bodies, we must size the enclosure according to which of the following requirements?
    (a) The minimum distance for straight pull calculations from where the conductors enter to the opposite wall must not be less than eight times the trade size of the largest raceway.
    (b) The distance for angle pull calculations from the raceway entry to the opposite wall must not be less than six times the trade diameter of the largest raceway, plus the sum of the diameters of the remaining raceways on the same wall and row.
    (c) The distance between raceways enclosing the same conductor(s) shall not be less than six times the trade diameter of the largest raceway.
    (d) all of the above are correct.

29. When conductors enter an enclosure opposite a removable cover, the distance from where the conductors enter to the removable cover shall not be less than _____ .
    (a) six times the largest raceway   (b) eight times the largest raceway
    (c) a or b   (d) none of these

The following information applies to the next three questions.
    A junction box contains two $2^{1}/_{2}$ inch raceways on the left side and one $2^{1}/_{2}$ inch raceway on the right side. The conductors from one $2^{1}/_{2}$ inch raceway (left wall) are pulled through the raceway on the right wall. The other $2^{1}/_{2}$ inch raceway conductors (on the side) are pulled through a $2^{1}/_{2}$ inch raceway at the bottom of the pull box.

30. What is the distance from the left wall to the right wall?
    (a) 18 inches   (b) 21 inches   (c) 24 inches   (d) 20 inches

31. • What is the distance from the bottom wall to the top wall?
    (a) 18 inches   (b) 21 inches   (c) 24 inches   (d) 15 inches

32. What is the distance from between the raceways that contain the same conductors?
    (a) 18 inches   (b) 21 inches   (c) 24 inches   (d) 15 inches

The following information applies to the next three questions.
    A junction box contains two 2 inch raceways on the left side and two 2 inch raceways on the top.

33. What is the distance from the left wall to the right wall?
    (a) 28 inches   (b) 21 inches   (c) 24 inches   (d) 14 inches

34. What is the distance from the bottom wall to the top wall?
    (a) 18 inches   (b) 21 inches   (c) 24 inches   (d) 14 inches

35. What is the distance between the 2 inch raceways that contain the same conductors?

    (a) 18 inches   (b) 21 inches   (c) 24 inches   (d) 12 inches

☆ Challenge Questions

Part A – Raceway Fill Calculations

36. • A 3 inch schedule 40 PVC raceway contains seven No. 1 RHW conductors without outer cover. How many No. 2 THW conductors may be installed in this raceway with the existing conductors?

    (a) 11   (b) 15   (c) 20   (d) 25

Part B – Outlet Box Fill Calculations

37. • Determine the minimum cubic inches required for two No. 10 TW passing through a box, four No. 14 THHN terminating, two No. 12 TW terminating to a receptacle, and one No 12 equipment bonding jumper from the receptacle to the box.

    (a) 18.5 cubic inches   (b) 22 cubic inches   (c) 20 cubic inches   (d) 21.75 cubic inches

38. • When determining the number of conductors in a box fill two No. 18 fixture wires, one 14/3 nonmetallic-sheathed cable with ground, one duplex switch, and two cable clamps would count as _____ .

    (a) 9 conductors   (b) 8 conductors   (c) 10 conductors   (d) 6 conductors

Part C – Pull, Junction Boxes, and Conduit Bodies

The Figure 5–32 applies to the next four questions:

39. The minimum horizontal dimension for the junction box shown in the diagram is _____ (Figure 5–32).

    (a) 21 inches   (b) 18 inches

    (c) 24 inches   (d) 20 inches

40. • The minimum vertical dimension for the junction box is _____ (Figure 5–32).

    (a) 16 inches   (b) 18 inches

    (c) 20 inches   (d) 24 inches

41. The minimum distance between the two 2 inch raceways that contain the same conductor "C" would be _____ (Figure 5–32).

    (a) 12 inches   (b) 18 inches

    (c) 24 inches   (d) 30 inches

42. • If a 3 inch raceway entry (250 kcmil) is in the wall opposite to a removable cover, the distance from that wall to the cover must not be less than _____ (Figure 5–32)?

    (a) 4 inches   (b) 4¹/₂ inches

    (c) 5 inches   (d) 6 inches

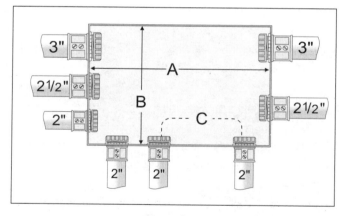

Figure 5–32

NEC® Questions from Section 90-1 through Sections 250-62

The following *National Electrical Code*® questions are in consecutive order. Questions with • indicate that 75 percent or less of exam takers get the question correct.

Article 90 – Introduction

43. Compliance with the provisions of the *Code* will result in _____ .
(a) good electrical service   (b) an efficient electrical system
(c) an electrical system free from hazard   (d) all of these

44. • The *Code* applies to the installation _____ .
(a) of electrical conductors and equipment within or on public and private buildings
(b) of outside conductors and equipment on the premises
(c) of optical fiber cable
(d) all these

45. Installations of communications equipment under the exclusive control of communications utilities located outdoors or in building spaces used exclusively for such installations _____ covered by the *Code*.
(a) are   (b) are sometimes   (c) are not   (d) might be

46. Material identified by the superscript letter "x" includes text extracted from _____ .
(a) other *Code* articles and Sections   (b) other NFPA documents
(c) other IOWA publications   (d) none of these

47. Explanatory material, such as references to other standards, references to related Sections of the *Code*, or information related to a *Code* rule is included in the *NEC*® in the form of a FPN.
(a) True   (b) False

48. Plans and specifications that provide ample space in raceways, spare raceways, and additional spaces will allow for future increases in the use of electricity. Distribution centers located in readily accessible locations will provide convenience and safety of operation. See Sections 110-26 and 240-24 for clearances and accessibility.
(a) True   (b) False

NEC® Chapter 1 – General Requirements

Article 100 – Definition

49. A junction box above a suspended ceiling that has removable panels is considered to be _____ .
(a) concealed   (b) accessible   (c) readily accessible   (d) recessed

50. Acceptable to the authority having jurisdiction means that the equipment has been _____ .
(a) identified   (b) listed   (c) approved   (d) labeled

51. According to the *Code*, automatic is self-acting, operating by its own mechanism when actuated by some impersonal influence such as _____ .
(a) change in current strength   (b) temperature
(c) mechanical configuration   (d) all of these

52. _____ branch circuits supply energy to one or more outlets to which appliances are to be connected.
(a) General purpose   (b) Multiwire   (c) Individual   (d) Appliance

53. • For a circuit to be considered a multiwire branch circuit, it must have _____ .
(a) two or more ungrounded conductors with a potential difference between them
(b) a grounded conductor having equal potential difference between it and each ungrounded conductor of the circuit
(c) a grounded conductor connected to the neutral (grounded) conductor of the system
(d) all of these

82. The *Code* prohibits damage to the internal parts of electrical equipment by foreign material such as paint, plaster, cleaners, etc. Precautions must be taken to provide protection from the detrimental effects of paint, plaster, cleaners, etc. on the internal parts of _____ .
     (a) panelboards   (b) receptacles   (c) lampholders   (d) a and b

83. Connection of conductors to terminal parts shall ensure a thoroughly good connection without damaging the conductors and shall be made by means of _____ .
     (a) solder lugs
     (b) pressure connectors
     (c) splices to flexible leads
     (d) all of these

84. The temperature rating associated with the ampacity of a _____ shall be so selected and coordinated so as to not exceed the lowest temperature rating of any connected termination, conductor, or device.
     (a) terminal   (b) conductor   (c) device   (d) all of these

85. Design B, C, D, or E motors shall be wired with conductors having an insulation rating of at least _____ °C, and the conductor shall be sized based on the 75°C ampacity as listed in Table 310-16.
     (a) 60   (b) 75   (c) 90   (d) none of these

86. The *NEC*® required series-rated installations to be field-marked to indicate the maximum level of fault-current for which the system has been installed.
     (a) True   (b) False

87. • Working space distances shall be measured from the _____ of equipment or apparatus if such are enclosed.
     (a) front   (b) opening   (c) a or b   (d) none of these

88. Concrete, brick, or tile walls shall be considered as _____ , as it applies to working space requirements..
     (a) inconsequential   (b) in the way   (c) grounded   (d) none of these

89. The working space in front of the electric equipment shall not be less than _____ inches wide or the width of the equipment, whichever is greater.
     (a) 15   (b) 30   (c) 40   (d) 60

90. When normally enclosed live parts are exposed for inspection or servicing, the working space, if in a passageway or general open space, shall be suitably _____ .
     (a) accessible   (b) guarded   (c) open   (d) enclosed

91. The minimum headroom of working spaces about motor control centers shall be _____ .
     (a) 3´   (b) 5´   (c) 6´   (d) 6´ 6″

92. • Ventilated heating or cooling equipment (including ducts) that service the electrical room or space cannot be installed in the dedicated space above a panelboard or switchboard.
     (a) True   (b) False

93. To guard live parts over 50 volts but less than 600 volts, the equipment can be _____ .
     (a) isolated in a room accessible to qualified persons only
     (b) located on a balcony
     (c) elevated 8 feet or more above the floor
     (d) any of these

94. Entrances to rooms and other guarded locations containing exposed live parts shall be marked with conspicuous _____ forbidding unqualified persons to enter.
     (a) warning signs   (b) locks   (c) a and b   (d) neither a nor b

95. Openings in ventilated dry-type _____ or similar openings in other equipment over 600 volts shall be designed so that foreign objects inserted through these openings will be deflected from energized parts.
     (a) lamp holders   (b) motors   (c) fuse holders   (d) transformers

96. At least one entrance not less than 24 inches wide and 6 feet high shall be provided to give access to the working space about electrical equipment that operates at over 600 volts, nominal. On switchboard and control panels exceeding 6 feet in width, there shall be one entrance at each end of such equipment, except where the work space required in Section 110-34(a) is doubled.
(a) True   (b) False

97. Where switches, cutouts, or other equipment operating at 600 volts, nominal, or less are installed in a room or enclosure where there are exposed live parts or exposed wiring operating at over 600 volts, nominal, the high-voltage equipment shall be effectively separated from the space occupied by _____ by a suitable partition, fence, or screen.
(a) the access area
(b) the low-voltage equipment
(c) unauthorized persons
(d) motor-control equipment

98. • Warning signs for over 600 volts shall read "Warning – High Voltage – Keep Out".
(a) True   (b) False

Chapter 2 – Wiring and Protection

Article 200 – Use and Identification of Grounded Conductor

99. Article 200 contains the requirements for _____ .
(a) identification of terminals
(b) grounded conductors in premises wiring systems
(c) identification of grounded conductors
(d) all of these

100. An insulated grounded conductor of No. _____ or smaller shall be identified by a continuous white or natural gray outer finish along its entire length.
(a) 3   (b) 4   (c) 6   (d) 8

101. When different wiring systems are in the same raceway or enclosure, the grounded conductor of one system can be white and the grounded conductor of the other system can be gray.
(a) True   (b) False

102. Identification of the grounded conductor shall be by a metal or metal coating substantially white in color or the word _____ located adjacent to the identified terminal.
(a) grounded   (b) white   (c) connected   (d) danger

Article 210 – Branch Circuit

103. • Multiwire branch circuits shall _____ .
(a) supply only line-to-neutral loads
(b) provide a means to disconnect simultaneously all ungrounded conductors if the multiwire branch circuit supplies more than one device or equipment on the same yoke for commercial installations
(c) have their conductors originate from different panelboards
(d) none of these

104. Each system _____ conductor, wherever accessible, shall be identified by separate color coding, marking tape, tagging, or other equally effective means. This only applies to multiwire branch circuits when there is more than one system voltage in a building.
(a) grounded   (b) ungrounded   (c) grounding   (d) all of these

105. Receptacles and cord connectors having grounding terminals must have those terminals effectively _____ .
(a) grounded   (b) bonded   (c) labeled   (d) listed

106.   All 125-volt, single-phase 15 and 20 ampere receptacles installed in bathrooms of _____ shall have ground-fault circuit-interrupter protection for personnel.
(a) guest rooms in hotels/motels
(b) dwelling units
(c) office buildings
(d) all of these

107.   Grade-level portions of _____ accessory buildings used for storage or work areas shall have ground-fault circuit-interrupter protection for all 15 and 20 ampere, 125-volt receptacles.
(a) finished   (b) unfinished   (c) a or b   (d) none of these

108.   • Dwelling unit GFCI protection is required for all 125-volt, 15 and 20 ampere, receptacles installed to supply countertop appliances in kitchens.
(a) True   (b) False

109.   GFCI protection is required for all 125-volt, 15 and 20 ampere, receptacles installed on rooftops, including those for fixed electric snow-melting or de-icing equipment.
(a) True   (b) False

110.   A dedicated 20-ampere circuit is permitted to supply power to a dwelling unit bathroom for receptacle outlet(s) and other equipment within the same bathroom.
(a) True   (b) False

111.   The recommended maximum total voltage drop on both the feeders and branch circuits is _____ percent.
(a) 3   (b) 2   (c) 5   (d) 3.6

112.   • A single receptacle installed on an individual branch circuit must be rated at least _____ percent of the rating of the circuit.
(a) 50   (b) 60   (c) 90   (d) 100

113.   • A branch circuit rated 20 amperes serves four receptacles. The rating of the receptacles must not be less than _____ amperes.
(a) 10   (b) 15   (c) 25   (d) 30

114.   The total rating of utilization equipment fastened in place shall not exceed _____ percent of the branch circuit ampere rating, where the circuit supplies receptacles for cord- and plug-connected equipment.
(a) 50   (b) 75   (c) 100   (d) 125

115.   A permanently installed cord connector on a cord pendant shall be considered a receptacle outlet.
(a) True   (b) False

116.   In dwelling units, when determining the spacing of general use receptacles, _____ on exterior walls are not considered wall space.
(a) fixed panels   (b) fixed glass   (c) sliding panels   (d) all of these

117.   • In dwelling units, outdoor receptacles can be connected to the 20 ampere small appliance branch circuit.
(a) True   (b) False

118.   The 24 inch dimension required in a dwelling unit countertop is measured along the wall line or center line, and the intent is that there be a receptacle outlet for every _____ linear feet or fraction thereof of counter length.
(a) 2   (b) 3   (c) 4   (d) 5

119.   When breaks occur in dwelling unit kitchen countertop spaces for ranges, refrigerators, sinks, etc., each countertop surface is considered a separate counter space for determining receptacle placement.
(a) True   (b) False

120.   Receptacle outlets in a dwelling unit bathroom must be installed within ___ inches of the outside edge of each basin and the receptacle must be located on a wall that is adjacent to the basin location.
(a) 12   (b) 18   (c) 24   (d) 36

121. For a one-family dwelling at least one receptacle outlet is required in each _____ .
     (a) basement
     (b) attached garage
     (c) detached garage with electric power
     (d) all of these

122. Receptacles installed behind a bed in the guest rooms in hotels and motels shall be located so as to prevent the bed from contacting an attachment plug, or the receptacle shall be provided with a suitable guard.
     (a) True   (b) False

123. At least ___ wall switch-controlled lighting outlet(s) shall be installed in every dwelling-unit habitable room and bathroom.
     (a) one   (b) three   (c) six   (d) none of these

124. • Which rooms in a dwelling unit must have a switch-controlled lighting outlet?
     (a) every habitable room   (b) bathrooms
     (c) hallways and stairways   (d) all of these

125. At least one lighting outlet controlled _____ located at the point of entry to the attic, underfloor space, utility room, and basement shall be installed where these spaces are used for storage or contain equipment requiring servicing.
     (a) by a switch   (b) containing a switch   (c) by a wall switch   (d) b and c

Article 215 – Feeders

126. The feeder conductor ampacity shall not be less than that of the service-entrance conductors where the feeder conductors carry the total load supplied by service-entrance conductors with an ampacity of _____ amperes or less.
     (a) 100   (b) 60   (c) 55   (d) 30

127. Where a feeder supplies _____ in which equipment grounding conductors are required, the feeder shall include or provide a grounding means to which the equipment grounding conductors of the branch circuits shall be connected.
     (a) equipment disconnecting means
     (b) electrical systems
     (c) branch circuits
     (d) electric discharge lighting equipment

Article 220 – Branch Circuit and Feeder Calculations

128. When a calculation results in a fraction of ampere of _____ or more, we "round up" to the next ampere.
     (a) 0.49   (b) 0.50   (c) 0.51   (d) none of these

129. Receptacle outlets in dwelling units are part of the _____ VA per square foot general lighting load.
     (a) 3   (b) 5   (c) 2   (d) 4

130. • The demand factors listed in Table 220-11 shall not be applied in determining the number of _____ for general illumination.
     (a) receptacles   (b) light outlets   (c) branch circuits   (d) switches

131. For other than dwelling units, the feeder and service load calculation for track lighting is to be determined at 150 volt-amperes for every ___ feet of track installed.
     (a) 4   (b) 6   (c) 2   (d) none of these

132. When sizing a feeder, the appliance loads in dwelling units can apply a demand factor of 75 percent of nameplate ratings for _____ or more appliances fastened in place on the same feeder.
     (a) two   (b) three   (c) four   (d) five

133. The feeder demand load for four 6 kW cooktops is _____ kW.
     (a) 17   (b) 4   (c) 12   (d) 24

134. • The maximum unbalanced feeder load for household electric dryers shall be calculated at _____ percent of the demand load as determined by Table 220-18.
(a) 50   (b) 70   (c) 85   (d) 115

135. Feeder and service-entrance conductors with demand load determined by the use of Table 220-30 shall be permitted to have the _____ load determined by Section 220-22.
(a) feeder   (b) circuit   (c) neutral   (d) none of these

136. The connected load to which the demand factors of Table 220-32 apply shall include 3 volt-amperes per _____ for general lighting and general-use receptacles.
(a) inch   (b) foot   (c) square inch   (d) square foot

137. Feeder conductors for new restaurants shall not be required to be of _____ ampacity than the service-entrance conductors.
(a) greater   (b) lesser   (c) equal   (d) none of these

138. • For each farm building or load supplied by _____ or more branch circuits, the load for feeders, service-entrance conductors, and service equipment shall be computed in accordance with demand factors not less than indicated in Table 220-40.
(a) one   (b) two   (c) three   (d) four

Article 225 – Outside Branch Circuits and Feeders

139. Conductors for _____ lighting shall be of the rubber-covered or thermoplastic type.
(a) indirect   (b) festoon   (c) show window   (d) temporary

140. If festoon lighting exceeds a span of _____ feet, messenger wires shall support the conductors.
(a) 20   (b) 30   (c) 40   (d) 50

141. • Communications conductors shall be mounted not less than _____ inches below power conductors rated over 300 volts supported on poles.
(a) 6   (b) 10   (c) 20   (d) 30

142. Outside branch circuits and feeder conductors shall have a vertical clearance of _____ feet from the surface of a roof.
(a) 3   (b) 6   (c) 8   (d) 10

143. Overhead branch circuit and feeder conductors shall not be installed _____ openings through which materials may be moved, such as openings in farm and commercial buildings, and shall not be installed where they will obstruct entrance to these building openings.
(a) above   (b) beneath   (c) inside   (d) outside

144. Where buildings exceed three stories, overhead lines shall be arranged so that a clear space at least 6 feet wide will be left either adjacent to the buildings or beginning not over 8 feet from them to facilitate the_____ when necessary for fire fighting.
(a) safety of persons (b) entrance of water   (c) raising of ladders (d) all of these

145. Lamps for outdoor lighting shall be below all _____, transformers, or other electrical utilization equipment.
(a) switches   (b) feeder circuits   (c) branch circuits   (d) energized conductors

Article 230 – Services

146. A building or structure shall be supplied by a maximum of _____ service(s).
(a) one   (b) two   (c) three   (d) as many as desired

147. • Branch circuit and feeder conductors shall not be installed in the same _____ with service conductors.
(a) raceway   (b) cable   (c) enclosure   (d) a or b

148.   Overhead service conductors to a building shall maintain a vertical clearance of 10 feet above platforms, projections, or surfaces from which they might be reached. This vertical clearance shall extend _____ feet measured horizontally from the platform, projections, or surfaces from which they might be reached.

  (a) 3 (b) 6 (c) 8 (d) 10

149.   The _____ clearances of all service drop conductors shall be based on a conductor temperature of 60ºF, no wind; with final unloaded sag in the wire, conductor, or cable.

  (a) horizontal (b) lateral (c) vertical (d) final

150.   The requirement for maintaining a 3 foot vertical clearance from the edge of the roof shall not apply to the final conductor span where the service drop is attached to _____ .

  (a) a service pole (b) the side of a building
  (c) an antenna (d) the base of a building

151.   • Overhead service drop conductors shall have a horizontal clearance of _____ feet from a pool.

  (a) 6 (b) 10 (c) 8 (d) 4

152.   Service-lateral conductors are required to be insulated (except the grounded conductor) when it is _____ .
  (a) bare copper in a raceway
  (b) bare copper and part of a cable assembly that is identified for underground use
  (c) copper-clad aluminum with individual insulation
  (d) a and b

153.   To supply power to a single-phase water heater, the size of the underground service lateral conductors shall not be smaller than _____ .

  (a) No. 4 copper (b) No. 8 aluminum (c) No. 12 copper (d) none of these

154.   Service-entrance conductors are required to be insulated except when they are _____ .
  (a) bare copper installed in a raceway
  (b) bare copper and part of the cable assembly identified for underground use
  (c) copper-clad aluminum with individual insulation
  (d) a and b

155.   Service entrance conductors can be spliced or tapped by clamped or bolted connections at anytime as long as _____ .
  (a) The free ends of conductors are covered with an insulation that is equivalent to that of the conductors or with an insulating device identified for the purpose [*Section 110-14*]
  (b) Wire connectors or other splicing means installed on conductors that are buried in the earth shall be listed for direct burial [*Section 300-5(e)*].
  (c) No splice is made in a raceway [*Section 300-13(a)*].
  (d) all of these

156.   Individual open conductors and cables other than service-entrance cables shall not be installed within _____ feet of grade level or where exposed to physical damage.

  (a) 8 (b) 10 (c) 12 (d) 15

157.   Where individual open conductors are not exposed to the weather, the conductors shall be mounted on _____ knobs.

  (a) door (b) insulated metal (c) glass or porcelain (d) none of these

158.   Service heads for service raceways shall be _____ .

  (a) raintight (b) weatherproof (c) rainproof (d) watertight

159.   Service heads must be located _____ .
  (a) above the point of attachment
  (b) below the point of attachment
  (c) within 36 inches from the point of attachment
  (d) none of these

160.   • Where the service disconnecting means is mounted on a switchboard having exposed busbars on the back, a raceway shall be permitted to terminate at a _____ .
    (a) connector   (b) box   (c) bushing   (d) cabinet

161.   Each service disconnect must be permanently marked to identify it as a service disconnecting means.
    (a) True   (b) False

162.   There shall be no more than ___ disconnects installed for each service, or for each set of service-entrance conductors as permitted in Section 230-2 and 230-40.
    (a) two   (b) four   (c) six   (d) none of these

163.   In a multiple-occupancy building, each occupant shall have access to its own service _____ .
    (a) conductor   (b) disconnecting means   (c) cable   (d) overcurrent protection

164.   • The service disconnecting means shall plainly indicate whether it is in the _____ position.
    (a) open or closed   (b) on or off   (c) up or down   (d) correct

165.   Service conductors shall be connected to the disconnecting means by _____ or other approved means.
    (a) pressure connectors   (b) clamps
    (c) solder   (d) a or b

166.   Each _____ service conductor shall have overload protection.
    (a) overhead   (b) underground   (c) ungrounded   (d) none of these

167.   Where the service overcurrent devices are locked or sealed, or not readily accessible to the _____, the branch circuit overcurrent devices shall be located in a readily accessible location.
    (a) inspector   (b) electrician   (c) occupant   (d) all of these

168.   Circuits used only for the operation of fire alarm, other protective signaling systems, or the supply to fire pump equipment shall be permitted to be connected on the _____ of the service overcurrent protection device where separately provided with overcurrent protection.
    (a) base   (b) load side   (c) supply side   (d) top

169.   As used in Section 230-95, the rating of the service disconnecting means is considered to be the rating of the largest _____ that can be installed or the highest continuous current trip setting for which the actual overcurrent protection device installed in a circuit breaker is rated or can be adjusted.
    (a) fuse   (b) circuit   (c) wire   (d) all of these

170.   • Where ground-fault protection is provided for the _____ disconnecting means and interconnection is made with another supply system by a transfer device, means or devices may be needed to ensure proper ground-fault sensing by the ground-fault protection equipment.
    (a) circuit
    (b) service
    (c) switch and fuse combination
    (d) b and c

Article 240 – Overcurrent Protection

171.   Overcurrent protection for conductors and equipment is provided to _____ the circuit if the current reaches a value that will cause an excessive or dangerous temperature in conductors or conductor insulation.
    (a) open   (b) close   (c) interrupt   (d) blow

172.   Unless specifically permitted in 240-3(e) through 240-3(g), the overcurrent protection shall not exceed _____ after any correction factors for ambient temperature and number of conductors have been applied.
    (a) 15 amperes for No. 14 copper
    (b) 20 amperes for No. 12 copper
    (c) 30 amperes for No. 10 copper
    (d) all of these

173. Flexible cords approved for use with listed appliances or portable lamps shall be considered as protected by a 20 ampere circuit breaker if they are _____ .
     (a) not more than 6 feet in length
     (b) No. 20 and larger
     (c) No. 18 and larger
     (d) No. 16 and larger

174. When can breakers or fuses be used in parallel?
     (a) Factory assembled in parallel
     (b) Listed as a unit
     (c) a and b
     (d) a or b

175. Where an orderly shutdown is required to minimize hazard(s) to personnel and equipment, a system of coordination based on two conditions shall be permitted. Those two conditions are (1) _____ short-circuit protection, and (2) _____ indication based on monitoring systems or devices.
     (a) uncoordinated, overcurrent
     (b) coordinated, overcurrent
     (c) coordinated, overload
     (d) none of these

176. A _____ shall be considered equivalent to an overcurrent trip unit.
     (a) current transformer
     (b) overcurrent relay
     (c) a and b
     (d) a or b

177. Single-pole breakers with approved handle ties can be used for the protection for each ungrounded conductor for line-to-line connected loads.
     (a) True   (b) False

178. Conductors supplying a transformer shall be permitted to be tapped without overcurrent protection at the tap where the conductors supplying the _____ of a transformer have an ampacity at least 1/3 of the rating of the overcurrent device protecting the feeder conductors.
     (a) primary   (b) secondary   (c) tertiary   (d) none of these

179. No overcurrent protection device shall be connected in series with any conductor that is intentionally grounded, except where the overcurrent protection device opens all conductors of the circuit, including the _____ conductor, and is so designed that no pole can operate independently.
     (a) ungrounded   (b) grounding   (c) grounded   (d) none of these

180. Overcurrent protection devices are not permitted to be located _____ .
     (a) where exposed to physical damage
     (b) near easily ignitable materials, such as clothes closets
     (c) in bathrooms of dwelling units
     (d) all of these

181. Plug fuses of 15 amperes and lower ratings shall be identified by a _____ configuration of the window, cap, or other prominent part to distinguish them from fuses of higher ampere ratings.
     (a) octagonal   (b) rectangular   (c) hexagonal   (d) triangular

182. Fuseholders of the Edison-base type shall be installed only where they are made to accept _____ fuses by the use of adapters.
     (a) Edison base   (b) medium base   (c) heavy-duty base   (d) Type S

183. Type S fuses, fuse holders, and adapters are required to be designed so that _____ would be difficult.
     (a) installation   (b) tampering   (c) shunting   (d) b and c

184. Fuseholders for cartridge fuses shall be so designed that it is difficult to put a fuse of any given class into a fuseholder that is designed for a _____ lower, or _____ higher, than that of the class to which the fuse belongs.
    (a) voltage, amperage
    (b) amperage, voltage
    (c) voltage, current
    (d) current, voltage

185. Cartridge fuses and fuseholders shall be classified according to _____ ranges.
    (a) voltage   (b) amperage   (c) voltage or amperage   (d) voltage and amperage

186. Markings on circuit breakers required by the *Code* shall be permitted to be made visible by removal of a _____ or cover.
    (a) plate   (b) trim   (c) cover   (d) a or b

187. Where used as switches in 120-volt and 277-volt fluorescent lighting circuits, circuit breakers shall be marked _____ .
    (a) UL   (b) SWD   (c) Amps   (d) VA

Article 250 – Grounding and Bonding

188. Conductive materials enclosing electrical conductors or equipment are grounded to limit the voltage to ground and are bonded to facilitate the operation of overcurrent protection devices under ground-fault conditions.
    (a) True   (b) False

189. • Currents that introduce noise or data errors in electronic equipment shall be considered objectionable currents.
    (a) True   (b) False

190. _____ on equipment to be grounded shall be removed from contact surfaces to ensure good electrical continuity.
    (a) Conductive coatings
    (b) Nonconductive coatings
    (c) Manufacturer's instructions
    (d) all of these

191. A grounding connection shall not be made to any grounded circuit conductor on the _____ side of the service disconnecting means except as permitted for separately derived systems or separate buildings.
    (a) supply   (b) power   (c) line   (d) load

192. Where service-entrance conductors are installed in parallel, the grounded conductor in each raceway can be sized, based on the size of the ungrounded service-entrance (phase)conductor in the raceway, but not smaller than _____ .
    (a) No. 6   (b) No. 1   (c) No. 1/0   (d) none of these

193. A grounding electrode is required if a building or structure is supplied by a feeder, or by more than one branch circuit.
    (a) True   (b) False

194. • What size copper grounding electrode conductor is required for a service that has three sets of 500 kcmil copper per phase?
    (a) No. 1   (b) No. 1/0   (c) No. 2/0   (d) No. 3/0

195. An additional electrode shall supplement a metal underground water pipe. Where the supplemental electrode is a made electrode as in Section 250-83(c) or (d), that portion of the bonding jumper that is the sole connection to the supplemental grounding electrode shall not be required to be larger than _____ wire.
    (a) No. 8   (b) No. 6   (c) No. 4   (d) No. 1

196. If none of the electrodes specified in Section 250-50 are available, then _____ can be used as the required grounding electrode.
    (a) local metal underground systems or structures (not gas)
    (b) ground rods or pipes
    (c) plate electrodes
    (d) any of these

197. • Stainless steel ground rods shall be listed and shall not be less than _____ in diameter.
(a) $1/2$ inch   (b) $3/4$ inch   (c) 1 inch   (d) $1 1/4$ inches

198. Supplementary grounding electrodes can supplement the equipment grounding conductor, but they cannot be used as the sole equipment grounding conductor.
(a) True   (b) False

199. When a building has two or more electrodes for different services, the separate electrodes must be _____ . This shall be considered as a single electrode in this sense.
(a) effectively bonded together   (b) spaced no more than 6 feet apart
(c) a and b   (d) none of these

200. The grounding electrode conductor shall be made of which of the following materials?
(a) copper   (b) aluminum   (c) copper-clad aluminum   (d) any of these

Unit 6

Conductor Sizing and Protection Calculations

OBJECTIVES

After reading this unit, the student should be able to briefly explain the following concepts:

Conductor allowable ampacity	Conductors in parallel	Overcurrent protection of equipment
Conductor ampacity	Current-carrying conductors	conductors
Conductor bundling derating factor	Equipment conductors size and	Overcurrent protection
Conductor sizing summary	protection examples	Overcurrent protection of conductors
Conductor insulation property	Minimum conductor size	Terminal ratings
Conductor size – voltage drop		

After reading this unit, the student should be able to briefly explain the following terms:

Ambient temperature	Continuous load	Parallel conductors
American Wire Gauge	Current-carrying conductors	Temperature correction factors
Ampacity	Fault current	Terminal ratings
Ampacity derating factors	Interrupting rating	Voltage drop
Conductor bundling	Overcurrent protection device	
Conductor properties	Overcurrent protection	

PART A – GENERAL CONDUCTOR REQUIREMENTS

6–1 CONDUCTOR INSULATION PROPERTY [Table 310-13]

Table 310-13 of the *NEC*® provides information on conductor properties such as permitted use, maximum operating temperature, and other insulation details (Figure 6–1).

The following abbreviations and explanations should be helpful in understanding Table 310-13 as well as Table 310-16.

-2   Conductor is permitted to be used at a continuous 90°C operation temperature

F   Fixture wire (solid or 7 strand) [Table 402-3]

FF   Flexible fixture wire (19 strands) [Table 402-3]

H   75°C insulation rating

HH   90°C insulation

N   Nylon outer cover

T   Thermoplastic insulation

W   Wet or damp

Fixture wires, see Article 402, Tables 402-3 and 402-5.

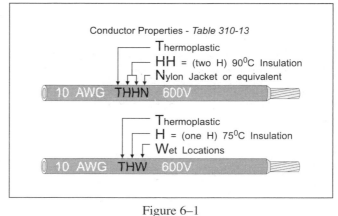

Figure 6–1

Conductor Properties

TABLE 310-13 CONDUCTOR INFORMATION					
	Column 2	**Column 3**	**Column 4**	**Column 5**	**Column 6**
Type Letter	Trade Name	Maximum Operating Temperature	Applications Provisions	Sizes Available	Outer Covering
THHN	Heat resistant thermoplastic	90°C	Dry and damp locations	14 – 1000	Nylon jacket or equivalent
THHW	Moisture- & heat-resistant thermoplastic	75°C / 90°C	Wet locations / Dry and damp locations	14 – 1000	None
THW	Moisture- & heat-resistant thermoplastic	75°C / 90°C	Dry, damp, and wet locations / Within electrical discharge lighting equipment *See Section 410–31*	14 – 2000	None
THWN	Moisture- & heat-resistant thermoplastic	75°C	Dry and wet locations	14 – 1000	Nylon jacket or equivalent
TW	Moisture-resistant thermoplastic	60°C	Dry and wet locations	14 – 2000	None
XHHW	Moisture- & heat-resistant cross-linked synthetic polymer	90°C / 75°C	Dry and damp locations / Wet locations	14 – 2000	None

❑ **Table 310-13**

TW can be described as _____ (Figure 6–2).
(a) thermoplastic insulation
(b) suitable for dry or wet locations
(c) maximum operating temperature of 60°C
d) all of the above
  • Answer: (d) all of the above

❑ **Table 402-3**

TFFN can be described as _____ .
(a) stranded fixture wire
(b) thermoplastic insulation with a nylon outer cover
(c) suitable for dry and wet locations
(d) all of the above except (c)
  • Answer: (d) all of the above except (c)

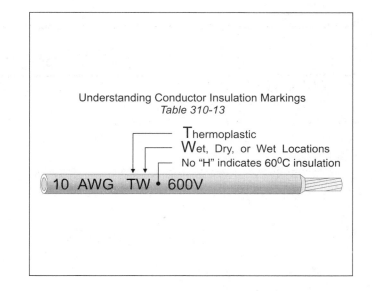

Understanding Conductor Insulation Markings
Table 310-13

Thermoplastic
Wet, Dry, or Wet Locations
No "H" indicates 60°C insulation

10 AWG TW 600V

Figure 6–2
Understanding Conductor Insulation Markings

6–2 CONDUCTOR ALLOWABLE AMPACITY [*Section 310-15*]

The ampacity of a conductor is the current the conductor can carry continuously under specific conditions of use [Article 100 definition]. The ampacity of a conductor is listed in Table 310-16 under the condition of no more than three current-carrying conductors bundled together in an ambient temperature of 86°F (Figure 6–3). The ampacity of a conductor changes if the ambient temperature is not 86°F or if more than three current-carrying conductors are bundled together for more than two feet.

Section 310-15 lists two ways of determining conductor ampacity:

(a)   General Requirements

(1)   Tables or Engineering Supervision. There are two ways to determine conductor ampacity:

• Tables 310-16 [*Section 310-15(b)*]
• Engineering formula

Note: For all practical purposes, use the ampacities listed in Table 310-16.

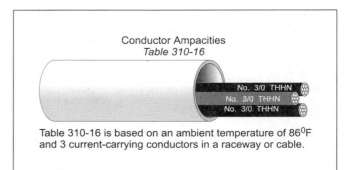

Conductor Ampacities
Table 310-16

Table 310-16 is based on an ambient temperature of 86⁰F and 3 current-carrying conductors in a raceway or cable.

Figure 6–3
Conductor Ampacities

TABLE 310–16. ALLOWABLE AMPACITIES OF INSULATED CONDUCTORS
Based On Not More Than Three Current-Carrying Conductors and Ambient Temperature of 30°C (86°F)

Size	Temperature Rating of Conductor, See Table 310-13						Size
	60°C (40°F)	75°C (167°F)	90°C (194°F)	60°C (40°F)	75°C (167°F)	90°C (194°F)	
AWG kcmil	TW	THHN THW THWN XHHW Wet Location	THHN THHW XHHW Dry Location	TW	THHN THW THWN XHHW Wet Location	THHN THHW WHHN Dry Location	AWG kcmil
	COPPER			**ALUMINUM/COPPER-CLAD ALUMINUM**			
14*	20	29	25				12
12*	25	25	30	20	20	25	10
10*	30	35	40	25	30	35	8
8	40	50	55	30	40	45	8
6	55	65	75	40	50	60	6
4	70	85	95	55	65	75	4
3	85	100	110	65	75	85	3
2	95	115	130	75	90	100	2
1	110	130	150	85	100	115	1
1/0	125	150	170	100	120	135	1/0
2/0	145	175	195	115	135	150	2/0
3/0	165	200	225	130	155	175	3/0
4/0	195	230	260	150	180	205	4/0
250	215	255	290	170	205	230	250
300	240	285	320	190	230	255	300
350	260	310	350	210	250	280	350
400	280	335	380	225	270	305	400
500	320	380	430	260	310	350	500

*See 240-3(d)

FPN: The ampacities listed in Table 310-16 are based on temperature alone and do not take voltage drop into consideration. Voltage drop considerations are for efficiency of operation and not safety; therefore, sizing conductors for voltage drop is not a *Code* requirement; see Sections 215-2(d) FPN No. 2 for more details.

Ampacity Definition. The ampacity of a conductor is the current in ampere that a conductor can carry continuously without exceeding its temperature rating [Article 100].

(b) Table Ampacity

The allowable ampacities of a conductor is listed in Table 310-16. These ampacities are based on the condition where no more than 3 current-carrying conductors are bundled together at an ambient temperature of 86°F.

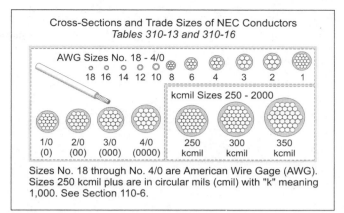

Figure 6–4
Cross-Section and Trade Sizes of *NEC®* Conductors

CAUTION: Section 240-3(d) specifies that the maximum overcurrent protection device permitted for copper No. 14 is 15 amperes, No. 12 is 20 amperes, and No. 10 is 30 amperes. The maximum overcurrent protection device permitted for aluminum No. 12 is 15 amperes and No. 10 is 25 amperes. The overcurrent protection device limitations of Section 240–3(d) does not apply to the following:

Air-conditioning conductor......................................Section 440–22
Capacitor conductors ....Article 460
Class I remote control conductors ....Article 725
Cooking equipment taps ....Article 725
Feeder taps...Section 240–21
Fixture wires and taps ....210–19(c) Exception
Motor branch circuit conductors............................Section 430–52
Motor feeder conductor ....Section 430–62
Motor control conductors...Section 430–72
Motor taps branch circuits.......................................Section 430–53(d)
Motor taps feeder conductors ....Section 430-28
Power loss hazard conductors ....Section 240–3(a)
Two-wire transformers..Section 240–3(i)
Transformer tap conductor ....Section 240–21(b)(d) and (m)
Welders...Article 630

6–3 CONDUCTOR SIZING [*SECTION 110-6*]

Conductors are sized according to the American Wire Gage (AWG) from Number 40 (No. 40) through Number 0000 (No. 4/0). The smaller the AWG size, the larger the conductor. Conductors larger than No. 4/0 are identified according to their circular mil area, such as 250,000, 300,000, 500,000. The circular mil size is often expressed in kcmil, such as 250 kcmil, 300 kcmil, 500 kcmil etc. (Figure 6–4).

Smallest Conductor Size

The smallest size conductor permitted by the *National Electrical Code®* for branch circuits, feeders, or services is No. 14 copper or No. 12 aluminum [Table 310-5]. Some local codes require a minimum No. 12 for commercial and industrial installations. Conductors smaller than No. 14 are permitted for:

Class I remote control circuits [*Section 402-11, Exception, and 725-27*]
Fixture wire [*Section 402-5 and 410-24*]
Flexible cords [*Section 400-12*]
Motor control circuits [*Section 430-72*]
Nonpower-limited fire alarm circuits [*Section 760-16*]
Power-limited fire alarm circuits [*Section 760-71(a)*].

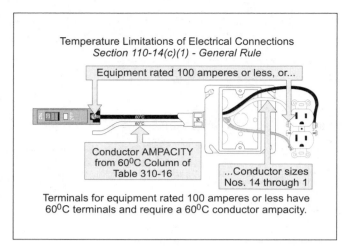

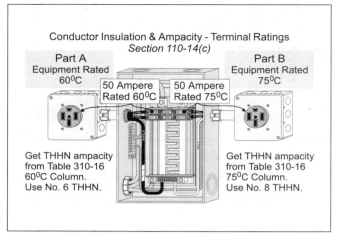

Figure 6–5
Temperature Limitations of Electrical Connections

Figure 6–6
Conductor Insulation & Ampacity – Terminal Ratings

6–4 TERMINAL RATINGS [*SECTION 110-14(c)*]

When selecting a conductor for a circuit, always select the conductors not smaller than the terminal ratings of the equipment.

Circuits Rated 100 Ampere and Less [*Section 110-14(c)(1)*]

Equipment terminals rated 100 ampere or less (and pressure connector terminals for No. 14 through No. 1 conductors), shall have the conductor sized no smaller than the 60°C temperature rating listed in Table 310-16, unless the terminals are marked otherwise (Figure 6–5).

❏ **Terminal Rated 60°C [*Section 110-14(c)(1)(b)*]**

What size THHN conductor is required for a 50 ampere circuit, listed for use at 60°C (Figure 6–6, Part A)?

(a) No. 10   (b) No. 8   (c) No. 6   (d) any of these

• Answer: (c) No. 6

Conductors must be sized to the lowest temperature rating of either the equipment or the conductor. THHN insulation can be used, but the conductor size must be selected based on the 60°C terminal rating of the equipment, not the 90°C rating of the insulation. Using the 60°C column of Table 310-16, this 50 ampere circuit requires a No. 6 THHN conductor (rated 55 ampere at 60°C).

❏ **Terminal Rated 75°C [*Section 110-14(c)(1)(c)*]**

What size THHN conductor is required for a 50 ampere circuit, listed for use at 75°C (Figure 6–6, Part B)?

(a) No. 10   (b) No. 8   (c) No. 6   (d) any of these

• Answer: (b) No. 8

Conductors must be sized according to the lowest temperature rating of either the equipment or the conductor. THHN conductors can be used, but the conductor size must be selected according to the 75°C terminal rating of the equipment, not the 90°C rating of the insulation. Using the 75°C column of Table 310-16, this installation would permit No. 8 THHN (rated 50 ampere 75°C) to supply the 50 ampere load.

Circuits over 100 Ampere [*Section 110-14(c)(2)*]

Terminals for equipment rated over 100 ampere and pressure connector terminals for conductors larger than No. 1 shall have the conductor sized according to the 75°C temperature rating listed in Table 310-16 (Figure 6–7).

❏ **Over 100 Ampere [*Section 110-14(c)(2)(b)*]**

What size THHN conductor is required to supply a 225 ampere feeder?

(a) No. 1/0   (b) No. 2/0   (c) No. 3/0   (d) No. 4/0

• Answer: (d) No. 4/0

The conductors in this example must be sized to the lowest temperature rating of either the equipment or the conductor. THHN conductors can be used, but the conductor size must be selected according to the 75°C terminal rating of the equipment.

Using the 75°C column of Table 310-16, this would require a No. 4/0 THHN (rated 230 ampere at 75°C) to supply the 225 ampere load. No. 3/0 THHN is rated 225 ampere at 90°C, but we must size the conductor to the terminal rating at 75°C.

Minimum Conductor Size Table

When sizing conductors, the following table must always be used to determine the minimum size conductor:

Terminal Size and Matching Cooper Conductor		
Terminal ampacity	**60°C Terminals Wire Size**	**75° Terminals Wire Size**
15	14	14
20	12	12
30	10	10
40	8	8
50	6	8
60	4	6
70	4	4
100	1	3
125	1/0	1
150	–	1/0
200	–	3/0
225	–	4/0
250	–	250 kcmil
300	–	350 kcmil
400	–	2 – 3/0
500	–	2 – 250 kcmil

CAUTION: When sizing conductors, we must consider conductor voltage drop, ambient temperature correction, and conductor bundle adjustment factors. These subjects are covered later in this book.

What is the purpose of THHN if we can not use its higher ampacity?

In general, 90°C rated conductor ampacities cannot be used for sizing circuit conductors. However, THHN offers the opportunity of having a greater conductor ampacity for conductor ampacity derating. The higher ampacity of THHN can permit a conductor to be used without having to increase its size because of conductor ampacity derating. Remember, the advantage of THHN is not to permit a smaller conductor, but it might prevent you from having to install a larger conductor because of ampacity adjustments (Figure 6–8).

Note: This is explained in detail in Part B of this unit.

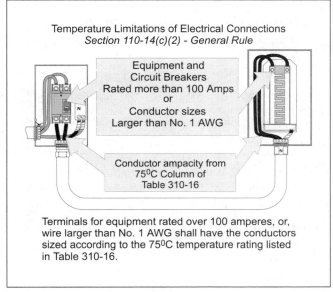

Temperature Limitations of Electrical Connections
Section 110-14(c)(2) - General Rule

Equipment and Circuit Breakers Rated more than 100 Amps or Conductor sizes Larger than No. 1 AWG

Conductor ampacity from 75⁰C Column of Table 310-16

Terminals for equipment rated over 100 amperes, or, wire larger than No. 1 AWG shall have the conductors sized according to the 75⁰C temperature rating listed in Table 310-16.

Figure 6–7
Temperature Limitations of Electrical Connections

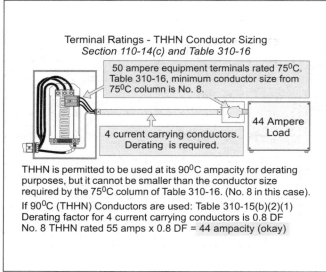

Terminal Ratings - THHN Conductor Sizing
Section 110-14(c) and Table 310-16

50 ampere equipment terminals rated 75⁰C.
Table 310-16, minimum conductor size from
75⁰C column is No. 8.

44 Ampere
Load

4 current carrying conductors.
Derating is required.

THHN is permitted to be used at its 90⁰C ampacity for derating
purposes, but it cannot be smaller than the conductor size
required by the 75⁰C column of Table 310-16. (No. 8 in this case).

If 90⁰C (THHN) Conductors are used: Table 310-15(b)(2)(1)
Derating factor for 4 current carrying conductors is 0.8 DF
No. 8 THHN rated 55 amps x 0.8 DF = 44 ampacity (okay)

Figure 6–8
Terminal Ratings – THHN Conductor Sizing

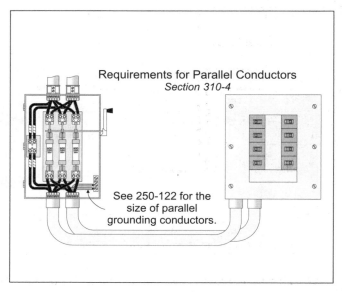

Requirements for Parallel Conductors
Section 310-4

See 250-122 for the
size of parallel
grounding conductors.

Figure 6–9
Requirements for Parallel Conductors

6–5 CONDUCTORS IN PARALLEL [*SECTION 310-4*]

Parallel conductors (electrically joined at both ends) permit a smaller cross-sectional area per ampere. This can result in significant cost saving for circuits over 300 ampere (Figure 6–9). The following table demonstrates the increase circular mil area required per ampere with larger conductors.

Conductor Size	Circular Mils Chapter 9, Table 8	Ampacity 75°C	Circular Mils Per Ampere
No. 1/0	105,600 cm	150 ampere	704 cm/per amp
No. 3/0	167,600 cm	200 ampere	838 cm/per amp
250 kcmil	250,000 cm	255 ampere	980 cm/per amp
500 kcmil	500,000 cm	380 ampere	1,316 cm/per amp
750 kcmil	750,000 cm	475 ampere	1,579 cm/per amp

❏ Sizing Parallel Conductors

What size 75°C conductors are required for a 600-ampere service that has a calculated demand load of 550 ampere (Figure 6–10)?

(a) 500 kcmil   (b) 750 kcmil   (c) 1,000 kcmil   (d) 1,250 kcmil

• Answer: (d) One 1,250 kcmil conductor, rated 590 ampere [Table 310-16]

❏ Sizing Parallel Conductors

What size conductors would be required in one raceway (nipple) for a 600-ampere service? The calculated demand load is 550 ampere (Figure 6–11)?

(a) two – 300 kcmil   (b) two – 250 kcmil   (c) two – 500 kcmil   (d) two – 750 kcmil

• Answer: (a) Two–300 kcmil conductors, each rated 285 ampere

285 ampere × 2 conductors = 570 ampere [Table 310-16]

If we parallel the conductors, we can use two 300 kcmil conductors (total 600,000) instead of one 1,250 kcmil conductor.

Grounding Conductors in Parallel [*Section 310-4*]

When equipment grounding conductors are installed with circuit conductors that are run in parallel, each raceway must have an equipment grounding conductor sized according to the overcurrent protection device rating that protects the circuit [*Section 250-122*] (Figure 6–12).

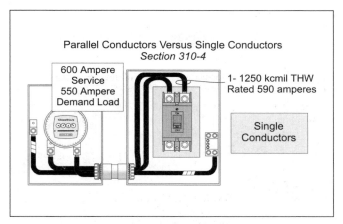

Figure 6–10
Parallel Conductors versus Single Conductors

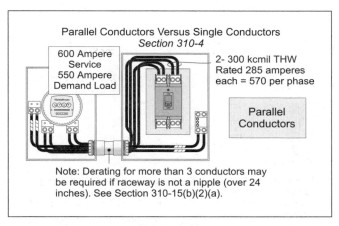

Figure 6–11
Parallel Conductors versus Single Conductors

❏ **Sizing Grounding Conductors in Parallel [*Section 250-122(f)*]**

What size equipment grounding conductor is required in each of two raceways for a 600-ampere feeder (Figure 6–13)?

(a) No. 3   (b) No. 2   (c) No. 1   (d) No. 1/0

• Answer: (c) No. 1

There must be a No. 1 equipment grounding conductor in each of the two raceways.

6–6 CONDUCTOR SIZE – VOLTAGE DROP [*SECTION 210-19(a) FPN No. 4, and 215-2(d) FPN No. 2*]

Conductor voltage drop is the result of the conductor's resistance (R) opposing the current (I) flow, $E_{vd} = R \times I$.

The *NEC*® generally does not require that voltage drop be limited on conductors, but the FPN suggest that we consider its effects. Voltage drop is covered in detail in Unit 8 of this book.

6–7 OVERCURRENT PROTECTION [Article 240]

Overcurrent protection devices are intended to open the circuit to prevent damage to persons or property due to excessive or dangerous heat. Overcurrent protection devices have two ratings, overcurrent and ampere interrupting current (AIC).

Note: Overcurrent protection devices are also used to open the circuit to clear ground-faults.

Overcurrent Rating [*Section 240-1 FPN*]. This is the actual ampere rating of the protection device, such as 15, 20, or 30 ampere. If the current flowing through the protection device exceeds the device setting for a significant period of time, the protection device will open to remove dangerous heat (Figure 6–14).

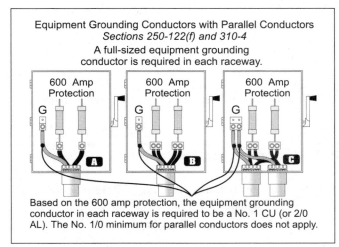

Figure 6–12
Equipment Grounding Conductors with Parallel Conductors

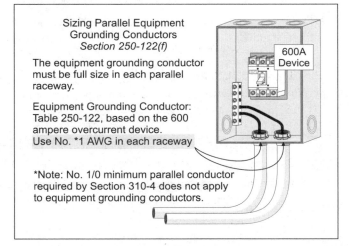

Figure 6–13
Sizing Parallel Equipment Grounding Conductors

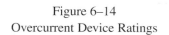

Figure 6–14
Overcurrent Device Ratings

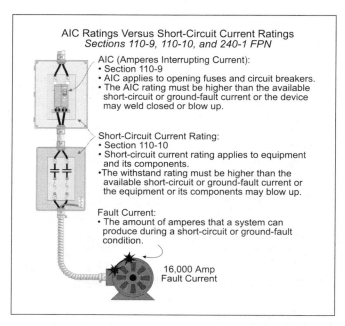

Figure 6–15
A/C Ratings versus Short-Circuit Current Ratings

Interrupting Rating (Short-circuit) [*Section 110-9*]. Overcurrent protection devices, such as circuit breakers and fuses, are intended to open the circuit at fault levels. Accordingly, they are required to have an interrupting rating sufficient for the maximum possible fault current available on the line side terminals of the equipment.

If the overcurrent protection device is not rated for the available fault current, it could explode while attempting to clear the fault, and/or the downstream equipment could suffer serious damage causing possible hazards to people. UL, ANSI, IEEE, NEMA, manufacturers, and other organizations have considerable literature on how to properly size overcurrent protection and methods for calculating available short-circuit current (Figure 6–15).

Note: The minimum interruption rating for a circuit breaker is 5,000 ampere [Section 240-83(c)] and 10,000 ampere for fuses [Section 240-60(c)] Figure 6–16).

Standard Size Protection Devices [*Section 240-6(a)*]. The *National Electrical Code*® list standard sized overcurrent protection devices: 15, 20, 25, 30, 35, 40, 45, 50, 60, 70, 80, 90, 100, 110, 125, 150, 175, 200, 225, 250, 300, 350, 400, 450, 500, 600, 700, 800, 1,000, 1,200, 1,600, 2,000, 2,500, 3,000, 4,000, 5,000, and 6,000 ampere. Exception: Additional standard ratings for fuses include : 1, 3, 6, 10, and 601 ampere.

Continuous Load. Overcurrent protection devices are sized no less than 125 percent of the continuous load, plus 100 percent of the noncontinuous load. [*Section 210-20(a)*, *215-3*, and *384-16(d)*] (Figure 6–17).

❑ **Continuous Load**
   What size protection device is required for a 100 ampere continuous load (Figure 6–18)?
   (a) 150 ampere   (b) 100 ampere   (c) 125 ampere   (d) any of these
      • Answer: (c) 125 ampere, 100 ampere × 1.25 = 125 ampere [*Section 240-6(a)*]

6–8 OVERCURRENT PROTECTION OF CONDUCTORS – GENERAL REQUIREMENTS [*SECTION 240-3*]

There are many different rules for sizing and protecting conductors and equipment. It is not simply No. 12 wire and a 20 ampere breaker. The general rule is that conductors must be protected according to their ampacity as listed in Table 310-16. Other methods of protection are permitted or required as listed in subsections (b) through (g) of this Section.

   (b) Next Higher Overcurrent Device Rating. The next higher protection device is permitted if all of the following conditions are met (Figure 6–19):

   (1) Conductors do not supply multioutlet receptacle branch circuits for portable cord- and plug-connected loads.
   (2) The ampacity of a conductor does not correspond with the standard ampere rating of a fuse or circuit breaker as listed

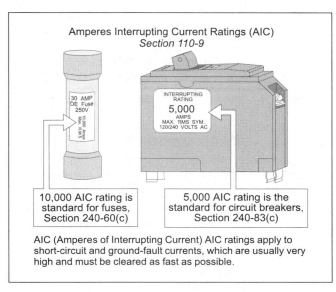

Figure 6–16
Amperes Interrupting Current Ratings (A/C)

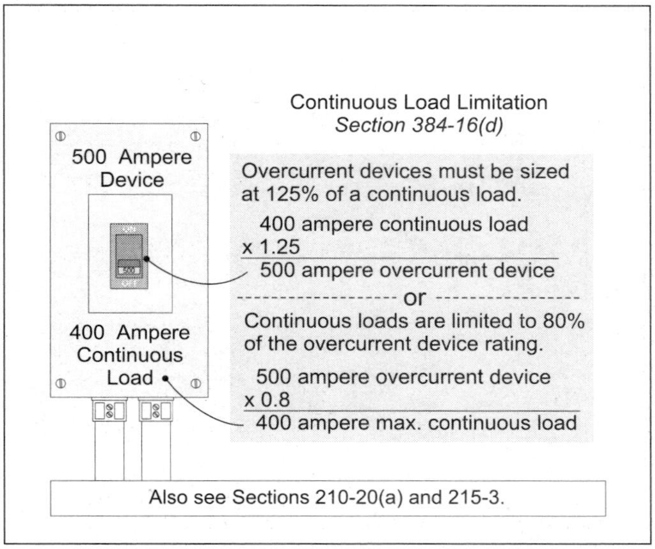

Figure 6–17
Continuous Load Limitation

in Section 240-6(a).

(3) The next size up breaker or fuse does not exceed 800 ampere.

❏ **Overcurrent Protection of Conductors**

What size conductor is required for a 104-ampere continuous load that is protected with a 150-ampere breaker (Figure 6–20)?

(a) No. 1/0   (b) No. 1   (c) No. 2   (d) any of these

   • Answer: (b) No. 1

The conductor must be sized no less than 125% of the continuous load: 104 amperes × 1.25 = 130 amperes

No. 1 THHN is rated 30 amperes at 75°C [*Section 110-14(c)(2)(b)*] and can be protected by a 150-ampere protection device.

(c) Circuits with Overcurrent Protection over 800 Ampere. If the circuit overcurrent protection device exceeds 800 ampere, the circuit conductor ampacity must not be less than the rating of the overcurrent protection device as listed in Section 240-6(a) (Figure 6–21).

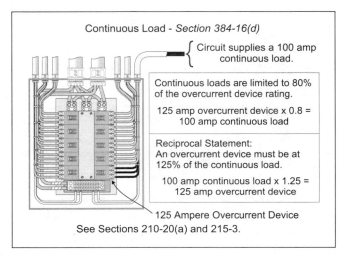

Figure 6–18
Continuous Loads

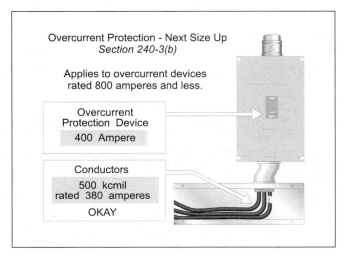

Figure 6–19
Overcurrent Protection – Next Size Up

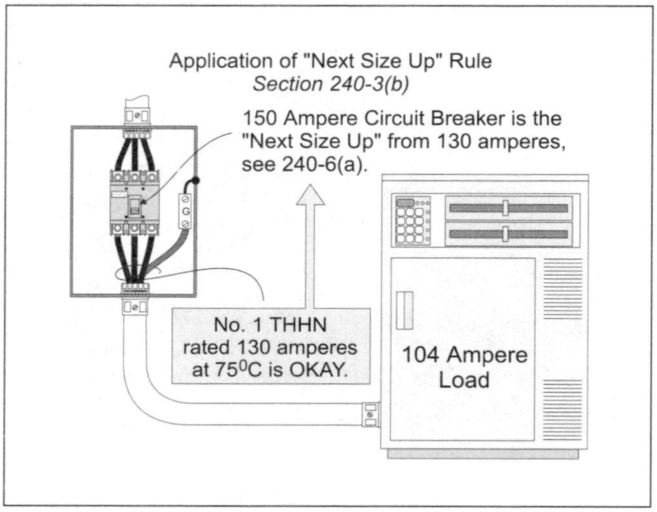

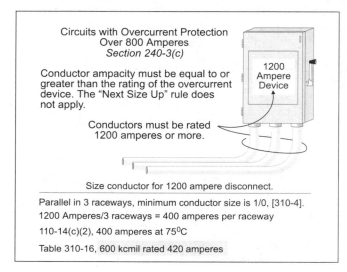

Figure 6–20
Application of "Next Size Up" Rule

Figure 6–21
Circuits with Overcurrent Protection over 800 Amperes

(d) Small Conductors. Unless specifically permitted in 240-3(e) through 240-3(g), overcurrent protection shall not exceed 15 ampere for No. 14, 20 ampere for No. 12, and 30 ampere for No. 10 copper, or 15 ampere for No. 12, and 25 ampere for No. 10 aluminum and copper-clad aluminum after ampacity correction (Figure 6–22).

6–9 OVERCURRENT PROTECTION OF CONDUCTORS – SPECIFIC REQUIREMENTS

When sizing and protecting conductors for equipment, be sure to apply the specific *NEC*® requirement.

Equipment

Air-Conditioning [*Section 440-22,* and *440-32*]

Appliances [*Section 422-10* and *422-11*]

Cooking Appliances [*Section 210-19(c), 210-21(b)(4),* Note 4 of Table 220-19]

Electric Heating Equipment [*Section 424-3(b)*]

Fire Protective Signaling Circuits [*Section 760-23*]

Motors:

Branch Circuits [*Section 430-22(a),* and *430-52*]

Feeders [*Section 430-24,* and *430-62*]

Remote Control [*Section 430-72*]

Panelboard [*Section 384-16(a)*]

Transformers [*Section 240-21* and *450-3*]

Feeders and Services

Dwelling Unit Feeders and Neutral [*Section 215-2, 310-15(b)(6)*]

Feeder Conductor [*Section 215-2* and *215-3*]

Service Conductors [*Section 230-42* and *230-90(a)*]

Temporary Conductors [*Section 305-4*]

Grounded (neutral) Conductor

Neural Calculations [*Section 220-22*]

Grounded Service Size [*Section 230-24(b)*]

Tap Conductors

Tap - Ten foot [*Section 240-21(b)(1)*]

Tap - Twenty-five foot [*Section 240-21(b)(2)*]

Tap - One hundred foot [*Section 240-21(b)(4)*]

Tap - Outside Feeder [*Section 240-21(b)(5)*]

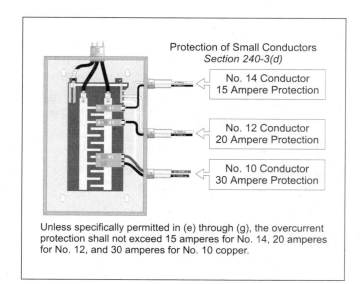

Figure 6–22
Protection of Small Conductors

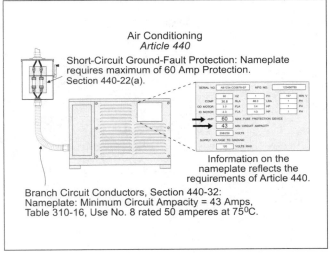

Air Conditioning
Article 440

Short-Circuit Ground-Fault Protection: Nameplate requires maximum of 60 Amp Protection. Section 440-22(a).

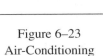

Information on the nameplate reflects the requirements of Article 440.

Branch Circuit Conductors, Section 440-32: Nameplate: Minimum Circuit Ampacity = 43 Amps, Table 310-16, Use No. 8 rated 50 amperes at 75°C.

Figure 6–23
Air-Conditioning

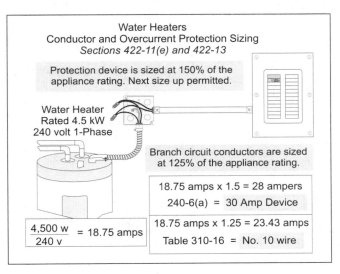

Water Heaters
Conductor and Overcurrent Protection Sizing
Sections 422-11(e) and 422-13

Protection device is sized at 150% of the appliance rating. Next size up permitted.

Water Heater
Rated 4.5 kW
240 volt 1-Phase

Branch circuit conductors are sized at 125% of the appliance rating.

18.75 amps x 1.5 = 28 ampers
240-6(a) = 30 Amp Device

$\dfrac{4{,}500\ w}{240\ v}$ = 18.75 amps

18.75 amps x 1.25 = 23.43 amps
Table 310-16 = No. 10 wire

Figure 6–24
Water Heaters

6–10 EQUIPMENT CONDUCTORS SIZE AND PROTECTION EXAMPLES

Air-Conditioning

An air-conditioner nameplate indicates the minimum circuit ampacity of 43 ampere and maximum fuse size of 60 ampere. What is the minimum size branch circuit conductor and the maximum size overcurrent protection device (Figure 6–23)?

   (a) No. 6 with a 60 ampere fuse   (b) No. 8 with a 60 ampere fuse
   (c) No. 8 with a 50 ampere fuse   (d) No. 10 with a 30 ampere fuse

     • Answer: (b) No. 8 with a 60 ampere fuse

    <u>Conductor:</u> The conductors must be sized based on the 75°C column of Table 310-16: No. 8 rated 50 amperes.

    <u>Overcurrent Protection:</u> The protection device must not be greater than a 60-ampere fuse, either one-time or dual-element.

❏ Water Heater [*Section 422-11(e) and 422-13*]

What size conductor and protection device is required for a 4,500 VA, 240 volt water heater (Figure 6–24)?

   (a) No. 10 wire with 20-ampere protection   (b) No. 10 wire with 25-ampere protection
   (c) No. 10 wire with 30-ampere protection   (d) b or c

     • Answer: (d) b or c

    $I = VA/E = 4{,}500\ VA/240\ volts = 18.75$ ampere

    <u>Conductor Size:</u> The conductor is sized at 125 percent of the water heater rating [*Section 422-13*].
    Minimum conductor = 18.75 ampere $\times$ 1.25 = 23.4 ampere
    Conductor is sized according to the 60°C column of Table 310-16 = No. 10 rated 30 ampere
    <u>Overcurrent Protection:</u> Overcurrent protection device sized no more than 150 percent of appliance rating, [*Section 422-11(e)*].
    8.75 ampere $\times$ 1.50 = 28.1 ampere, next size up = 30 ampere

❏ Motor

What size branch circuit conductor and short-circuit protection (circuit breaker) is required for a 2 horsepower (12 ampere) motor rated 230 volts (Figure 6–25)?

   (a) No. 14 with a 15-ampere breaker   (b) No. 12 with a 20-ampere breaker
   (c) No. 12 with a 30-ampere breaker   (d) No. 14 with a 30-ampere breaker

     • Answer (d) No. 14 with a 30-ampere protection device

    <u>Conductors:</u> Conductors are sized no less than 125 percent of the motor full-load current [*Section 430-6(a)* and *430-22(a)*].
    12 ampere $\times$ 1.25 = 15 ampere, Table 310-16, No. 14 is rated 20 ampere.
    <u>Overcurrent Protection:</u> The short-circuit protection (circuit breaker) is sized at 250 percent of motor full-load current.
    12 ampere $\times$ 2.5 = 30 ampere [*Section 240-6(a)* and *430-52(c)(1)* Exception No. 1]

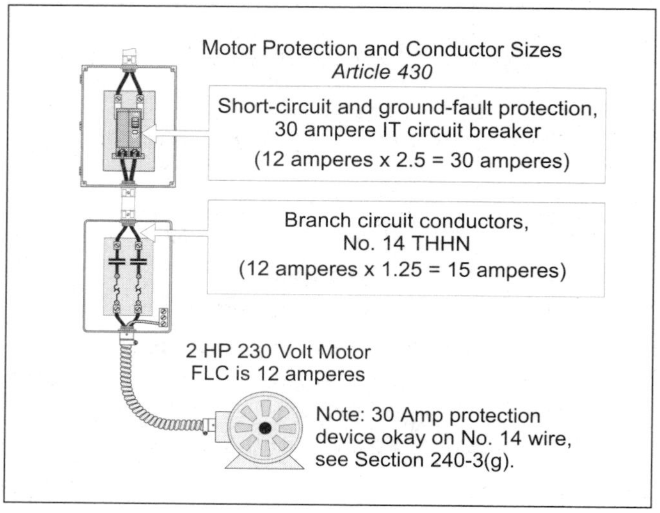

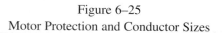

Motor Protection and Conductor Sizes
Article 430

Short-circuit and ground-fault protection,
30 ampere IT circuit breaker
(12 amperes x 2.5 = 30 amperes)

Branch circuit conductors,
No. 14 THHN
(12 amperes x 1.25 = 15 amperes)

2 HP 230 Volt Motor
FLC is 12 amperes

Note: 30 Amp protection
device okay on No. 14 wire,
see Section 240-3(g).

Figure 6–25
Motor Protection and Conductor Sizes

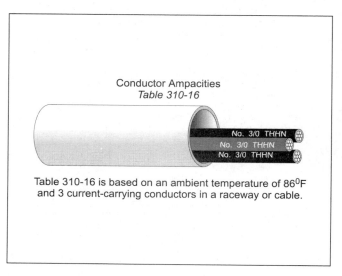

Conductor Ampacities
Table 310-16

Table 310-16 is based on an ambient temperature of 86°F
and 3 current-carrying conductors in a raceway or cable.

Figure 6–26
Conductor Ampacities

PART B – CONDUCTOR AMPACITY CALCULATIONS

6–11 CONDUCTOR AMPACITY [*SECTION 310-10*]

The insulation temperature rating of a conductor is limited to an operating temperature that prevents serious heat damage to the conductor's insulation. If the conductor carries excessive current, the I^2R heating within the conductor can destroy the conductor insulation. To limit elevated conductor operation temperatures, the current flow (ampacity) in the conductors must be limited.

Allowable Ampacities

The ampacity of a conductor is the current the conductors can carry continuously under the specific condition of use [Article 100 definition]. The ampacity of a conductor is listed in Table 310-16 under the condition of no more than three current-carrying conductors bundled together in an ambient temperature of 86°C. The ampacity of a conductor changes if the ambient temperature is not 86°F or if more than three current-carrying conductors are bundled together in any way (Figure 6–26).

6–12 AMBIENT TEMPERATURE DERATING FACTOR [Table 310-16]

The ampacity of a conductor as listed on Table 310-16 is based on the conductor operating at an ambient temperature of 86°F (30°C). When the ambient temperature is different than 86°F (30°C) for a prolonged period of time, the conductor ampacity listed in Table 310-16 must be adjusted to a new ampacity.

In general, 90°C rated conductor ampacities cannot be used for sizing circuit conductors. However, higher insulation temperature rating offers the opportunity of having a greater conductor ampacity for conductor ampacity derating and a reduced temperature correction factor. The temperature correction factors used to determine the new conductor ampacity are listed at the bottom of Table 310-16. The following formula can be used to determine the conductors new ampacity when the ambient temperature is not 86°F (30°C) (Figure 6–27):

Ampacity = Allowable Ampacity × Temperature Correction Factor

Note: Conductor ampacity adjustment does not apply if the different ambient is 10 feet or less and does not exceed 10 percent of the total length of the conductor [*Section 310-15(a)(2)* Exception].

❏ **Ambient Temperature below 86°F (30°C)**

The ampacity of No. 12 THHN when installed in a walk-in cooler that has an ambient temperature of 50°F (Figure 6–28)?

(a) 31 ampere   (b) 35 ampere   (c) 30 ampere   (d) 20 ampere

• Answer: (a) 31 ampere

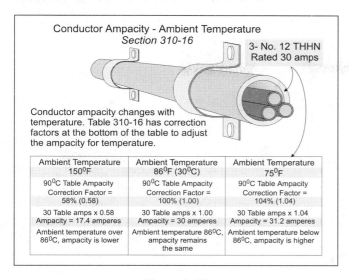

Figure 6–27
Conductor Ampacity – Ambient Temperature

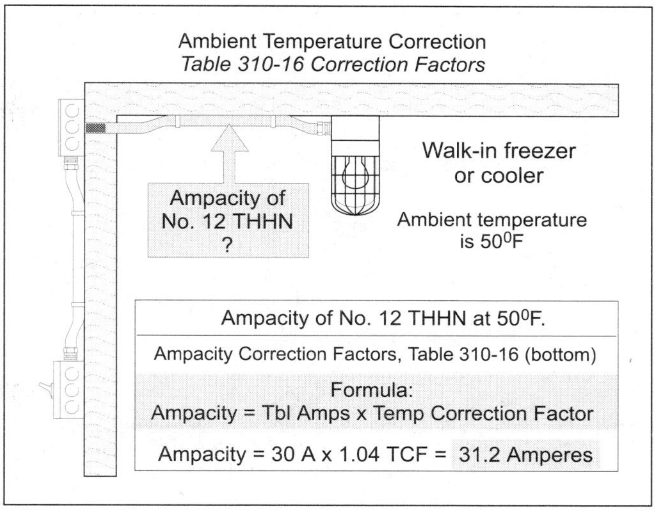

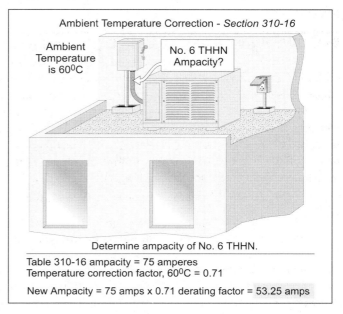

Figure 6–28
Ambient Temperature Correction

New Ampacity = Table 310-16 Ampacity × Temperature Correction Factor

Table 310-16 ampacity for No. 12 THHN is 30 ampere at 90°C.

Temperature Correction Factor for 90°C conductor installed at 50°C is 1.04.

New Ampacity = 30 ampere × 1.04 = 31.2 ampere

Note: Ampacity increases when the ambient temperature is less than 86°F (30°C).

❏ **Ambient Temperature above 86°F (30°C)**

What is the ampacity of No. 6 THHN when installed on a roof that has an ambient temperature of 60°C (Figure 6–29)?

(a) 53 ampere   (b) 35 ampere   (c) 75 ampere   (d) 60 ampere

• Answer: (a) 53 ampere

New Ampacity = Table 310-16 Ampacity × Ambient Temperature Correction Factor

Table 310-16 ampacity for No. 6 THHN is 75 ampere at 90°C.

Temperature Correction Factor for 90°C conductor rating installed at 60°C is 0.71.

New Ampacity = 75 ampere × 0.71 = 53.25 ampere

Note: Ampacity decreases when the ambient temperature is more than 86°F (30°C).

❏ **Conductor Size**

What size conductor is required to supply a 40-ampere load? The conductors pass through a room where the ambient temperature is 100°F (Figure 6–30).

(a) No. 10 THHN   (b) No. 8 THHN
(c) No. 6 THHN   (d) any of these

• Answer: (b) No. 8 THHN

The conductor to the load must have an ampacity of 40 ampere after applying the ambient temperature derating factor.

New Ampacity = Table 310-16 Ampere × Ambient Temperature Correction

No. 10 THHN = 40 ampere × 0.91 = 36.4 ampere

No. 8 THHN = 55 ampere × 0.91 = 50.0 ampere

Figure 6–29
Ambient Temperature Correction

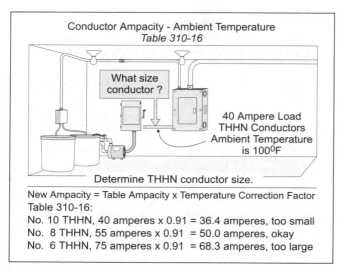

Figure 6–30
Conductor Ampacity – Ambient Temperature

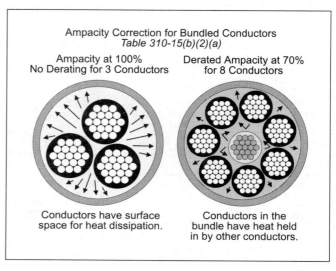

Figure 6–31
Ampacity Correction for Bundled Conductors

6–13 CONDUCTOR BUNDLING DERATING FACTOR [Table 310-15(b)(2)]

When many conductors are bundled together, the ability of the conductors to dissipate heat is reduced. The *National Electrical Code*® requires that the ampacity of a conductor be reduced whenever four or more current-carrying conductors are bundled together (Figure 6–31). In general, 90°C rated conductor ampacities cannot be used for sizing a circuit conductor. However, higher insulation temperature rating offers the opportunity of having a greater conductor ampacity for conductor ampacity derating.

The ampacity derating factors used to determine the new ampacity is listed in Table 310-15(b)(2)(a). The following formula can be used to determine the new conductor ampacity when more than three current-carrying conductors are bundled together:

New Ampacity = Table 310-16 Ampacity × Correction Factor

Note: Conductor bundle ampacity adjustment factors do not apply to conductors in a nipple that does not exceeding 24 inches, see Exception 3 to Section 310-15(b)(2)(a) (Figure 6–32).

❑ **Conductor Ampacity**

What is the ampacity of four current-carrying No. 10 THHN conductors installed in a raceway or cable (Figure 6–33)?

(a) 20 ampere   (b) 24 ampere   (c) 32 ampere   (d) none of these

• Answer: (c) 32 ampere

Ampacity = Table 310-16 Ampacity × Adjustment

Table 310-16 ampacity for No. 10 THHN is 40 ampere at 90°C.

Bundle adjustment factor for four current-carrying conductors is 0.8.

New Ampacity = 40 ampere × 0.8 = 32 ampere

❑ **Conductor Size**

A raceway contains four current-carrying conductors. What size conductor is required to supply a 40-ampere non-continuous load (Figure 6–34)?

(a) No. 10 THHN   (b) No. 8 THHN
(c) No. 6 THHN   (d) none of these

• Answer: (b) No. 8 THHN

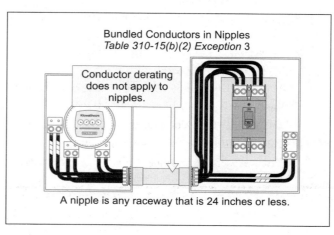

Figure 6–32
Bundled Conductors in Nipples

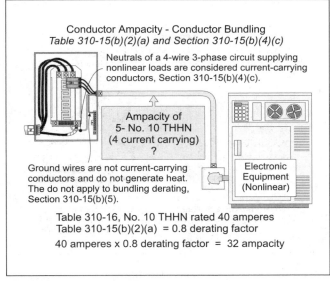

Figure 6–33
Conductor Ampacity – Conductor Bundling

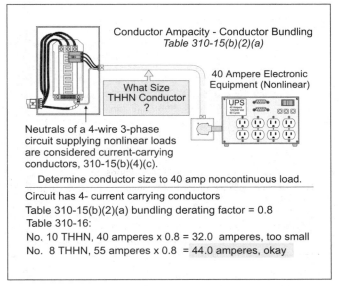

Figure 6–34
Ampacity Correction for Bundled Conductors

The conductor must have an ampacity of 40 ampere after applying bundle adjustment factor.

Ampacity = 310-16 Ampacity × Bundle Adjustment Factor

No. 10 THHN = 40 ampere × 0.8 = 32 ampere, too small

No. 8 THHN = 55 ampere × 0.8 = 44 ampere, just right

No. 6 THHN = 75 ampere × 0.8 = 60 ampere, larger than required by the *NEC*®

6–14 AMBIENT TEMPERATURE AND CONDUCTOR BUNDLING DERATING FACTORS

If the ambient temperature is different than 86°F (30°C) and there are more than three current-carrying conductors bundled together, then the ampacity listed in Table 310-16 must be adjusted for both conditions (Figure 6–35).

The following formula can be used to determine the new conductor ampacity when both ambient temperature and bundle adjustment factors apply:

Ampacity = Table 310-16 Ampacity × Temperature Correction Factor × Bundle Adjustment Factor

Note: Ampacity adjustment does not apply if the length of conductors or ambient temperature is 10 feet or less and does not exceed 10 percent of the conductor length [*Section 310-15(a)(2)* Exception].

❑ **Conductor Ampacity**

What is the ampacity of four current-carrying No. 8 THHN conductors installed in ambient temperature of 100°F (Figure 6–36)?

    (a) 25 ampere   (b) 40 ampere   (c) 55 ampere   (d) 60 ampere

      • Answer: (b) 40 ampere

New Ampacity = Table 310-16 Ampacity × Temperature Factor × Bundle Factor

Table 310-16 ampacity of No. 8 THHN is 55 ampere at 90°C.

Temperature Correction factor for 90°C conductor insulation at 100°F is 0.91.

Bundle adjustment factor for four conductors is 0.8.

New Ampacity = 55 ampere × 0.91 × 0.8 = 40 ampere

6–15 CURRENT-CARRYING CONDUCTORS

Table 310-15(b)(2)(a) adjustment factors only apply when there is more than three current-carrying conductors bundled. Naturally, all phase conductors are considered current-carrying, and the following should be helpful in determining which other conductors are considered current-carrying:

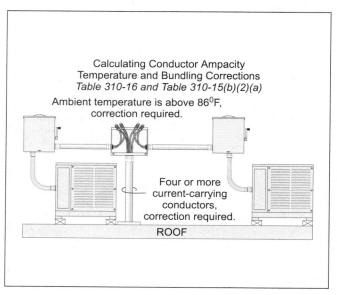

Figure 6–35
Calculating Conductor Ampacity with Temperature
and Bunching

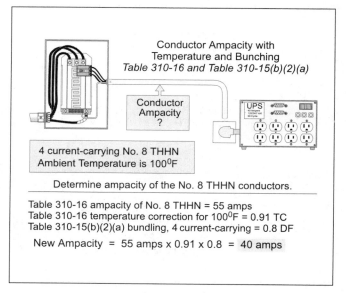

Figure 6–36
Conductor Ampacity with Temperature and Bunching

Grounded (neutral) Conductor – Balanced Circuits, Section 310-15(b)(4)(a)

The grounded (neutral) conductor of a balanced 3-wire circuit, or a balanced 4-wire wye circuit is not considered a current-carrying conductor (Figure 6–37).

Grounded (neutral) Conductor – Unbalanced 3-wire Wye Circuit, Section 310-15(b)(4)(b)

The grounded (neutral) conductor of balanced 3-wire wye circuit is considered a current-carrying conductor (Figure 6–38).

This can be proven with the following formula:

$$I_{Neutral} = \sqrt{(L_1^2 + L_2^2)\text{-}(L_1 \times L_2)} \quad L_1 = \textbf{Current of one phase} \quad L_2 = \textbf{Current of the other phase}$$

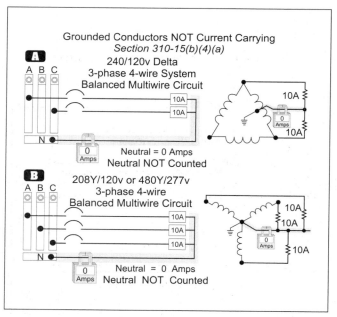

Figure 6–37
Grounded Conductors Not Current Carrying

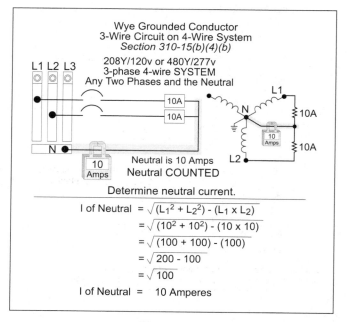

Figure 6–38
Wye Grounded Conductor

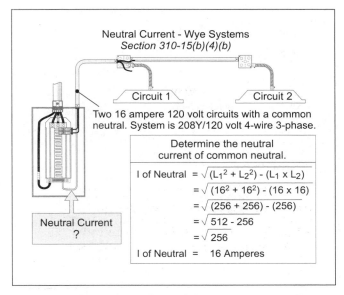

Figure 6–39
Neutral Current – Wye Systems

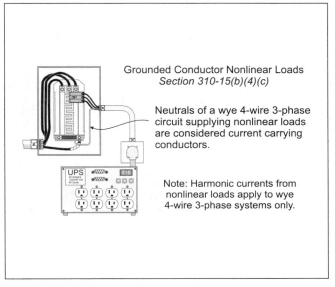

Figure 6–40
Grounded Conductor Nonlinear Loads of a
4-Wire, 3-Phase System

❏ Grounded (neutral) Conductor

What is the neutral current for a balanced 16 ampere, 3-wire, 208Y/120-volt branch circuit of a 4-wire, 3-phase wye system that supplies fluorescent lighting (Figure 6–39)?

(a) 8 ampere   (b) 16 ampere   (c) 32 ampere   (d) none of these

• Answer: (b) 16 ampere

$$I_{Neutral} = \sqrt{(L_1^2 + L_2^2) - (L_1 \times L_2)}, = \sqrt{(16^2 + 16^2) - (16 \times 16)}, = \sqrt{512 - 256} = \sqrt{256} = 16 \text{ ampere}$$

Grounded (neutral) Conductor – Nonlinear Loads, Section 310-15(b)(4)(c)

The grounded (neutral) conductor of a balanced 4-wire wye circuit that is at least 50 percent loaded with nonlinear loads (computers, electric discharge lighting, etc.) is considered a current-carrying conductor (Figure 6–40).

CAUTION: Nonlinear loads produce harmonic currents that add on the neutral conductor, and the current on the neutral can be doubled (Figure 6–41).

Two-wire Circuits

Both the grounded and ungrounded conductor of a two-wire circuit carries current and both are considered current-carrying (Figure 6–42).

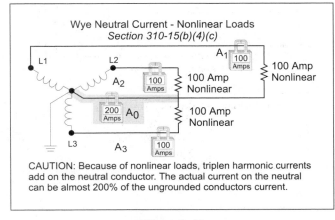

Figure 6–41
Wye-Neutral Current – Nonlinear Loads

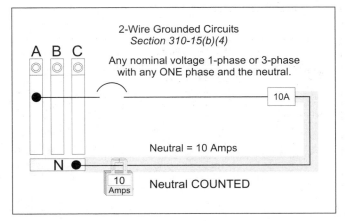

Figure 6–42
2-Wire Grounded Circuits

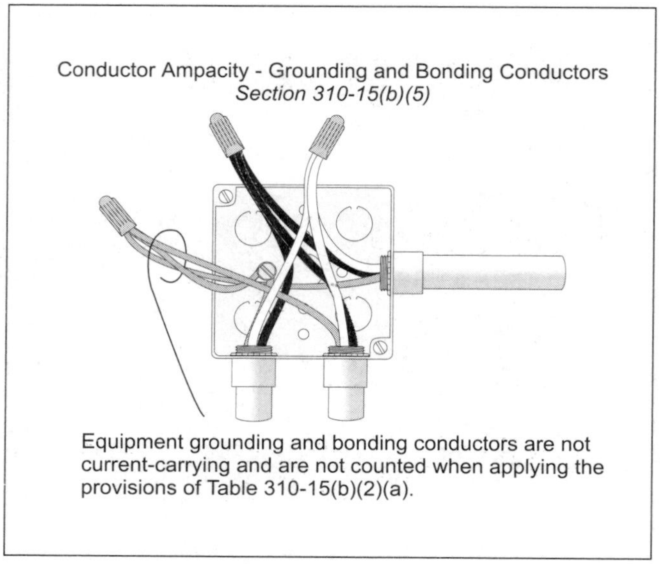

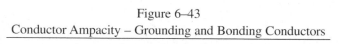

Conductor Ampacity - Grounding and Bonding Conductors
Section 310-15(b)(5)

Equipment grounding and bonding conductors are not current-carrying and are not counted when applying the provisions of Table 310-15(b)(2)(a).

Figure 6–43
Conductor Ampacity – Grounding and Bonding Conductors

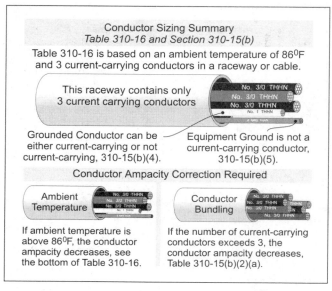

Conductor Sizing Summary
Table 310-16 and Section 310-15(b)

Table 310-16 is based on an ambient temperature of 86°F and 3 current-carrying conductors in a raceway or cable.

This raceway contains only 3 current carrying conductors

Grounded Conductor can be either current-carrying or not current-carrying, 310-15(b)(4).

Equipment Ground is not a current-carrying conductor, 310-15(b)(5).

Conductor Ampacity Correction Required

Ambient Temperature

If ambient temperature is above 86°F, the conductor ampacity decreases, see the bottom of Table 310-16.

Conductor Bundling

If the number of current-carrying conductors exceeds 3, the conductor ampacity decreases, Table 310-15(b)(2)(a).

Figure 6–44
Conductor Sizing Summary

Grounding and Bonding Conductors, Section 310-15(b)(b)

Grounding and bonding conductors do not normally carry current and are not considered current-carrying (Figure 6–43).

Note: Grounding and bonding conductors are not counted when adjusting conductor ampacity for the effects of conductor bunching.

6–16 CONDUCTOR SIZING SUMMARY

The ampacity of a conductor changes with changing conditions. The factors that affect conductor ampacity are (Figure 6–44):

1. The allowable ampacity as listed in Table 310-16.
2. The ambient temperature correction factors if the ambient temperature is not 86°F.
3. Conductor ampacity derating factors apply if three or more current-carrying conductors are bundled together.

Terminal Ratings, Section 110-14(c)

Equipment rated 100 ampere or less must have the conductor sized no smaller than the 60°C column of Table 310-16. Equipment rated over 100 ampere must have the conductors sized no smaller than for the 75°C column of Table 310-16. However, higher insulation temperature rating offers the opportunity of having a greater conductor ampacity for conductor ampacity derating.

Unit 6 – Conductor Sizing and Protection Summary Questions

6–1 Conductor Insulation Property [Table 310-13]

1. THHN can be described as _____ .
    (a) thermoplastic insulation with a nylon outer cover
    (b) suitable for dry and wet locations
    (c) having a maximum operating temperature of 90ºC
    (d) a and c

6–2 Conductor Allowable Ampacity [*Section 310-15*]

2. • The maximum overcurrent protection device size for No. 14 is 15 ampere, No. 12 is 20 ampere, and No. 10 is 30 ampere. This is a general rule, but it does not apply to motors or air-conditioners according to Section 240-3.
    (a) True   (b) False

6–3 Conductor Sizing

3. • Conductor sizes are expressed in American Wire Gauge (AWG) from No. 40 through No. 4/0. Conductors larger than _____ are express in circular mils.
    (a) No. 1/0   (b) No. 1   (c) No. 3/0   (d) No. 4/0

4. The smallest size conductor permitted for branch circuits, feeders, and services for residential, commercial, and industrial locations is _____ .
    (a) No. 14 copper   (b) No. 12 aluminum   (c) No. 12 copper   (d) a and b

6–4 Terminal Rating [*Section 110-14(c)*]

5. Equipment terminals rated 100 ampere or less (circuit breakers, fuses, etc.) and pressure connector terminals for No. 14 through No. 1 conductors shall have the conductor sized according to 60ºC temperature rating as listed in Table 310-16.
    (a) True   (b) False

6. • What is the minimum size THHN conductor that is permitted to terminate on a 70-ampere circuit breaker of fuse? Be sure to comply with the requirements of Section 110-14(c)(1).
    (a) No. 8   (b) No. 6   (c) No. 4   (d) none of these

7. • What size THHN conductor is required for a 70-ampere branch circuit if the circuit breaker and equipment is listed for 75ºC terminals and the load does not exceed 65 ampere?
    (a) No. 10   (b) No. 8   (c) No. 6   (d) No. 4

8. Terminals for equipment rated over 100 ampere and pressure connector terminals for conductors larger than No. 1 shall have the conductor sized according to 75ºC temperature rating as listed in Table 310-16.
    (a) True   (b) False

9. • What size THHN conductor is required for an air-conditioning unit if the nameplate requires a conductor ampacity of 34 ampere? Terminals of all the equipment and circuit breakers are rated 75ºC.
    (a) No. 12   (b) No. 10   (c) No. 8   (d) No. 14

10. What is the minimum size THHN conductor required for a 150-ampere circuit breaker or fuse? Be sure to comply with the requirements of Section 110-14(c)(2).
    (a) No. 1/0   (b) No. 2/0   (c) No. 3/0   (d) No. 4/0

11. In general, THHN (90°C) conductor ampacities cannot be used when sizing conductors; but when more than three current-conductors conductors are bundled together, or if the ambient temperature is greater than 86°F, the allowable conductor ampacity must be decreased. THHN offers the opportunity of having a greater ampacity for conductor derating purposes, thereby permitting the same conductor to be used without having to increase the conductor size.
    (a) True   (b) False

12. • What size conductor is required to supply a 190-ampere load in a dry location? Terminals are rated 75°C.
    (a) 300 kcmil   (b) No. 4/0   (c) No. 3/0   (d) none of these

6–5 Conductors In Parallel [*Section 310-4*]

13. • Phase and grounded (neutral) conductors sized No. 1 AWG and larger are permitted to be connected in parallel.
    (a) True   (b) False

14. To ensure that the currents are evenly distributed between the parallel conductors, each conductor within a parallel set must be installed in the same type of raceway (metallic or nonmetallic) and must be the same length, material, circular mils, insulation type, and must terminate in the same method.
    (a) True   (b) False

15. • Paralleling of conductors is done by sets.
    (a) True   (b) False

16. When an electric relay (coil) is energized, the initial current can be very high, causing significant voltage drop. The reduced voltage at the coil (because of voltage drop) can cause the coil contacts to chatter (open and close like a buzzer) or not close at all. Paralleling of control wiring conductors is often necessary and permitted by the *NEC*® to reduce the effects of voltage drop for long control runs.
    (a) True   (b) False

17. When equipment grounding conductors are installed in parallel, each raceway must have a full-size equipment grounding conductor sized according to the overcurrent protection device rating of that circuit.
    (a) True   (b) False

18. All parallel equipment grounding conductors are required to be a minimum No. 1/0.
    (a) True   (b) False

19. What size equipment grounding conductor is required in each raceway for an 800 ampere, 500 kcmil feeder parallel in two raceways?
    (a) No. 3   (b) No. 2   (c) No. 1   (d) No. 1/0

20. If we have an 800 ampere service with a calculated demand load of 750 ampere, what size 75°C conductors would be required if parallel in two raceways?
    (a) No. 4/0   (b) 250 kcmil   (c) 500 kcmil   (d) 750 kcmil

21. • What are the conductors required for a 250-ampere feeder paralleled in two raceways?
    (a) No. 3   (b) No. 2   (c) No. 2/0   (d) No. 1/0

6–6 Conductor Size – Voltage Drop [*Section 210–196) FPN No. 4, and 215–2(b)* FPN No. 2]

22. There is no mandatory rule in the *NEC*® limiting the voltage drop on conductors, but the *Code* recommends that we consider its effect.
    (a) True   (b) False

6–7 Overcurrent Protection [Article 240]

23. One of the purposes of conductor overcurrent protection is to protect the conductors against excessive or dangerous heat.
     (a) True   (b) False

24. Overcurrent devices must be designed and rated to clear fault current and must have a short-circuit interrupting rating sufficient for the available fault levels. The minimum interruption rating for circuit breakers is _____ ampere and _____ ampere for fuses.
     (a) 10,000; 10,000   (b) 5,000; 5,000   (c) 5,000; 10,000   (d) 10,000; 5,000

25. The following are the standard sized circuit breakers and fuses: _____ .
     (a) 25 ampere   (b) 90 ampere   (c) 350 ampere   (d) any of these

26. The maximum continuous load permitted on a overcurrent protection device is limited to _____ of the device rating.
     (a) 80%   (b) 100%   (c) 125%   (d) 150%

6–8 Overcurrent Protection of Conductors – General Requirements [*Section 240-3*]

27. If the ampacity of a conductor does not correspond with the standard ampere rating of a fuse or circuit breaker, the next size up protection device is permitted. This applies only if the conductors supply multioutlet receptacles for portable cord- and plug-connected loads.
     (a) True   (b) False

28. What size conductor (75°C) is required for a 70-ampere breaker that supplies a 70-ampere load?
     (a) No. 8   (b) No. 6   (c) No. 4   (d) any of these

6–11 Conductor Ampacity [*Section 310-10* and *310-15*]

29. The temperature rating of a conductor is the maximum operating temperature the conductor insulation can withstand (without serious damage) over a prolonged period of time. The _____ provide guidance for adjusting the conductor's ampacities for the different conditions.
     (a) conductor allowable ampacities   (b) ambient temperature correction factors
     (c) correction factors   (d) all of these

6–12 Ambient Temperature Derating Factor [Table 310-16]

30. The ampacities listed in Table 310-16 apply only when the ambient temperature is 40°C and there are no more than two current-carrying conductors bundled together. If the ambient temperature is not 40°C, or there are more than two current-carrying conductors in a raceway, the allowable ampacities must be adjusted to reflect the ampacity under the condition of use.
     (a) True   (b) False

31. • What is the ampacity of No. 8 THHN conductors when installed in a walk-in cooler if the ambient temperature is 50°F?
     (a) 40 ampere   (b) 50 ampere   (c) 55 ampere   (d) 57 ampere

32. • What size conductor is required to feed a 16-ampere load when the conductors are in an ambient temperature of 100°F? The circuit is protected with a 20-ampere overcurrent protection device.
     (a) No. 14 THHN   (b) No. 12 THHN   (c) No.10 THHN   (d) No. 8 THHN

6–13 Conductor Bundle Derating Factor [Table 310-15(b)(2)(a)]

33. When four or more current-carrying conductors are bundled together for more than _____ , the conductor allowable ampacity must be reduced according to the factors listed in Table 310-15(b)(2)(a).
     (a) 12 inches   (b) 24 inches   (c) 36 inches   (d) 48 inches

34. What is the ampacity of four No. 1/0 THHN conductors?
     (a) 111 ampere   (b) 136 ampere   (c) 153 ampere   (d) 171 ampere

35.   • A raceway contains eight current-carrying conductors. What size conductor is required to feed a 21 ampere noncontinuous lighting load? The overcurrent protection device is rated 30 ampere.
(a) No. 14 THHN   (b) No. 12 THHN   (c) No. 10 THHN   (d) any of these

6–14 Ambient Temperature and Conductor Bundling Derating Factors

36.   What is the ampacity of eight current-carrying No. 10 THHN conductors installed in ambient temperature of 100°F?
(a) 21 ampere   (b) 26 ampere   (c) 32 ampere   (d) 40 ampere

6–15 Current-Carrying Conductors

37.   • The neutral conductor of a balanced 3-wire, delta circuit, or 4-wire wye circuit is considered a current-carrying conductor for the purpose of applying the derating factors of Table 310-15(b)(2)(a).
(a) True   (b) False

38.   The neutral conductor of a balanced 4-wire wye circuit that is at least 50 percent loaded with nonlinear loads (electric-discharge lighting, electronic ballast, dimmers, controls, computers, laboratory test equipment, medical test equipment, recording studio equipment, etc.) is not considered a current-carrying conductor for the purpose of applying bundle adjustment factors.
(a) True   (b) False

39.   • The neutral conductor of a balanced 3-wire wye circuit is not considered a current-carrying conductor for the purpose of applying bundle adjustment factors.
(a) True   (b) False

6–16 Conductor Sizing Summary

40.   • The ampacity of a conductor can be different along the length of the conductor. The higher calculated ampacity can be used if the length of the lower ampacity is no more than 10 feet or no more than 10 percent of the length of the circuit conductors.
(a) True   (b) False

41.   Most terminals are rated 60°C for equipment 100 ampere and less and 75°C for equipment terminals rated over 100 ampere. Regardless of the conductor ampacity, conductors must be sized no smaller than the terminal temperature rating.
(a) True   (b) False

☆ Challenge Questions

6–7 Overcurrent Protection [Article 240]

42.   • A continuous load of 27 ampere requires the circuit overcurrent protection device to be sized at _____ ampere.
(a) 20   (b) 30   (c) 40   (d) 35

43.   • What size overcurrent protection device is required for a 45 ampere continuous load? The circuit is in a raceway with 14 current-carrying conductors.
(a) 45 ampere   (b) 50 ampere   (c) 60 ampere   (d) 70 ampere

44.   • A 65 ampere continuous load requires a _____ ampere overcurrent protection device.
(a) 60   (b) 70   (c) 75   (d) 90

45.   • A department store (continuous load) feeder supplies a lighting load of 103 ampere. The minimum size overcurrent protection device permitted for this feeder is _____ ampere.
(a) 110   (b) 125   (c) 150   (d) 175

6–12 Ambient Temperature Derating Factor [Table 310-16]

46.   • A No. 2 TW conductor is installed in a location where the ambient temperature is expected to be 102°F. The temperature correction factor for conductor ampacity in this location is _____ .
    (a) .96   (b) .88   (c) .82   (d) .71

47.   • If the ambient temperature is 71°C, the minimum insulation that a conductor must have and still have the capacity to carry current is _____ .
    (a) 60°C   (b) 105°C   (c) 90°C   (d) any of these

6–13 Conductor Bundling Derating Factor [Table 310-15(b)(2)(a)]

48.   • The ampacity of six current-carrying No. 4/0 XHHW aluminum conductors installed in a ground floor slab (wet location) is _____ .
    (a) 135 ampere   (b) 185 ampere   (c) 144 ampere   (d) 210 ampere

6–14 Ambient Temperature and Conductor Bundle Derating Factors

49.   • The ampacity of 15 current-carrying No. 10 RHW aluminum conductors in an ambient temperature of 75°F would be _____ .
    (a) 30 ampere   (b) 22 ampere   (c) 16 ampere   (d) 12 ampere

50.   • A No. _____ THHN conductor is required for a 19.7-ampere load if the ambient temperature is 75°F and there are nine current-carrying conductors in the raceway.
    (a) 14   (b) 12   (c) 10   (d) 8

51.   • The ampacity of the nine current-carrying No. 10 THW conductors installed in a 20 inch-long raceway is _____ .
    (a) 25 ampere   (b) 30 ampere   (c) 35 ampere   (d) none of these

52.   • The ampacity of 10 current-carrying No. 6 THHW conductors installed in an 18 inch-long raceway with an ambient temperature of 39°C is _____ .
    (a) 47 ampere   (b) 68 ampere   (c) 66 ampere   (d) 75 ampere

6–15 Current-Carrying Conductors

53.   A raceway contains the following: One 4-wire multiwire branch circuit that supplies a balanced incandescent 120-volt lighting load; one 4-wire multiwire branch circuit that supplies a balanced 120-volt fluorescent lighting load; two conductors that supply a receptacle; and there is one equipment grounding conductor. The system is 3-phase 208Y/120 volts. Taking all of these factors into consideration, how many of these conductors are considered current-carrying?
    (a) 7 conductors   (b) 8 conductors   (c) 9 conductors   (d) 11 conductors

54.   • There is a total of nine No. 10 THW conductors in a raceway. The system voltage is 3-phase, 208Y/120 volts. One conductor is an equipment grounding conductor; four conductors supply a 4-wire multiwire branch circuit for balanced electric-discharge lighting, and the remaining conductors supply a 4-wire multiwire branch circuit for balanced incandescent lighting. Taking all of these factors into consideration, how many of these conductors are considered current-carrying?
    (a) 6 conductors   (b) 7 conductors   (c) 9 conductors   (d) 10 conductors

NEC® Questions from Section 230-66 through Sections 370-16

The following *National Electrical Code*® questions are in consecutive order. Questions with • indicate that 75 percent or less of exam takers get the question correct.

Article 250 – Grounding and Bonding

55.   • A service that contains No. 12 service-entrance conductors, shall require a No. _____ grounding electrode conductor.
    (a) 6   (b) 12   (c) 8   (d) 10

56.   The connection (attachment) of the grounding electrode conductor to a grounding electrode shall _____ .
       (a) be readily accessible
       (b) be made in a manner that will ensure a permanent and effective ground
       (c) not require bonding around insulated joints of a metal piping system
       (d) none of these

57.   The connection of the grounding electrode conductor to a buried grounding electrode (driven ground rod) shall be made
       with a listed terminal device that is _____ .
       (a) suitable for direct burial   (b) accessible
       (c) readily accessible   (d) concrete proof

58.   • Metal enclosures for conductors added to existing installations of _____ , which do not provide an equipment ground,
       are not required to be grounded if they are less than 25 feet long and free from probable contact with grounded
       conductive material.
       (a) nonmetallic-sheathed cable   (b) open wiring
       (c) knob-and-tube wiring   (d) all of these

59.   An accessible point must be available for bonding of communications systems to the building grounding electrode. This
       Section requires the accessible bonding point to be _____ .
       (a) exposed metallic service raceways.
       (b) exposed grounding electrode conductor.
       (c) 6 inches of No. 6 copper conductor connected to the service equipment or raceway.
       (d) any one of these

60.   • Metal parts that serve as the grounding conductor must be _____ together to ensure electrical continuity and have the
       capacity to conduct safely any fault current likely to be imposed.
       (a) grounded   (b) effectively bonded   (c) attached   (d) any of these

61.   Where required for the reduction of electric noise for electronic equipment, electrical continuity of the metal raceway is
       not required, and the metal raceway can terminate to a(n) _____ nonmetallic fitting(s) or spacer on the electronic
       equipment.
       (a) listed   (b) labeled   (c) identified   (d) marked

62.   The bonding jumper for service raceways must be sized according to the _____ .
       (a) calculated load
       (b) service-entrance conductor size
       (c) service drop size
       (d) load to be served

63.   What is the minimum size equipment grounding conductor required for equipment connected to a 40 ampere rated
       circuit?
       (a) No. 12   (b) No. 14   (c) No. 8   (d) No. 10

64.   A building or structure, which is supplied by a feeder, must have the interior metal water piping system bonded with a
       conductor sized from _____ .
       (a) Table 250-66   (b) Table 250-122   (c) Table 310-16   (d) none of these

65.   Exposed noncurrent-carrying metal parts likely to become energized must be grounded where _____ .
       (a) within 8 feet vertically or 5 feet horizontally of ground or grounded objects
       (b) located in wet or damp locations
       (c) in electrical contact with metal
       (d) all of these

66.   Which of the following appliances installed in residential occupancies need not be grounded?
       (a) a toaster   (b) an aquarium   (c) a dishwasher   (d) a refrigerator

67. The equipment grounding conductor shall be identified by _____ .
   (a) a continuous outer green finish
   (b) being bare
   (c) a continuous outer green finish with one or more yellow stripes
   (d) any of these

68. Where a single equipment grounding conductor is used for multiple circuits in the same raceway, the single equipment grounding conductor must be sized according to _____ .
   (a) the combined rating of all the overcurrent protection devices
   (b) the largest overcurrent protection device of the multiple circuits
   (c) the combined rating of all the loads
   (d) any of these

69. Where necessary to comply with Section 250-2(d), the equipment grounding conductors shall be sized larger than specified in _____ .
   (a) Table 250-66   (b) Table 250-122   (c) Table 310-16   (d) none of these

70. When considering whether equipment is effectively grounded, the structural metal frame of a building shall be permitted to be used as the required equipment grounding conductor for ac equipment.
   (a) True   (b) False

71. • An _____ shall be used to connect the grounding terminal of a grounding-type receptacle to a grounded box.
   (a) equipment bonding jumper
   (b) equipment grounding jumper
   (c) a or b
   (d) a and b

72. Equipment bonding jumper shall be used to connect the grounding terminal of a grounding-type receptacle to a grounded box. Where the box is surface mounted, direct metal-to-metal contact between the device yoke and the box shall be permitted to ground the receptacle to the box.
   (a) True   (b) False

73. Where there are multiple equipment grounding conductors present in a receptacle outlet box, all equipment grounding conductors must be spliced and a pigtail brought out for the receptacle.
   (a) True   (b) False

74. DC systems to be grounded shall have the grounding connection made at _____ .
   (a) one supply station only
   (b) one or more supply stations
   (c) the individual services
   (d) any point on the premise wiring

75. • Cases or frames of instrument transformers are not required to be grounded _____ .
   (a) when accessible to qualified persons only
   (b) for current transformers where the primary is not over 150 volts to ground and which are used exclusively to supply current to meters
   (c) for potential transformers where the primary is less than 150 volts to ground
   (d) a or b

Article 280 – Surge Arresters

76. The conductor used to connect the surge arrester to _____ shall not be any longer than necessary.
   (a) ground   (b) bus   (c) a or b   (d) none of these

Chapter 3 – Wiring Methods and Material

Article 300 – Wiring Methods

77. Unless specified elsewhere in the *Code*, Chapter 3 shall be used for voltages of _____ .
    (a) 600 volts to ground or less   (b) 300 volts between conductors or less
    (c) 600 volts, nominal, or less   (d) 600 volts RMS

78. • Circuit conductors that operate at 277 volts (with 600 volt insulation) may occupy the same enclosure or raceway with 48-volt dc conductors that have an insulation rating of 300 volts.
    (a) True   (b) False

79. Where nonmetallic-sheathed cables pass through cut or drilled slots or holes in metal members, the cable needs to be protected by _____ securely covering all metal edges fastened in the opening prior to installation of the cable.
    (a) bushings   (b) grommets   (c) a or b   (d) a and b

80. UF cable is used to supply power for a 120-volt, 30-ampere circuit. If the cable is installed outdoors underground, the minimum cover requirement is _____ inches.
    (a) 6   (b) 12   (c) 18   (d) 24

81. • UF cable is used to supply power to a 120-volt, 15-ampere GFCI-protected circuit. If the cable is installed outdoors under the driveway of a one-family dwelling, the minimum cover requirement is _____ inches.
    (a) 12   (b) 24   (c) 16   (d) 6

82. Where underground conductors and cables emerge from underground, they shall be protected to a point _____ feet above finished grade. In no case shall the protection be required to exceed 18 inches below grade.
    (a) 3   (b) 6   (c) 8   (d) 10

83. Backfill used for underground wiring must not _____ .
    (a) damage the wiring method   (b) prevent compaction of the fill
    (c) contribute to the corrosion of the raceway   (d) all of these

84. All conductors of a circuit are required to be _____ .
    (a) in the same raceway   (b) in close proximity in the same trench
    (c) the same size   (d) a and b

85. • The provisions required for mounting conduits on indoor walls or in rooms that must be hosed down frequently is _____ between the mounting surface and the electrical equipment.
    (a) a ¹/₄-inch air space   (b) separated by insulated bushings
    (c) separated by noncombustible tubing   (d) none of these

86. • All metal raceways, cable armor, boxes, fittings, cabinets, and other metal enclosures for conductors must be _____ joined together to form a continuous electrical conductor.
    (a) electrically   (b) permanently   (c) metallically   (d) none of the above

87. Conductors in raceways must be _____ between outlet devices and there shall be no splice or tap within a raceway itself.
    (a) continuous   (b) installed   (c) copper   (d) in conduit

88. An 8″ x 8″ x 4″ deep junction/splice box would only require 6 inches of free conductor measured from the point in the box where the conductors enter the enclosure. The 3 inch outside-the-box rule _____ apply.
    (a) does   (b) does not   (c) sometimes   (d) none of these

89. A bushing can be used instead of a box where conductors emerge from a raceway and enter or terminate at equipment, such as open switchboards, unenclosed control equipment, or similar equipment.
    (a) True   (b) False

90. The *NEC*® permits prewired raceway assemblies where specifically permitted elsewhere in the *Code*.
    (a) True   (b) False

91. • A 100-foot vertical run of No. 4/0 copper requires the conductors to be supported at _____ locations.
    (a) 4   (b) 5   (c) 2   (d) none of these

92. Electrical installations in hollow spaces, vertical shafts, and ventilation or air-handling ducts shall be so made that the possible spread of fire or products of combustion will not be _____ .
    (a) substantially increased   (b) allowed
    (c) increased   (d) possible

93. Wiring methods permitted in ducts or plenums used for environmental air are _____ .
    (a) flexible metal conduit of any length
    (b) electrical metallic tubing
    (c) armor cable (BX)
    (d) nonmetallic sheath cable

94. The space above a dropped ceiling used for environmental air is considered as _____ and the wiring limitations of Section _____ apply.
    (a) a plenum, 300-22(b)
    (b) other spaces, 300-22(c)
    (c) a duct, 300-22(b)
    (d) none of these

95. Wiring methods and equipment installed behind panels designed to permit access (such as drop ceilings) shall be so arranged and secured to permit the removal of panels to give access to electrical equipment.
    (a) True   (b) False

Article 305 – Temporary Wiring

96. Temporary wiring must be _____ immediately upon the completion of the purpose for which it was installed.
    (a) disconnected   (b) removed   (c) deenergized   (d) any of these

97. Type NM and NMC cables can be used for temporary wiring in structures of a height of _____ .
    (a) 18 feet   (b) 6 feet   (c) 12 feet   (d) any of these

98. Multiwire branch circuits for temporary wiring shall be provided with a means to disconnect simultaneously all _____ conductors at the power outlet or panelboard where the branch circuit originated.
    (a) underground   (b) overhead   (c) ungrounded   (d) grounded

99. Cable assemblies, as well as flexible cords and cables, shall be supported at intervals that ensure protection from physical damage. Support shall be in the form of _____ or similar fittings designed as not to damage the cable or cord assembly.
    (a) staples   (b) cable ties   (c) straps   (d) all of these

100. All 125-volt, single-phase, 15, 20, and 30 ampere receptacle outlets in use by personnel for temporary power shall have ground-fault circuit-interrupter protection for personnel, except receptacles on a two-wire, single-phase portable or vehicle-mounted generator rated not more than _____ kW, where the circuit conductors of the generator are insulated from the generator frame and all other grounded surfaces.
    (a) 3   (b) 4   (c) 5   (d) 6

Article 310 – Conductors for General Wiring

101. Where installed in raceways, conductors No. _____ and larger shall be stranded.
    (a) 10   (b) 6   (c) 8   (d) 4

102. In general, the minimum size phase, neutral, or grounded conductor permitted for use in parallel is No. _____ .
    (a) 10   (b) 1   (c) 1/0   (d) 4

103. The minimum size conductor permitted in a commercial building for branch circuits under 600 volts is No. _____ .
    (a) 14   (b) 12   (c) 10   (d) 8

104.   • Where conductors of different insulation are associated together, the limiting temperature of any conductor shall not be exceeded.
(a) True   (b) False

105.   Suffixes to designate the number of conductors within a cable are _____ .
(a) D – Two insulated conductors laid parallel
(b) M – Two or more insulated conductors twisted spirally
(c) T – Two or more insulated conductors twisted in parallel
(d) a and b

106.   TFE-insulated conductors are manufactured in sizes from No. 14 through No. _____ .
(a) 2   (b) 1   (c) 2/0   (d) 4/0

107.   Where six current-carrying conductors are run in the same conduit or cable, the ampacity of each conductor shall be adjusted by a factor of _____ percent.
(a) 90   (b) 60   (c) 40   (d) 80

108.   On a 4-wire, 3-phase wye circuit, where the major portion of the load consists of electrical discharge lighting, data processing equipment, or other harmonic current inducting loads, the neutral conductor shall be counted when applying Section 310-15(b)(2) adjustment factors.
(a) True   (b) False

109.   In designing circuits, the current-carrying capacity of conductors should be corrected for heat at room temperatures above _____°F.
(a) 30   (b) 86   (c) 94   (d) 75

110.   Thermal resistively is the reciprocal of thermal conductivity; it is designated Rho and expressed in units of _____ .
(a) °F-cm/volt   (b) °F-cm/watt   (c) °C-cm/volt   (d) °C-cm/watt

111.   Table 310-71 provides ampacities of an insulated three-conductor copper cable isolated in air, based on conductor temperature of 90°C (194°F) and ambient air temperature of 40°C (104°F). If the conductor size is No. 4/0 AWG, MV-105, and the voltage range is 2,001 to 5,000, then the ampacity is _____ amperes.
(a) 250   (b) 285   (c) 320   (d) 325

Article 318 through 342

112.   Cable tray systems shall not be used _____ .
(a) in hoistways
(b) where subject to severe physical damage
(c) in hazardous locations
(d) a and b

113.   Each run of cable tray shall be _____ before the installation of cables.
(a) inspected   (b) tested   (c) completed   (d) all of these

114.   Steel cable trays shall not be used as equipment grounding conductors for circuits with ground-fault protection above _____ amperes.
(a) 200   (b) 300   (c) 600   (d) 800

115.   Cable _____ made and insulated by approved methods shall be permitted to be located within a cable tray provided they are accessible and do not project above the side rails.
(a) connections   (b) jumpers   (c) splices   (d) conductors

116.   Where single conductor cables comprising each phase or neutral of a circuit are connected in parallel in a cable tray, the conductors shall be installed _____ to prevent current unbalance in the paralleled conductors due to inductive reactance.
(a) in groups consisting of not more than three conductors per phase or neutral
(b) in groups consisting of not more than one conductor per phase or neutral
(c) as individual conductors securely bound to the cable tray
(d) in separate groups

117. Open wiring on insulators is a(n) _____ wiring method using cleats, knobs, tubes, and flexible tubing for the protection and support of single insulated conductors run in or on buildings, and not concealed by the building structure.
    (a) temporary   (b) acceptable   (c) enclosed   (d) exposed

118. Where screws are used to mount knobs for the support of open wiring on insulators, they shall be of a length sufficient to penetrate the wood to a depth equal to at least _____ the height of the knob and the full thickness of the cleat.
    (a) one-eighth   (b) one-quarter   (c) one-third   (d) one-half

119. Open conductors shall be separated at least _____ inches from metal raceways, piping, or other conducting material, and from any exposed lighting, power, or signaling conductor, or shall be separated by a continuous and firmly fixed nonconductor in addition to the insulation of the conductor.
    (a) 2   (b) $2^1/_2$ (c) 3   (d) $3^1/_2$

120. Where open conductors cross ceiling joists and wall studs and are exposed to physical damage, they shall be protected by a substantial running board. Running boards shall extend at least _____ inch outside the conductors, but not more than _____ inches, and the protecting sides shall be at least 2 inches high and at least 1 inch nominal in thickness.
    (a) $^1/_2$, 1   (b) $^1/_2$, 2   (c) 1, $2^1/_2$ (d) 1, 2

121. The messenger shall be supported at dead ends and at intermediate locations so as to eliminate _____ on the conductors.
    (a) static   (b) magnetism   (c) tension   (d) induction

122. Concealed knob-and-tube wiring shall be permitted to be used in commercial garages, theaters and similar locations, motion picture studios, hazardous (classified) locations, or in the hollow spaces of walls, ceilings, and attics where such spaces are insulated by loose, rolled, or foamed-in-place insulating material that envelops the conductors.
    (a) True   (b) False

123. Where knob-and-tube conductors pass through wood cross members in plastered partitions, conductors shall be protected by noncombustible, nonabsorbent, insulating tubes extending not less then _____ inches beyond the wood member.
    (a) 2   (b) 3   (c) 4   (d) 6

124. Type IGS cable is a factory assembly of one or more conductors, each individually insulated and enclosed in a loose fit nonmetallic flexible conduit as an integrated gas spacer cable rated _____ volts.
    (a) 0 – 6,000   (b) 600 – 6,000   (c) 0 – 600   (d) 3,000 – 6,000

125. Type FCC cable consists of _____ conductors.
    (a) three or more square   (b) two or more round
    (c) three or more flat   (d) two or more flat

126. Use of FCC systems in damp locations shall be _____ .
    (a) restricted
    (b) permitted
    (c) permitted provided the system is encased in concrete
    (d) approved by special permission

127. General use branch circuits using flat conductor cable shall not exceed _____ amperes.
    (a) 15   (b) 20   (c) 30   (d) 40

128. All bare FCC cable end fittings shall be _____ .
    (a) sealed   (b) insulated   (c) listed   (d) all of these

129. Receptacles, receptacle housings, and self-contained devices used with flat conductor cable systems shall be _____ .
    (a) rated a minimum of 20 amperes
    (b) rated a minimum of 15 amperes
    (c) identified for this use
    (d) none of these

130.   Each FCC transition assembly shall incorporate means for _____ .
       (a) facilitating the entry of the Type FCC cable into the assembly
       (b) connecting the Type FCC cable to grounded conductors
       (c) electrically connecting the assembly to the metal cable shields and grounding conductors
       (d) all of these

131.   The radius of the inner edge of any bend shall not be less than _____ times the MI cable diameter for a cable that has a diameter of over $^3/_4$ inch.
       (a) 6   (b) 3   (c) 8   (d) 10

132.   Where MI cable terminates, a _____ shall be provided immediately after stripping to prevent the entrance of moisture into the insulation.
       (a) bushing   (b) connector   (c) fitting   (d) seal

133.   The conductor insulation in Type MI cable shall be a highly compressed refractory mineral that will provide proper _____ for the conductors.
       (a) covering   (b) spacing   (c) resistance   (d) none of these

134.   Electrical nonmetallic tubing is composed of a material that is resistant to moisture, chemical atmospheres, and is _____ .
       (a) flame-resistant (b) flame-retardant (c) fireproof   (d) nonflammable

135.   Electrical nonmetallic tubing and fittings shall be permitted to be _____ in a concrete slab on grade where the tubing is placed upon sand or approved screenings.
       (a) encased   (b) embedded   (c) either a or b   (d) both a and b

136.   Electrical nonmetallic tubing is not permitted in hazardous (classified) locations.
       (a) True   (b) False

137.   All cut ends of electrical nonmetallic tubing shall be trimmed inside and _____ to remove rough edges.
       (a) outside   (b) outside, smoothing all surfaces,
       (c) the surface smoothed   (d) none of these

138.   Bushings or adapters are required at electrical nonmetallic tubing terminations to protect the conductors from abrasion.
       (a) True   (b) False

139.   AC cable shall be supported at intervals not exceeding $4^1/_2$ feet and the cable shall be secured within ___ inches of cable termination.
       (a) 4   (b) 8   (c) 9   (d) 12

140.   Although exposed Type AC cable is required to closely follow the surface of the building, it can be used on the underside of floor joists in basements where supported at each joist where not subject to damage.
       (a) True   (b) False

141.   Where armor cable is run on the side of rafters, studs, or floor joists in an accessible attic, protection is required for the cable with running boards.
       (a) True   (b) False

142.   Type MC cable shall be supported and secured at intervals not exceeding _____ feet.
       (a) 3   (b) 6   (c) 4   (d) 2

143.   The metallic sheath of metal-clad cable shall be continuous and _____ .
       (a) flame-retardant (b) weatherproof   (c) close fitting   (d) all of these

144.   Type NM, Type NMC cables shall be permitted to be used in one- and two-family dwellings of a height of 24 feet.
       (a) True   (b) False

145.   The definition of a first floor, as it applies to nonmetallic-sheath cable, is the level where _____ percent or more of the outside wall area is at or above finish grade.
       (a) 25   (b) 50   (c) 75   (d) 100

146. Where NMC cable is run at angles with joists in unfinished basements it shall be permissible to secure cables not smaller than _____ conductors directly to the lower edges of the joist.

    (c) 2- No. 6   (b) 3- No. 8   (c) 3- No. 10   (d) all of these

147. Nonmetallic-sheathed cables run through framing members are considered to be adequately supported.

    (a) True   (b) False

148. The ampacity of nonmetallic-sheath cable shall be that of 60°C as listed in Table 310-16. However the 90°C ampacity listed in Table 310-16 can be used for ampacity adjustment purposes, provided the final adjusted ampacity does not exceed that of a _____ rated conductor.

    (a) 120°C   (b) 60°C   (c) 90°C   (d) none of these

149. _____ is a type of multiconductor cable permitted for use as an underground service entrance cable.

    (a) SE   (b) NMC   (c) UF   (d) USE

150. SE cable installed outdoor for branch circuits and feeder shall be installed in accordance with the requirements of Article 225, the cable shall be secured within _____ inches of termination and $4^1/_2$ feet, and when installed underground the cable shall comply with the underground requirements in Article 339 - Underground Feeder Cable.

    (a) 6   (b) 8   (c) 12   (d) 24

151. Type UF cable is permitted for _____ .

    (a) interior use   (b) wet locations   (c) direct burial   (d) all of these

152. Type UF cable shall not be used where subject to physical damage. When this cable is subject to physical damage, it shall be protected by a suitable method such as a raceway.

    (a) True   (b) False

153. The classification of nonmetallic extensions includes _____ .

    (a) surface extensions intended for mounting directly on the surface of walls
    (b) surface extensions intended for mounting directly on the surface of ceilings
    (c) aerial cable containing a supporting messenger cable as part of the assembly
    (d) all of these

154. Receptacle-type tap connectors for nonmetallic extensions shall be of the _____ .

    (a) protected-type   (b) grounding-type   (c) locking-type   (d) all of these

155. Nonmetallic extensions shall be secured in place by approved means at intervals not exceeding _____ inches.

    (a) 6   (b) 8   (c) 10   (d) 16

Article 343 through Article 351

156. The use of nonmetallic conduit with conductors shall be permitted _____ .

    (a) for direct burial underground installation   (b) encased or embedded in concrete
    (c) in cinder fill   (d) all of these

157. For _____ , nonmetallic underground conduit with conductors shall be trimmed away from the conductors or cables using an approved method that will not damage the conductor or cable insulation or jacket.

    (a) convenience   (b) appearances   (c) termination   (d) safety to persons

158. Materials such as straps, bolts, screws, etc., associated with the installation of intermediate metal conduit in concrete, in direct contact with the earth are required to be _____ .

    (a) weatherproof   (b) weathertight   (c) corrosion resistant   (d) none of these

159. Threadless couplings approved for use with intermediate metal conduit in wet locations shall be of the _____ type.

    (a) rainproof   (b) raintight   (c) moistureproof   (d) concrete-tight

160. Intermediate metal conduit shall be firmly fastened within _____ of each outlet box, junction box, device box, fitting, cabinet, or other conduit termination.

    (a) 12 inches   (b) 18 inches   (c) 2 feet   (d) 3 feet

161. When IMC is installed through bored or punched holes in framing members, additional support requirements are not necessary. This applies to both wood and metal framing members.
    (a) True   (b) False

162. Rigid metal conduit shall be permitted to be installed in concrete, in direct contact with the earth, or in areas subject to severe corrosive influences where protected by _____ and judged suitable for the condition.
    (a) ceramic   (b) corrosion protection   (c) PVC   (d) orangeburg

163. Where threadless couplings and connectors used with rigid metal conduit are buried in masonry or concrete, they shall be of the _____ type.
    (a) raintight   (b) wet and damp location   (c) nonabsorbent   (d) concrete-tight

164. The minimum radius for a bend of 1 inch rigid conduit with three No. 10 TW conductors is _____ inches, if a one-shot bender is used.
    (a) 13   (b) 11   (c) 5   (d) 9

165. Straight runs of 1 inch rigid metal conduit using threaded couplings may be secured at intervals not exceeding _____ feet.
    (a) 5   (b) 10   (c) 12   (d) 14

166. Rigid nonmetallic conduit and fittings used above ground shall be resistant to _____ .
    (a) moisture and chemical atmospheres
    (b) low temperatures and sunlight
    (c) distortion from heat
    (d) all of these

167. Rigid nonmetallic conduit can be installed exposed in buildings _____ .
    (a) three floors   (b) twelve floors   (c) six floors   (d) of any height

168. An equipment grounding conductor is not required in the conduit if the grounded (neutral) conductor is used to ground equipment as permitted in Section 250-142.
    (a) True   (b) False

169. 1 inch rigid nonmetallic conduit must be supported every _____ feet.
    (a) 2   (b) 3   (c) 4   (d) 6

170. The minimum and maximum size of EMT is _____ inches, except for special installations.
    (a) $^5/_{16}$ to 3 (b) $^3/_8$ to 4   (c) $^1/_2$ to 3   (d) $1^1/_2$ to 4

171. A run of electrical metallic tubing between outlet boxes shall not exceed _____ offsets close to the box.
    (a) 360° plus   (b) 360° total including
    (c) four quarter bends plus   (d) 180° total including

172. • Flexible metallic tubing shall be permitted to be used _____ .
    (a) in dry and damp locations   (b) for direct burial
    (c) in lengths over six feet   (d) for a maximum of 1,000 volts

173. Where flexible metal conduit is installed in a wet location, which of the following conductor types are required to be used?
    (a) THWN   (b) XHHW   (c) THW   (d) any of these

174. The maximum run length of $^3/_8$ inch flexible metal conduit for any circuit is _____ feet.
    (a) 3   (b) 8   (c) 12   (d) no limit

175. • $^3/_8$ inch flexible metal conduit used for the supply to recessed lighting fixtures must have the raceway secured every _____ inches.
    (a) 18
    (b) 36
    (c) 48
    (d) is not required to be secured

176. The use of listed and marked liquidtight flexible metal conduit shall be permitted for _____ .
   (a) direct burial where listed and marked for the purpose
   (b) exposed work
   (c) concealed work
   (d) all of these

177. Liquidtight flexible conduit where used as a fixed raceway, must be secured within _____ inch(es) on each side of the box and at intervals not exceeding _____ feet.
   (a) 12, 4$^1/_2$ (b) 18, 3 (c) 12, 3 (d) 18, 4

178. • Where liquidtight flexible conduit is used to connect to equipment and flexibility where required, a separate _____ conductor must be installed.
   (a) bond jumper (b) bonding
   (c) equipment grounding conductor (d) none of these

Article 352 through Article 370

179. In general, the voltage limitation between conductors in a surface metal raceway shall not exceed _____ volts unless the metal has a thickness listed for higher voltage.
   (a) 300 (b) 150 (c) 600 (d) 1,000

180. It is permissible to run unbroken lengths of surface metal raceways through dry _____ .
   (a) walls (b) partitions (c) floors (d) all of these

181. Surface metal raceways and their fittings shall be so designed that the sections can be _____ .
   (a) electrically coupled together
   (b) mechanically coupled together
   (c) installed without subjecting the wires to abrasion
   (d) all of these

182. Underfloor raceways spaced less than 1 inch apart shall be covered with concrete to a depth of _____ inch(es).
   (a) $^1/_2$ (b) 4 (c) 1 $^1/_2$ (d) 2

183. When an outlet from an underfloor raceway is discontinued, the circuit conductors supplying the outlet _____ .
   (a) may be spliced
   (b) may be reinsulated
   (c) may be handled like abandoned outlets on loop wiring
   (d) shall be removed from the raceway

184. A _____ is defined as a single, enclosed tubular space in a cellular metal floor member.
   (a) cellular metal floor raceway
   (b) cell
   (c) header
   (d) none of these

185. A junction box used with a cellular metal floor raceway shall be _____ .
   (a) level with the floor grade
   (b) sealed against the entrance of water or concrete
   (c) metal and electrically continuous
   (d) all of these

186. In cellular concrete floor raceways, a "cell" shall be defined as a single, enclosed _____ space in a floor made of precast cellular concrete slabs, the direction of the cell being parallel to the direction of the floor member.
   (a) circular (b) oval (c) tubular (d) hexagonal

187. Connections from cellular concrete floor raceway headers to cabinets shall be made by means of _____ .
   (a) listed metal raceways (b) PVC raceways
   (c) listed fittings (d) a and c

188.   When an outlet is _____ from a cellular concrete floor raceway, the sections of circuit conductors supplying the outlet shall be removed from the raceway.
(a) discontinued   (b) abandoned   (c) removed   (d) all of these

189.   Conductors larger than that for which the wireway is designed shall be permitted to be installed in any wireway.
(a) True   (b) False

190.   Wireways where run horizontally shall be supported at each end and the distance between supports shall not exceed _____ feet.
(a) 5   (b) 10   (c) 3   (d) 6

191.   Flat cable assemblies shall be permitted only as branch circuits to supply suitable tap devices for _____ loads.
(a) lighting   (b) small power   (c) small appliance   (d) all of these

192.   Flat cable assembly shall consist of _____ conductors.
(a) 2   (b) 3   (c) 4   (d) any of these

193.   All extensions from flat cable assemblies shall be made by approved wiring methods, within the _____ , installed at either end of the flat cable assembly runs.
(a) end-caps   (b) junction boxes
(c) surface metal raceway   (d) underfloor metal raceway

194.   Busways shall be securely supported, unless otherwise designed, and marked at intervals not to exceed _____ feet.
(a) 10   (b) 5   (c) 3   (d) 8

195.   • Where devices or plug-in connections for tapping off feeder or branch circuits from busways consist of an externally operable fusible switch, _____ shall be provided for operation of the disconnecting means from the floor.
(a) ropes   (b) chains   (c) hook sticks   (d) any of these

196.   Cablebus is ordinarily assembled at the _____ .
(a) point of installation   (b) manufacturer's location
(c) distributor's location   (d) none of these

197.   Cablebus framework, where _____ , shall be permitted as the equipment grounding conductor for branch circuits and feeders.
(a) bonded   (b) welded   (c) protected   (d) galvanized

198.   • A cablebus system shall include approved fittings for _____ .
(a) bends   (b) U-turns   (c) dead ends   (d) all of these

199.   • Nonmetallic boxes are permitted for use with _____ .
(a) flexible nonmetallic conduit
(b) liquidtight nonmetallic conduit
(c) nonmetallic cables and raceways
(d) all of these

200.   According to the *NEC®*, the volume of a 3 inch x 2 inch x 2 inch device box is _____ cubic inches.
(a) 12   (b) 14   (c) 10   (d) 8

Unit 7

Motor Calculations

OBJECTIVES

After reading this unit, the student should be able to briefly explain the following concepts:

Branch circuit short-circuit ground fault protection	Highest rated motor	Motor calculation steps
Feeder protection	Motor overcurrent protection	Overload protection
Feeder conductor size	Motor branch circuit conductors	
	Motor VA calculations	

After reading this unit, the student should be able to briefly explain the following terms:

Dual-element fuse	Motor nameplate current rating	Overload
Ground-fault	Next smaller device size	Service factor (SF)
Inverse time breaker	One-time fuse	Short-circuit
Heaters	Overcurrent protection	Temperature rise
Motor full-load current	Overcurrent	

INTRODUCTION

When sizing conductors and overcurrent protection for motors, we must comply with the requirements of the *National Electrical Code®*, specifically Article 430 [*Section 240-3(g)*].

7–1 MOTOR BRANCH CIRCUIT CONDUCTORS [*SECTION 430-22(a)*]

Branch circuit conductors to a single motor must have an ampacity of not less than 125 percent of the motor's full-load current as listed in Tables 430-147 through 430-150 [*Section 430-6(a)*]. The actual conductor size must be selected from Table 310-16 according to the terminal temperature rating (60° or 75°C) of the equipment [*Section 110-14(c)*] (Figure 7–1).

❏ **Motor Branch Circuit Conductors**

What size THHN conductor is required for a 2 horsepower 230-volt, single-phase motor (Figure 7–2)?

(a) No. 14   (b) No. 12

(c) No. 10   (d) No. 8

  • Answer: (a) No. 14

Motor Full-Load Current – Table 430-148:

2 horsepower 230 volts single-phase = 12 ampere

Conductor Sized no more than 125 percent of Motor Full-Load Current:

12 ampere × 1.25 = 15 ampere, Table 310-16, No. 14 THHN rated 20 ampere at 60°C

Note: The minimum size conductor permitted for building wiring is No. 14 [*Section 310-5*]; however,

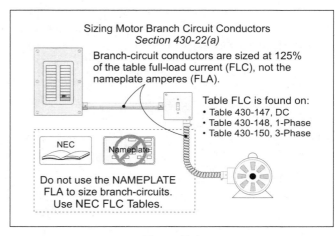

Figure 7–1
Sizing Motor Branch Circuit Conductors

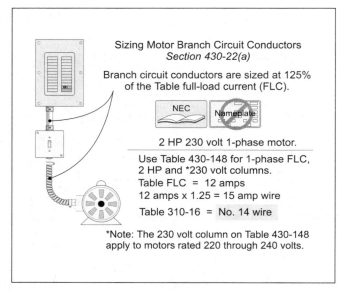

Figure 7–2
Sizing Motor Branch Circuit Conductors

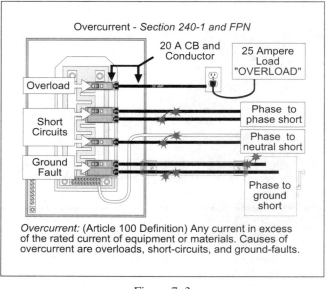

Figure 7–3
Overcurrent

local codes and many industrial facilities have requirements that No. 12 be used as the smallest branch circuit wire.

7–2 MOTOR OVERCURRENT PROTECTION

Motors and their associated equipment must be protected against overcurrent (overload, short-circuit, or ground-fault) [Article 100] (Figure 7–3). Due to the special characteristics of induction motors, overcurrent protection is generally accomplished by having the overload protection separated from the short-circuit and ground-fault protection device (Figure 7–4). Article 430, Part C contains the requirements for motor overload protection, and Part D of Article 430 contains the requirements for motor short-circuit and ground-fault protection.

Overload Protection

Overload is the condition in which current exceeds the equipment ampere rating, which can result in equipment damage due to dangerous overheating [Article 100]. Overload protection devices, sometimes called heaters, are intended to protect the motor, the motor-control equipment, and the branch circuit conductors from excessive heating due to motor overload [*Section 430-31*].

Overload protection is not intended to protect against short-circuits or ground-fault currents (Figure 7–5).

If overload protection is to be accomplished by the use of fuses, a fuse must be installed to protect each ungrounded conductor [*Section 430-36* and *430-55*].

Short-Circuit and Ground-Fault Protection

Branch circuit short-circuit and ground-fault protection devices are intended to protect the motor, the motor control apparatus, and the conductors against short-circuit or ground-faults, but they are not intended to protect against an overload [*Section 430-51*] (Figure 7–6).

Note: The ground-fault protection device required for motor circuits is not the type required for personnel [*Section 210-8*], feeders [*Section 215-9* and *240-13*], services [*Section 230-95*], or temporary wiring for receptacles [*Section 305-6*].

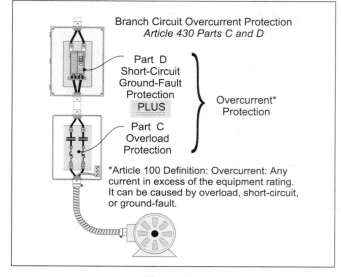

Figure 7–4
Branch Circuit Overcurrent Protection

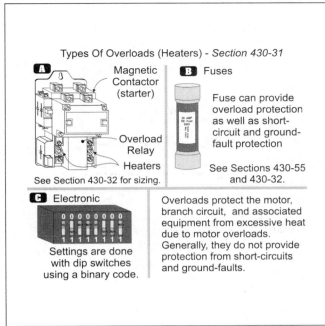

Figure 7–5
Types of Overloads

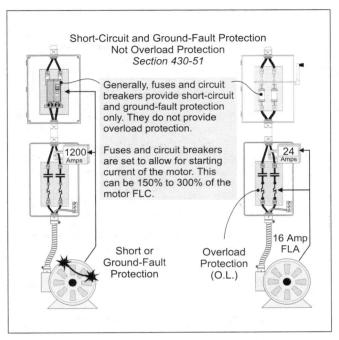

Figure 7–6
Short-Circuit and Ground-Fault Protection

7–3 OVERLOAD PROTECTION [*SECTION 430-32(a)*]

In addition to short-circuit and ground-fault protection, motors must be protected against overload. Generally, the motor overload device is part of the motor starter; however, a separate overload device like a dual-element fuse can be used. Motors rated more than 1 horsepower without integral thermal protection, and motors 1 horsepower or less (automatically started) [*Section 430-32(c)*], must have an overload device sized in response to the motor nameplate current rating [*Section 430-6(a)*]. The overload device must be sized no larger than the requirements of Section 430-32.

Service Factor

Motors with a nameplate service factor (SF) rating of 1.15 or more shall have the overload protection device sized no more than 125 percent of the motor nameplate current rating.

Service Factor

If a dual-element fuse is used for overload protection, what size fuse is required for a 5 horsepower 230-volt, single-phase motor, service factor 1.16, if the motor nameplate current rating is 28 ampere (Figure 7–7)?

(a) 25 ampere (b) 30 ampere (c) 35 ampere (d) 40 ampere

• Answer: (c) 35 ampere

Overload protection is sized to motor nameplate current rating [*Section 430-6(a), 430-32(a)(1),* and *430-55*].

28 ampere $\times$ 1.25 = 35 ampere [*Section 240-6(a)*]

Temperature Rise

Motors with a nameplate temperature rise rating not over 40°C shall have the overload protection device sized no more than 125 percent of motor nameplate current rating.

Temperature Rise

If a dual-element fuse is used for the overload protection, what size fuse is required for a 50 horsepower 460-volt, 3-phase motor, temperature rise 39°C, motor nameplate current rating of 60 ampere (FLA) (Figure 7–8)?

(a) 40 ampere (b) 50 ampere (c) 60 ampere (d) 70 ampere

• Answer: (d) 70 ampere

Overloads are sized according to the motor nameplate current rating, not the motor full-load current rating.

60 ampere $\times$ 1.25 = 75 ampere, 70 ampere [*Section 240-6(a)* and *430-32(a)(1)*]

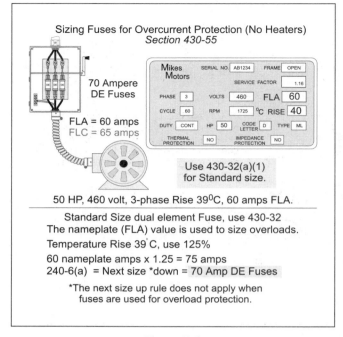

Sizing Fuses for Overcurrent Protection (No Heaters)
Section 430-55

35 Ampere DE Fuses

Mikes Motors — SERIAL NO. AB1234 — FRAME OPEN — SERVICE FACTOR 1.16 — PHASE 1 — VOLTS 230 — FLA 28.0 — CYCLE 60 — RPM 1725 — C RISE 40 — DUTY CONT — HP 5 — CODE LETTER D — TYPE ML — THERMAL PROTECTION NO — IMPEDANCE PROTECTION NO

Use 430-32(a)(1) for Standard size.

5 HP, 230 volt, SF 1.16, 28 amps nameplate.

Standard Size Dual Element Fuse, Use 430-32.

The nameplate value is used to size overloads.
Service Factor of 1.16, use 125%
28 nameplate amps x 1.25 = 35 amps

240-6(a) = 35 Amp Dual Element Fuses

Figure 7–7
Sizing Fuses for Overcurrent Protection

Sizing Fuses for Overcurrent Protection (No Heaters)
Section 430-55

70 Ampere DE Fuses

FLA = 60 amps
FLC = 65 amps

Mikes Motors — SERIAL NO. AB1234 — FRAME OPEN — SERVICE FACTOR 1.16 — PHASE 3 — VOLTS 460 — FLA 60 — CYCLE 60 — RPM 1725 — °C RISE 40 — DUTY CONT — HP 50 — CODE LETTER D — TYPE ML — THERMAL PROTECTION NO — IMPEDANCE PROTECTION NO

Use 430-32(a)(1) for Standard size.

50 HP, 460 volt, 3-phase Rise 39°C, 60 amps FLA.

Standard Size dual element Fuse, use 430-32
The nameplate (FLA) value is used to size overloads.
Temperature Rise 39°C, use 125%
60 nameplate amps x 1.25 = 75 amps
240-6(a) = Next size *down = 70 Amp DE Fuses

*The next size up rule does not apply when fuses are used for overload protection.

Figure 7–8
Sizing Fuses for Overcurrent Protection

All Other Motors

Motors that do not have a service factor rating of 1.15 and up, or a temperature rise rating of 40°C and less, must have the overload protection device sized at not more than 115 percent of the motor nameplate ampere rating.

Number of Overloads [*Section 430-37*]

An overload protection device must be installed in each ungrounded conductor according to the requirements of Table 430-37. If a fuse is used for overload protection, a fuse must be installed in each ungrounded conductor [*Section 430-36*].

7–4 BRANCH CIRCUIT SHORT-CIRCUIT GROUND-FAULT PROTECTION [*SECTION 430-52(c)(1)*]

In addition to overload protection, each motor and its accessories requires short-circuit and ground-fault protection. *NEC®* Section 430-52(c) requires the motor branch circuit short-circuit and ground-fault protection (except torque motors) to be sized no greater than the percentages listed in Table 430-152. When the short-circuit ground-fault protection device value determined from Table 430-152 does not correspond with the standard rating or setting of overcurrent protection devices as listed in Section 240-6(a), the next higher protection device size may be used [*Section 430-52(c)(1)* Exception No. 1] (Figure 7–9).

To determine the percentage from Table 430-152 to be used to size the motor branch circuit, short-circuit, and ground-fault protection device, the following steps should be helpful:

Step 1: ➝ Locate the motor type on Table 430-152: Wound rotor, direct-current, or all other motors.

Step 2: ➝ Select the percentage from Table 430-152 according to the type of protection device such as nontime delay (one-time) fuse, dual element fuse, or inverse time circuit breaker.

NEC® Table 430–152			
Percent of Full-Load Current (FLC) Tables 430–147, 148 and 150			
Type of Motor	One-Time Fuse	Dual-Element Fuse	Circuit Breaker
Direct Current and Wound-Rotor Motors	150	150	150
All Other Motors	300	175	250

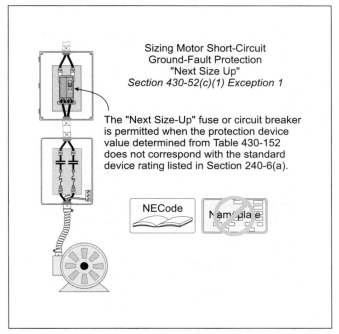

Figure 7–9
Sizing Motor Short-Circuit Ground-Fault Protection

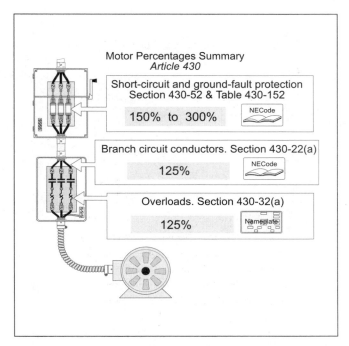

Figure 7–10
Motor Percentages Summary

Note: Where the protection device rating determined by Table 430-152 does not correspond with the standard size or rating of fuses or circuit breakers listed in Section 240-6(a), use of the next higher standard size or rating shall be permitted [*Section 430-52(c)(1)* Exception No. 1].

❏ **Branch Circuit**

Which of the following statements are true?

(a) The branch circuit short-circuit protection (nontime delay fuse) for a 3 horsepower 115-volt motor, single-phase shall not exceed 110 ampere.

(b) The branch circuit short-circuit protection (dual element fuse) for a 5 horsepower Design E, 230-volt motor, single-phase shall not exceed 50 ampere.

(c) The branch circuit short-circuit protection (inverse time breaker) for a 25 horsepower synchronous, 460-volt motor, 3-phase shall not exceed 70 ampere.

(d) all of these are true

• Answer: (d) all of these are true

Short-circuit and ground-fault protection, 430-53(c)(1) Exception No. 1 and Table 430-152:

Table 430-148 – 34 ampere × 3.00 = 102 ampere, next size up permitted, 110 ampere

Table 430-148 – 28 ampere × 1.75 = 49 ampere, next size up permitted, 50 ampere

Table 430-150 – 26 ampere × 2.50 = 65 ampere, next size up permitted, 70 ampere

WARNING: Conductors are sized at 125 percent of the motor full-load current [*Section 430-22(a)*], overloads are sized from 115 percent to 125 percent of the motor nameplate current rating [*Section 430-32(a)(1)*], and the short-circuit ground-fault protection device is sized from 175 percent to 300 percent of the motor full-load current [Table 430-152]. There is no relationship between the branch circuit conductor ampacity (125 percent) and the short-circuit ground-fault protection device (175 percent up to 300 percent) (Figure 7–10).

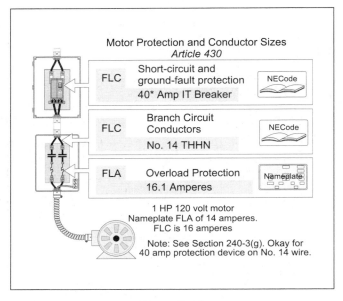

Motor Protection and Conductor Sizes
Article 430

FLC   Short-circuit and ground-fault protection
      40* Amp IT Breaker   NECode

FLC   Branch Circuit Conductors
      No. 14 THHN   NECode

FLA   Overload Protection
      16.1 Amperes   Nameplate

1 HP 120 volt motor
Nameplate FLA of 14 amperes.
FLC is 16 amperes

Note: See Section 240-3(g). Okay for 40 amp protection device on No. 14 wire.

Figure 7–11
Motor Protection and Conductor Sizes

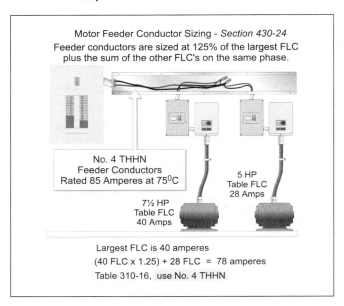

Motor Feeder Conductor Sizing - *Section 430-24*
Feeder conductors are sized at 125% of the largest FLC plus the sum of the other FLC's on the same phase.

No. 4 THHN
Feeder Conductors
Rated 85 Amperes at 75°C

5 HP
Table FLC
28 Amps

7½ HP
Table FLC
40 Amps

Largest FLC is 40 amperes
(40 FLC x 1.25) + 28 FLC = 78 amperes
Table 310-16, use No. 4 THHN

Figure 7–12
Motor Feeder Conductor Sizing

❏ **Branch Circuit**

Which of the following statements are true for a 1 horsepower 120-volt motor, nameplate current rating of 14 ampere (Figure 7–11)?

(a) The branch circuit conductors can be No. 14 THHN.

(b) Overload protection is from 16.1 ampere to 18.2 ampere.

(c) Short-circuit and ground-fault protection is permitted to be a 40-ampere circuit breaker.

(d) all of these are true
   • Answer: (d) all of these are true
   <u>Conductor Size [*Section 430-22(a)*]</u>
   16 ampere $\times$ 1.25 = 20 ampere = No. 14 at 60°C, Table 310-16
   <u>Overload Protection Size</u>
   Standard [*Section 430-32(a)(1)*] – 14 ampere (nameplate) $\times$ 1.15 = 16.1 ampere
   <u>Short-Circuit and Ground-Fault Protection [*Section 430-52(c)(1), Tables 430-152* and *240-6*]</u>
   16 ampere $\times$ 2.50 = 40-ampere circuit breaker

This bothers many electrical people, but the No. 14 THHN conductors and motor are protected against overcurrent by the 16-ampere overload protection device and the 40-ampere short-circuit protection device.

7–5 FEEDER CONDUCTOR SIZE [*SECTION 430-24*]

Conductors that supply several motors must have an ampacity of not less than:

(1) 125 percent of the highest rated motor full-load current (FLC) [*Section 430-17*], plus

(2) The sum of the full-load currents of the other motors (on the same phase) [*Section 430-6(a)*].

❏ **Feeder Conductor Size**

What size feeder conductor (in ampere) is required for two motors. Motor 1 – 7½ horsepower single-phase, 230 volts (40 ampere) and motor 2–5 horsepower single-phase 230 volts (28 ampere)? Terminals rated for 75°C (Figure 7–12).

(a) 50 ampere   (b) 60 ampere   (c) 70 ampere   (d) 80 ampere
   • Answer: (d) 80 ampere, (40 ampere $\times$ 1.25) + 28 ampere = 78 ampere

   Note: No. 4 conductor at 75°C, Table 310-16 is rated for 85 ampere.

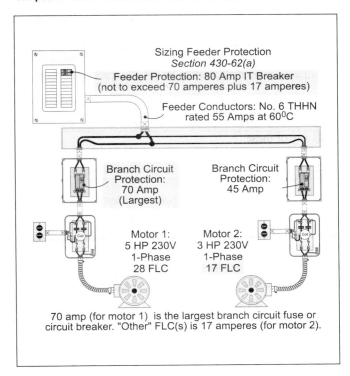

Figure 7–13
Sizing Feeder Protection

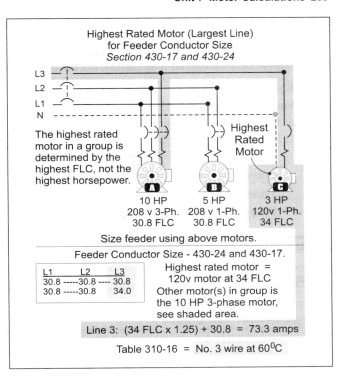

Figure 7–14
Highest Rated Motor

7–6 FEEDER PROTECTION [SECTION 430-62(a)]

Motor feeder conductors must have protection against short-circuits and ground-faults but not overload. The protection device must be sized not greater than the largest branch circuit short-circuit ground-fault protection device [*Section 430-52(c)*] of any motor of the group, plus the sum of the full-load currents of the other motors (on the same time and phase).

❏ **Feeder Protection**

What size feeder protection (inverse time breaker) is required for a 5 horsepower 230-volts, single-phase motor and a 3 horsepower 230-volt, single-phase motor (Figure 7–13)?

(a) 30-ampere breaker   (b) 40-ampere breaker   (c) 50-ampere breaker   (d) 80-ampere breaker

    • Answer: (d) 80 ampere breaker

Motor Full-Load Current [Table 430-148]

5 horsepower motor FLC = 28 ampere [Table 430-148]

3 horsepower motor FLC = 17 ampere [Table 430-148]

Branch Circuit Protection [*Section 430-52(c)(1), Table 430-152* and *240-6(a)*]

5 horsepower: 28 ampere $\times$ 2.5 = 70 ampere

3 horsepower: 17 ampere $\times$ 2.5 = 42.5 ampere: Next size up, 45 ampere

Feeder Conductor [*Section 430-24(a)*]

28 ampere $\times$ 1.25 + 17 ampere = 52 ampere, No. 6 rated 55 ampere at 60°C, Table 310-16

Feeder Protection [*Section 430-62*]

Not greater than 70-ampere protection, plus 17 ampere = 87 ampere: Next size down, 80 ampere

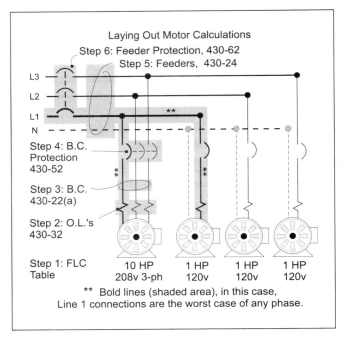

Figure 7–15
Laying Out Motor Calculations

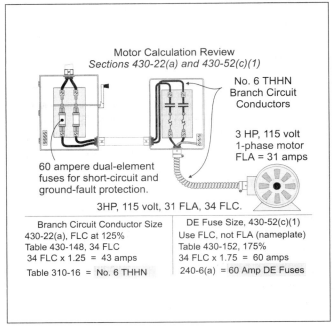

Figure 7–16
Motor Calculation Review

7–9 MOTOR CALCULATION REVIEW

❏ Branch Circuit

Size the branch circuit conductors (THHN) and short-circuit ground-fault protection device for a 3 horsepower 115-volt, single-phase motor. The motor nameplate full-load ampere is 31 ampere and dual-element fuses are to be used for short-circuit and ground-fault protection (Figure 7–16).

Branch Circuit Conductors [*Section 430-22(a)*]

Branch circuit conductors to a single motor must have an ampacity of not less than 125 percent of the motor full-load current as listed in Tables 430-147 through 430-150 [*Section 430-6(a)*].

34 ampere $\times$ 125 percent = 43 ampere

Table 310-16, 60°C Terminals: The conductor must be a No. 6 THHN rated 55 ampere [*Section 110-14(c)(1)*].

Branch Circuit Short-Circuit Protection [*Section 430-52(c)(1)*]

The branch circuit short-circuit and ground-fault protection device protects the motor, the motor control apparatus, and the conductors against overcurrent due to short-circuits or ground-faults, but not overload [*Section 430-51*]. The branch circuit short-circuit and ground-faults protection devices sized by considering the type of motor and the type of protection device according to the motor full-load current listed in Table 430-152. When the protection device values determined from Table 430-152 do not correspond with the standard rating of overcurrent protection devices as listed in Section 240-6(a), the next higher overcurrent protection device must be installed.

34 ampere $\times$ 175 percent = 60 ampere dual-element fuse [*Section 240-6(a)*]

See Example No. D8 in Appendix D of the *NEC*®.

❏ Feeder

Size the feeder conductor (THHN) and protection device (inverse time breakers 75°C terminal rating) for the following motors: Three 1 horsepower 115-volt single-phase motors; three 5 horsepower 208-volt, single-phase motors; and one 15 horsepower wound-rotor 208-volt, 3-phase motor (Figure 7–17).

Branch circuit short-circuit protection [*Section 240-6(a), 430-52(c)(1),* and *Table 430-152*]

15 horsepower: 46.2 ampere $\times$ 150 percent (wound-rotor) = 69 ampere: Next size up = 70 ampere

5 horsepower: 30.8 ampere $\times$ 250 percent = 77 ampere: Next size up = 80 ampere

1 horsepower: 16 ampere $\times$ 250 percent = 40 ampere

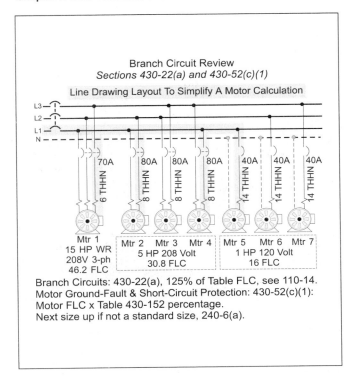

Figure 7–17
Branch Circuit Review

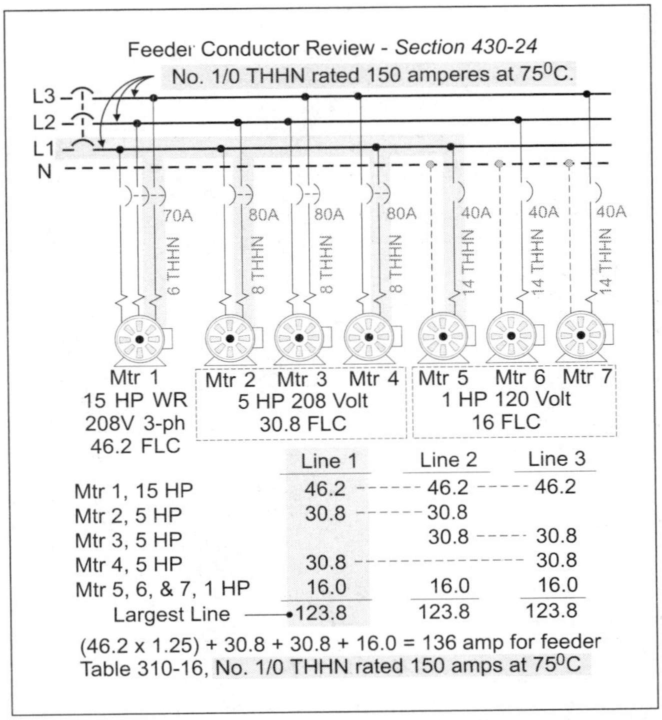

Figure 7–18
Feeder Conductor Review

Feeder Conductor [*Section 430-24*]

Conductors that supply several motors must have an ampacity of not less than 125 percent of the highest rated motor full-load current [*Section 430-17*], plus the sum of the other motor full-load currents [*Section 430-6(a)*].

(46.2 ampere × 1.25) + 30.8 ampere + 30.8 ampere + 16 ampere = 136 ampere

Table 310-16, No. 1/0 THHN, rated 150 ampere (Figure 7–18)

Note: When sizing the feeder conductor, be sure to only include the motors that are on the same phase. For that reason, only four motors are used for this feeder calculation.

Feeder Protection [*Section 430-62*]

Feeder conductors must be protected against short-circuits and ground-faults sized not to be greater than the maximum branch circuit short-circuit ground-fault protection device [*Section 430-52(c)(1)*] plus the sum of the full-load currents of the other motors (on the same phase).

80 ampere + 30.8 ampere + 46.2 ampere + 16 ampere = 174 ampere

Next size down, 150 ampere circuit breaker [*Section 240-6(a)*]. See Example No. D8 in Appendix D of the *NEC*® (Figure 7–19).

Note: When sizing the feeder protection, be sure to only include the motors that are on the same phase. For that reason, only four motors are used for the feeder calculation.

7–10 MOTOR VA CALCULATIONS

The input VA of a motor is determined by multiplying the motor volts by the motor ampere. To determine the *motor VA rating*, the following formulas can be used:

Motor VA (single-phase) = Volts × Motor Ampere

Motor VA (three-phase) = Volts × Motor Ampere × $\sqrt{3}$

Note: Many people believe that a 230-volt motor consumes less power that a 115-volt motor, but both motors consume the same amount of power.

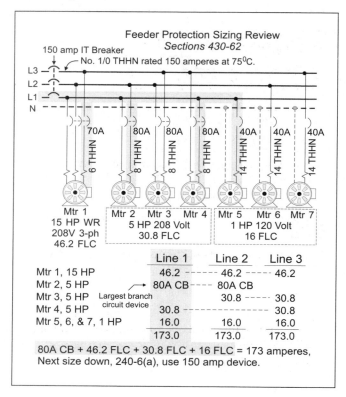

Figure 7–19
Feeder Protection Sizing Review

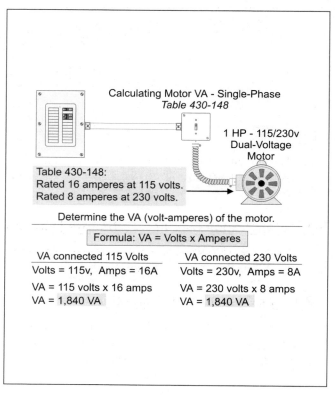

Figure 7–20
Calculating Motor VA – Single Phase

❏ Motor VA – Single-Phase

What is the motor input VA of a 1 horsepower motor rated 115/230 volts (Figure 7–20)?

(a) 1,840 VA at 115 volts   (b) 1,840 VA at 230 volts   (c) a and b   (d) none of these

• Answer: (c) a and b, Table 430-148

Motor VA = Volts × Full-Load Current

VA at 230 volts = 230 volts × 8 ampere = 1,840 VA

VA at 115 volts = 115 volts × 16 ampere = 1,840 VA

❏ Motor VA – Three-Phase

What is the input VA of a 5 horsepower 230-volt, 3-phase motor (Figure 7–21)?

(a) 6,055 VA   (b) 3,730 VA

(c) 6,440 VA   (d) 8,050 VA

• Answer: (a) 6,055 VA, Table 430-150

Motor VA = Volts × Full-Load Current × $\sqrt{3}$

Table 430-150 Full-Load Current = 15.2 ampere

Motor VA = 230 volts × 15.2 ampere × 1.732

Motor VA = 6,055 VA

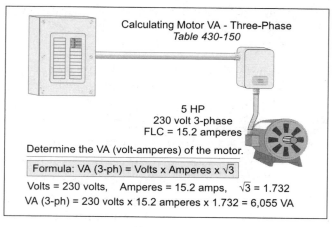

Figure 7–21
Calculating Motor VA – Three-Phase

Unit 7 – Motor Calculations Summary Questions

Introduction

1. When sizing conductors and overcurrent protection for motors, you must comply with the requirements of *NEC*® Article 430, not Article 240.
   (a) True   (b) False

7–1 Motor Branch Circuit Conductors [*Section 430-22(a)*]

2. What size THHN conductor is required for a 5 horsepower 230-volt, single-phase motor? Terminals are rated 75°C.
   (a) No. 14   (b) No. 12   (c) No. 10   (d) No. 8

7–2 Motor Overcurrent Protection

3. Motors and their associated equipment must be protected against overcurrent (overload, short-circuit, or ground-fault), but because of the special characteristics of induction motors, overcurrent protection is generally accomplished by having the overload protection separate from the short-circuit and ground-fault protection.
   (a) True   (b) False

4. • Which parts of Article 430 contain the requirements for motor overcurrent protection?
   (a) Overload protection – Part C   (b) Short-circuit ground-fault protection — Part D
   (c) a and b   (d) none of these

5. • Overload is the condition where current is greater than the equipment ampacity rating resulting in equipment damage due to dangerous overheating [Article 100]. Overload protection devices, sometimes called heaters, are intended to protect the _____ from dangerous overheating.
   (a) motor   (b) motor-control equipment
   (c) branch circuit conductors   (d) all of these

6. The branch-circuit short-circuit and ground-fault protection device is intended to protect the motor, the motor control apparatus, and the conductors against overcurrent due to _____ .
   (a) short-circuits   (b) ground-faults   (c) overloads   (d) a and b

7–3 Overload Protection [*Section 430-32(a)*]

7. • The *NEC*® requires motor overload protection devices to be sized according to the motor full-load current rating as listed in Tables 430-147, 148, or 150.
   (a) True   (b) False

8. The standard size overload protection device must be sized according to the requirements of Section 430-32 of the *NEC*®. If the overload protection relay sized according to Section 430-32 is not capable of carrying the motor starting and running current, the next size up overload can be used if sized according to the requirements of Section 430-34.
   (a) True   (b) False

9. Motors with a nameplate service factor (SF) rating of 1.15 or more must have the overload protection device sized at no more than _____ percent of the motor nameplate current rating.
   (a) 100   (b) 115   (c) 125   (d) 135

10. Motors with a nameplate temperature rise rating not over 40°C must have the overload protection device sized at no more than _____ percent of motor nameplate current rating.
    (a) 100   (b) 115   (c) 125   (d) 135

11. Motors that have a service factor of 1.10 must have the overload protection device sized at not more than _____ percent of the motor nameplate ampere rating.
     (a) 100   (b) 115   (c) 125   (d) 135

12. If a dual-element fuse is used for overload protection, what size fuse is required for a 5 horsepower 208-volt, 3-phase motor, service factor 1.16, motor nameplate current rating of 16 ampere (FLA)?
     (a) 20 ampere   (b) 25 ampere   (c) 30 ampere   (d) 35 ampere

13. If a dual-element fuse is used for the overload protection, what size fuse is required for a 30 horsepower 460-volt, 3-phase synchronous motor, temperature rise 39°C?
     (a) 20 ampere   (b) 25 ampere   (c) 30 ampere   (d) 40 ampere

7–4 Branch Circuit Short-Circuit Ground-Fault Protection [*Section 430-52(c)(1)*]

14. In addition to overload protection, each motor and its accessories requires short-circuit and ground-fault protection according to the requirements of Section 430-52. When sizing the branch circuit protection device, we must consider which of the following factors?
     (a) the motor type, such as induction, synchronous, wound-rotor, etc.
     (b) the motor *Code* letter starting characteristics
     (c) the type of protection device to be used, fuse or breaker
     (d) all of these

15. The *NEC®* requires motor branch circuit short-circuit and ground-fault protection to be sized not greater than the percentages listed in Table 430-152. When the short-circuit ground-fault protection device value determined from Table 430-152 does not correspond with the standard rating of overcurrent protection devices as listed in Section 240-6(a), the next _____ device size must be used.
     (a) smaller   (b) larger   (c) a or b   (d) none of these

16. To determine the percentage of the motor FLC from Table 430-152 that is to be used to size the motor branch circuit short-circuit ground-fault protection device, which of the following steps should be used:
     (a) Locate the motor type on Table 430-152, such as direct current, wound rotor, high-reactance, autotransformer start, or all other motors.
     (b) Locate the motor starting conditions, such as *Code* letter or no *Code* letter.
     (c) Select the percentage from Table 430-152 according to the type of protection device, such as one-time fuse, dual-element fuse, or circuit breaker.
     (d) all of these

17. If the branch circuit short-circuit ground-fault protection dual-element fuse selected (sized not greater than the percentages listed in Table 430-152,) is not capable of carrying the load, the next larger size dual-element fuse can be used. The next size dual-element fuse cannot exceed _____ of the motor full-load current rating.
     (a) 125%   (b) 150%   (c) 175%   (d) 225%

18. Conductors are sized at _____ percent of the motor full-load currents [*Section 430-6* and *430-22*], overloads from _____ percent, and the motor short-circuit ground-fault protection device (inverse time circuit breaker) is sized up to _____ . (There is no relationship between the branch circuit conductor ampacity and the short-circuit ground-fault protection device!)
     (a) 125%, 115%, 250%   (b) 100%, 125%, 150%
     (c) 125%, 125%, 125%   (d) 100%, 100%, 100%

19. Which of the following statements are true for a 10 horsepower 208-volt, 3-phase motor, nameplate current 29 ampere?
     (a) The branch circuit conductors can be No. 8 THHN
     (b) Overload protection is from 33 ampere to 38 ampere
     (c) Short-circuit and ground-fault protection is an 80-ampere circuit breaker
     (d) all of these

7–5 Feeder Conductor Size [*Section 430-24*]

20. Feeder conductors that supply several motors must have an ampacity of not less than:
    (a) 125 percent of the highest rated motor FLC.
    (b) the sum of the full-load currents of the other motors on the same phase [*Section 430-6(a)*].
    (c) a or b
    (d) a and b

21. Motor feeder conductors (sized according to 430-24) must have a feeder protection device to protect against short-circuits and ground-faults (not overloads), sized not greater than:
    (a) the largest branch circuit short-circuit ground-fault protection device [*Section 430-52*] of any motor of the group.
    (b) the sum of full-load currents of the other motors on the same time phase.
    (c) a or b
    (d) a and b

7–6 Feeder Protection [*Section 430-62(a)*]

22. • Which of the following statements about a 30 horsepower 460-volt, 3-phase synchronous motor and a 10 horsepower 460-volt, 3-phase motor are true?
    (a) The 30 horsepower motor has No. 8 THHN with a 80-ampere breaker.
    (b) The 10 horsepower motor has No. 14 THHN with a 35-ampere breaker.
    (c) The feeder conductors must be No. 6 THHN with a 90-ampere breaker.
    (d) all of these.

7–7 Highest-Rated Motor [*Section 430-17*]

23. When selecting the feeder conductors and short-circuit ground-fault protection device, the highest rated motor shall be the highest rated _____ .
    (a) horsepower   (b) full-load current   (c) nameplate current   (d) any of these

24. • Which is the highest rated motor of the following?
    (a) 25 horsepower synchronous, 3-phase, 460 volt
    (b) 20 horsepower 3-phase, 460 volt
    (c) 15 horsepower 3-phase, 460 volt
    (d) 3 horsepower 120 volt

7–10 Motor VA Calculations

25. What is the VA input of a dual voltage 5 horsepower 3-phase motor rated 460/230 volts?
    (a) 3,027 VA at 460 volts
    (b) 6,055 VA at 230 volts
    (c) 6,055 VA at 460 volts
    (d) b and c

26. What is the input VA of a 3 horsepower 208-volt, single-phase motor?
    (a) 3,890 VA   (b) 6,440 VA   (c) 6,720 VA   (d) none of these

☆ Challenge Questions

7–1 Motor Branch Circuit Conductors [*Section 430-22(a)*]

27. • The branch circuit conductors of a 5 horsepower 230-volt motor with a nameplate rating of 25 ampere shall have an ampacity of not less than _____ . Note: The motor is used for intermittent duty and cannot run for more than 5 minutes at any one time due to the nature of the apparatus it drives.
    (a) 33 ampere   (b) 37 ampere   (c) 21 ampere   (d) 23 ampere

7–3 Overload Protection [*Section 430-32(a)(1)*]

28. The standard overload protection device for a 2 horsepower 115-volt motor that has a full-load current rating of 24 ampere, and a nameplate rating of 21.5 ampere shall not exceed _____ .
     (a) 20.6 ampere   (b) 24.7 ampere   (c) 29.9 ampere   (d) 33.8 ampere

29. • The maximum overload protective device relay for a 2 horsepower 115-volt motor with a nameplate rating of 22 ampere is _____ . The Service factor is 1.2.
     (a) 30.8 ampere   (b) 33.8 ampere   (c) 33.6 ampere   (d) 22.6 ampere

Ultimate Trip Setting [*Section 430-32(a)(2)*]

30. The ultimate trip overload device of a thermally protected 1 horsepower 120-volt motor would be rated no more than _____ .
     (a) 31.2 ampere   (b) 26 ampere   (c) 28 ampere   (d) 23 ampere

7–4 Branch Circuit Short-Circuit Ground-Fault Protection [*Section 430-52* and *Table 430-152*]

31. A 2 horsepower 120-volt wound rotor motor requires a _____ branch circuit short-circuit protection device.
     (a) 15 ampere   (b) 20 ampere   (c) 25 ampere   (d) 40 ampere

32. • The branch circuit short-circuit protection device for a 10 horsepower 230-volt, single-phase wound rotor motor shall not exceed _____ . Note: Use an inverse-time breaker for protection.
     (a) 125 ampere   (b) 50 ampere   (c) 75 ampere   (d) 80 ampere

33. The branch circuit protection for a 125 horsepower 240-volt, direct current motor is _____ .
     (a) 400 ampere   (b) 600 ampere   (c) 700 ampere   (d) 800 ampere

7–5 Feeder Conductor Size [*Section 430-24*]

34. • The motor controller (below) requires a No. _____ THHN for the feeder, if the motor terminals are rated for 75°C.
     (a) 2   (b) 3/0   (c) 4/0   (d) 250 kcmil

7–6 Feeder Protection [*Section 430-62(a)*]

35. • There are three motors: one 5 horsepower 230-volt motor with service factor of 1.2 and two 1 horsepower 120-volt motors. The three motors are fed with a 3-wire cable. Using an inverse time breaker, the feeder conductor protection device after balancing all three motors would be _____ .
     (a) 60 ampere   (b) 70 ampere   (c) 80 ampere   (d) 90 ampere

36. • If an inverse-time breaker is used for the feeder short-circuit protection, what size protection is required for the following 3-phase motors?
     Motor 1 = 40 horsepower 52 FLC
     Motor 2 = 20 horsepower 27 FLC
     Motor 3 = 10 horsepower 14 FLC
     Motor 4 = 5 horsepower 7.6 FLC
     (a) 225 ampere   (b) 200 ampere   (c) 125 ampere   (d) 175 ampere

37. • If dual-element fuses are used to protect a three-wire 115/230-volt feeder conductor for twenty-two $1/2$ horsepower single-phase, 115-volt motors, the fuse size selected should not be greater than _____ .
     (a) 125 ampere   (b) 90 ampere   (c) 100 ampere   (d) 110 ampere

38. • The feeder protection for a 25 horsepower 208-volt, 3-phase and three 3 horsepower 120-volt motors would be _____ , after balancing. Note: Use Inverse time breakers (ITB).
     (a) 225 ampere   (b) 200 ampere   (c) 300 ampere   (d) 250 ampere

NEC® Questions from Section 370-16 through Sections 630-12

Article 370 through Article 384

39. Equipment grounding conductor(s) and not more than _____ fixture wires (smaller than No. 14) shall be permitted to be omitted from the calculations where they enter the box from a domed fixture or similar canopy.
    (a) 2　　　　　　　(b) 3　　　　　　　(c) 4　　　　　　　(d) none of these

40. A box has three equipment grounding conductors. When determining the number of conductors in a box for box fill calculations; the three grounding conductors are counted as _____ conductor(s).
    (a) 3　　　　　　　(b) 6　　　　　　　(c) 1　　　　　　　(d) 0

41. • Metal plugs or plates used with nonmetallic boxes shall be recessed _____ inch.
    (a) $\frac{1}{8}$　　　　　　　(b) $\frac{3}{8}$　　　　　　　(c) $\frac{1}{4}$　　　　　　　(d) $\frac{3}{4}$

42. • Only a _____ wiring method can be used for a surface extension from a cover, and the wiring method must include an equipment grounding conductor.
    (a) solid　　　　　　　(b) flexible　　　　　　　(c) rigid　　　　　　　(d) cord

43. Outlet boxes can be secured to suspended ceiling framing members by mechanical means such as _____, or other means identified for the suspended ceiling framing member(s).
    (a) bolts　　　　　　　(b) screws　　　　　　　(c) rivets　　　　　　　(d) all of these

44. • Enclosures not over 100 cubic inches that have threaded entries that support lighting fixtures or contain devices shall be considered adequately supported where two or more conduits are threaded wrenchtight into the enclosure where each conduit is supported within _____ inches.
    (a) 12　　　　　　　(b) 18　　　　　　　(c) 24　　　　　　　(d) 30

45. • The internal depth of outlet boxes intended to enclose flush devices shall be at least _____ inch.
    (a) $\frac{1}{2}$　　　　　　　(b) $\frac{7}{8}$　　　　　　　(c) $\frac{5}{8}$　　　　　　　(d) none of these

46. Paddle fans cannot be supported directly by an outlet box.
    (a) True　　　　　　　(b) False

47. Which of the following are required to be accessible?
    (a) outlet boxes　　　　(b) junction boxes　　　　(c) pull boxes　　　　(d) all of these

48. Metal boxes over _____ cubic inches in size shall be constructed so as to be of ample strength and rigidity.
    (a) 50　　　　　　　(b) 75　　　　　　　(c) 100　　　　　　　(d) 125

49. For straight pulls, the length of the box shall be not less than _____ times the outside diameter, over sheath, of the largest conductor or cable entering the box on systems over 600 volts.
    (a) 18　　　　　　　(b) 16　　　　　　　(c) 36　　　　　　　(d) 48

50. Cabinets or cutout boxes installed in wet locations shall be _____ .
    (a) waterproof　　　　(b) raintight　　　　(c) weatherproof　　　　(d) watertight

51. • Cables entering a cutout box shall _____ .
    (a) be secured independently to the cutout box
    (b) can be sleeved through a chase
    (c) have a maximum of two cables per connector
    (d) all of these

52. An auxiliary gutter is permitted to contain _____ .
    (a) conductors　　　　(b) overcurrent devices　　　　(c) busways　　　　(d) a or c

53. • Auxiliary gutters shall not contain more than _____ conductors at any cross-section.
    (a) 30　　　　(b) 40 current-carrying　　　　(c) 20　　　　(d) 30 current-carrying

54.   • The continuous current-carrying capacity of 1 square-inch copper busbar mounted in an unventilated enclosure is
       _____ amperes.
       (a) 500   (b) 750   (c) 650   (d) 1,500

55.   • Auxiliary gutters shall be constructed and installed so as to maintain _____ continuity.
       (a) mechanical   (b) electrical   (c) a or b   (d) a and b

56.   Switches or circuit breakers shall not disconnect the grounded conductor of a circuit unless the switch or circuit breaker
       _____ .
       (a) can be opened and closed by hand levers only
       (b) simultaneously disconnects all conductors of the circuit
       (c) open the grounded conductor before it disconnects the ungrounded conductors
       (d) none of these

57.   All switches and circuit breakers used as switches must be installed so they can be operated from a readily accessible
       place.
       (a) True   (b) False

58.   A faceplate for a flush-mounted snap switch shall not be less than _____ inches thick when made of a nonferrous metal.
       (a) 0.03   (b) 0.04   (c) 0.003   (d) 0.004

59.   Alternating current general use snap switches can control _____ .
       (a) resistive loads that do not exceed the ampere and voltage rating of the switch
       (b) inductive and tungsten filament (120 volt) loads that do not exceed the ampere and voltage rating of the switch
       (c) motor loads (2 horsepower or less) that do not exceed 80 percent of the ampere rating of the switch
       (d) all of these

60.   A fused switch shall not have fuses _____ .
       (a) in series   (b) in parallel   (c) less than 100 amperes   (d) over 15 amperes

61.   Barriers shall be placed in all service switchboards to isolate the service _____ and terminals from the remainder of the
       switchboard.
       (a) busbars   (b) conductors   (c) cables   (d) none of these

62.   Switchboards that have any exposed live parts shall be located in permanently _____ locations and then only where
       under competent supervision and accessible only to qualified persons.
       (a) dry   (b) mounted   (c) supported   (d) all of these

63.   An insulated conductor used within a switchboard shall be _____ .
       (a) listed   (b) flame-retardant
       (c) rated for the highest voltage it may contact   (d) all of these

64.   All panelboard circuit breaker _____ shall be legibly identified as to purpose or use on a circuit directory located on the
       face or inside of the panel doors.
       (a) manufacturers   (b) conductors   (c) feeders   (d) modifications

65.   A lighting and appliance branch circuit panelboard contains six 3-pole breakers and eight 2-pole breakers. The
       maximum allowable number of single-pole breakers that can be added to this panelboard is _____ .
       (a) 8   (b) 16   (c) 28   (d) 42

66.   When equipment grounding conductors are installed in panelboards, a _____ is required for the proper termination of
       the equipment grounding conductors.
       (a) neutral
       (b) grounded terminal bar
       (c) grounding terminal bar
       (d) none of these

Article 400 through Article 410

67. Type HPD cord is permitted for _____ usage.
    (a) not hard   (b) hard   (c) extra hard   (d) all of these

68. A three conductor No. 16, SJE cable (one conductor is used for grounding) shall have a maximum ampacity of _____ amperes for each conductor.
    (a) 13   (b) 12   (c) 15   (d) 8

69. Flexible cords shall not be used as a substitute for _____ wiring.
    (a) temporary   (b) fixed   (c) overhead   (d) none of these

70. Flexible cords shall be connected to devices and to fittings so that tension will not be transmitted to joints or terminal screws. This shall be accomplished by _____ .
    (a) knotting the cord   (b) winding with tape
    (c) fittings designed for the purpose   (d) all of these

71. One conductor of flexible cords intended to be used as a _____ circuit conductor shall have a continuous marker readily distinguishing it from the other conductor or conductors.
    (a) grounded   (b) equipment   (c) ungrounded   (d) all of these

72. The maximum operating temperature of asbestos-covered, type AF heat resistant wire is _____ °C.
    (a) 124   (b) 150   (c) 194   (d) 220

73. • Fixture wires shall be permitted (1) for installation in lighting fixtures and in similar equipment where enclosed or protected and not subject to _____ in use, or (2) for connecting lighting fixtures to the branch circuit conductors supplying the lighting fixtures.
    (a) bending or twisting
    (b) knotting
    (c) stretching or straining
    (d) none of these

74. Interior locations protected from weather but subject to moderate degrees of moisture such as some basements, some barns, some cold-storage warehouses and the like, the partially protected locations under canopies, marquees, roofed open porches, and the like, shall be considered to be _____ locations.
    (a) damp   (b) wet   (c) hazardous   (d) dry

75. No part of cord-connected lighting fixtures, hanging lighting fixtures, lighting track, pendants, or suspended ceiling fans shall be located within a zone measured 3 feet horizontally and _____ feet vertically from the top of the bathtub rim or shower stall threshold.
    (a) 4   (b) 6   (c) 8   (d) none of these

76. The *NEC*® requires a lighting outlet in clothes closets.
    (a) True   (b) False

77. In clothes closets recessed incandescent lighting fixtures with a completely enclosed lamp shall be permitted to be installed in the wall or on the ceiling, provided there is a minimum clearance of _____ inches between the fixture and the nearest point of a storage space.
    (a) 3   (b) 6   (c) 9   (d) 12

78. The maximum weight of a light fixture that may be mounted by the screw shell of a brass socket is _____ pounds.
    (a) 2   (b) 6   (c) 3   (d) 50

79. Metal poles used for the support lighting fixtures must be bonded to an _____ .
    (a) grounding electrode   (b) grounded conductor
    (c) equipment grounding conductor   (d) any of these

80. Where the outlet box provides adequate support, a fixture weighing up to _____ pounds may be supported by the outlet box.
    (a) 6   (b) 30   (c) 50   (d) 75

81.   Trees can be used to support lighting fixtures.
       (a) True   (b) False

82.   Lighting fixtures shall be wired with conductors having insulation suitable for the environmental conditions and _____
       to which the conductors will be subjected.
       (a) temperature   (b) voltage   (c) current   (d) all of these

83.   No _____ splices or taps shall be made within or on a fixture.
       (a) unapproved   (b) untested   (c) uninspected   (d) unnecessary

84.   • Lighting fixtures that require adjustment or aiming can be cord-connected without an attachment plug.
       (a) True   (b) False

85.   Lighting fixtures shall not be used as a raceway for circuit conductors, except lighting fixtures designed for end-to-end
       assembly to form a continuous raceway, or lighting fixtures connected together by recognized wiring methods shall be
       permitted to carry through conductors of 2-wire or _____ branch circuit supplying the lighting fixtures.
       (a) small appliance   (b) appliance   (c) multiwire   (d) industrial

86.   Portable lamps shall be wired with _____, recognized by Section 400-4, and have an attachment plug of the polarized or
       grounding type.
       (a) flexible cable   (b) flexible cord
       (c) nonmetallic flexible cable   (d) nonmetallic flexible cord

87.   A 1,000-watt incandescent lamp requires a _____ base.
       (a) mogul   (b) standard   (c) admedium   (d) copper

88.   • Receptacles mounted in boxes that are set back of the wall surface shall be installed so that the mounting _____ of the
       receptacle is held rigidly at the surface of the wall.
       (a) screws or nails   (b) yoke or strap   (c) face plate   (d) none of these

89.   A receptacle shall be considered to be in a location protected from the weather where located under roofed open
       porches, canopies, marquees, and the like, and will not be subjected to _____ .
       (a) spray from a hose
       (b) a direct lightning hit
       (c) beating rain or water run-off
       (d) falling or wind-blown debris

90.   An enclosure that is weatherproof only when the receptacle cover is closed can be used for receptacles in a wet location,
       when the receptacle is used for _____ while attended.
       (a) portable equipment (b) portable tools   (c) fixed equipment   (d) a and b

91.   Grounding-type attachment plugs shall be used only where an _____ ground is provided.
       (a) equipment   (b) isolated   (c) computer   (d) branch circuit

92.   The minimum distance an outlet box containing fixture tap supply conductors can be placed from a recessed fixture is
       _____ feet.
       (a) 1   (b) 2   (c) 3   (d) 4

93.   Lighting fixtures shall be so constructed that adjacent combustible material will not be subjected to temperature in
       excess of _____°C.
       (a) 75   (b) 90   (c) 185   (d) 140

94.   Surface-mounted lighting fixtures with ballast must have a minimum clearance of _____ inch(es) from combustible low
       density fiberboard, unless the fixture is marked "Suitable for Surface Mounting on Combustible Low-Density Cellulose
       Fiberboard."
       (a) $1/2$ (b) 1   (c) $1^1/2$ (d) 2

95.   • Electric-discharge lighting fixtures having an open-circuit voltage exceeding _____ volts shall not be installed in or on
       dwelling occupancies.
       (a) 120   (b) 250   (c) 600   (d) 1,000

96. Lighting track fittings shall be permitted to be equipped with general-purpose receptacles.
    (a) True   (b) False

97. Track lighting shall not be installed within the zone measured 3 feet horizontally and _____ feet vertically from the top of the bathtub rim.
    (a) 2   (b) 3   (c) 4   (d) 8

Article 422 through Article 427

98. • Individual appliances that are continuously loaded shall have the branch circuit rating sized no less than _____ percent of the appliance marked ampere rating.
    (a) 150   (b) 100   (c) 125   (d) 80

99. A(n) _____ branch circuit must supply central heating equipment, other than fixed electrical space heating equipment.
    (a) multiwire   (b) individual
    (c) multipurpose   (d) small appliance branch circuit

100. A disposal can be cord-and-plug connected. The cord must not be less than 18 inches or more than _____ inches and must be protected from physical damage.
    (a) 30   (b) 36   (c) 42   (d) 48

101. Ceiling fans that do not exceed _____ pounds in weight, with or without accessories, must be supported by outlet boxes identified for such use and supported in accordance with Sections 370-23 and 370-27.
    (a) 20   (b) 25   (c) 30   (d) 35

102. For permanently connected appliances rated over _____ volts, amperes, or horsepower, the branch circuit switch or circuit breaker shall be permitted to serve as the disconnecting means where the switch or circuit breaker is within sight from the appliance or is capable of being locked in the open position.
    (a) 200   (b) 300   (c) 400   (d) 500

103. Electrically heated smoothing irons shall be equipped with an identified _____ means.
    (a) disconnecting   (b) temperature limiting   (c) current limiting   (d) none of these

104. The branch circuit conductor and overcurrent protection device for fixed electric space heating equipment loads shall not be smaller than _____ percent of the total load.
    (a) 80   (b) 100   (c) 125   (d) 150

105. Fixed electric space heating equipment shall be installed to provide the _____ spacing between the equipment and adjacent combustible material, unless it has been found to be acceptable where installed in direct contact with combustible material.
    (a) required   (b) minimum   (c) maximum   (d) safest

106. • Resistance-type heating elements in electric space heating equipment shall be protected at not more than _____ .
    (a) 95 percent of the nameplate value   (b) 60 amperes
    (c) 48 amperes   (d) 150 percent of the rated current

107. Conductors located above a heated ceiling shall be considered as operating in an ambient temperature of _____ °C.
    (a) 86   (b) 30   (c) 50   (d) 20

108. • The minimum clearance between an electric space heating cable and an outlet box used for surface lighting fixtures shall not be less than _____ inches.
    (a) 8   (b) 14   (c) 18   (d) 6

109. A boiler employing resistance-type immersion heating elements contained in an ASME rated and stamped vessel and rated more than 120 amperes shall have the heating elements subdivided into loads not exceeding _____ amperes.
    (a) 70   (b) 100   (c) 120   (d) 150

110. A heating panel is completely assembly provided with a junction box or a length of flexible conduit for connection to a _____ .
    (a) wiring system   (b) service   (c) branch circuit   (d) approved conductor

111. Examples of snow melting resistance heaters could be _____ .
 (a) tubular heaters and strip heaters
 (b) immersion heaters and heating blankets
 (c) heating cables or heating tape
 (d) all of these

112. Exposed elements of impedance heating systems shall be physically guarded, isolated, or thermally insulated with a _____ jacket to protect against contact by personnel in the area.
 (a) corrosion resistant   (b) waterproof   (c) weatherproof   (d) flame retardant

113. All fixed outdoor de-icing and snow-melting equipment shall be provided with a means for disconnection from all _____ conductors.
 (a) grounded   (b) grounding   (c) ungrounded   (d) neutral

114. The ampacity of branch circuit conductors and the rating or setting of overcurrent protective devices supplying fixed electric heating equipment for pipelines and vessels shall be not less than _____ percent of the total load of the heaters.
 (a) 75   (b) 100   (c) 125   (d) 150

Article 430 through Article 480

115. • For general motor application, the motor branch circuit short-circuit ground-fault protection device must be sized based on the _____ amperes.
 (a) motor nameplate   (b) NEMA standards   (c) *NEC®* Table   (d) Factory Mutual

116. Conductors for motors, controllers, and terminals of control circuit devices are required to be of copper wire unless identified with different conductors.
 (a) True   (b) False

117. A motor terminal housing with rigidly mounted motor terminals shall have a minimum of _____ inch between line terminals for a 230-volt motor.
 (a) $1/4$ (b) $3/8$ (c) $1/2$ (d) $5/8$

118. When selecting the conductors according to Section 430-24, the motor with the highest _____ shall be the highest-rated motor.
 (a) FLC   (b) horsepower   (c) nameplate   (d) code letter

119. Motor tap conductors shall not exceed _____ feet.
 (a) 10   (b) 15   (c) 20   (d) 25

120. An overload device used to protect continuous-duty motors (rated more than 1 horsepower) shall be selected to trip or rated at no more than _____ percent of the motor nameplate full-load current rating for motors with a marked service factor not less than 1.15.
 (a) 110   (b) 115   (c) 120   (d) 125

121. Where an overload relay selected by Section 430-32(a)(1) and (c)(1) is not sufficient to start the motor or carry the load, the next higher size overload relay is permitted. Provided the trip current does not exceed _____ percent of motor full-load current rating if the motor is marked with a service factor of 1.12.
 (a) 100   (b) 110   (c) 120   (d) 130

122. • The minimum number of overload unit(s) required for a 3-phase alternating current motor shall be _____ .
 (a) one   (b) two   (c) three   (d) any of these

123. • If _____ shutdown is necessary to reduce hazards to persons, the overload sensing devices shall be permitted to be connected to a supervised alarm instead of causing the motors to shut down.
 (a) emergency   (b) normal   (c) orderly   (d) none of these

124. • Feeder conductors that supply motor and other loads must have overcurrent protection against short-circuit, ground-fault, and overload.
 (a) True   (b) False

125. Motor control circuit conductors that extend beyond the motor control equipment enclosure shall be required to have short-circuit and ground-fault protection sized not greater than _____ percent of the conductor ampacity as listed in Table 310-16 for 60°C conductors.

    (a) 100   (b) 150   (c) 300   (d) 500

126. If the control circuit transformer is located in the controller enclosure, the transformer must be connected to the _____ side of the control circuit disconnect.

    (a) line   (b) load   (c) adjacent   (d) none of these

127. Each motor shall be provided with an individual controller.

    (a) True   (b) False

128. The motor disconnecting means is not required to be within sight of the motor, if the required controller disconnect is capable of being individually locked in the open position.

    (a) True   (b) False

129. Where more than one motor disconnecting means is provided in the same motor branch circuit, only one of the disconnecting means is required to be readily accessible.

    (a) True   (b) False

130. The branch circuit overcurrent protection device such as a plug fuse may serve as the disconnecting means for a stationary motor of 1/8 horsepower or less.

    (a) True   (b) False

131. An oil switch used for both a controller and disconnect is permitted on a circuit whose rating _____ volts or 100 amperes.

    (a) is 600   (b) exceeds 600   (c) does not exceed 600   (d) none of these

132. Article 440 applies to electric-driven air-conditioning and refrigeration equipment that has a hermetic refrigerant motor compressor.

    (a) True   (b) False

133. Which of the following statement(s) about an air-conditioning disconnect is true?
    (a) The disconnecting means is required to be within sight of the unit.
    (b) The disconnecting means can be mounted directly on the unit.
    (c) The disconnecting means can be inside the unit.
    (d) All of these

134. The attachment plug and receptacle shall not exceed _____ amperes at 250 volts for a cord- and-plug-connected air-conditioning motor compressor.

    (a) 15   (b) 20   (c) 30   (d) 40

135. Each generator shall be provided with a _____ giving the maker's name, the rated frequency, power factor, number of phases if of alternating current, and the rating in kilowatts or kilovolt-amperes.

    (a) list   (b) faceplate   (c) nameplate   (d) sticker

136. The phase converter disconnecting means shall be _____ and located in sight from the phase converter.
    (a) protected from physical damage
    (b) readily accessible
    (c) easily visible
    (d) clearly identified

137. A transformer not over 600 volts requires overcurrent protection:
    (a) Individual overcurrent device on the primary side rated or set at not more than 125 percent of the rated primary current of the transformer.
    (b) Secondary overcurrent device set at not more than 125 percent of the rated secondary current of the transformer, if the primary overcurrent device is set at not more than 250 percent of the rated primary current of the transformer.
    (c) a or b
    (d) none of these

166. Each circuit leading to or through a dispensing pump shall be provided with a switch or other acceptable means to disconnect simultaneously from the source of supply all conductors of the circuit, including the _____ conductor, if any.
   (a) grounding   (b) grounded   (c) bonding   (d) all of these

167. Single-pole breakers utilizing approved handle ties cannot be used for the circuit disconnect for gasoline dispensing equipment.
   (a) True   (b) False

168. In a gas station, an approved seal shall be _____ .
   (a) provided in each conduit run entering a dispenser
   (b) provided in each conduit run leaving a dispenser
   (c) the first fitting after the conduit emerges from the ground or concrete
   (d) all of these

169. An aboveground tank in a bulk storage plant is a Class I, Group D, Division 1 location within _____ feet of open end of vent, extending in all directions.
   (a) 12   (b) 10   (c) 6   (d) 5

Article 517 through Article 555

170. The authority having jurisdiction may judge a location utilized for _____ as a "non-hazardous location" providing the conditions of the *Code* are met.
   (a) drying or curing   (b) dipping and coating   (c) spraying operations   (d) none of these

171. The patient care area is any portion of a health care facility such as business offices, corridors, lounges, day rooms, dining rooms, or similar areas.
   (a) True   (b) False

172. In health care facilities, patient bed receptacles located in general care areas shall be supplied by at least two branch circuits. The branch circuits for these receptacles can originate from two separate transfer switches on the emergency system.
   (a) True   (b) False

173. In critical care areas of health care centers, each patient bed location shall be provided with a minimum of _____ duplex receptacles.
   (a) 10   (b) 3   (c) 1   (d) 4

174. • The essential electrical systems in a health care facility shall have sources of power from:
   (a) a normal source generally supplying the entire electrical system.
   (b) one or more alternate sources for use when the normal source is interrupted.
   (c) a or b
   (d) a and b

175. In anesthetic areas, low-voltage equipment that is frequently in contact with the bodies of persons or has exposed current-carrying elements shall _____ .
   (a) operate on an electrical potential of 10 volts or less
   (b) be moisture resistant
   (c) be intrinsically safe or double-insulated
   (d) all of these

176. Examples of places of assembly could include (but are not limited to) _____ .
   (a) restaurants   (b) conference rooms   (c) pool rooms   (d) all of these

177. The following wiring methods can be installed in a place of assembly:
   (a) metal raceways
   (b) MC cable
   (c) Type AC cable containing an insulated equipment grounding conductor
   (d) all of these

178. On fixed stage equipment, portable or strip light fixtures and connector strips shall be wired with conductors having insulation rated _____°C or suitable for the conditions.
(a) 75   (b) 90   (c) 125   (d) 200

179. All receptacles in dressing rooms of theaters shall be _____ .
(a) controlled by wall switches with pilot lights
(b) controlled by wall switches located within the room
(c) limited to a 30-ampere branch circuit
(d) a and b

180. Cord connectors for carnival, circuses, and fairs can be laid on the ground when the connectors are _____ for a wet location, but they must not be placed in audience traffic paths or within areas accessible to the public, unless guarded.
(a) listed   (b) labeled   (c) approved   (d) all of these

181. GFCI protection is required for all _____ ampere, 125-volt receptacle outlets used by personnel at carnivals, circuses, and fairs (not used for cooking or refrigeration equipment). The GFCI protection can be by circuit breaker, receptacle, an integral part of the attachment plug, or a listed cord set incorporating GFCI protection.
(a) 15   (b) 20   (c) 30   (d) all of these

182. A professional-type projector employs _____-millimeter film.
(a) 35   (b) 70   (c) a or b   (d) none of these

183. The plans, specifications, and other building details for construction of manufactured buildings are included in the _____ details.
(a) manufactured buildings   (b) building components
(c) building structure   (d) building system

184. A _____ is a factory-assembled transportable structure that bears a label identifying it as a manufactured home built on a permanent chassis and designed to be used as a dwelling with or without a permanent foundation.
(a) manufactured home   (b) mobile home
(c) dwelling unit   (d) all of these

185. • A mobile home that is factory-equipped with gas or oil-fired heating and cooking appliances shall be permitted to be supplied with a listed mobile home power-supply cord rated _____ amperes.
(a) 30   (b) 35   (c) 40   (d) 50

186. Ground-fault circuit-interrupter (GFCI) protection in a mobile home is required for _____ .
(a) outlets installed outdoors
(b) receptacles within 6 feet of a lavatory
(c) a receptacle within a bathroom light fixture
(d) all of these

187. • A small mobile home park has six mobile homes. What is the total park electrical wiring system load after applying the demand factors permitted in Article 550?
(a) 4,640 VA   (b) 27,840 VA   (c) 96,000 VA   (d) none of these

188. • The working clearance for a distribution panelboard located in a recreational vehicle shall be no less than _____ .
(a) 24 inches wide   (b) 30 inches deep   (c) 30 inches wide   (d) a and b

189. • Electrical service and feeders of a recreational vehicle park shall be calculated at a minimum of _____VA per site (not including tent sites).
(a) 1,200   (b) 2,400   (c) 3,600   (d) 9,600

190. In locating receptacles in a marina, consideration should be given to _____ .
(a) the maximum tide level
(b) the minimum tide level
(c) wave action
(d) a and c

191. The demand percentage used to calculate 45 receptacles on a boat yard feeder is _____ percent.
(a) 90   (b) 80   (c) 70   (d) 50

Article 600 through Article 630

192. Each commercial building and occupancy with ground floor access for pedestrians shall have at least one outside sign outlet supplied by a _____-ampere branch circuit that supplies no other load.
(a) 15   (b) 20   (c) a or b   (d) none of these

193. Each sign and outline lighting system, or feeder circuit or branch circuit supplying a sign or outline lighting system shall be controlled by an externally operable switch or circuit breaker that opens all _____ conductors.
(a) ungrounded   (b) grounded   (c) grounding   (d) all of these

194. Sign and outline lighting enclosures for live parts other than lamps and neon tubing shall _____ .
(a) have ample structural strength and rigidity.
(b) be constructed of metal or shall be listed.
(c) if of sheet steel be at least 0.016 inches thick.
(d) all of the above

195. Ballasts, transformers, and electronic power supplies shall be located where accessible and shall be securely fastened in place and a working space at least _____ shall be provided.
(a) 3 feet high, 3 feet wide, by 3 feet deep
(b) 4 feet high, 3 feet wide, by 3 feet deep
(c) 6 feet high, 3 feet wide, by 3 feet deep
(d) none of these

196. Each section of a manufactured wiring system shall be marked to identify _____ .
(a) its location   (b) its type of cable or conduit
(c) the size of the wires installed   (d) suitability for wet or damp locations

197. The conductors to the hoistway door interlocks from the hoistway riser shall be suitable for a temperature of not less than _____°C and shall be type SF or equivalent.
(a) 200   (b) 60   (c) 90   (d) 110

198. A separate _____ is required for the elevator car lights, receptacle(s), auxiliary lighting power source, and ventilation.
(a) branch circuit   (b) disconnect   (c) connection   (d) none of these

199. All 125-volt, 15 and 20 ampere receptacle installed in machine rooms and spaces for elevators and escalators shall be GFCI protected by a _____ .
(a) GFCI receptacle
(b) GFCI circuit breaker
(c) GFCI protection not required
(d) a or b

200. Each arc welder shall have overcurrent protection rated or set at not more than _____ percent of the rated primary current of the welder.
(a) 100   (b) 125   (c) 150   (d) 200

Unit 8

Voltage Drop Calculations

OBJECTIVES

After reading this unit, the student should be able to briefly explain the following concepts:

Alternating current resistance as compared to direct current

Conductor resistance

Conductor resistance – alternating current circuits

Conductor resistance – direct current circuits

Determining circuit voltage drop

Extending circuits

Limiting current to limit voltage drop

Limiting conductor length to limit voltage drop

Resistance – alternating current

Sizing conductors to prevent excessive voltage drop

Voltage drop considerations

Voltage drop recommendations

After reading this unit, the student should be able to briefly explain the following terms:

American wire gauge

CM = circular mils

Conductor

Cross-sectional area

D = distance

Eddy currents

E_{VD} = Conductor voltage drop expressed in volts

I = load in ampere at 100 percent

K = direct current constant

Ohm's Law method

Q = alternating current adjustment factor

R = resistance

Skin effect

Temperature coefficient

VD = volts dropped

PART A – CONDUCTOR RESISTANCE CALCULATIONS

8–1 CONDUCTOR RESISTANCE

Metals that carry electric current are called conductors or wires and oppose the flow of electrons. Conductors can be solid, stranded, copper, or aluminum. The conductor's opposition to the flow of current (resistance) is determined by the material type (copper/aluminum), the cross-sectional area (wire size), and the conductor's length and operating temperature. The *resistance* of a conductor is expressed in ohms.

Material

Silver is the best conductor because it has the lowest resistance, but its high cost limits its use to special applications. *Aluminum* is often used when weight or cost are important considerations, but *copper* is the most common type of metal used for electrical conductors (Figure 8–1).

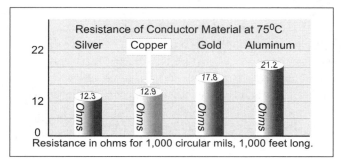

Figure 8–1
Resistance of Conductor Material at 75°C

231

Cross-Sectional Area

The *cross-sectional area* of a conductor is the conductor's surface area of a cross-section expressed in circular mils. The greater the conductor cross-sectional area, the greater the number of available electron paths and the lower the conductor resistance. Conductors are sized according to the *American Wire Gauge,* which ranges from a No. 40 to the largest, No. 4/0. Conductor resistance varies inversely with the conductor's diameter; that is, the smaller the wire size, the greater the resistance, and the larger the wire size, the lower the resistance (Figure 8–2).

Conductor Length

The resistance of a conductor is directly proportional to its length. The following table provides examples of conductor resistance and *circular mils area* for conductor lengths of 1,000 feet. Naturally, longer or shorter lengths will result in different conductor resistances.

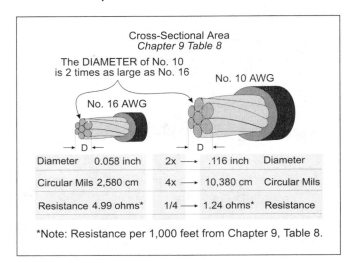

Figure 8–2
Cross-Sectional Area

Conductor Properties – *NEC*® Chapter 9, Table 8			
Conductor Size American Wire Gauge	Conductor Resistance Per 1,000 Feet at 75°C	Conductor Diameter	Conductor Area Circular Mils
No. 14	3.140 ohm (solid)	0.073	4,110
No. 12	1.980 ohm (stranded)	0.092	6,530
No. 10	1.240 ohm (stranded)	0.116	10,380
No. 8	0.778 ohm (stranded)	0.146	16,510
No. 6	0.491 ohm (stranded)	0.184	26,240

Temperature

The resistance of conductors changes with changing temperature; this is called *temperature coefficient.* Temperature coefficient describes the effect that temperature has on the resistance of a conductor. Positive temperature coefficient indicates that as the temperature rises, the conductor resistance will also rise. Examples of conductors that have a positive temperature coefficient are silver, copper, gold, and aluminum conductors. Negative temperate coefficient means that as the temperature increases, the conductor resistance decreases.

The conductor resistances listed in the *National Electrical Code*® Table 8 and 9 of Chapter 9 are based on an operating temperature of 75°C. A 3-degree change in temperature results in a 1 percent change in conductor resistance for both copper and aluminum conductors. The formula to determine the change in conductor resistance with changing temperature is listed at the bottom of Table 8. For example, the resistance of copper at 90°C is about 5 percent more than at 75°C.

8–2 CONDUCTOR RESISTANCE – DIRECT CURRENT CIRCUITS, [Chapter 9, Table 8]

The *National Electric Code*® lists the resistance and area circular mils for both direct current and alternating current circuit conductors. Direct current circuit conductor resistances are listed in Chapter 9, Table 8, and alternating current circuit conductor resistances are listed in Chapter 9, Table 9.

The *direct current conductor resistances* listed in Chapter 9, Table 8 apply to conductor lengths of 1,000 feet. The following formula can be used to determine the conductor resistance for conductor lengths other than 1,000 feet:

$$\text{Direct Current Conductor Resistance} = \frac{\text{Conductor Resistance Ohms}}{1,000 \text{ Feet}} \times \text{Conductor Length}$$

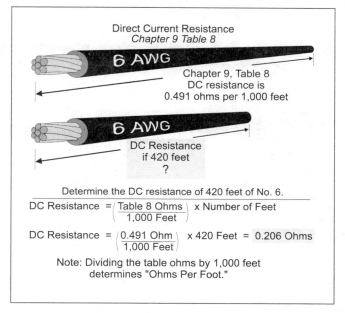

Figure 8–3
Direct Current Resistance

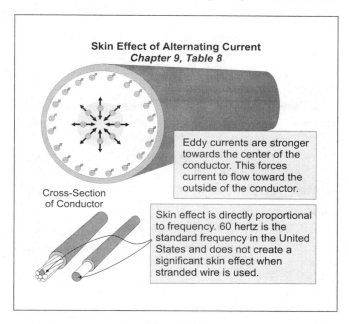

Figure 8–4
Skin Effect of Alternating Current

❏ Conductor Resistance Copper

What is the resistance of 420 feet of No. 6 copper (Figure 8–3)?

(a) 0.49 ohm   (b) 0.29 ohm   (c) 0.72 ohm   (d) 0.21 ohm

 • Answer: (d) 0.21 ohm

The resistance of No. 6 copper 1,000 feet long is 0.491 ohm, Chapter 9, Table 8.

The resistance of 420 feet is: (0.491 ohm/1,000 feet) $\times$ 420 feet = 0.206 ohm, rounded to 0.21

❏ Conductor Resistance Aluminum

What is the resistance of 1,490 feet of No. 3 aluminum?

(a) 0.60 ohm   (b) 0.29 ohm   (c) 0.72 ohm   (d) 0.21 ohm

 • Answer: (a) 0.60 ohm

The resistance of No. 3 aluminum 1,000 feet long is 0.403 ohm, Chapter 9, Table 8.

The resistance of 1,490 feet is: (0.403 ohm/1,000 feet) $\times$ 1,490 feet = 0.60 ohm

8–3 CONDUCTOR RESISTANCE – ALTERNATING CURRENT CIRCUITS

In direct current circuits, the only property that opposes the flow of electrons is resistance. In alternating current circuits, the expanding and collapsing magnetic field within the conductor induces an electromotive force that opposes the flow of alternating current. This opposition to the flow of alternating current is called *inductive reactance* which is measured in ohms.

In addition, alternating current flowing through a conductor generates small, erratic, independent currents called *eddy currents*. Eddy currents are greatest in the center of the conductors and repel the flowing electrons toward the conductor surface; this is known as *skin effect* (Figure 8–4).

Because of skin effect, the effective cross-sectional area of an alternating current conductor is reduced, which results in an increase of the conductor resistance. The total opposition to the flow of alternating current (resistance and inductive reactance) is called *impedance* and is measured in ohms.

8–4 ALTERNATING CURRENT RESISTANCE AS COMPARED TO DIRECT CURRENT

The opposition to current flow is greater for alternating current as compared to direct current circuits, because of inductive reactance, eddy currents, and skin effect. The following two tables give examples of the difference between alternating current circuits as compared to direct current circuits (Figure 8–5).

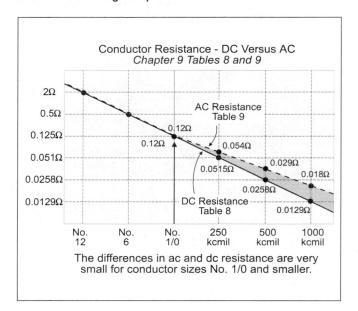

Figure 8–5
Conductor Resistance

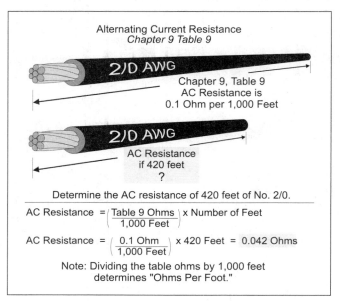

Figure 8–6
Alternating Current Resistance

COPPER – Alternating Current versus Direct Current Resistance at 75°C			
Conductor Size	Alternating Current Chapter 9, Table 9	Direct Current Chapter 9, Table 8	AC resistance greater than DC resistance by %
250,000	0.054 ohms per 1,000 feet	0.0515 ohms per 1,000 feet	4.85%
500,000	0.029 ohms per 1,000 feet	0.0258 ohms per 1,000 feet	12.40%
1,000,000	0.018 ohms per 1,000 feet	0.0129 ohms per 1,000 feet	39.50%
ALUMINUM – Alternating Current versus Direct Current Resistance at 75°C			
Conductor Size	Alternating Current Chapter 9, Table 9	Direct Current Chapter 9, Table 8	AC resistance greater than DC resistance by %
250,000	0.086 ohms per 1,000 feet	0.0847 ohms per 1,000 feet	1.5%
500,000	0.045 ohms per 1,000 feet	0.0424 ohms per 1,000 feet	6.13%
1,000,000	0.025 ohms per 1,000 feet	0.0212 ohms per 1,000 feet	17.12%

8–5 RESISTANCE ALTERNATING CURRENT [Chapter 9, Table 9 of the *NEC*®]

Alternating current conductor's resistances are listed in Chapter 9, Table 9 of the *NEC*®. The alternating current resistance of a conductor is dependent on the conductors material (copper or aluminum) and on the magnetic property of the raceway.

❏ Alternating Current Resistance

What is the alternating current resistance for a 250,000 circular mils conductor 1,000 feet long?

Copper conductor in nonmetallic raceway	= 0.052 ohms per 1,000 feet
Copper conductor in aluminum raceway	= 0.057 ohms per 1,000 feet
Copper conductor in steel raceway	= 0.054 ohms per 1,000 feet
Aluminum conductor in nonmetallic raceway	= 0.085 ohms per 1,000 feet
Aluminum conductors in aluminum raceway	= 0.090 ohms per 1,000 feet
Aluminum conductors in steel raceway	= 0.086 ohms per 1,000 feet

Alternating Current Conductor Resistance Formula

The following formula can be used to determine the resistance of different lengths of conductors:

$$\text{Alternating Current Resistance} = \frac{\text{Conductor Resistance Ohms}}{\text{1,000 Feet}} \times \text{Conductor Length}$$

What is the alternating current resistance of 420 feet of No. 2/0 copper installed in a metal raceway (Figure 8–6)?

(a) 0.069 ohm   (b) 0.042 ohm   (c) 0.072 ohm   (d) 0.021 ohm

   • Answer: (b) 0.042 ohm

   The resistance of No. 2/0 copper is 0.1 ohm per 1,000 feet, Chapter 9, Table 9.

   The resistance of 420 feet of No. 2/0 is: (0.1 ohm/1,000 feet) × 420 feet = 0.042 ohm

❑ Resistance Aluminum

What is the alternating current resistance of 169 feet of 500 kcmil installed in aluminum conduit?

(a) 0.0049 ohm   (b) 0.0029 ohm   (c) 0.0054 ohm   (d) 0.0021 ohm

   • Answer: (c) 0.0054 ohm

   The resistance of 500 kcmil installed in aluminum conduit is 0.032 ohm per 1,000 feet.

   Resistance of 169 feet of 500 kcmil in aluminum conduit: (0.032 ohm/1,000 feet) × 169 feet = 0.0054 ohm

Converting Copper to Aluminum or Aluminum to Copper

When requested to determine the replacement conductor for copper or aluminum the following steps should be helpful:

Step 1: ➤ Determine the resistance of the existing conductor using Table 9, Chapter 9 for 1,000 feet.

Step 2: ➤ Using Table 9, Chapter 9, locate a replacement conductor that has a resistance of not more than the existing conductors.

Step 3: ➤ Verify that the replacement conductor has an ampacity [Table 310-16] sufficient for the load.

❑ Aluminum to Copper

A 100-ampere, 240-volt, single-phase load is wired with No. 2/0 aluminum conductors in a steel raceway. What size copper wire can we use to replace the aluminum wires and not have a greater voltage drop?

Note: The wire selected must have an ampacity of at least 100 ampere (Figure 8–7).

(a) No. 1/0   (b) No. 1   (c) No. 2   (d) No. 3

   • Answer: (b) No. 1 copper

The resistance of No. 2/0 aluminum (steel raceway) is 0.16 ohms per 1,000 feet.

The resistance of No. 1 copper (steel raceway) is 0.16 ohms per 1,000 feet (ampacity of 130 ampere at 75°C).

Determining the Resistance of Parallel Conductors

The resistance total in a parallel circuit is always less than the smallest resistor. The equal resistors formula can be used to determine the resistance total of parallel conductors:

$$\text{Resistance Total} = \frac{\text{Resistance of One Conductor*}}{\text{Number of Parallel Conductors}}$$

*Resistance according to Chapter 9, Table 8 or Table 9 of the *NEC*®, assuming 1,000 feet unless specified otherwise.

❑ Resistance of Parallel Conductors

What is the direct current resistance for two 500 kcmil conductors in parallel (Figure 8–8)?

(a) 0.0129 ohms   (b) 0.0258 ohms   (c) 0.0518 ohms   (d) 0.0347 ohms

   • Answer: (a) 0.0129 ohm

$$\text{Resistance Total} = \frac{\text{Resistance of One Conductor}}{\text{Number of Parallel Conductors}} = \frac{0.0258 \text{ ohm}}{2 \text{ conductors}} = 0.0129 \text{ ohm}$$

Note: The resistance of 1,000 kcmil in Chapter 9, Table 8 is 0.0129 ohms.

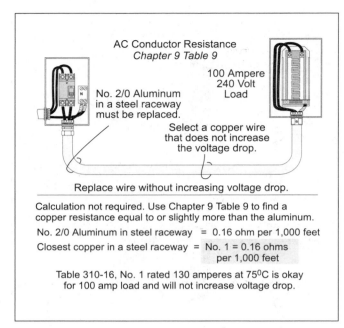

Figure 8–7
AC Conductor Resistance

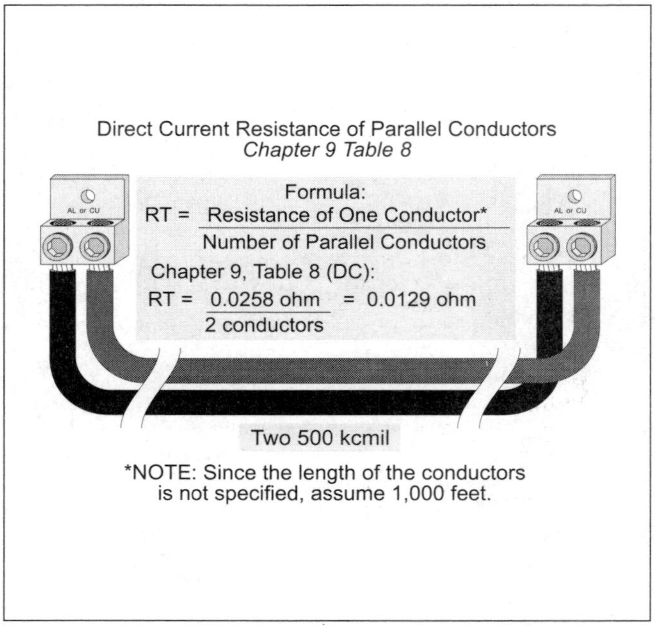

Figure 8–8
Direct Current Resistance of Parallel Conductors

PART B – VOLTAGE DROP CALCULATIONS

8–6 VOLTAGE DROP CONSIDERATIONS

The voltage drop of a circuit is in direct proportion to the conductors resistance and the magnitude (size) of the current. The longer the conductor, the greater the conductor resistance, the greater the conductor voltage drop; or, the greater the current, the greater the conductor voltage drop.

Note: Undervoltage for inductive loads can cause overheating, inefficiency, and a shorter life span for electrical equipment. This is especially true in such solid-state equipment as TVs, data processing equipment (e.g., computers), and similar equipment. When a conductor resistance causes the voltage to drop below an acceptable point, the conductor size should be increased (Figure 8–9).

8–7 *NEC*® VOLTAGE DROP RECOMMENDATIONS

Contrary to many beliefs, the *NEC*® does not contain any requirements for sizing ungrounded conductors for voltage drop. It does recommend in many areas of the *NEC*® that we consider the effects of conductor voltage drop when sizing conductors.

See some of these recommendations in the FPN to sections 210-19(a), 215-2, 230-31(c), and 310-15(a)(1). Please be aware that FPN in the *NEC*® are recommendations, *not* requirements [*Section 90-5(c)*]. The *NEC*® recommends that the maximum combined voltage drop for both the feeder and branch circuit should not exceed 5 percent, and the maximum on the feeder or branch circuit should not exceed 3 percent (Figure 8–10).

❑ *NEC*® **Voltage Drop Recommendation**

What are the minimum *NEC*® recommended operating volts for a 115-volt rated load that is connected to a 120/240 volt source (Figure 8–11)?

(a) 120 volts   (b) 115 volts   (c) 114 volts   (d) 116 volts

• Answer: (c) 114 volts

The maximum conductor voltage drop recommended for both the feeder and branch circuit is 5 percent of the voltage source (120 volts). The total conductor voltage drop (feeder and branch circuit) should not exceed 120 volts × 0.05 = 6 volts. The operating voltage at the load is calculated by subtracting the conductors voltage drop from the voltage source:
120 volts – 6 volts dropped = 114 volts.

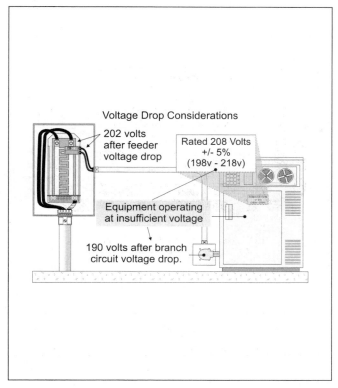

Figure 8–9
Voltage Drop Considerations

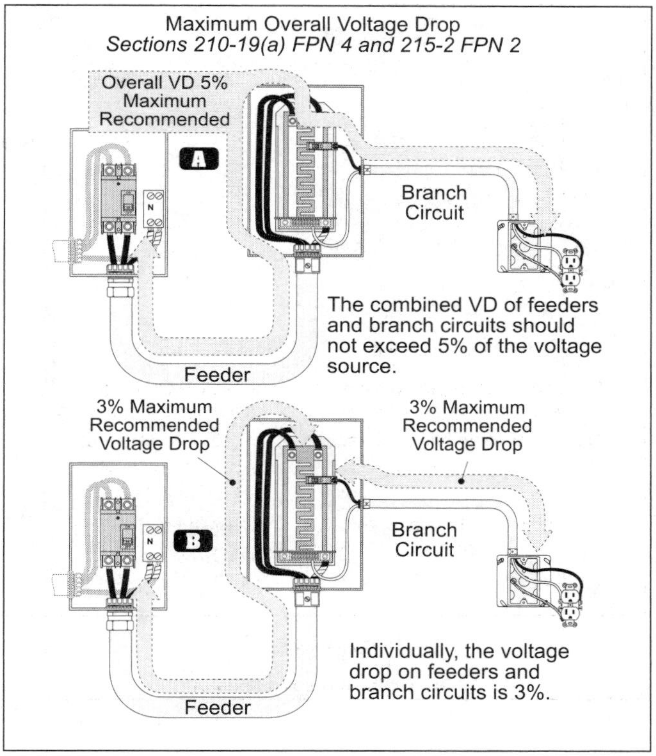

Figure 8–10
Maximum Overall Voltage Drop

8–8 DETERMINING CIRCUIT CONDUCTORS VOLTAGE DROP

When the circuit conductors have already been installed, the voltage drop of the conductors can be determined by the Ohm's Law method or by the formula method.

Ohm's Law Method – Single-Phase Only

VD = I × R

VD = Conductor voltage drop expressed in volts.

I = The load in ampere at 100 percent, not at 125 percent, for motors or continuous loads.

R* = Conductor Resistance, Chapter 9, Table 8 direct current; or Chapter 9, Table 9 for alternating current.

*For conductors No. 1/0 and smaller, the difference in resistance between direct current and alternating current circuits is so little that it can be ignored. In addition, you can ignore the small difference in resistance between stranded and solid wires.

❑ **Voltage Drop 120 Volt**

What is the voltage drop of two No. 12 THHN conductors that supply a 16-ampere, 120-volt load located 100 feet from the power supply (Figure 8–12)?

(a) 3.2 volts (b) 6.4 volts (c) 9.6 volts (d) 12.8 volts

• Answer: (b) 6.4 volts

VD = I × R

I = 16 ampere

R = 2 ohm per 1,000 feet, Chapter 9, Table 9: (2 ohm/1,000 feet) × 200 feet = 0.4 ohm

VD = 16 ampere × 0.4 ohm = 6.4 volts

❑ **Voltage Drop 240 Volt**

A single-phase, 24-ampere, 240-volt load is located 160 feet from the panelboard and is wired with No. 10 THHN. What is the voltage drop of the circuit conductors (Figure 8–13)?

(a) 4.53 volts (b) 9.22 volts (c) 3.64 volts (d) 5.54 volts

• Answer: (b) 9.22 volts

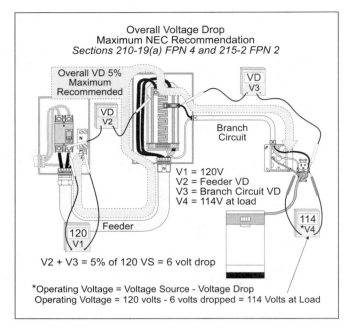

Figure 8–11
Overall Voltage Drop

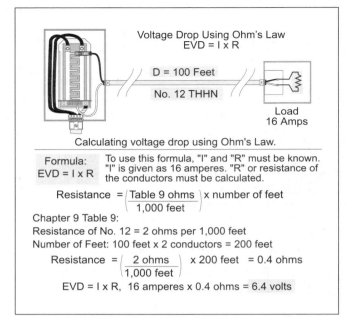

Figure 8–12
Voltage Drop Using Ohm's Law

VD = I × R

I = 24 ampere

R = 1.2 ohm per 1,000 feet, Chapter 9, Table 9: (1.2 ohm/1,000 feet) × 320 feet = 0.384 ohm

VD = 24 ampere × 0.384 ohm = 9.216 volts dropped

Voltage Drop Using the Formula Method

In addition to the Ohm's Law method, the following formula can be used to determine the conductor voltage drop:

$$\text{VD (single-phase)} = \frac{2 \times (K \times Q) \times I \times D}{CM}$$ $$\text{VD (three-phase)} = \frac{\sqrt{3} \times (K \times Q) \times I \times D}{CM}$$

Note: $\sqrt{3} = 1.732$

VD = Volts Dropped: The voltage drop of the circuit expressed in volts. The *NEC®* recommends a maximum 3 percent voltage drop for either the branch circuit or feeder.

K = Direct Current Constant: This constant K represents the direct current resistance for a 1,000 circular mils conductor that is 1,000 feet long, at an operating temperature of 75°C. The constant K value is 12.9 ohm for copper, and 21.2 ohm for aluminum.

I = Ampere: The load in ampere at 100 percent, (not at 125 percent for motors or continuous loads).

Q = Alternating Current Adjustment Factor: For alternating current circuits with conductors No. 2/0 and larger, the direct current resistance constant K must be adjusted for the effects of self-induction (eddy currents). The Q-Adjustment Factor is calculated by dividing the alternating current resistance listed in Chapter 9, Table 9 by the direct current resistance listed in Chapter 9, Table 8 in the *NEC®*. For all practical exam purposes, this resistance adjustment factor can be ignored because exams rarely give alternating current voltage drop questions with conductors larger than No. 1/0.

D = Distance: The distance the load is from the power supply.

CM = Circular-Mils: The circular mils of the circuit conductor as listed in *NEC®* Chapter 9, Table 8.

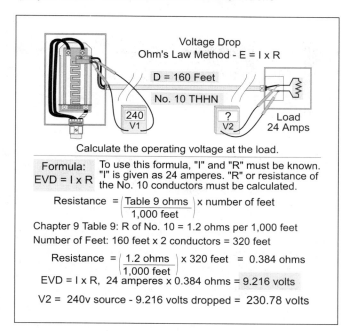

Voltage Drop
Ohm's Law Method - E = I x R

D = 160 Feet

No. 10 THHN

240 V1   ? V2   Load 24 Amps

Calculate the operating voltage at the load.

Formula: To use this formula, "I" and "R" must be known.
EVD = I x R "I" is given as 24 amperes. "R" or resistance of the No. 10 conductors must be calculated.

Resistance = $\left(\dfrac{\text{Table 9 ohms}}{1,000 \text{ feet}}\right)$ x number of feet

Chapter 9 Table 9: R of No. 10 = 1.2 ohms per 1,000 feet
Number of Feet: 160 feet x 2 conductors = 320 feet

Resistance = $\left(\dfrac{1.2 \text{ ohms}}{1,000 \text{ feet}}\right)$ x 320 feet = 0.384 ohms

EVD = I x R, 24 amperes x 0.384 ohms = 9.216 volts

V2 = 240v source - 9.216 volts dropped = 230.78 volts

Figure 8–13
Voltage Drop Ohm's Law Method

Voltage Drop
Formula Method

D = 160 Feet

No. 10 THHN

240 V1   ? V2   Load 24 Amps

Calculate the conductor voltage drop.

Formula: VD = ?
VD = $\dfrac{2 \times K \times I \times D}{CM}$ K = 12.9 ohms, copper
I = 24 amps
D = 160 feet
CM = 10,380

VD = $\dfrac{2 \text{ wires x } 12.9\,\Omega \text{ x } 24 \text{ amps x } 160 \text{ ft}}{10,380 \text{ circular mils}}$ = 9.54 volts dropped on circuit

Load = 240v Source - 9.54 Volts Dropped = 230.46 volts at load

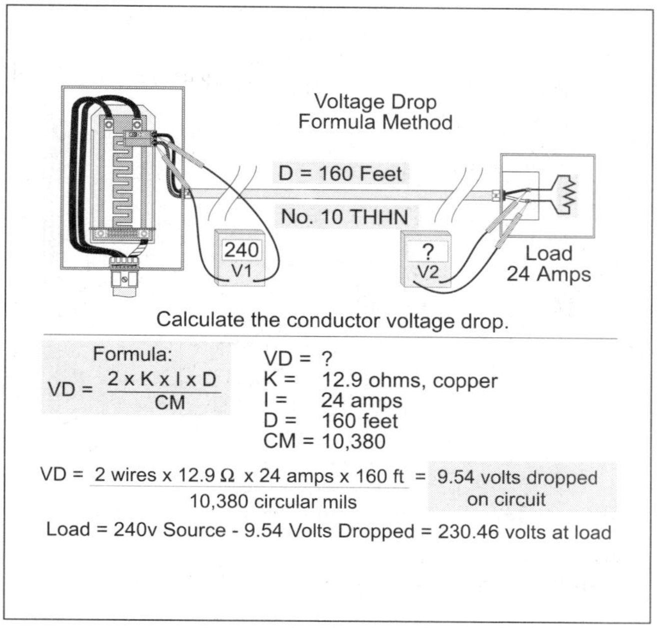

Figure 8–14
Voltage Drop Formula Method

❏ **Voltage Drop – Single-Phase**

A 24-ampere, 240-volt load is located 160 feet from a panelboard and is wired with No. 10 THHN. What is the approximate voltage drop of the branch circuit conductors (Figure 8–14)?

 (a) 4.25 volts   (b) 9.5 volts   (c) 3 percent   (d) 5 percent

 • Answer: (b) 9.5 volts is the closest

$$VD = \frac{2 \times K \times I \times D}{CM}$$

K = 12.9 ohm, copper

I = 24 ampere

D = 160 feet

CM = 10,380 Chapter 9, Table 8 (No. 10)

$$VD = \frac{2 \text{ wires} \times 12.9 \text{ ohms} \times 24 \text{ amps} \times 160 \text{ feet}}{10,380 \text{ circular mils}} = 9.54 \text{ volts dropped}$$

❏ **Voltage Drop – Three-Phase**

A 3-phase, 36 kVA load rated 208 volts is located 80 feet from the panelboard and is wired with No. 1 THHN aluminum. What is the approximate voltage drop of the feeder circuit conductors (Figure 8–15)?

 (a) 3.5 volts   (b) 7 volts   (c) 3 percent   (d) 5 percent

 • Answer: (a) 3.5 volts is the closest

$$VD = \frac{\sqrt{3} \times K \times I \times D}{CM}$$

$\sqrt{3} = 1.732$

K = 21.2 ohm, aluminum

$$I = \frac{VA}{(E \times \sqrt{3})} = \frac{36,000 \text{ VA}}{(208 \text{ volts} \times 1.732)} = 100 \text{ ampere}$$

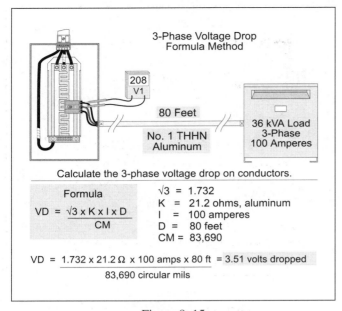

Figure 8–15
3-Phase Voltage Drop Formula Method

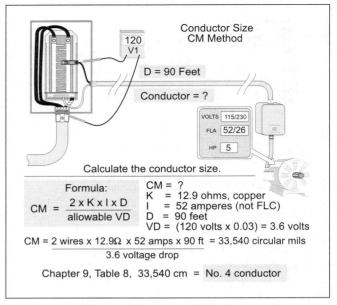

Figure 8–16
Conductor Size CM Method

D = 80 feet

CM = 83,690 Chapter 9, Table 8

$$VD = \frac{1.732 \times 21.2 \text{ ohms} \times 100 \text{ amps} \times 80 \text{ feet}}{83,690 \text{ circular mils}} = 3.51 \text{ volts dropped}$$

8–9 SIZING CONDUCTORS TO PREVENT EXCESSIVE VOLTAGE DROP

The size of a conductor (actually its resistance) affects voltage drop. If we want to decrease the voltage drop of a circuit, we can increase the size of the conductor (reduce its resistance). When sizing conductors to prevent excessive voltage drop, use the following formulas:

$$\textbf{CM (single-phase)} = \frac{\textbf{2} \times \textbf{K} \times \textbf{I} \times \textbf{D}}{\textbf{Volts Dropped}} \qquad \textbf{CM (three-phase)} = \frac{\sqrt{\textbf{3}} \times \textbf{K} \times \textbf{I} \times \textbf{D}}{\textbf{Volts Dropped}}$$

❏ **Size Conductor – Single-Phase**

A 5 horsepower motor is located 90 feet from a 120/240-volt panelboard. What size conductor should be used if the motor nameplate indicates 52 ampere at 115 volts? Terminals rated for 75°C (Figure 8–16).

(a) No. 10 THHN   (b) No. 8 THHN   (c) No. 6 THHN   (d) No. 4 THHN

• Answer: (d) No. 4 THHN

$$CM = \frac{2 \times K \times I \times D}{VD}$$

K = 12.9 ohm, copper

I = 52 ampere at 115 volts

D = 90 feet

VD = 120 volts × 0.03 = 3.6 volts

$$CM = \frac{2 \text{ wires} \times 12.9 \text{ ohms} \times 52 \text{ ampere} \times 90 \text{ feet}}{3.6 \text{ volts}} = 33,540 \text{ (No. 4, Chapter 9, Table 8)}$$

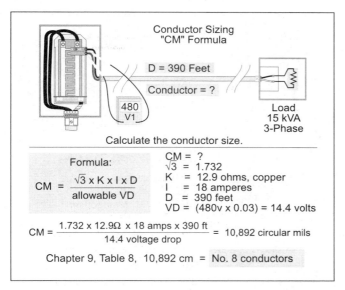

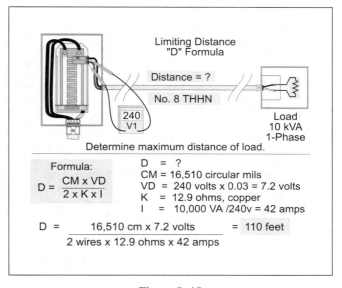

Figure 8–17
Conductor Sizing "CM" Formula

Figure 8–18
Limiting Distance "D" Formula

Note: *NEC®* Section 430-22(a), requires that the motor conductors be sized not less than 125 percent of the motor full-load currents as listed in Table 430-148. The motor full-load current is 56 ampere, and the conductor must be sized at 56 ampere × 1.25 = 70 ampere. The No. 4 THHN required for voltage drop is rated for 75 ampere at 75°C according to Table 310-16 and Section 110-14(c).

❏ **Size Conductor – Three-Phase**

A 3-phase, 15 kVA load rated 480 volts is located 390 feet from the panelboard. What size conductor is required to prevent the voltage drop from exceeding 3 percent (Figure 8–17)?

    (a) No. 10 THHN   (b) No. 8 THHN   (c) No. 6 THHN   (d) No. 4 THHN

    • Answer: (b) No. 8 THHN

$$CM = \frac{\sqrt{3} \times K \times I \times D}{VD}$$

K = 12.9 ohm, copper

$$I = 18 \text{ ampere } \frac{VA}{(E \times 1.732)} = \frac{15,000 \text{ VA}}{(480 \text{ volts} \times 1.732)} = 18 \text{ ampere}$$

D = 390 feet

VD = 480 volts × 0.03 = 14.4 volts

$$CM = \frac{1.732 \times 12.9 \text{ ohms} \times 18 \text{ ampere} \times 390 \text{ feet}}{14.4 \text{ volts}} = 10,892 \text{ (No. 8, Chapter 9, Table 8)}$$

8–10 LIMITING CONDUCTOR LENGTH TO LIMIT VOLTAGE DROP

Voltage drop can also be reduced by limiting the length of the conductors. The following formulas can be used to help determine the maximum conductor length to limit the voltage drop to *NEC®* suggestions:

$$\textbf{D (single-phase)} = \frac{\textbf{CM} \times \textbf{VD}}{2 \times \textbf{K} \times \textbf{I} \sqrt{3}} \qquad\qquad \textbf{D (three-phase)} = \frac{\textbf{CM} \times \textbf{VD}}{\sqrt{3} \times \textbf{K} \times \textbf{I}}$$

❏ **Distance – Single-Phase**

What is the maximum distance a single-phase, 10 kVA, 240-volt load can be located from the panelboard so the voltage drop does not exceed 3 percent? The load is wired with No. 8 THHN (Figure 8–18).

    (a) 55 feet   (b) 110 feet   (c) 165 feet   (d) 220 feet

    • Answer: (b) 110 feet

$$D = \frac{CM \times VD}{2 \times K \times I}$$

CM = 16,510 (No. 8), Chapter 9, Table 8

VD = 240 volts × 0.03 = 7.2 volts

K = 12.9 ohm, copper

I = VA/E = 10,000 VA/240 volts = 42 amperes

$$D = \frac{16,510 \text{ circular mils} \times 7.2 \text{ volts}}{2 \text{ wires} \times 12.9 \text{ ohms} \times 42 \text{ ampere}} = 110 \text{ feet}$$

❏ **Distance – Three-Phase**

What is the maximum distance a 3-phase, 37.5 kVA, 480-volt transformer, wired with No. 6 THHN can be located from the panelboard so that the voltage drop does not exceed 3 percent (Figure 8–19)?

(a) 275 feet   (b) 325 feet

(c) 375 feet   (d) 425 feet

   • Answer: (c) 375 feet

$$D = \frac{CM \times VD}{\sqrt{3} \times K \times I}$$

CM = 26,240 (No. 6), Chapter 9, Table 8

K = 12.9 ohm, copper

VD = 480 volts × 0.03 = 14.4 volts

$$I = \frac{VA}{(\text{Volts} \times 1.732)} = \frac{37,500 \text{ VA}}{(480 \text{ volts} \times 1.732)} = 45 \text{ ampere}$$

$$D = \frac{26,240 \text{ circular mils} \times 14.4 \text{ volts}}{1.732 \times 12.9 \text{ ohms} \times 45 \text{ ampere}} = 376 \text{ feet}$$

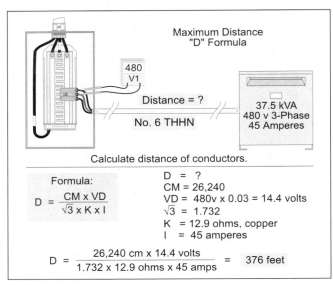

Maximum Distance "D" Formula

480 V1

Distance = ?

No. 6 THHN

37.5 kVA
480 v 3-Phase
45 Amperes

Calculate distance of conductors.

Formula:

$$D = \frac{CM \times VD}{\sqrt{3} \times K \times I}$$

D = ?
CM = 26,240
VD = 480v x 0.03 = 14.4 volts
√3 = 1.732
K = 12.9 ohms, copper
I = 45 amperes

$$D = \frac{26,240 \text{ cm} \times 14.4 \text{ volts}}{1.732 \times 12.9 \text{ ohms} \times 45 \text{ amps}} = \boxed{376 \text{ feet}}$$

Figure 8–19
Maximum Distance "D" Formula

8–11 LIMITING CURRENT TO LIMIT VOLTAGE DROP

Sometimes the only method of limiting the circuit voltage drop is to limit the load on the conductors. The following formulas can be used to determine the maximum load.

$$I \text{ (single-phase)} = \frac{CM \times VD}{2 \times K \times D}$$ $$I \text{ (three-phase)} = \frac{CM \times VD}{\sqrt{3} \times K \times D}$$

❏ **Maximum Load – Single-Phase**

An existing installation contains No. 1/0 THHN aluminum conductors in a nonmetallic raceway to a panelboard located 220 feet from a 230-volt power source. What is the maximum load that can be placed on the panelboard so that the *NEC*® recommendations for voltage drop are not exceeded (Figure 8–20)?

(a) 51 ampere   (b) 78 ampere   (c) 94 ampere   (d) 115 ampere

   • Answer: (b) 78 ampere

$$I = \frac{CM \times VD}{2 \times K \times D}$$

CM = 105,600 (No. 1/0), Chapter 9, Table 8

VD = 230 volts × 0.03 = 6.9 volts

K = 21.2 ohm, aluminum

D = 220 feet

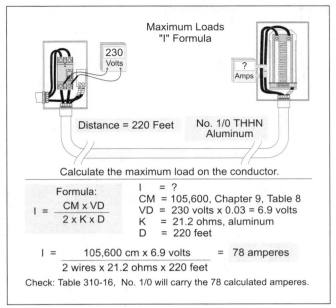

Calculate the maximum load on the conductor.

Formula:

$$I = \frac{CM \times VD}{2 \times K \times D}$$

I = ?
CM = 105,600, Chapter 9, Table 8
VD = 230 volts x 0.03 = 6.9 volts
K = 21.2 ohms, aluminum
D = 220 feet

$$I = \frac{105,600 \text{ cm} \times 6.9 \text{ volts}}{2 \text{ wires} \times 21.2 \text{ ohms} \times 220 \text{ feet}} = 78 \text{ amperes}$$

Check: Table 310-16, No. 1/0 will carry the 78 calculated amperes.

Figure 8–20
Maximum Loads "I" Formula

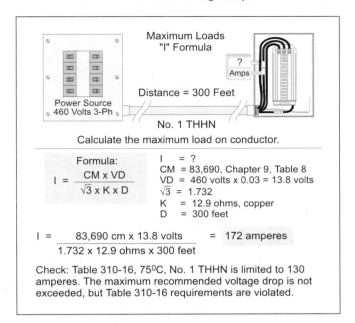

Calculate the maximum load on conductor.

Formula:

$$I = \frac{CM \times VD}{\sqrt{3} \times K \times D}$$

I = ?
CM = 83,690, Chapter 9, Table 8
VD = 460 volts x 0.03 = 13.8 volts
√3 = 1.732
K = 12.9 ohms, copper
D = 300 feet

$$I = \frac{83,690 \text{ cm} \times 13.8 \text{ volts}}{1.732 \times 12.9 \text{ ohms} \times 300 \text{ feet}} = 172 \text{ amperes}$$

Check: Table 310-16, 75°C, No. 1 THHN is limited to 130 amperes. The maximum recommended voltage drop is not exceeded, but Table 310-16 requirements are violated.

Figure 8–21
Maximum Loads "I" Formula

$$I = \frac{105,600 \text{ circular mils} \times 6.9 \text{ volts drop}}{2 \text{ wires} \times 21.2 \text{ ohms} \times 220 \text{ feet}} = 78 \text{ ampere}$$

Note: The maximum load permitted on 1/0 THHN Aluminum at 75°C is 120 ampere [Table 310-16].

❏ **Maximum Load – Three-Phase**

An existing installation contains No. 1 THHN conductors in an aluminum raceway to a panelboard located 300 feet from a 3-phase 460/230-volt power source. What is the maximum load the conductors can carry so that the *NEC*® recommendation for voltage drop is not exceeded (Figure 8–21)?

(a) 170 ampere (b) 190 ampere (c) 210 ampere (c) 240 ampere

• Answer: (a) 170 ampere

$$I = \frac{CM \times VD}{\sqrt{3} \times K \times D}$$

CM = 83,690 (No. 1), Chapter 9, Table 8

VD = 460 volts × 0.03 = 13.8 volts

K = 12.9 ohm, copper

D = 300 feet

$$I = \frac{(83,690 \text{ circular mils} \times 13.8 \text{ volts})}{(1.732 \times 12.9 \text{ ohms} \times 300 \text{ feet})} = 172 \text{ ampere}$$

Note: The maximum load permitted on No. 1 THHN at 75°C is 130 ampere [*Section 110-14(c)* and *Table 310-16*].

8–12 EXTENDING CIRCUITS

If you want to extend an existing circuit and you want to limit the voltage drop, follow these steps:

Step 1: ➥ Determine the voltage drop of the existing conductors.

$$VD \text{ (single-phase)} = \frac{2 \times K \times I \times D}{CM}$$

$$VD \text{ (three-phase)} = \frac{\sqrt{3} \times K \times I \times D}{CM}$$

Step 2: ⇒ Determine the voltage drop permitted for the extension by subtracting the voltage drop of the existing conductors from the permitted voltage drop.

Step 3: ⇒ Determine the extended conductor size.

$$CM \text{ (single-phase)} = \frac{2 \times K \times I \times D}{VD}$$

$$CM \text{ (three-phase)} = \frac{\sqrt{3} \times K \times I \times D}{VD}$$

❏ **Extending Circuits**

An existing junction box is located 55 feet from the panelboard and contains No. 4 THW aluminum. This circuit is to be extended 65 feet and supply a 50-ampere, 240-volt load. What size copper conductors must be used for the extension (Figure 8–22)?

(a) No. 8 THHN   (b) No. 6 THHN
(c) No. 4 THHN   (d) No. 5 THHN

• Answer: (b) No. 6 THHN

Step 1: ⇒ Determine the voltage drop of the existing conductors:

$$\text{Single-Phase VD} = \frac{2 \times K \times I \times D}{CM}$$

K = 21.2 ohm, aluminum
D = 55 feet
I = 50 ampere
CM = No. 4 (41,740 circular mils, Chapter 9, Table 8)

$$VD = \frac{2 \text{ wires} \times 21.2 \text{ ohms} \times 50 \text{ amps} \times 55 \text{ feet}}{41,740 \text{ circular mils}} = 2.79 \text{ volts}$$

Step 2: ⇒ Determine the voltage drop permitted for the extension by subtracting the voltage drop of the existing conductors from the total permitted voltage drop.

Total permitted voltage drop = 240 volts × 0.03 = 7.2 volts less 2.79 volts = 4.41 volts

Step 3: ⇒ Determine the extended conductor size.

$$CM = \frac{2 \times K \times I \times D}{VD}$$

K = 12.9 ohm, copper
I = 50 ampere
D = 65 feet
VD = 7.2 volts total less 2.79 volts = 4.41 volts

$$CM = \frac{2 \text{ wires} \times 12.9 \text{ ohms} \times 50 \text{ amps} \times 65 \text{ feet}}{4.41 \text{ volts drop}} = 19,014 \text{ (No. 6, Chapter 9, Table 8)}$$

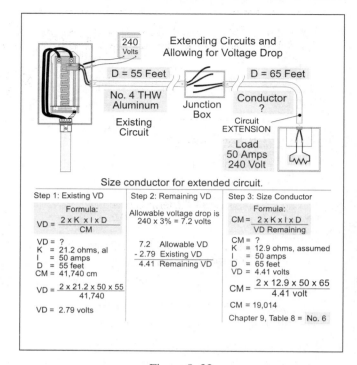

Size conductor for extended circuit.

Step 1: Existing VD	Step 2: Remaining VD	Step 3: Size Conductor
Formula: $VD = \frac{2 \times K \times I \times D}{CM}$	Allowable voltage drop is 240 x 3% = 7.2 volts	Formula: $CM = \frac{2 \times K \times I \times D}{VD \text{ Remaining}}$
VD = ? K = 21.2 ohms, al I = 50 amps D = 55 feet CM = 41,740 cm	7.2 Allowable VD - 2.79 Existing VD 4.41 Remaining VD	CM = ? K = 12.9 ohms, assumed I = 50 amps D = 65 feet VD = 4.41 volts
$VD = \frac{2 \times 21.2 \times 50 \times 55}{41,740}$		$CM = \frac{2 \times 12.9 \times 50 \times 65}{4.41 \text{ volt}}$
VD = 2.79 volts		CM = 19,014
		Chapter 9, Table 8 = No. 6

Figure 8–22
Extending Circuits

Unit 8 – Voltage Drop Summary Questions

8–1 Conductor Resistance

1. The _____ the number of free electrons, the better the conductivity of the conductor.
   (a) greater   (b) fewer

2. Conductor resistance is determined by the _____ .
   (a) material type   (b) cross-sectional area   (c) conductor length   (d) all of these

3. _____ is the best conductor, better than gold, but the high cost limits its use to special applications, such as fuse elements and some switch contacts.
   (a) Silver   (b) Copper   (c) Aluminum   (d) none of these

4. _____ conductors are often used when weight or cost are an important consideration.
   (a) Silver   (b) Copper   (c) Aluminum   (d) none of these

5. • Conductor cross-sectional area is expressed in _____ .
   (a) square inches   (b) mils   (c) circular mils   (d) none of these

6. The resistance of a conductor is directly proportional to its length.
   (a) True   (b) False

7. The resistance of a conductor changes with temperature. Temperature coefficient describes the effect that temperature has on the resistance of a conductor. Conductors with a _____ temperature coefficient have an increase in resistance with an increase in temperature.
   (a) positive   (b) negative   (c) neutral   (d) none of these

8–2 Conductor Resistance – Direct Current Circuits [Chapter 9, Table 8]

8. The *National Electrical Code®* lists the resistance and area in circular mils for both direct current and alternating current conductors. Direct current conductor resistances are listed in Chapter 9, Table _____ and alternating current conductor resistance are listed in Chapter 9, Table _____ .
   (a) 1, 5   (b) 3, 4   (c) 9, 8   (d) 8, 9

9. What is the direct current resistance of 400 feet of No. 6?
   (a) 0.2 ohm   (b) 0.3 ohm   (c) 0.4 ohm   (d) 0.5 ohm

10. What is the direct current resistance of 150 feet of No. 1?
    (a) 0.023 ohm   (b) 0.031 ohm   (c) 0.042 ohm   (d) 0.056 ohm

11. What is the direct current resistance of 1,400 feet of No. 3 AL?
    (a) 0.23 ohm   (b) 0.31 ohm   (c) 0.42 ohm   (d) 0.56 ohm

12. What is the direct current resistance of 800 feet of No. 1/0 AL?
    (a) 0.23 ohm   (b) 0.16 ohm   (c) 0.08 ohm   (d) 0.56 ohm

13. What is the direct current resistance of 120 feet of No. 1?
    (a) 0.23 ohm   (b) 0.16 ohm   (c) 0.02 ohm   (d) 0.56 ohm

14. What is the direct current resistance of 1,249 feet of No. 3 AL?
    (a) 0.23 ohm   (b) 0.16 ohm   (c) 0.02 ohm   (d) 0.5 ohm

8–3 Conductor Resistance – Alternating Current Circuits

15. The intensity of the magnetic field is dependent on the intensity of alternating current. The greater the current flow, the greater the overall magnetic field.
    (a) True   (b) False

16. • The expanding and collapsing magnetic field within the conductor exerts a force on the moving electrons. This force is called counter-electromotive force (CEMF).
    (a) True   (b) False

17. _____ currents are small independent currents that are produced as a result of the expanding and collapsing magnetic field. They flow erratically within the conductor opposing current flow and consuming power.
    (a) Lenz   (b) Ohm's   (c) Eddy   (d) Kerchoff's

18. The expanding and collapsing magnetic field induces a counter voltage within the conductors, which repels the flowing electrons towards the conductor surface. This is known as _____ effect.
    (a) inductive   (b) skin   (c) surface   (d) watt

8–4 Alternating Current Resistance As Compared to Direct Current

19. The opposition to current flow is greater for alternating current circuits because of _____ than for direct current circuits.
    (a) eddy currents   (b) skin effect   (c) CEMF   (d) all of these

8–5 Conductor Resistance – Alternating Current [Chapter 9, Table 9 of The *NEC®*]

20. The alternating current conductor resistances listed in Chapter 9, Table 9 of the *NEC®* are different for copper and aluminum and for nonmagnetic and magnetic raceways.
    (a) True   (b) False

21. What is the alternating current resistance of 320 feet of No. 2/0?
    (a) 0.03 ohm   (b) 0.04 ohm   (c) 0.05 ohm   (d) 0.06 ohm

22. What is the alternating current resistance of 1,000 feet of 500 kcmil when installed in an aluminum raceway?
    (a) 0.032 ohm   (b) 0.027 ohm   (c) 0.029 ohm   (d) 0.030 ohm

23. What is the alternating current resistance of 220 feet of No. 2/0?
    (a) 0.012 ohm   (b) 0.022 ohm   (c) 0.33 ohm   (d) 0.43 ohm

24. What is the alternating current resistance of 369 feet of 500 kcmil installed in nonmetallic conduit?
    (a) 0.01 ohm   (b) 0.02 ohm   (c) 0.03 ohm   (d) 0.04 ohm

25. A 36-ampere load is located 80 feet from the panelboard and is wired with No. 1 THHN aluminum. What is the total resistance of the circuit conductors?
    (a) 0.04 ohm   (b) 0.25 ohm   (c) 0.5 ohm   (d) all of these

26. What size copper conductors can be used to replace No. 1/0 aluminum that supplies a 110-ampere load? Note: We do not want to increase the circuit voltage drop.
    (a) No. 3   (b) No. 2   (c) No. 1   (d) No. 1/0

27. What is the direct current resistance in ohms for three 300 kcmil conductors in parallel, 1,000 feet long?
    (a) 0.014 ohm   (b) 0.026 ohm   (c) 0.052 ohm   (d) 0.047 ohm

28. What is the ac resistance in ohms for three 1/0 THHN aluminum conductors in parallel?
    (a) 0.1 ohm   (b) 0.2 ohm   (c) 0.4 ohm   (d) 0.067 ohm

8–6 Voltage Drop Consideration

29. Because of the great demand for electricity, utilities sometimes are required to reduce their output voltage. In addition, the utility and customer transformers, services, feeders, and branch circuit conductors oppose the flow of current. The opposition to current flow results in voltage drop. All circuits have voltage drop, simply because all conductors have resistance.
     (a) True   (b) False

30. When sizing conductors for feeders and branch circuits, the *NEC*® _____ that we take voltage drop into consideration.
     (a) permits   (b) suggests   (c) requires   (d) demands

31. _____ equipment such as motors and electromagnetic ballast can overheat at reduced voltage. This results in reduced equipment operating life and inconvenience to the customer.
     (a) Inductive   (b) Electronic   (c) Resistive   (d) all of these

32. • _____ equipment such as computers, laser printers, copy machines, etc., can suddenly power down because of reduced voltage, resulting in data losses.
     (a) Inductive   (b) Electronic   (c) Resistive   (d) all of these

33. Resistive loads such as incandescent lighting and electric space heating can have their power output decreased by the square of the voltage.
     (a) True   (b) False

34. What is the power consumed of a 4.5 kW, 230-volt water heater operating at 200 volts?
     (a) 2,700 watts   (b) 3,400 watts   (c) 4,500 watts   (d) 5,500 watts

35. How can conductor voltage drop be reduced?
     (a) reduce conductor resistance
     (b) increase conductor size
     (c) decrease conductor length
     (d) all of these

8–7 *NEC*® Voltage Drop Recommendations

36. If the branch circuit supply voltage is 208 volts, the maximum recommended voltage drop of the circuit should not be more than _____ .
     (a) 3.6 volts   (b) 6.24 volts   (c) 6.9 volts   (d) 7.2 volts

37. If the feeder supply voltage is 240 volts, the maximum recommended voltage drop of the feeder should not be more than _____ .
     (a) 3.6 volts   (b) 6.24 volts   (c) 6.9 volts   (d) 7.2 volts

8–8 Determining Circuit Conductors Voltage Drop

38. What is the voltage drop of two No. 12 THHN conductors supplying a 12-ampere continuous load? Note: The continuous load is located 100 feet from the power supply.
     (a) 3.2 volts   (b) 4.75 volts   (c) 6.4 volts   (d) 12.8 volts

39. A single-phase, 24-ampere, 240-volt load is located 160 feet from the panelboard. The load is wired with No. 10 THHN. What is the approximate voltage drop of the branch circuit conductors?
     (a) 4.25 volts   (b) 9.5 volts   (c) 3 percent   (d) 5 percent

40. A 3-phase, 36 kVA load, rated 208 volts is located 100 feet from the panelboard and is wired with No. 1 THHN aluminum. What is the approximate voltage drop of the circuit conductors?
     (a) 3.5 volts   (b) 5 volts   (c) 3 volts   (d) 4.4 volts

8–9 Sizing Conductors to Prevent Excessive Voltage Drop

41. A single-phase, 5 horsepower motor is located 110 feet from a panelboard. The nameplate indicates that the voltage is 115/230 and the FLA is 52/26 ampere. What size conductor is required if the motor winding are connected in parallel and operates at 115 volts?
     (a) No. 10 THHN   (b) No. 8 THHN   (c) No. 6 THHN   (d) No. 3 THHN

42. A single-phase, 5 horsepower motor is located 110 feet from a panelboard. The nameplate indicates that the voltage is 115/230 and the FLA is 52/26 ampere. What size conductor is required if the motor winding are connected in series and operates at 230 volts?
     (a) No. 10 THHN   (b) No. 8 THHN   (c) No. 6 THHN   (d) No. 4 THHN

43. A 3-phase, 15 kW, 480-volt load is located 300 feet from the panelboard. What size conductor is required to prevent the voltage drop from exceeding 3 percent?
     (a) No. 10 THHN   (b) No. 8 THHN   (c) No. 6 THHN   (d) No. 4 THHN

8–10 Limiting Conductor Length to Limit Voltage Drop

44. What is the approximate distance a single-phase, 7.5 kVA, 240-volt load can be located from the panelboard so the voltage drop does not exceed 3 percent? The load is wired with No. 8 THHN.
     (a) 55 feet   (b) 110 feet   (c) 145 feet   (d) 220 feet

45. What is the approximate distance a 3-phase, 37.5 kVA, 460-volt transformer, wired with No. 6 THHN, can be located from the panelboard so the voltage drop does not exceed 3 percent?
     (a) 250 feet   (b) 300 feet   (c) 345 feet   (d) 400 feet

8–11 Limiting Current to Limit Voltage Drop

46. An existing installation consists of No. 1/0 THHN copper conductors in a nonmetallic raceway to a panelboard located 200 feet from a 240-volt power source. What is the maximum load that can be placed on the panelboard so that the *NEC*® recommendations for voltage drop are not exceeded?
     (a) 94 ampere   (b) 109 ampere   (c) 71 ampere   (d) 147 ampere

47. An existing installation contains No. 2 THHN aluminum conductors in an aluminum raceway to a panelboard located 300 feet from a 3-phase 460/230-volt power source. What is the maximum load the conductors can carry without exceeding the *NEC*® recommendation for voltage drop?
     (a) 83 ampere   (b) 85 ampere   (c) 64 ampere   (d) 49 ampere

8–12 Extending Circuits

48. An existing junction box is located 65 feet from the panelboard and contains No. 4 THHN aluminum conductors. What size copper conductors can be used to extend this circuit 85 feet and supply a 50-ampere, 208-volt load?
     (a) No. 8 THHN   (b) No. 6 THHN   (c) No. 4 THHN   (d) No. 5 THHN

8–13 Miscellaneous

49. What is the circuit voltage if the conductor voltage drop is 3.3 volts? Assume 3 percent voltage drop.
     (a) 110 volts   (b) 115 volts   (c) 120 volts   (d) none of these

☆ Challenge Questions

8–1 Conductor Resistance

50. The resistance of a conductor is affected by temperature change. This is called the _____ .
     (a) temperature correction factor   (b) temperature coefficient
     (c) ambient temperature factor   (d) none of these

8–3 Conductor Resistance – Alternating Current Circuits

51.   The total opposition to current flow in an alternating current circuit is expressed in ohms and is called _____ .
       (a) impedance   (b) conductance   (c) reluctance   (d) resistance

8–5 Resistance Alternating Current [Chapter 9, Table 9 of the *NEC®*]

52.   A 40-ampere, 240-volt, single-phase load is located 150 feet from an existing junction box. The junction box is located 50 feet from the panelboard and is wired with No. 4 THHN aluminum wire. The total resistance of the two No. 4 conductors from the panelboard to the junction box is approximately _____ ohm.
       (a) 0.03   (b) 0.09   (c) 0.05   (d) 0.04

53.   A load is located 100 feet from a 230-volt power supply and is wired with No. 4 THHN aluminum conductors. What size copper conductor can be used to replace the aluminum conductors and not increase the conductor voltage drop?
       (a) No. 6   (b) No. 8   (c) No. 1   (d) No. 1/0

8–7 *NEC®* Voltage Drop Recommendations

54.   A 40-ampere, 240-volt rated, single-phase load is wired 150 feet from a junction box, and the junction box is located 50 feet from a panelboard (for a total of 200 feet). If the voltage at the panelboard is 240 volts, what is the minimum voltage recommended by the *NEC®* at the 40-ampere load?
       (a) 228.2 volts   (b) 232.8 volts   (c) 236.2 volts   (d) 117.7 volts

8–8 Determining Circuit Conductors Voltage Drop

55.   What is the voltage drop of 2 No. 4 aluminum conductors that supply a 5 horsepower single-phase, 208-volt motor that has a nameplate rating of 55 ampere? The motor is located 95 feet from the power supply.
       (a) 3.25 volts   (b) 5.31 volts   (c) 6.24 volts   (d) 7.26 volts

8–9 Sizing Conductors to Prevent Excessive Voltage Drop

56.   A 40-ampere, 240-volt, single-phase load is located 150 feet from an existing junction box. The junction box is located 50 feet from the panelboard. When the 40-ampere load is on, the voltage at the junction box would be calculated to be 236 volts. The *NEC®* recommends voltage drop for this branch circuit not exceed 3 percent of the 240 voltage source (7.2 volts). What size conductor could be installed from the junction box to the load and still meet *NEC®* recommendations?
       (a) No. 3   (b) No. 1   (c) No. 1/0   (d) No. 6

8–10 Limiting Conductor Length to Limit Voltage Drop

57.   How far can a 50-ampere, 3-phase, 230-volt load be located from the panel, if fed with No. 3 THHN, and still meet *NEC®* recommendations for voltage drop?
       (a) 275 feet   (b) 300 feet   (c) 325 feet   (d) 350 feet

8–11 Limiting Current to Limit Voltage Drop

58.   Two No. 8 THHN supply a 120-volt load that is located 225 feet from the panelboard. What is the maximum load, in ampere, that can be applied to these conductors without exceeding the *NEC®* recommendation on conductor voltage drop?
       (a) 0 ampere   (b) 5 ampere   (c) 10 ampere   (d) 15 ampere

8–13 Miscellaneous Voltage Drop Questions

59.   An 8 ohm resistor is connected to a 120-volt power supply. Using a voltmeter, we measure 112 volts across the resistor. What is the current of the 8 ohm resistor in ampere?
       (a) 14 ampere   (b) 13 ampere   (c) 15 ampere   (d) 19 ampere

60.   A 480-volt, single-phase feeder carries 400 ampere and has a 7.2 voltage drop. What is the total resistance of the conductors in this circuit?

(a) .1880 ohm   (b) .1108 ohm   (c) .0190 ohm   (d) .0180 ohm

61.   An 8 ohm resistor operates at 112 volts and is connected to a 115-volt power supply. The percent of voltage drop for the circuit is _____ percent.

(a) 2.6   (b) 3.0   (c) 5.0   (d) 7.0

NEC® Questions from Section 90 through Chapter 9

Article 90 – Introduction

62.   In the event the *Code* requires new products, constructions, or materials that are not yet available at the time the *Code* is adopted, the _____ may permit the use of the products, constructions, or materials that comply with the most recent previous edition of this *Code* adopted by the jurisdiction.

(a) architect   (b) master electrician
(c) authority having jurisdiction   (d) supply house

NEC® Chapter 1 – General Requirements

63.   • Admitting close approach, not guarded by locked doors, elevation, or other effective means, is commonly referred to as _____ .

(a) accessible (equipment)   (b) accessible (wiring methods)
(c) accessible, readily   (d) all of these

64.   The conductor between the final overcurrent protection device and the outlet(s) are known as the _____ conductors.

(a) feeder   (b) branch circuit   (c) home run   (d) none of these

65.   The *Code* defines a _____ as a structure that stands alone or that is cut off from adjoining structures by fire walls, with all openings therein protected by approved fire doors.

(a) unit   (b) apartment   (c) building   (d) utility

66.   • The _____ of any system is the ratio of the maximum demand for a system, or part of a system, to the total connected load of a system under consideration.

(a) load   (b) demand factor   (c) minimum load   (d) computed factor

67.   Continuous duty is defined as _____ .
(a) when the load is expected to continue for three hours or more
(b) operation at a substantially constant load for an indefinite length of time
(c) operation at loads and for intervals of time, both of which may be subject to wide variations
(d) operation at which the load may be subject to maximum current for six hours or more

68.   _____ is defined as intentionally connected to earth through a ground connection or connections of sufficiently low impedance and having sufficient current-carrying capacity to prevent the build-up of voltages that may result in undue hazards to connected equipment or to persons.
(a) Effectively grounded
(b) A proper wiring system
(c) A lighting rod
(d) A grounded conductor

69.   • In a grounded system, the conductor that connects the circuit grounded (neutral) conductor at the service to the grounding electrode is called the _____ conductor.
(a) main grounding
(b) common main
(c) equipment grounding
(d) grounding electrode

70. A hoistway is any _____ in which an elevator or dumbwaiter is designed to operate.
    (a) hatchway or well hole   (b) vertical opening or space
    (c) shaftway   (d) all of these

71. The *Code* defines a _____ as one familiar with the construction and operation of the equipment and the hazards involved.
    (a) inspector   (b) master electrician
    (c) journeyman electrician   (d) qualified person

72. • When one electrical circuit controls another circuit through a relay, that first circuit is called a _____ .
    (a) control circuit   (b) remote-control circuit   (c) signal circuit   (d) controller

73. The service conductor between the terminals of the service equipment and a point usually outside the building, clear of building walls, where joined by tap or splice to the service drop are service entrance conductors, is an _____ system.
    (a) underground   (b) complete   (c) overhead   (d) grounded

74. The voltage of a circuit is defined by the *Code* as the _____ root-mean-square (effective) difference of potential between any two conductors of the circuit.
    (a) lowest   (b) greatest   (c) average   (d) nominal

75. _____ means so constructed or protected that exposure to the weather will not interfere with successful operation.
    (a) Weatherproof   (b) Weathertight   (c) Weather-resistant   (d) All weather

76. Only wiring methods recognized as _____ are included in the *Code*.
    (a) identified   (b) efficient   (c) suitable   (d) cost effective

77. Many equipment terminations are marked with _____ .
    (a) electrical tape   (b) a nameplate   (c) a tightening torque   (d) red paint

78. All services, feeders, and branch circuits must be legibly marked to indicate their intended purpose unless located and arranged in such a manner so that the purpose is clearly evident.
    (a) True   (b) False

79. The dimension of working clearance for access to live parts operating at 300 volts, nominal, to ground where there are exposed live parts on both sides of the work space (not guarded) with the operator between, is _____ feet, according to Table 110-26(a).
    (a) 3   (b) $3^1/_2$ (c) 4   (d) $4^1/_2$

80. • A 6-foot wide control panel operating at over 600 volts requires a minimum of _____ entrances to the required working space about electrical equipment.
    (a) one   (b) two   (c) three   (d) four

Chapter 2 – Wiring and Protection

81. An insulated conductor larger than No. 6 that is not white or gray may be reidentified and used as a grounded conductor.
    (a) True   (b) False

82. Receptacles connected to circuits having different voltages, frequencies, or types of current (alternating or direct current) on the _____ shall be of such design that the attachment plugs used on these circuits are not interchangeable.
    (a) building   (b) interior   (c) same premises   (d) exterior

83. Branch circuit conductors that supply a continuous load shall have an ampacity of not less than 100 percent of the continuous load, plus 125 percent of the noncontinuous load.
    (a) True   (b) False

84. What is the maximum cord-and plug-connected load permitted on a 15-ampere multioutlet receptacle that is supplied by a 20-ampere circuit?
    (a) 12   (b) 16   (c) 20   (d) 24

85.   The small appliance circuit applies to the _____ as well as the kitchen.
      (a) dining room   (b) refrigerator   (c) breakfast room   (d) all of these

86.   • Heating, air-conditioning, and refrigeration equipment on rooftops or in attics and crawl spaces, shall have receptacle outlets for use in servicing these units. The receptacle shall be _____ .
      (a) GFCI-protected   (b) at an accessible location
      (c) within 25 feet of the equipment   (d) all of the above

87.   In _____ rooms, other than kitchens and bathrooms of dwelling units, one or more receptacles controlled by a wall switch shall be permitted in lieu of lighting outlets.
      (a) habitable   (b) finished   (c) all   (d) a and b

88.   Feeder ground-fault protection of equipment shall not be required where ground-fault protection of equipment is provided on the _____ side of the feeder.
      (a) load   (b) supply   (c) service   (d) none of these

89.   The minimum feeder allowance for show window lighting expressed in volt-amps per linear foot shall be _____VA.
      (a) 400   (b) 200   (c) 300   (d) 180

90.   A dwelling unit served by a single 3-wire, 120/240 volt set of service-entrance or feeder conductors with an ampacity of _____ or greater, it shall be permissible to compute the feeder and service loads in accordance with Table 220-30 instead of the method specified in Part B of Article 220.
      (a) 100   (b) 125   (c) 150   (d) 175

91.   Overhead conductors for festoon lighting shall not be smaller than No. _____ .
      (a) 10   (b) 12   (c) 14   (d) 18

92.   If the voltage between conductors does not exceed 300 volts, the outside conductor clearance over the roof overhang can be reduced from 8 feet to 18 inches. But no more than _____ feet of conductors can pass over the roof overhang.
      (a) 6   (b) 8   (c) 10   (d) none of these

93.   The disconnecting means for a remote building or structure shall be installed either inside or outside of the building or structure served or where the conductors pass through the building or structure. The disconnecting means shall be at a _____ location "nearest" the point of entrance of the conductors.
      (a) accessible   (b) readily accessible   (c) convenient   (d) any of these

94.   Service conductors supplying one building shall not _____ of another building.
      (a) be installed on the exterior walls
      (b) pass through the interior
      (c) a and b
      (d) none of these

95.   Service drop conductors shall have _____ .
      (a) sufficient ampacity to carry the load
      (b) adequate mechanical strength
      (c) a or b
      (d) a and b

96.   Reasonable efficiency of operation can be provided when voltage drop is taken into consideration in sizing the _____ conductors.
      (a) service-vertical   (b) grounding   (c) service-lateral   (d) none of these

97.   For services exceeding 600 volts, nominal, the _____ switch shall be accessible to qualified persons only.
      (a) power   (b) isolating   (c) emergency shut-off   (d) all of these

98.   The standard ampere ratings for fuses and fixed trip circuit breakers are listed in Article 240; additional standard ratings for fuses shall be _____ ampere(s).
      (a) 1   (b) 6   (c) 601   (d) all of these

99. • Circuit breaker handle tie-bars are required for multiwire branch circuits in commercial and industrial buildings if they supply line-to-neutral loads.
(a) True   (b) False

100. Type _____ fuse adapters shall be so designed that once inserted in a fuseholder they cannot be easily removed.
(a) A   (b) E   (c) S   (d) P

101. The earth can be used as the sole equipment grounding conductor.
(a) True   (b) False

102. The grounding electrode conductor at the service is permitted to terminate on an equipment grounding terminal bar if a (main) bonding jumper is installed between the neutral bus and the equipment grounding terminal.
(a) True   (b) False

103. Where livestock is housed, that portion of the equipment grounding conductor run underground to the separate building or structure disconnecting means shall be _____ .
(a) insulated or covered
(b) continuous and unbroken
(c) copper
(d) a and c

104. When a metal underground water pipe is to be used as the grounding electrode, it must be _____ .
(a) supplemented by an additional electrode
(b) protected from physical damage
(c) in direct contact with the earth for a minimum of 10 feet
(d) a and c

105. Air terminal conductors or electrodes used for grounding air terminals _____ be used as the required grounding electrode for the building/structure grounding electrode system.
(a) shall   (b) shall not   (c) can   (d) any of these

106. When an underground metal water piping system is used as a grounding electrode, effective bonding shall be provided around insulated joints and sections around any equipment that is likely to be disconnected for repairs. _____ conductors shall be of sufficient length to permit removal of such equipment while retaining the integrity of the bond.
(a) Main bonding jumper
(b) Equipment grounding
(c) Bonding
(d) none of these

107. Isolated metal elbows installed underground are permitted if located so that no part is less than _____ inches below finish grade.
(a) 6   (b) 12   (c) 18   (d) according to Table 300-5

108. For circuits over 250 volts to ground (480Y/277 volt), electrical continuity can be maintained between a box or enclosure (with no ringed knockouts) and a metal conduit by _____ .
(a) threadless fitting for cables with metal sheath
(b) double locknuts on threaded conduit (one inside and one outside the box or enclosure)
(c) set screw conduit fittings that have shoulders that seat firmly against the box or enclosure with a single locknut on the inside.
(d) all of these

109. Liquidtight flexible metal conduit listed for grounding is permitted to serve as the equipment grounding conductor (with specific limitations) if the ground return path of the liquidtight flexible metal conduit does not exceed _____ feet.
(a) 3   (b) 4   (c) 6   (d) 8

110.   A grounding-type receptacle can replace a nongrounding type receptacle at an outlet box that does not contain an equipment grounding conductor if the equipment grounding conductor is connected to _____ .

(a) a grounding electrode system as described in Section 250-50

(b) a grounding electrode conductor

(c) equipment grounding terminal bar within the enclosure where the branch circuits for the receptacle or branch circuit originates

(d) any of these

111.   When equipment grounding conductor(s) are installed in a metal box, an electrical connection is required between the equipment grounding conductor and the metal box enclosure by means of a _____ .

(a) grounding screw   (b) listed grounding clip

(c) listed grounding fittings   (d) any of these

112.   The grounding conductor for secondary circuits of instrument transformers and for instrument cases shall not be smaller than No. _____ copper.

(a) 18   (b) 16   (c) 14   (d) 12

Chapter 3 – Wiring Methods and Material

113.   Cables laid in wood notches require protection against nails or screws by a steel plate at least _____ inch thick.

(a) $1/16$ (b) $1/8$ (c) $1/2$ (d) none of these

114.   The air handling area beneath raised floors for data processing systems is not a plenum and is not required to comply with the requirements of Section 300-22 but shall be permitted in accordance with Article 645.

(a) True   (b) False

115.   It is permitted to use individual open conductors for a period not to exceed 90 days for Christmas decorative lighting and similar purposes where the circuit voltage to ground does not exceed _____ volts.

(a) 50   (b) 125   (c) 150   (d) 277

116.   It is not the intent of Section 310-4 to require that conductors of one phase, neutral, or grounded circuit conductor be the same as those of another phase, neutral, or grounded circuit conductor to achieve _____ .

(a) polarity   (b) balance   (c) grounding   (d) none of these

117.   Which conductor has a temperature rating of 90ºC?

(a) RH   (b) RHW   (c) THHN   (d) TW

118.   • Aluminum and copper-clad aluminum of the same circular mil size and insulation have _____ .

(a) the same physical characteristics   (b) the same termination

(c) the same ampacity   (d) different ampacities

119.   Where a solid-bottom cable tray, having a usable inside depth of 6 inches or less, contains multiconductor control and/or signal cables only, the sum of the cross-sectional areas of all cables at any cross section shall not exceed _____ percent of the interior cross-sectional area of the cable tray.

(a) 25   (b) 30   (c) 35   (d) 40

120.   Conductors No. _____ or larger supported on solid knobs shall be securely tied thereto by tie wires having an insulation equivalent to that of the open wire.

(a) 14   (b) 12   (c) 10   (d) 8

121.   The conductors supported by messenger shall be permitted to come into contact with the messenger supports or any structural members, walls, or pipes.

(a) True   (b) False

122.   Use of FCC systems shall be permitted on wall surfaces in _____ .

(a) surface metal raceways

(b) cable trays

(c) busways

(d) any of these

123. No more than _____ layers of flat conductor cable can cross at any one point.
    (a) 2   (b) 3   (c) 4   (d) none of these

124. Where single-conductor Type MI cables are used, all phase conductors and, where used, the _____ conductor shall be grouped together to minimize induced voltage on the metal sheath.
    (a) grounded   (b) neutral   (c) grounding   (d) largest

125. • Electrical nonmetallic tubing is permitted for direct earth burial when used with fittings listed for this purpose.
    (a) True   (b) False

126. Where Type AC cable is run across the top of floor joists, or within 7 feet of floor or floor joists across the face of rafters or studding, in attics and roof spaces that are accessible, the cable shall be protected by substantial guard strips that have a height of _____ .
    (a) 3 inches   (b) 1 inch   (c) 2 inches   (d) as high as the cable

127. • Smooth-sheath MC cables that have external diameters of not greater than 1 inch, shall have a bending radius of not more than _____ times the cable external diameter.
    (a) 5   (b) 10   (c) 12   (d) 13

128. Two conductor nonmetallic sheath cables cannot be stapled on edge.
    (a) True   (b) False

129. Switch, outlet, and tap devices of insulating material shall be permitted to be used without boxes in exposed Type NMC cable.
    (a) True   (b) False

130. The overall covering of UF cable shall be _____ .
    (a) suitable for direct burial in the earth
    (b) flame-retardant
    (c) moisture, fungus, and corrosion resistant
    (d) all of these

131. Nonmetallic surface extensions with one or more extensions shall be permitted to be run in any direction from an existing outlet, but not on the floor or within _____ inches from the floor.
    (a) 6   (b) 4   (c) 3   (d) 2

132. Nonmetallic underground conduit with conductors shall be capable of being supplied on reels without damage or _____ and shall be of sufficient strength to withstand abuse, such as impact or crushing, in handling and during installation without damage to conduit or conductors.
    (a) distortion   (b) breakage   (c) shattering   (d) all of these

133. A run of intermediate metal conduit shall not contain more than the equivalent of _____ quarter bends including the offsets located immediately at the outlet or fitting.
    (a) 1   (b) 2   (c) 3   (d) 4

134. • Two-inch rigid metal conduit shall be supported every _____ feet.
    (a) 10   (b) 12   (c) 14   (d) 15

135. Rigid metal conduit as shipped shall _____ .
    (a) be in standard lengths of 10 feet   (b) include a coupling on each length
    (c) be reamed and threaded on each end   (d) all of these

136. When installing rigid nonmetallic conduit _____ .
    (a) all cut ends shall be trimmed inside and outside to remove rough edges
    (b) there shall be a support within 2 feet of each box and cabinet
    (c) all joints shall be made by an approved method
    (d) a and c

137. • The *Code* requires that EMT be supported within 3 feet of each termination and each coupling.
    (a) True   (b) False

138. How many No. 12 XHHW conductors, not counting a bare ground wire, are allowed in $^3/_8$ inch flexible metal conduit (maximum of six feet) with outside fittings?
      (a) 4   (b) 3   (c) 2   (d) 5

139. Where flexibility is necessary, securing of the raceway is not required for lengths not exceeding _____ feet at terminals.
      (a) 2   (b) 3   (c) 4   (d) 6

140. The *Code* permits listed manufactured, prewired, liquidtight, flexible, nonmetallic assemblies, in sizes from _____ inch through 1 inch.
      (a) $^1/_4$ (b) $^3/_8$ (c) $^1/_2$ (d) all of these

141. When an extension is made from a surface metal enclosure, the enclosure must contain a means for terminating a(n) _____ .

      (a) grounded conductor   (b) ungrounded conductor
      (c) equipment grounding conductor   (d) all of these

142. The combined area of all conductors installed at any point of an underfloor raceway shall not exceed _____ percent of the internal cross-sectional area.
      (a) 75   (b) 60   (c) 40   (d) 30

143. For cellular concrete floor raceways, junction boxes shall be _____ the floor grade and sealed against the free entrance of water or concrete.
      (a) leveled to   (b) above   (c) below   (d) perpendicular to

144. Where conductors No. 4 and larger enter then exit a wireway, the distance between the conductor entries shall not be less than _____ times the trade diameter of the raceway or cable connector.
      (a) two   (b) four   (c) six   (d) none of these

145. Flat cable assemblies shall have conductors of No. _____ special stranded copper.
      (a) 14   (b) 12   (b) 10   (d) all of these

146. The maximum branch circuit rating of a flat conductor cable is _____ amperes.
      (a) 15   (b) 20   (c) 30   (d) 50

147. • Each section of cablebus shall be marked with the manufacturer's name or trade designation and the minimum diameter, number, voltage rating, and ampacity of the conductors to be installed. Markings shall be so located as to be visible after installation.
      (a) True   (b) False

148. When determining the number of conductors in a box, one or more internal cable clamps, whether factory or field supplied, are present in the box, a double volume allowance in accordance with Table 370-16(b) shall be made based on the largest conductor present in the box.
      (a) True   (b) False

149. Boxes can be supported from a multiconductor cord or cable, provided the conductors are protected from _____ .
      (a) strain   (b) temperature   (c) sunlight   (d) swaying

150. A switch enclosure shall not be used as a junction box, except where adequate space is provided so that the conductors do not fill the wiring space at any cross-section to more than 40 percent of the cross-sectional area of the space, and so that _____ do not fill the wiring space at any cross-section to more than 75 percent of the cross-sectional area of the space.
      (a) splices   (b) taps   (c) conductors   (d) all of these

151. When the derating factors of Section 310-15(b)(2) are used, auxiliary gutters shall not contain more than _____ at any cross-section.
      (a) 25 conductors
      (b) 40 current-carrying conductors
      (c) 20 conductors
      (d) no limit on the number of conductors

152. Knife switches rated for more than 1,200 amperes at 250 volts _____ .
     (a) are used only as isolating switches
     (b) may be opened under load
     (c) should be placed so that gravity tends to close them
     (d) should be connected in parallel

153. Switchboards shall be so placed as to reduce to a minimum the probability of communicating _____ to adjacent combustible materials.
     (a) sparks   (b) backfeed   (c) fire   (d) all of these

154. A lighting and appliance branch circuit panelboard is considered protected by the _____ conductor protection device if the protection device rating is not greater than the panelboard rating.
     (a) grounded   (b) feeder   (c) branch circuit   (d) none of these

Chapter 4 – Equipment for General Use

155. • Flexible cords and cables shall be protected by _____ where passing through holes in covers, outlet boxes, or similar enclosures.
     (a) bushings   (b) fittings   (c) a and b   (d) a or b

156. The enclosure for a receptacle installed in an outlet box flush-mounted on a wall surface, in a damp or wet location, shall be made weatherproof by means of a weatherproof faceplate assembly that provides a _____ connection between the plate and the wall surface.
     (a) sealed   (b) weathertight   (c) sealed and protected   (d) watertight

157. Dishwashers and trash compactors can be cord and plug connected. The cord must not be less than 36 inches or more than _____ inches and must be protected from physical damage. The receptacle for the appliances must be accessible and located in the space occupied by the appliances or adjacent thereto.
     (a) 30   (b) 36   (c) 42   (d) 48

158. GFCI protection for personnel shall be provided for electrically heated floors in _____ locations.
     (a) bathroom   (b) spa   (c) hot tub   (d) all of these

159. • To prevent skin effect heating, the provisions of Section 300-20 shall apply to the installation of a single conductor in a ferromagnetic envelope (metal enclosure).
     (a) True   (b) False

160. Torque requirements for a motor-control circuit device shall be a minimum of _____ pounds inches (unless otherwise indicated) for No. 14 and smaller screw-type pressure terminals.
     (a) 7   (b) 9   (c) 10   (d) 15

161. The motor branch circuit, short-circuit, and ground-fault protective device shall be capable of carrying the _____ current of the motor.
     (a) varying   (b) starting   (c) running   (d) continuous

162. • Motor disconnecting means shall be a motor-circuit switch rated in _____, be a molded case switch, or a circuit breaker.
     (a) horsepower   (b) watts   (c) amperes   (d) locked-rotor current

163. • Exposed live parts of motors and controllers operating at _____ volts or more between terminals shall be guarded against accidental contact by enclosure or by location.
     (a) 24   (b) 50   (c) 150   (d) 60

164. • The rating of the branch circuit short circuit ground-fault protection device for an individual motor-compressor shall not exceed _____ percent of the rated-load current or branch circuit selection current, whichever is greater, if the protection device will carry the starting current of the motor.
     (a) 100   (b) 125   (c) 175   (d) none of these

165. The ampacity of phase conductors from the generator terminals to the first overcurrent protection device shall not be less than _____ percent of the nameplate rating of the generator.

    (a) 75   (b) 115   (c) 125   (d) 140

166. Transformer vaults shall be located where they can be ventilated to the outside air without using flues or ducts wherever such an arrangement is _____ .

    (a) permitted   (b) practicable   (c) required   (d) all of these

167. Means shall be provided to disconnect simultaneously all ungrounded single-phase _____ conductors to the phase converter.

    (a) supply   (b) load   (c) service   (d) neutral

168. A vented alkaline-type battery operating at less than 250 volts shall be installed with not more than _____ cells in the series circuit in any one tray.

    (a) 10   (b) 12   (c) 18   (d) 20

Chapter 5 – Special Occupancies

169. • A Class III, Division _____ location is where easily ignitable fibers or combustible flying material are stored or handled but not manufactured.

    (a) 1   (b) 2   (c) 3   (d) all of these

170. Where seals are required for Class I locations, they shall comply with the following:

    (a) They shall be approved for Class I locations and shall be accessible.
    (b) The sealing compound shall be approved and must not be less than the trade size of the conduit, but in no case less than 5/8 inch.
    (c) Splices and taps shall not be made in seal-off fittings.
    (d) All of these

171. Raceways embedded in a masonry wall or buried beneath a floor shall be considered _____ .

    (a) outside the hazardous area if any extensions pass through such areas
    (b) within the hazardous area if any connections lead into such areas
    (c) hazardous if beneath a hazardous area
    (d) outside any hazardous location

172. Aboveground bulk-storage tanks shall be classified as _____ for the space between 5 feet and 10 feet from the open end of vent, extending in all directions.

    (a) Class I, Division 1   (b) Class I, Division 2
    (c) Class II, Division 1   (d) Class II, Division 2

173. An insulated No. 12 ground wire must be installed in a metal raceway for the redundant grounding of the metal parts of _____ located in the patient care area of health care facilities.

    (a) receptacles   (b) switches   (c) dimmers   (d) all of these

174. Flexible conductors used to supply portable stage equipment shall be _____ .

    (a) listed   (b) extra-hard usage (c) hard-usage   (d) a and b

175. Each receptacle of dc plugging boxes shall be rated at not _____ amperes when used on a stage or set of a motion picture studio.

    (a) more than 30   (b) less than 20   (c) less than 30   (d) more than 20

176. Service-entrance conductors for a manufactured building shall be installed _____ .

    (a) after erection at the building site
    (b) before erection where the point of attachment is known prior to manufacture
    (c) before erection at the building site
    (d) a or b

177. • The coupler of a mobile home is included when determining the number of branch circuits for a mobile home.

    (a) True   (b) False

178. Feeders to floating buildings can be installed in _____ if listed for wet locations and sunlight resistance.
(a) rigid nonmetallic conduit
(b) liquidtight metallic conduit
(c) liquidtight nonmetallic conduit
(d) portable power cables

179. The *NEC*® requires a(n) _____ disconnecting means by which each boat can be isolated from its supply circuit.
(a) accessible   (b) readily accessible   (c) remote   (d) any of these

Chapter 6 – Special Equipment

180. Each sign and outline lighting system, feeder circuit, or branch circuit supplying a sign or outline lighting system, shall be controlled by an externally operable switch and circuit breaker that opens all ungrounded conductors. In addition, the disconnecting means must be _____ .
(a) accessible   (b) readily accessible
(c) above grade level   (d) none of these

181. • A crane has two motors: one rated 106 amperes for 30 minutes and the other rated 72 amperes for 60 minutes. The minimum calculated motor load for the two motors is _____ amperes.
(a) 150   (b) 204.5   (c) 142   (d) 110

182. • An elevator shall have a single means for disconnecting all ungrounded main power supply conductors for each unit.
(a) This does not include the emergency power service.
(b) This includes the emergency power service.
(c) This does not include the emergency power service if the system is automatic.
(d) No elevators are to operate on emergency power systems.

183. The ampacity for the supply conductors for a resistance welder with a duty cycle of 15 percent and a primary current of 21 amperes is _____ amperes.
(a) 9.45   (b) 8.19   (c) 6.72   (d) 5.67

184. Electric pipe organ circuits shall be so arranged that all conductors shall be protected from overcurrent by an overcurrent protection device rated at not more than _____ amperes.
(a) 20   (b) 15   (c) 6   (d) none of these

185. No. 18 and No. 16 fixture wires can be used for the control and operating circuits of X-ray equipment when protected by not larger than a _____ ampere overcurrent protection device.
(a) 15   (b) 20   (c) 25   (d) 30

186. A generator output circuit shall be isolated from ground, except where the capacitive coupling inherent in the generator causes the generator terminals to have voltages from terminal to ground that are equal.
(a) True   (b) False

187. A wet-niche lighting fixture is intended to be installed in a _____ .
(a) transformer   (b) forming shell   (c) hydromassage bathtub   (d) all of these

188. • At dwelling units, a 125-volt receptacle is required to be installed within 20 feet of the inside wall of the pool.
(a) True   (b) False

189. • Underground outdoor pool or spa equipment rooms or pits must have adequate drainage to prevent water accumulation during abnormal operation.
(a) True   (b) False

190. A forming shell shall be provided with a number of grounding terminals that shall be _____ the number of conduit entries.
(a) one more than   (b) two more than   (c) the same as   (d) none of these

191.   Stainless steel, copper, or copper alloy clamps approved for direct burial use can be used to connect the bonding conductor to the common bonding grid of a swimming pool.
(a) True   (b) False

192.   A clearly labeled emergency shutoff or control switch for the purpose of stopping the motors(s) that provide power to the recirculation system and jet system shall be installed. The emergency shutoff control switch must be _____ to the user, located at least 5 feet away, and within sight of the spa or hot tub. This requirement does not apply to single-family dwelling units.
(a) accessible   (b) readily accessible   (c) available   (d) none of these

Chapter 7 – Special Conditions

193.   Emergency lighting, emergency power, or both in a building or group of buildings will be available within the time period required for the application, but not to exceed _____ seconds.
(a) 5   (b) 10   (c) 30   (d) 60

194.   Optional standby systems shall have adequate capacity and rating for the supply of _____ .
(a) all emergency lighting and power loads
(b) all equipment to be operated at one time
(c) 100 percent of the appliance loads and 50 percent of the lighting loads
(d) 100 percent of the lighting loads and 75 percent of the appliance loads

195.   Power supply and Class 1 circuit conductors shall be permitted in the same cable, enclosure, or raceway only where the _____ .
(a) equipment powered is functionally associated
(b) circuits involved are not a mixture of alternating and direct current
(c) Class 1 circuits are never permitted with the power supply
(d) none of these

196.   Class 2 control circuits and power circuits _____ .
(a) may occupy the same raceway
(b) shall be put in different raceways
(c) a and b
(d) none of these

197.   Electrical fire alarm installations in hollow spaces, vertical shafts, and ventilation or air-handling ducts shall be made so that the spread of fire or products of combustion (such as smoke or gases) will not be _____ .
(a) possible   (b) substantially increased   (c) permitted   (d) none of these

198.   Power-limited fire alarm cables installed as wiring within buildings shall be _____ as being resistant to the spread of fire.
(a) labeled   (b) listed   (c) identified   (d) marked

Chapter 8 – Communications Systems

199.   Soft-drawn or medium-drawn copper lead in conductors for receiving antenna systems shall be permitted where the maximum span between points of support is less than _____ feet.
(a) 35   (b) 30   (c) 20   (d) 10

200.   What length of nipple may utilize 60 percent conductor fill?
(a) 12 inch   (b) 18 inch   (c) 24 inch   (d) all of these

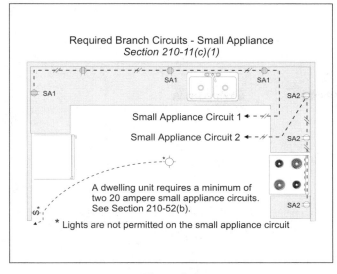

Figure 9–3
Required Branch Circuits – Small Appliance

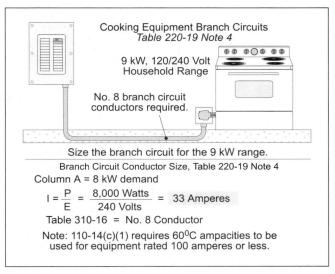

Figure 9–4
Cooking Equipment Branch Circuits

This can be converted to ampere by dividing the power by the voltage, $I = \dfrac{P}{E} = \dfrac{8 \text{ kW} \times 1,000}{240 \text{ volts*}} = 33$ ampere

*Assume 120/240 volt, single-phase for all calculations unless the question gives a specific voltage and system.

❑ **More Than 12 kW**

What is the branch circuit load (in ampere) for one 14 kW range (Figure 9–5)?

(a) 33 ampere   (b) 37 ampere   (c) 50 ampere   (d) 58 ampere

  • Answer: (b) 37 ampere

Step 1:➡ Since the range exceeds 12 kW, follow Note 1 of Table 220-19. The first step is to determine the demand load as listed in Column A of Table 220-19 for one unit = 8 kW.

Step 2:➡ We must increase the Column A value (8 kW) by 5 percent for each kW that the average range (in this case 14 kW) exceeds 12 kW. This results is an increase of the Column A value (8 kW) by 10 percent,

8 kW x 1.1 = 8.8 kW or 8,800 Watts.

Step 3: Convert the demand load in kW to ampere. $I = \dfrac{P}{E} = \dfrac{8,800 \text{ watts}}{240 \text{ volts}} = 36.7$ ampere

❑ **One Counter-Mounted Cooking Unit**

What is the branch circuit load (in ampere) for one 6 kW counter-mounted cooking unit (Figure 9–6)?

(a) 15 ampere   (b) 20 ampere   (c) 25 ampere   (d) 30 ampere

  • Answer: (c) 25 ampere

$I = \dfrac{P}{E} = \dfrac{6,000 \text{ watts}}{240 \text{ volts}} = 25$ ampere

---

One Counter-Mounted Cooking Unit and up to Two Ovens [Section 220-19 Note 4]

To calculate the load for one counter-mounted cooking unit and up to two wall-mounted ovens, complete the following steps:

Step 1:➡ Total Load. Add the nameplate ratings of the cooking appliances and treat this total as one range.

Step 2:➡ Table 220-19 Demand. Determine the demand load for one unit from Table 220-19, Column A.

Step 3:➡ Net Computed Load. If the total nameplate rating exceeds 12 kW, increase Column A (8 kW) 5 percent for each kW, or major fraction (0.5 kW), that the combined rating exceeds 12 kW.

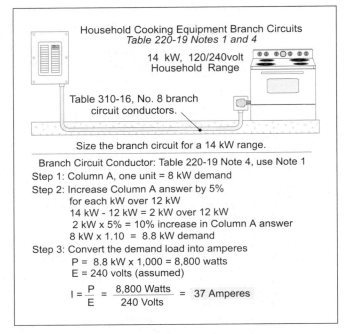

Figure 9–5
Household Cooking Equipment Branch Circuits

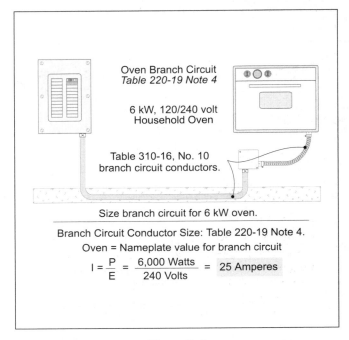

Figure 9–6
Oven Branch Circuit

❏ Cooktop and One Oven

What is the branch circuit load (in ampere) for one 6 kW counter-mounted cooking unit and one 3 kW wall-mounted oven (Figure 9–7)?

(a) 25 ampere   (b) 38 ampere   (c) 33 ampere   (d) 42 ampere

• Answer: (c) 33 ampere

Step 1:➛ Total connected load: 6 kW + 3 kW = 9 kW.

Step 2:➛ Demand load for one range, Column A of Table 220-19: 8 kW.

Step 3:➛ $I = \dfrac{P}{E} = \dfrac{8,000 \text{ watts}}{240 \text{ volts}} = 33.3$ ampere

❏ Cooktop and Two Ovens

What is the branch circuit load (in ampere) for one 6 kW counter-mounted cooking unit and two 4 kW wall-mounted ovens (Figure 9–8)?

(a) 22 ampere   (b) 27 ampere   (c) 33 ampere   (d) 37 ampere

• Answer: (d) 37 ampere

Step 1:➛ The total connected load: 6 kW + 4 kW + 4 kW = 14 kW.

Step 2:➛ Demand load for one range, Column A or Table 220-19: 8 kW.

Step 3:➛ Since the total (14 kW) exceeds 12 kW, we must increase Column A (8 kW) 5 percent for each kW that the total exceeds 12 kW.

14 kW exceeds 12 kW by 2 kW, which results in a 10 percent increase of the Column A value.

8 kW × 1.1 = 8.8 kW

Step 4:➛ $I = \dfrac{P}{E} = \dfrac{8,800 \text{ watts}}{240 \text{ volts}} = 36.7$ ampere

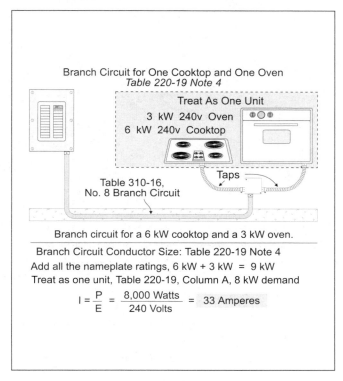

Figure 9–7
Branch Circuit for One Cooktop and One Oven

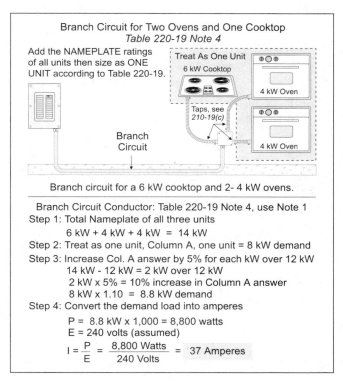

Figure 9–8
Branch Circuit for Two Ovens and One Cooktop

9–6 LAUNDRY RECEPTACLE(S) CIRCUIT [*SECTION 220-11(c)(2)*]

One 20-ampere branch circuit for the laundry receptacle outlet (or outlets) is required, and this laundry circuit cannot serve any other outlet such as the laundry room lights [*Section 210-52(f)*]. The *NEC*® does not require a separate circuit for the washing machine, but does require a separate circuit for the laundry room receptacle or receptacles (Figure 9–9). If the washing machine receptacle(s) are located in the garage or basement, GFCI protection might be required, see Section 210-8(a)(2) and (5) and their exceptions.

Feeder and Service

Each dwelling-unit shall have a feeder load consisting of 1,500 volt-ampere for the 20-ampere laundry circuit [*Section 220-16(b)*].

Other Related *Code* Rules

A laundry outlet must be within 6 feet of a washing machine [*Section 210-50(c)*].

Laundry area receptacle outlet required [*Section 210-52(f)*].

9–7 LIGHTING AND RECEPTACLES

General Lighting Volt-Ampere Load [*Section 220-3(a)*]

The *NEC*® requires a minimum 3 volt-ampere each square foot [Table 220-3(a)] for the *general lighting* and general-use receptacles. The dimensions for determining the area shall be computed from the outside dimensions of the building and shall not include open porches, garages, or spaces not adaptable for future use (Figure 9–10).

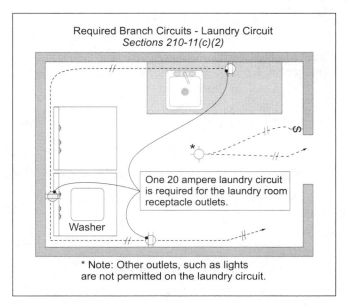

Figure 9–9
Required Branch Circuits – Laundry Circuit

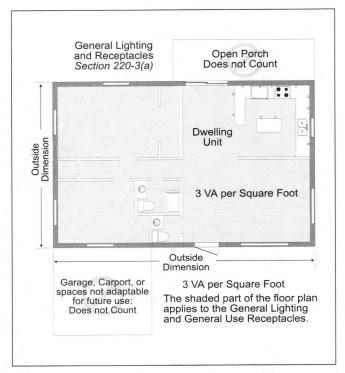

Figure 9–10
General Lighting and Receptacles

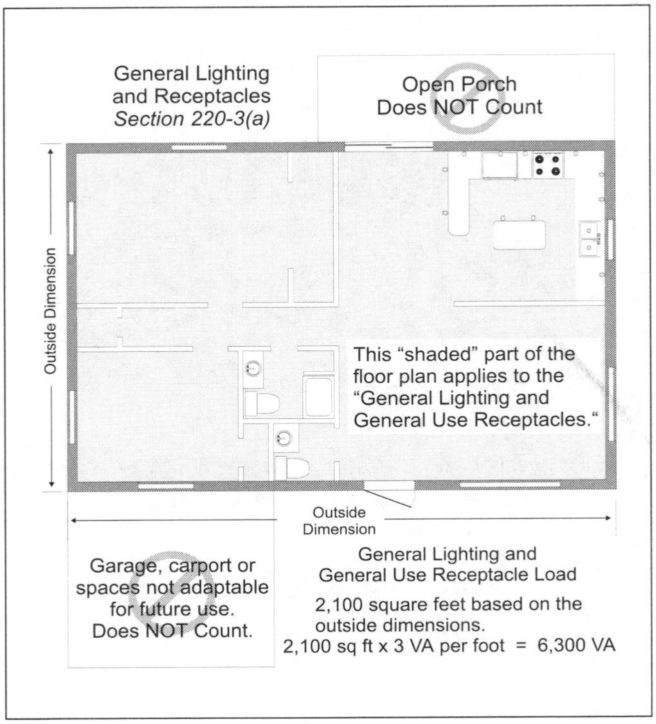

Figure 9–11
General Lighting and Receptacles

Note: The 3 volt-ampere each square foot rule for general lighting includes all 15- and 20-ampere general-use receptacles; but it does not include the appliance or laundry circuit receptacles, see *Section 220-3(10)* for details.

❏ **General Lighting Load [Table 220-3(a)]**

What is the general lighting and receptacle load for a 2,100 square-foot dwelling-unit that has 34 convenience receptacles and 12 lighting fixtures rated 100 watts each (Figure 9–11)?

(a) 2,100 VA   (b) 4,200 VA   (c) 6,300 VA   (d) 8,400 VA

    • Answer: (c) 6,300 VA

2,100 square feet $\times$ 3 VA = 6,300 VA

Note. No additional load is required for general-use receptacles and lighting outlets, see *Section 220-3(a)(10)*.

Number of Circuits Required [Chapter 9, Example No. D1(a)]

The number of branch circuits required for general lighting and receptacles shall be determined from the general lighting load and the rating of the circuits [*Section 210-11(a)*].

To determine the number of branch circuits for general lighting and receptacles, follow these steps:

Step 1:➔ Determine the general lighting VA load: Living area square footage $\times$ 3 VA.

Step 2:➔ Determine the general lighting ampere load: Ampere = $^{VA}/_{E}$.

Step 3:➔ Determine the number of branch circuits:

$$\frac{\text{General Light Ampere (Step 2)}}{\text{Circuit Ampere (100\%)}}$$

❏ **Number of 15-Ampere Circuits**

How many 15-ampere circuits are required for a 2,100 square-foot dwelling-unit (Figure 9–12)?

(a) 2 circuits   (b) 3 circuits   (c) 4 circuits   (d) 5 circuits

    • Answer: (c) 4 circuits

Step 1: ➛ General lighting VA:

2,100 square feet $\times$ 3 VA = 6,300 VA

Step 2: ➛ General lighting ampere: $I = \dfrac{VA}{E}$

$I = \dfrac{6,300 \text{ VA}}{120 \text{ volts*}} = 53$ ampere

*Use 120 volts single-phase unless specified otherwise.

Step 3: ➛ Determine the number of circuits:

$$\frac{\text{General Light Ampere}}{\text{Circuit Ampere}}$$

$$= \frac{53 \text{ ampere}}{15 \text{ ampere}} = 3.53 \text{ or 4 circuits}$$

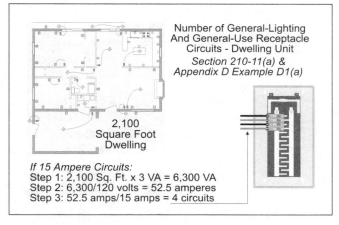

Number of General-Lighting
And General-Use Receptacle
Circuits - Dwelling Unit
Section 210-11(a) &
Appendix D Example D1(a)

2,100
Square Foot
Dwelling

If 15 Ampere Circuits:
Step 1: 2,100 Sq. Ft. x 3 VA = 6,300 VA
Step 2: 6,300/120 volts = 52.5 amperes
Step 3: 52.5 amps/15 amps = 4 circuits

Figure 9–12
Number of General Lighting and General
Use Receptacle Circuits

❑ **Number of 20-Ampere Circuits**

How many 20-ampere circuits are required for a 2,100 square-foot home?

(a) 2 circuits   (b) 3 circuits   (c) 4 circuits   (d) 5 circuits

   • Answer: (b) 3

Step 1: ➛ General lighting VA: 2,100 square foot $\times$ VA = 6,300 VA

Step 2: ➛ General lighting ampere: $I = \dfrac{VA}{E} = \dfrac{6,300 \text{ VA}}{120 \text{ volts}} = 53$ ampere

Step 3: ➛ Determine the number of circuits: $\dfrac{\text{General Lighting Ampere}}{\text{Circuit Ampere}} = \dfrac{53 \text{ ampere}}{20 \text{ ampere}} = 2.65 \text{ or 3 circuits}$

PART B – STANDARD METHOD – FEEDER/SERVICE LOAD CALCULATIONS

9–8 DWELLING UNIT FEEDER/SERVICE LOAD CALCULATIONS (Part B of Article 220)

When determining a dwelling-unit feeder or service using the standard method, use the following steps:

Step 1: ➛ **General Lighting and Receptacles, Small Appliance and Laundry Circuits [Table 220-11]**

The *NEC®* recognizes that the general lighting and receptacles, small appliance, and laundry circuits will not all be on, or loaded, at the same time and permits a demand factor to be applied [*Section 220-16*]. To determine the feeder demand load for these loads, follow these steps:

(a) Total Connected Load: Determine the total connected load for (1) general lighting and receptacles (3 VA per square foot), (2) two small appliance circuits each at 1,500 VA, and (3) one laundry circuit at 1,500 VA.

(b) Demand Factor: Apply Table 220-11 demand factors to the total connected load (Step 1).

First 3,000 VA at 100 percent demand

Remainder VA at 35 percent demand

Step 2: ➛ **Air-Conditioning versus Heat**

Because the air-conditioning and heating loads are not on at the same time, it is permissible to omit the smaller of the two loads [*Section 220-21*]. The air-conditioning demand load is sized at 125 percent of the largest air-conditioning VA, plus the sum of the other air-conditioning VA's [*Section 440-34*]. Fixed electric space-heating demand loads are calculated at 100 percent of the total heating load [*Section 220-15*].

Step 3:➡ **Appliances [*Section 220-17*]**

A demand factor of 75 percent is permitted to be applied when four or more appliances fastened in place, such as dishwasher, waste disposal, trash compactor, water heater, etc. are on the same feeder. This does not apply to motors [*Section 220-14*], space-heating equipment [*Section 220-15*], clothes dryers [*Section 220-18*], cooking appliances [*Section 220-19*], or air-conditioning equipment [*Section 440-34*].

Step 4:➡ **Clothes Dryer [*Section 220-18*]**

The feeder or service demand load for electric clothes dryers located in a dwelling-unit shall not be less than: (1) 5,000 watts or (2) the nameplate rating if greater than 5,000 watts. A feeder or service dryer load is not required if the dwelling-unit does not contain an electric dryer.

Step 5:➡ **Cooking Equipment [*Section 220-19*]**

Household cooking appliances rated over $1^3/_4$ kW can have the feeder and service loads calculated according to the demand factors of Section 220-19, Table and Notes 1, 2, and 3.

Step 6:➡ **Feeder and Service Conductor Size**

400 Ampere and Less: The feeder or service conductors are sized according to Table 310-15(b)(6) for 3-wire, single-phase, 120/240-volt systems up to 400 ampere. The grounded (neutral) conductor is sized based on the maximum unbalanced load [*Section 220-22*] according to conductors as listed in Table 310-16.

Over 400Ampere: The ungrounded and grounded (neutral) conductors are sized according to Table 310-16.

9–9 DWELLING UNIT FEEDER/SERVICE CALCULATIONS EXAMPLES

Step 1: General Lighting, Small Appliance, and Laundry Demand [Section 220-11]

➡ **General Lighting No. 1**

What is the general lighting, small appliance, and laundry demand load for a 2,700 square-foot dwelling-unit (Figure 9–13)?

(a) 8,100 VA   (b) 12,600 VA   (c) 2,700 VA   (d) 6,360 VA

• Answer: (d) 6,360 VA

General Lighting/Receptacles	(2,700 square feet × 3 VA)	8,100 VA
Small Appliance Circuits	(1,500 VA × 2)	3,000 VA
Laundry Circuit	(1,500 VA × 1)	1,500 VA
Total Connected Load		12,600 VA
First 3,000 VA at 100%		–3,000 VA × 1.00 = 3,000 VA
Remainder at 35%		9,600 VA × 0.35 = + 3,360 VA
		6,360 VA

➡ **General Lighting No. 2**

What is the general lighting, small appliance, and laundry demand load for a 6,540 square-foot dwelling-unit?

(a) 8,100 VA   (b) 12,600 VA   (c) 2,700 VA   (d) 10,392 VA

• Answer: (d) 10,392 VA

General Lighting/Receptacles	(6,540 square feet × 3 VA)	19,620 VA
Small Appliance Circuits	(1,500 VA × 2)	3,000 VA
Laundry Circuit	(1,500 VA × 1)	1,500 VA
Total Connected Load		24,120 VA
First 3,000 VA at 100%		–3,000 VA × 1.00 = 3,000 VA
Remainder at 35%		21,120 VA × 0.35 = + 7,392 VA
		10,392 VA

Step 2: Air-Conditioning versus Heat [Section 220-15]

When determining the cooling versus heating load, compare the air-conditioning load at 125 percent against the the heating load at 100 percent. Section 220-15 only refers to the heating load and Section 440-32 contains the requirement of the air-conditioning load. You are permitted to omit the smaller of the two loads according to Section 220-21.

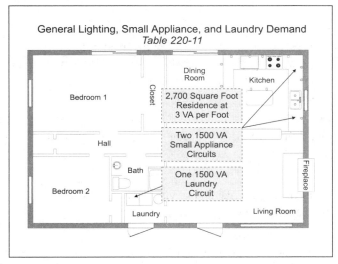

Figure 9–13
General Lighting, Small Appliance, and Laundry Demand

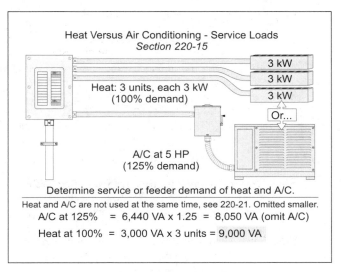

Figure 9–14
Heat versus Air Conditioning – Service Loads

➤ **Air-Conditioning versus Heat No. 1**

What is the service demand load for a air-conditioning (5 horsepower 230 volt) versus 3 baseboard heaters (each 3 kW) (Figure 9–14)?

(a) 6,400 VA   (b) 3,000 watts   (c) 8,050 VA   (d) 9,000 watts

• Answer: (d) 9,000 watts

A/C [*Section 440-34*] 230 volts × 28 ampere*
A/C = 6,440 × 1.25 = 8,050 VA, omit [*Section 220-21*]

Heat [*Section 220-15*] 3,000 watts × 3 units
Heat = 9,000 watts

*Table 430-148

➤ **Air-Conditioning versus Heat No. 2**

What is the service or feeder demand load for A/C (4 horsepower 230 volt) and electric space heater (10 kW)?

(a) 10,000 watts   (b) 3,910 watts   (c) 6,440 watts   (d) 10,350 watts

• Answer: (a) 10,000 kW heat

A/C: [*Section 440-34*] Since a 4 horsepower 230 volt motor is not listed in Table 430-148, we must determine the approximate VA rating by averaging the values for a 3 horsepower and 5 horsepower motor.

3 horsepower = 230 volts × 17 ampere* =   3,910 VA
5 horsepower = 230 volts × 28 ampere* =   <u>6,440 VA</u>
Total VA for 3 horsepower + 5 horsepower (8 horsepower) = 10,350 VA
VA for 4 horsepower = 10,350 VA (8 horsepower)/2 = 5,175 VA × 1.25 = 6,469 VA

Omit because it is less than 10 kW heat [*Section 220-21*] Note: This type of question is rarely given on an exam.

Step 3: Appliance Demand Load [Section 220-17]

➤ **Appliance No. 1**

What is the demand load for a waste disposal (940 VA), a dishwasher (1,250 VA), and a water heater (4,500 VA)?

(a) 5,018 VA   (b) 6,690 VA   (c) 8,363 VA   (d) 6,272 VA

• Answer: (b) 6,690 VA

Waste disposal	940 VA
Dishwasher	1,250 VA
Water heater	<u>4,500 VA</u>
	6,690 VA

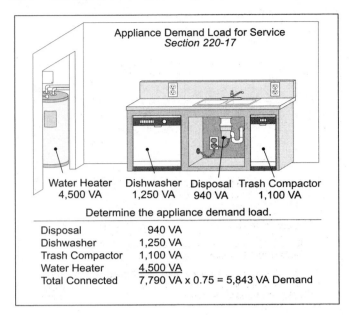

Appliance Demand Load for Service
Section 220-17

Water Heater   Dishwasher   Disposal   Trash Compactor
4,500 VA   1,250 VA   940 VA   1,100 VA

Determine the appliance demand load.

Disposal   940 VA
Dishwasher   1,250 VA
Trash Compactor   1,100 VA
Water Heater   4,500 VA
Total Connected   7,790 VA x 0.75 = 5,843 VA Demand

Figure 9–15
Appliance Demand Load for Service

Dryer Demand Load for Service
Section 220-18

Dryer:
4 kW 120/240 volt
Demand = 5 kW

4 kW
120/240 volt

Service demand load is
based on the nameplate
value but 5,000 VA is the
minimum value permitted.

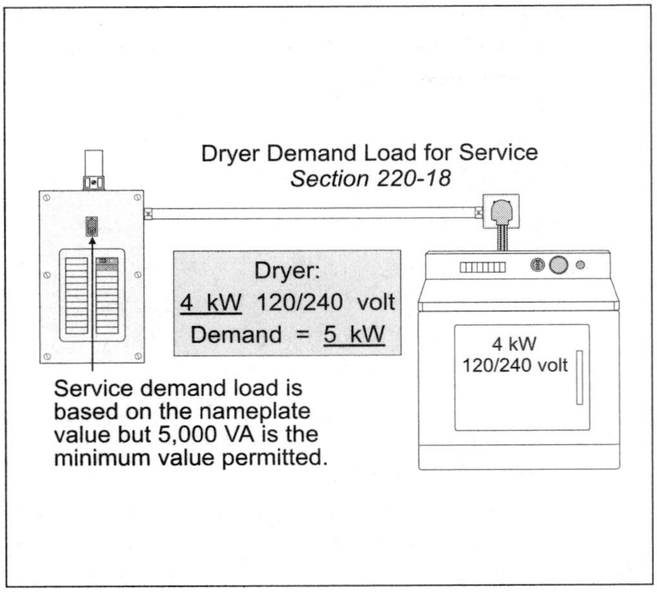

Figure 9–16
Dryer Demand Load for Service

➤ **Appliance No. 2**

What is the demand load for a waste disposal (940 VA), a dishwasher (1,250 VA), a trash compactor (1,100 VA), and a water heater (4,500 VA) (Figure 9–15)?

   (a) 7,790 VA   (b) 5,843 VA   (c) 7,303 VA   (d) 9,738 VA

     • Answer: (b) 5,843 VA

   Waste disposal   940 VA
   Dishwasher   1,250 VA
   Trash Compactor   1,100 VA
   Water Heater   4,500 VA
                     7,790 VA $\times$ 0.75 = 5,843 VA

Step 4: Dryer Demand Load [Section 220-18]

➤ **Dryer No. 1**

What is the service and feeder demand load for a 4 kW dryer (Figure 9–16)?

   (a) 4,000 watts   (b) 3,000 watts   (c) 5,000 watts   (d) 5,500 watts

     • Answer: (c) 5,000 VA

The dryer load must not be less than 5,000 VA.

➤ **Dryer No. 2**

What is the service and feeder demand load for a 5.5 kW dryer?

   (a) 4,000 watts   (b) 3,000 watts   (c) 5,000 watts   (d) 5,500 watts

     • Answer: (d) 5,500 VA

The dryer load must not be less than the nameplate rating if greater than 5,000 VA.

Step 5: Cooking Equipment Demand Load [Section 220-19]

➤ **Note 3 – Not over 3½ kW – Column B**

What is the service and feeder demand load for two 3 kW cooking appliances in a dwelling-unit?

   (a) 3 kW   (b) 4.8 kW   (c) 4.5 kW   (d) 3.9 kW

     • Answer: (c) 4.5 kW

   3 kW $\times$ 2 units = 6 kW $\times$ 0.75 = 4.5 kW

➡ Note 3 – Not over 8³/₄ – Column C

What is the service and feeder demand load for one 6 kW cooking appliance in a dwelling unit?

(a) 6 kW   (b) 4.8 kW
(c) 4.5 kW   (d) 3.9 kW

• Answer: (b) 4.8 kW

$6 \text{ kW} \times 0.8 = 4.8 \text{ kW}$

➡ Note 3 – Column B and C

What is the service and feeder demand load for two 3 kW ovens and one 6 kW cooktop in a dwelling unit (Figure 9–17)?

(a) 6 kW   (b) 4.8 kW
(c) 4.5 kW   (d) 9.3 kW

• Answer: (d) 9.3 kW

Column B demand: $(3 \text{ kW} \times 2) \times 0.75$	=	4.5 kW
Column C demand: $6 \text{ kW} \times 0.8$	=	<u>4.8 kW</u>
Total demand		9.3 kW

➡ Column A – Not over 12 kW

What is the service and feeder demand load for an 11.5 kW range in a dwelling unit?

(a) 11.5 kW   (b) 8 kW
(c) 9.2 kW   (d) 6 kW

• Answer: (b) 8 kW

Table 220-19 Column A: 8 kW

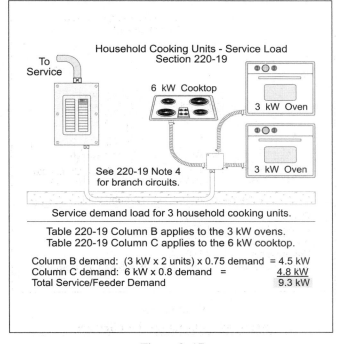

Figure 9–17
Household Cooking Units – Service Load

➡ Over 12 kW – Note 1

What is the service and feeder demand load for a 13.6 kW range in a dwelling unit?

(a) 8.8 kW   (b) 8 kW   (c) 9.2 kW   (d) 6 kW

• Answer: (a) 8.8 kW

Step 1:➡ 13.6 kW exceeds 12 kW by one kW and one major fraction of a kW. Column A value (8 kW) must be increased 5 percent for each kW or major fraction of a kW (0.5 kW or larger) over 12 kW.

Step 2:➡ $8 \text{ kW} \times 1.1 = 8.8 \text{ kW}$

Step 6: Service Conductor Size [Table 310-15(b)(6)]

➡ Service Conductor No. 1

What size THHN feeder or service conductor (120/240 volt, single-phase) is required for a 225-ampere service demand load?

(a) No. 1/0   (b) No. 2/0   (c) No. 3/0   (d) No. 4/0
• Answer: (c) No. 3/0

➡ Service Conductor No. 2

What size THHN feeder or service conductor (208Y/120 volt, single-phase) is required for a 225-ampere service demand load?

(a) No. 1/0   (b) No. 2/0   (c) No. 3/0   (d) No. 4/0

• Answer: (d) No. 4/0

Table 310-15(b)(6), does not apply to 208Y/120-volt systems because the grounded (neutral) conductor carries current, even if the loads are balanced.

9–10 DWELLING UNIT OPTIONAL FEEDER/SERVICE CALCULATIONS [*Section 220-30*]

Instead of sizing the dwelling-unit feeder and/or service conductors according to the standard method (Part B of Article 220), an optional method [*Section 220-30*] can be used. The following Steps can be used to determine the demand load:

Step 1:➡ **Section 220-30(b) General Loads**

The calculated load shall not be less 100 percent for the first 10 kW, plus 40 percent of the remainder of the following loads:

(1) Small Appliance and Laundry Branch Circuits: 1,500 volt-ampere for each 20-ampere small appliance and laundry branch circuit.

(2) General Lighting and Receptacles: 3 volt-ampere per square-foot.

(3) Appliances: The nameplate VA rating of all appliances and motors that are fastened in place (permanently connected), or located on a specific circuit.

Note: Be sure to use the range and dryer nameplate rating.

Step 2:➡ **Section 220-30(c) Heating and Air-Conditioning Load**

Include the largest of the following:

(1) 100 percent of the nameplate rating of the air-conditioning equipment.

(2) 100 percent of the heat pump compressors and supplemental heating unless the controller prevents both the compressor and supplemental heating from operating at the same time.

(3) 100 percent of the nameplate ratings of electric thermal storage and other heating systems where the usual load is expected to be continuous at the full nameplate value. Systems qualifying under this selection shall not be figured under any other selection in this table.

(4) 65 percent of the nameplate rating(s) of the central electric space heating, including integral supplemental heating in heat pumps where the controller prevents both the compressor and supplemental heating from operating at the same time.

(5) 65 percent of the nameplate rating(s) of electric space heating, if there are less than four separately controlled units.

(6) 40 percent of the nameplate rating(s) of electric space heating of four or more separately controlled units.

Step 3:➡ **Size Service/Feeder Conductors [Section 310-15(b)(6)].**

400 Ampere and Less: The ungrounded conductors are permitted to be sized to Table 310-15(b)(6) for 120/240-volt single-phase systems up to 400 ampere. The grounded (neutral) conductor must be sized to carry the unbalanced load, per Table 310-16.

Over 400 Ampere: The ungrounded and grounded neutral conductors are sized according to Table 310-16.

9–11 DWELLING UNIT OPTIONAL CALCULATION EXAMPLES

❏ **Optional Load Calculation No. 1**

What size service conductor is required for a 1,500 square foot dwelling-unit containing the following loads?

Dishwasher (1,200 VA) 115 volt	Water heater (4,500 VA) 230 volt
Disposal (900 VA) 115 volt	Dryer (4,000 VA) 230 volt
Cooktop (6,000 VA) 115/230 volt	Two ovens (each 3,000 VA) 115/230 volt
A/C (5 horsepower) 230 volt	Electric heating (10 kW), 230 volt

   (a) No. 6   (b) No. 4   (c) No. 3   (d) No. 2

   • Answer: (c) No. 3

Step 1: ➤ General Loads [*Section 220-30(b)*]

Small Appliance Circuits (1,500 VA × 2 circuits)	3,000 VA
General Lighting (1,500 square feet × 3 VA)	4,500 VA
Laundry Circuit	1,500 VA

Appliances (nameplate)

Dishwasher	1,200 VA		
Water Heater	4,500 VA		
Disposal	900 VA		
Dryer	4,000 VA		
Cooktop	6,000 VA		
Ovens 3,000 VA × 2 units	6,000 VA		
Total Connected Load	31,600 VA		
First 10,000 at 100%	10,000 VA × 1.00 =	10,000 VA
Remainder at 40%	21,600 VA × 0.40 =	8,640 VA
Demand load		18,640 VA

Step 2: ➤ Air-Conditioning versus Heat [*Section 220-30(c)*]
Air-conditioning at 100 percent versus 65 percent electric space-heating
Air Conditioner: 230 volts × 28 ampere = 6,440 VA (omit)
Electric heat: 10,000 VA × 0.65 = 6,500 VA

Step 3: ➤ Service/Feeder Conductors [*Section 310-15(c)(6)*]

General Loads	18,640 VA
Heat	6,500 VA
Total Demand Load	25,140 VA

I = VA/Volts

I = 25,140 VA/230 volts = 109 ampere

Note: The feeder/service conductor is sized to 110 ampere, No. 3 AWG.

❑ **Optional Load Calculation No. 2**

What size service conductor is required for a 2,330 square-foot residence with a 300 square-foot open porch and a 150 square-foot carport that contains the following:

Note: System voltage 120/240.

Dishwasher (1.5 kVA)	Waste disposal (1 kVA)	Trash compactor (1.5 kVA)
Water heater (6 kW)	Range (14 kW)	Dryer (4.5 kVA)
A/C (5 horsepower 230 volt)	Electric heat (four, each 2.5 kW)	

  (a) No. 5   (b) No. 4   (c) No. 3   (d) No. 2

    • Answer: (d) No. 2

Step 1: Total Connected Load

Totals

Step 1:➺ General Loads [*Section 220-30(b)*]

Small Appliance Circuits (1,500 VA × 2 circuits)	3,000 VA
General Lighting (2,330 square feet × 3 VA)	6,990 VA
Laundry circuit	1,500 VA

Appliances (nameplate)

Dishwasher	1,500 VA
Disposal	1,000 VA
Trash Compactor	1,500 VA
Water Heater	6,000 VA
Range	14,000 VA
Dryer (nameplate)	<u>4,500 VA</u>
Total Connected Load	39,900 VA
First 10,000 at 100%	<u>10,000 VA</u> × 1.00 = 10,000 VA
Remainder at 40%	29,000 VA × 0.40 = <u>11,996 VA</u>
Demand load	21,996 VA

Step 2:➺ Air-Conditioning versus Heat [*Section 220-30(c)*]

Air-conditioning at 100 percent versus 40 percent electric space-heating

Air-conditioner: 230 volts × 28 ampere [Table 430-148] = 6,440 VA

Electric heat: 10,000 VA × 0.40 = 4,000 VA, omit [*Section 220-21*]

Step 3:➺ Service/Feeder Conductors [*Section 310-15(c)(6)*]

General Loads	21,996 VA
A/C	<u>6,440 VA</u>
Total Demand Load	28,436 VA

I = VA/Volts,

I = 28,436 VA/240 volts

I = 118 ampere, No. 2

9–12 NEUTRAL CALCULATIONS – GENERAL [*SECTION 220-22*]

The feeder and service neutral load is the maximum unbalanced demand load between the grounded (neutral) conductor and any one ungrounded conductor as determined by Article 220 Part B (standard calculations). This means that since 240-volt loads are not connected to the neutral conductor, they are not considered for sizing the neutral conductor.

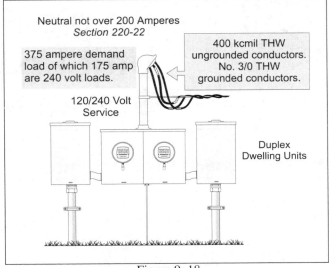

Figure 9–18
Neutral Not over 200 Amperes

❏ Neutral Not over 200 Ampere

What size 120/240-volt single-phase ungrounded and grounded (neutral) conductor is required for a 375 ampere, two-family dwelling-unit demand load, of which 175 ampere consist of 240-volt loads (Figure 9–18)?

    (a) Two – 400 kcmil and 1 – 350
    (b) Two – 350 kcmil and 1 – 350
    (c) Two – 500 kcmil and 1 – 500
    (d) Two – 400 kcmil and 1 – No. 3/0
      • Answer: (d) Two – 400 kcmil and 1 – No. 3/0

The ungrounded conductor is sized at 375 ampere (400 kcmil THHN) according to Table 310-15(b)(6). The grounded (neutral) conductor must be sized to carry the maximum unbalanced load of 200 ampere (120-volt loads only).

Cooking Appliance Feeder/Service Neutral Load [Section 220-22]

The feeder or service cooking appliance neutral load for household cooking appliances such as electric ranges, wall-mounted ovens, or counter-mounted cooking units shall be calculated at 70 percent of the demand load as determined by Section 220-19.

❏ Range Neutral

What is the neutral load (in ampere) for one 14 kW range?

    (a) 33 ampere   (b) 26 ampere   (c) 45 ampere   (d) 18 ampere
      • Answer: (b) 26 ampere

Step 1:➥ Since the range exceeds 12 kW, we must comply with Note 1 of Table 220-19. The first step is to determine the demand load as listed in Column A of Table 220-19 for one unit = 8 kW.

Step 2:➥ We must increase the Column A value (8 kW) by 5 percent for each kW that the average range (in this case 14 kW) exceeds 12 kW. This results is an increase of the Column A value (8 kW) by 10 percent.

    8 kW × 1.1 = 8.8 kW × 1,000 = 8,800 Watts

Step 3:➥ Convert the demand load to ampere.

$$I = \frac{P}{E} = \frac{8,800 \text{ VA}}{240 \text{ volts}} = 36.7 \text{ ampere}$$

Step 4:➥ Neutral load at 70 percent: 36.7 ampere × 0.7 = 26 ampere

---

Dryer Feeder/Service Neutral Load [Section 220-22]

The feeder and service dryer neutral demand load for electric clothes dryers shall be calculated at 70 percent of the demand load as determined by Section 220-18.

❏ Dryer over 5 kW

What is the feeder/service neutral load for one 5.5 kW dryer?

    (a) 3.85 kW   (b) 5.5 kW   (c) 4.6 kW   (d) none of these
      • Answer: (a) 3.85 kW
    5.5 kW × 0.7 = 3.85 kW

Unit 9 – Single-Family Dwelling Unit Load Calculations Summary Questions

Part A – General Requirements

9–2 Voltages

1. Unless other voltages are specified, branch circuit, feeder, and service loads shall be computed at nominal system voltages of _____ .
    (a) 120   (b) 120/240   (c) 208Y/120   (d) any of these

9–3 Fraction of an Ampere

2. When a calculation results in _____ of an ampere or larger, such fractions may be rounded up to the next ampere.
    (a) 0.15   (b) 0.5   (c) 0.49   (d) 0.051

9–4 Appliance (Small) Circuits [*Section 220–11(c)(1)*]

3. A minimum of _____ 20-ampere small appliance branch circuits are required for receptacle outlets in the kitchen, dining room, breakfast room, pantry, or similar areas of dining.
    (a) one   (b) two   (c) three   (d) none of these

4. When sizing the feeder or service, each dwelling-unit shall have a minimum feeder load of _____ VA for the two small appliance branch circuits.
    (a) 1,500   (b) 2,000   (c) 3,000   (d) 4,500

9–5 Cooking Equipment – Branch Circuit [Table 220-19, Note 4]

Table 220-19 Column A

5. What is the branch circuit demand load (in ampere) for one 11 kW range?
    (a) 21 ampere   (b) 27 ampere   (c) 38 ampere   (d) 33 ampere

6. What is the branch circuit load (in ampere) for one 13.5 kW range?
    (a) 33 ampere   (b) 37 ampere   (c) 50 ampere   (d) 58 ampere

Wall Mounted Oven and/or Counter-Mounted Cooking Unit

7. • What is the branch circuit load (in ampere) for one 6 kW wall-mounted oven, 240 volts, single-phase?
    (a) 15 ampere   (b) 25 ampere   (c) 30 ampere   (d) 40 ampere

8. • What is the branch circuit load (in ampere) for one 4.8 kW counter-mounted cooking unit, 240 volts, single-phase?
    (a) 15 ampere   (b) 20 ampere   (c) 25 ampere   (d) 30 ampere

9. What is the branch circuit load (in ampere) for one 6 kW counter-mounted cooking unit and one 3 kW wall-mounted oven, 240 volts, single-phase?
    (a) 25 ampere   (b) 38 ampere   (c) 33 ampere   (d) 42 ampere

10. What is the branch circuit load (in ampere) for one 6 kW counter-mounted cooking unit and two 4 kW wall-mounted ovens, 240 volts, single-phase?
    (a) 22 ampere   (b) 27 ampere   (c) 33 ampere   (d) 37 ampere

9–6 Laundry Receptacle(s) Circuit [*Section 220-11(c)(2)*]

11. The *NEC*® does not require a separate circuit for the washing machine, but requires a separate circuit for the laundry room receptacle or receptacles, of which one can be for the washing machine.
    (a) True   (b) False

12. Each dwelling-unit shall have a feeder load consisting of _____ VA for the 20-ampere laundry circuit.
    (a) 1,500   (b) 2,000   (c) 3,000   (d) 4,500

9–7 Lighting and Receptacles

13. The *NEC*® requires a minimum 3 volt-ampere each square foot for the required general lighting and receptacles . The dimensions for determining the area shall be computed from the outside of the building and shall not include _____ .
    (a) open porches   (b) garages
    (c) spaces not adaptable for future use   (d) all of these

14. The 3 volt-ampere each square foot for general lighting includes all 15 and 20 ampere general use receptacles, but not the appliance or laundry circuit receptacles.
    (a) True   (b) False

15. What is the general lighting load and receptacle load for a 2,100 square-foot home that has 14 extra convenience receptacles, and nine extra recessed fixtures rated 100 watts each?
    (a) 2,100 VA   (b) 4,200 VA   (c) 6,300 VA   (d) 8,400 VA

Number of General Lighting Circuits

16. The number of branch circuits required for general lighting and receptacles shall be determined from the general lighting load and the rating of the circuits.
    (a) True   (b) False

17. How many 15-ampere general lighting circuits are required for a 2,340 square-foot home?
    (a) 2   (b) 3   (c) 4   (d) 5

18. How many 20-ampere general lighting circuits are required for a 2,100 square-foot home?
    (a) 2   (b) 3   (c) 4   (d) 5

19. How many 20-ampere circuits are required for the general lighting and receptacles for a 1,500 square-foot dwelling-unit?
    (a) 2   (b) 3   (c) 4   (d) 5

Part B – Standard Method – Feeder/Service Load Calculation

9–8 Dwelling Unit Feeder/Service Load Calculations (Part B of Article 240)

20. The *NEC*® recognizes that the general lighting and receptacles and small appliance and laundry circuits will not all be on or loaded at the same time and permits a _____ to be applied to the total of these loads.
    (a) demand factor   (b) adjustment factor   (c) correction factor   (d) correction

21. Because the air-conditioning and heating loads are not on at the same time (simultaneously), it is permissible to omit the smaller of the two loads when determining the air-conditioning versus heat demand load.
    (a) True   (b) False

22. A load demand factor of 75 percent is permitted for _____ or more appliances fastened in place such as dishwasher, waste disposal, trash compactor, water heater, etc.
    (a) one   (b) two   (c) three   (d) four

23. The feeder or service demand load for electric clothes dryers located in dwelling-units shall not be less than _____ .
    (a) 5,000 watts   (b) nameplate rating   (c) a or b   (d) the greater of a or b

24.   • A feeder or service dryer load is required, even if the dwelling-unit does not contain an electric dryer.
      (a) True   (b) False

25.   Household cooking appliances rated 1³/₄ kW can have the feeder and service loads calculated according to the demand factors of Section 220-19, Table and Notes.
      (a) True   (b) False

26.   Dwelling-unit feeder and service ungrounded conductors are sized according to Note 3 of Table 310-16 for _____ .
      (a) 3-wire, single-phase, 120/240-volt systems up to 400 ampere
      (b) 3-wire, single-phase, 120/208-volt systems up to 400 ampere
      (c) 4-wire, single-phase, 120/240-volt systems up to 400 ampere
      (d) none of these

27.   What is the general lighting load for a 2,700 square-foot dwelling-unit?
      (a) 8,100 VA   (b) 12,600 VA   (c) 2,700 VA   (d) 6,840 VA

28.   What is the total connected load for general lighting and receptacles and small appliance and laundry circuits for a 6,540 square-foot dwelling-unit?
      (a) 8,100 VA   (b) 12,600 VA   (c) 2,700 VA   (d) 24,120 VA

29.   What is the feeder or service demand load for an air-conditioner (5 horsepower 230 volt) and three baseboard heaters (3 kW)?
      (a) 6,400 VA   (b) 3,000 VA   (c) 8,050 VA   (d) 9,000 VA

30.   What is the feeder or service demand load for an air-conditioner (4 horsepower 230 volt) and electric space heater (10 kW)?
      (a) 10,000 VA   (b) 3,910 VA   (c) 6,440 VA   (d) 10,350 VA

31.   What is the feeder or service demand load for a waste disposal (940 VA), dishwasher (1,250 VA), and a water heater (4,500 VA)?
      (a) 5,018 VA   (b) 6,690 VA   (c) 8,363 VA   (d) 6,272 VA

32.   • What is the feeder or service demand load for a waste disposal (940 VA), dishwasher (1,250 VA), water heater (4,500 VA), and a trash compactor (1,100 VA)?
      (a) 7,790 VA   (b) 5,843 VA   (c) 7,303 VA   (d) 9,738 VA

33.   What is the feeder or service demand load for a 4 kW dryer?
      (a) 4,000 VA   (b) 3,000 VA   (c) 5,000 VA   (d) 5,500 VA

34.   What is the feeder or service demand load for a 5.5 kW dryer?
      (a) 4,000 VA   (b) 3,000 VA   (c) 5,000 VA   (d) 5,500 VA

35.   • What is the feeder or service demand load for two 3 kW ranges?
      (a) 3 kW   (b) 4.8 kW   (c) 4.5 kW   (d) 3.9 kW

36.   What is the feeder or service demand load for one 6 kW cooking appliance?
      (a) 6 kW   (b) 4.8 kW   (c) 4.5 kW   (d) 3.9 kW

37.   What is the feeder or service demand load for one 6 kW and two 3 kW cooking appliances?
      (a) 6 kW   (b) 4.8 kW   (c) 4.5 kW   (d) 9.3 kW

38.   What is the feeder or service demand load for an 11.5 kW range?
      (a) 11.5 kW   (b) 8 kW   (c) 9.2 kW   (d) 6 kW

39.   What is the feeder or service demand load for a 13.6 kW range?
      (a) 8.8 kW   (b) 8 kW   (c) 9.2 kW   (d) 6 kW

40.   What size 120/240-volt single-phase service or feeder THHN conductors are required for a dwelling-unit that has a 190-ampere service demand load?
      (a) No. 1/0   (b) No. 2/0   (c) No. 3/0   (d) No. 4/0

41. • What is the net computed demand load for the general lighting and small appliance and laundry circuits for a 1,500 square foot-dwelling-unit?
     (a) 4,500 VA   (b) 9,000 VA   (c) 5,100 VA   (d) none of these

42. What is the feeder or service demand load for an air-conditioner (3 horsepower 230 volt) with a blower ($^1/_4$ horsepower) and electric space heating (4 kW)?
     (a) 3,910 VA   (b) 5,555 VA   (c) 4,000 VA   (d) none of these

43. What is the feeder or service demand load for a dwelling-unit that has one water heater (4 kVA) and one dishwasher (1.5 kVA)?
     (a) 4 kW   (b) 5.5 kW   (c) 4.13 kW   (d) none of these

44. What is the net computed load for a 4.5 kW dryer?
     (a) 4.5 kW   (b) 5 kW   (c) 5.5 kW   (d) none of these

45. What is the total net computed demand load for a 14 kW range?
     (a) 8 kW   (b) 8.4 kW   (c) 8.8 kW   (d) 14 kW

46. If the total demand load is 30 kW for a 120/240-volt dwelling-unit, what is the service and feeder conductor size?
     (a) No. 4   (b) No. 3/0   (c) No. 2   (d) No. 1/0

47. If the service raceway contains No. 2 service conductors, what size bonding jumper is required for the service raceway?
     (a) No. 8   (b) No. 6   (c) No. 4   (d) No. 3

48. If the service contains No. 2 conductors, what is the minimum size grounding electrode conductor required?
     (a) No. 8   (b) No. 6   (c) No. 4   (d) No. 3

Part C – Optional Method – Load Calculation Questions

9–10 Dwelling Unit Optional Feeder/Service Calculations [*Section 220-30*]

49. When sizing the feeder or service conductor according to the optional method we must:
     (a) Determine the total connected load of the general lighting and receptacles, small appliance and laundry branch circuits, and the nameplate VA rating of all appliances and motors.
     (b) Determine the demand load by applying the following demand factor to the total connected load: First 10 kVA at 100 percent, remainder at 40 percent.
     (c) Determine the air-conditioning versus heat demand load.
     (d) all of these.

Use this information for Question 50.
A 1,500 square-foot dwelling-unit containing the following loads:

Dishwasher (1.5 kVA)	Water heater (5 kVA)
Waste disposal (1 kVA)	Dryer (5.5 kVA)
Cooktop (6 kVA)	Two ovens (each 3 kVA)
A/C (3 horsepower 230 volts)	Electric heat (10 kW)

50. • Using the optional method, what size aluminum conductor is required for the 120/240-volt single-phase service?
     (a) No. 1   (b) No. 1/0   (c) No. 2/0   (d) No. 3/0

Use this information for Question 51.
A 2,330 square-foot residence with a 300 square-foot porch and 150 square-foot carport contains the following:

Dishwasher (1.5 kVA)	Waste disposal (1 kVA)
Trash compactor (1.5 kVA)	Water heater (6 kW)
Range (14 kW)	Dryer (4.5 kW)
A/C (5 horsepower 230 volts)	Baseboard heat (four each 2.5 kW)

51. Using the optional method, what size conductors are required for the 120/230-volt single-phase service?
     (a) No. 6   (b) No. 4   (c) No. 3   (d) No. 2

9–12 Neutral Calculations – General

52. The feeder and service neutral load is the maximum unbalanced demand load between the grounded (neutral) conductor and any one ungrounded conductor as determined by Article 220 Part B. Since 240-volt loads cannot be connected to the neutral conductor, 240-volt loads are not considered for sizing the feeder neutral conductor.
     (a) True   (b) False

53. • What size single-phase, 3-wire feeder is required for a 475-ampere demand load of which 275 ampere consists of 240-volt loads? Size the conductors based on 75°C terminals, in accordance with Section 110-14(c).
     (a) 2 – 750 and 1 – 350   (b) 2 – 500 and 1 – 350
     (c) 2 – 750 and 1 – 500   (d) 2 – 750 and 1 – No. 3/0

54. The feeder or service neutral load for household cooking appliances such as electric ranges, wall-mounted ovens, or counter-mounted cooking units shall be calculated at _____ of the demand load as determined by Section 220-19.
     (a) 50%   (b) 60%   (c) 70%   (d) 80%

55. What is the dwelling-unit service neutral load for one 9 kW ranges?
     (a) 5.6 kW   (b) 6.5 kW   (c) 3.5 kW   (d) 12 kW

56. The service neutral demand load for household electric clothes dryers shall be calculated at _____ of the demand load as determined by Section 220-18.
     (a) 50%   (b) 60%   (c) 70%   (d) 80%

57. What is the feeder neutral load for one 6 kW household dryers?
     (a) 4.2 kW   (b) 4.7 kW   (c) 5.4 kW   (d) 6.6 kW

☆ Challenge Questions

9–5 Cooking Equipment – Branch Circuit [Table 220-19 Note 4]

58. The branch circuit demand load for one 18 kW range is _____ .
     (a) 12 kW   (b) 8 kW   (c) 10.4 kW   (d) 18 kW

59. A dwelling-unit kitchen has the following appliances: one 9 kW cooktop and one wall-mounted oven rated 5.3 kW. The branch circuit demand load for these appliances is _____ .
     (a) 14.3 kW   (b) 12 kW   (c) 8.8 kW   (d) 8 kW

9–8 Dwellling Unit Feeder/Service Load Calculation (Part B of Article 220)

General Lighting [Table 220-11]

60. An apartment building contains 20 units, each 840 square feet. What is the general lighting feeder demand load for each dwelling-unit? Note: Laundry facilities are provided on the premises for all tenants and no laundry circuit is required in each unit, see 210-52(f) Exception No. 1.
     (a) 3,520 VA   (b) 3,882 VA   (c) 4,220 VA   (d) 6,300 VA

61. An 1,800 square-foot residence has a 300 square-foot open porch and a 450 square-foot carport. What is the demand load for the general lighting, receptacles, small appliance circuits, and laundry circuits?
     (a) 5,400 VA   (b) 7,900 VA   (c) 5,415 VA   (d) 6,600 VA

62. How many 15-ampere, 120-volt branch circuits are required for general use receptacle and lighting for a 1,800 square-foot dwelling-unit?
     (a) 1 circuit   (b) 2 circuits   (c) 3 circuits   (d) 4 circuits

63. How many 20-ampere, 120-volt branch circuits are required for general use receptacle and lighting for a 2,800 square-foot dwelling-unit?
     (a) 2 circuits   (b) 6 circuits   (c) 7 circuits   (d) 4 circuits

Appliance Demand Factors [*Section 220-17*]

64.   A dwelling-unit contains one of each of the following: washing machine (1.2 kW), water heater (4 kW), dishwasher (1.2 kW), trash compactor (1.5 kW). The appliance demand load added to the service is _____ .
      (a) 5.9 kW   (b) 6.7 kW   (c) 7.7 kW   (d) 8.8 kW

65.   • A dwelling-unit contains one of each of the following: water heater (4 kW), dishwasher ($^1/_2$ horsepower), dryer, pool pump ($^3/_4$ horsepower), cooktop (6 kW), oven (6 kW), A/C (4 horsepower 230 volts), heat (6 kW). What is the appliance demand load for the dwelling-unit?
      (a) 6.7 kW   (b) 4 kW   (c) 9 kW   (d) 11 kW

Dryer Calculation [*Section 220-18*]

66.   A dwelling-unit contains a 5.5 kW electric clothes dryer. What is the feeder demand load for the dryer?
      (a) 3.38 kW   (b) 4.5 kW   (c) 5 kW   (d) 5.5 kW

Cooking Equipment Demand Load [*Section 220-19*]

67.   The feeder load for twelve $1^3/_4$ kW cooktops is _____ .
      (a) 21 kW   (b) 10.5 kW   (c) 13.5 kW   (d) 12.5 kW

68.   The minimum neutral for a branch circuit to an 8 kW household range is _____ . See Article 210.
      (a) No. 10   (b) No. 12   (c) No. 8   (d) No. 6

69.   What is the maximum dwelling-unit feeder or service demand load for fifteen 8 kW cooking units?
      (a) 38.4 kW   (b) 33 kW   (c) 88 kW   (d) 27 kW

Note 2 of Table 220-19

70.   What is the feeder demand load for five 10 kW, five 14 kW, and five 16 kW household ranges?
      (a) 70 kW   (b) 23 kW   (c) 14 kW   (d) 33 kW

Note 3 to Table 220-19

71.   A dwelling-unit has one 6 kW cooktop and one 6 kW oven. What is the minimum feeder demand load for the cooking appliances?
      (a) 13 kW   (b) 8.8 kW   (c) 7.8 kW   (d) 8.2 kW

72.   The maximum feeder demand load for an 8 kW range is _____ .
      (a) 8.5 kW   (b) 8 kW   (c) 6.3 kW   (d) 12 kW

73.   The minimum feeder demand load for five 5 kW ranges, two 4 kW ovens, and four 7 kW cooking units is _____ .
      (a) 61 kW   (b) 22 kW   (c) 19.5 kW   (d) 18 kW

74.   The minimum feeder demand load for two 3 kW wall-mounted ovens and one 6 kW cooktop is _____ .
      (a) 9.3 kW   (b) 11 kW   (c) 8.4 kW   (d) 9.6 kW

9–9 Dwelling Unit Feeder/Service Calculations Examples

75.   • Both units of a duplex apartment require a 100-ampere main; the resulting 200-ampere service would require _____ THHN.
      (a) No. 1/0   (b) No. 2/0   (c) No. 3/0   (d) No. 4

76.   After all demand factors have been taken into consideration, the demand load for a dwelling-unit is 21,560 VA. The minimum service size for this residence would be _____ if the optional method of service calculations were used for a single-phase 120/240-volt system. Be sure to read Section 220-30 carefully.
      (a) 90 ampere   (b) 100 ampere   (c) 110 ampere   (d) 125 ampere

77.   After all demand factors have been taken into consideration, the net computed load for a service is 24,221 VA, 120/240-volt single-phase system. The minimum service size is _____ .
(a) 150 ampere   (b) 175 ampere   (c) 125 ampere   (d) 110 ampere

9–10 Optional Method – Load Calculation Questions [*Section 220-30*]

78.   Using the optional calculation method, the demand load for a 6 kW central space heating and a 4 kW air-conditioning unit would be _____ .
(a) 9,000 watts   (b) 3,900 watts   (c) 4,000 watts   (d) 5,000 watts

79.   The total connected load of a dwelling-unit is 25 kVA, not including heat or air-conditioning. If the heat is separately controlled in five rooms (10 kW), and the air-conditioning is 6 kW, then the total service demand load is _____ if the optional method is used.
(a) 22 kVA   (b) 29 kVA   (c) 35 kVA   (d) 36 kVA

80.   Using the optional method, the service size would be _____ for the following loads: 1,200 square feet first floor, plus 600 square feet on the second floor, and a 200 square-foot open porch, water heater (4 kW), dishwasher ($^1/_2$ horsepower), clothes dryer (4 kW), A/C (4 horsepower), heat (6 kW), pool pump ($^3/_4$ horsepower), oven (6 kW), and cooktop (6 kW). The service source voltage is 230/115-volt single-phase.
(a) 200 ampere   (b) 175 ampere   (c) 125 ampere   (d) 110 ampere

81.   An 1,800 square-foot residence contains the following: a 300 square-foot porch, a 150 square-foot carport, a 4 kW water heater, 10 kW heat separated in five rooms, one 1.5 kW dishwasher, one 6 kW range, one 4.5 kW dryer, two 3 kW ovens, and one 6 kW air conditioner. The 120/240-volt single-phase service size for the loads would be _____ . The optional method of service calculations were used.
(a) 175 ampere   (b) 110 ampere   (c) 125 ampere   (d) 150 ampere

82.   A dwelling-unit has 1,200 square feet on the first floor, 600 square feet upstairs (unfinished but adaptable for future use), and a 200 square-foot open porch with the following loads: pool pump ($^3/_4$ horsepower), range (13.9 kW), dishwasher (1.2 kW), water heater (4 kW), dryer (4 kW), A/C (5 horsepower), and heat (6 kW). Using the optional calculation, what size service and feeder conductor is required for a 120/240-volt single-phase service?
(a) No. 4   (b) No. 3   (c) No. 2   (d) No. 1

NEC® Questions from Section 90 through Section 551-71

Article 90 – Introduction

83.   The *National Electrical Code*® is _____ .
(a) intended to be a design manual
(b) meant to be used as an instruction guide for untrained persons
(c) for the practical safeguarding of persons and property
(d) published by the Bureau of Standards

84.   The *Code* covers underground installations in mines and self-propelled mobile surface mining machinery and its attendant electrical trailing cable.
(a) True   (b) False

NEC® Chapter 1 – General Requirements

85.   • A device that, by insertion in a receptacle, establishes connection between the conductors of the attached flexible cord and the conductors connected permanently to the receptacle is called an _____ .
(a) attachment plug   (b) plug cap   (c) cap   (d) all of these

86.   A solderless pressure connector is a device that _____ between two or more conductors or between one or more conductors and a terminal by means of mechanical pressure and without the use of solder.
(a) provides access   (b) protects the wiring
(c) is never needed   (d) establishes a connection

87. The *NEC*® term to define wiring that is not concealed is _____ .
    (a) open   (b) uncovered   (c) exposed   (d) bare

88. Conduit installed underground or encased in concrete slabs, which are in direct contact with the earth, shall be considered a _____ location.
    (a) dry   (b) damp   (c) wet   (d) moist

89. A large single panel, frame, or assembly of panels on which are mounted, on the face or back or both, switches, overcurrent and other protective devices, buses, and usually instruments is a _____ .
    (a) switchboard   (b) panel box   (c) switch box   (d) panelboard

90. The _____ of the circuit shall be so selected and coordinated as to permit the circuit protective devices used to clear a fault without extensive damage to the electrical components of the circuit.
    (a) overcurrent protective devices   (b) total circuit impedance
    (c) component short-circuit withstand ratings   (d) all of these

91. Enclosures housing electrical apparatus that are controlled by lock and key shall be considered _____ to qualified persons.
    (a) readily accessible   (b) accessible   (c) available   (d) none of these

92. Unless specified otherwise, live parts of electrical equipment operating at _____ volts or more shall be guarded.
    (a) 12   (b) 15   (c) 50   (d) 24

93. Unguarded live parts operating at 30,000 volts located above a working space shall be elevated _____ feet above the working space.
    (a) 24   (b) 18   (c) 12   (d) 9

Chapter 2 – Wiring and Protection

94. For the purpose of Article 200, connected so as to be capable of carrying current (as distinguished from connection through electromagnetic induction) defines the term _____ .
    (a) effectively grounded   (b) electrically connected
    (c) a grounded system   (d) none of these

95. 208Y/120- or 480Y/227-volt, 3-phase, 4-wire wye power systems used to supply nonlinear loads such as personal computers, energy efficient electronic ballasts, electronic dimming, etc. cause a distortion of the phase and neutral currents which produces high, unwanted and potentially hazardous harmonic neutral currents. The *Code* cautions us that the system design should allow for the possibility of high harmonic neutral currents.
    (a) True   (b) False

96. All 125-volt, single-phase 15- and 20-ampere receptacles installed in crawl spaces at or below grade level and in _____ of dwelling units shall have GFCI protection for personnel.
    (a) unfinished attics   (b) finished attics   (c) unfinished basements   (d) finished basements

97. _____ in dwelling units shall supply only loads within that dwelling unit or loads associated only with that dwelling unit.
    (a) Service-entrance conductors   (b) Ground-fault protection
    (c) Branch circuits   (d) none of these

98. Island or Peninsular countertop receptacle outlets can be installed below the countertop surface in dwelling units, when necessary for the physically impaired or if no wall space is available above the countertop.
    (a) True   (b) False

99. Dwelling unit or mobile home feeder conductors not over 400 amperes can be sized according to Section 310-15(b)(6).
    (a) True   (b) False

100. For feeder and service load calculations, equipment that contains four or more receptacles installed in an outlet box is to be considered at not less than ___ volt-amperes per receptacle.
    (a) 220   (b) 90   (c) 125   (d) none of these

101. • To determine the feeder demand load for ten 3 kW household cooking appliances, use _____ of Table 220-19.
    (a) Column A   (b) Column B   (c) Column C   (d) none of these

102. Where within _____ feet of any building or other structure, open wiring on insulators shall be insulated or covered.
    (a) 4   (b) 3   (c) 6   (d) 10

103. Where outdoor lampholders have terminals of a type that puncture the _____ and make contact with the conductors, they shall be attached only to conductors of the stranded type.
    (a) exterior of the building (b) paint   (c) insulation   (d) none of these

104. Service drops run over roofs shall have a minimum clearance of _____ feet.
    (a) 8   (b) 12   (c) 15   (d) 3

105. Service conductors shall be sized no less than _____ percent of the continuous load, plus 100 percent of the noncontinuous load.
    (a) 100   (b) 115   (c) 125   (d) 150

106. • The high-leg service entrance conductor of a 4-wire, delta-connected service is required to be permanently marked _____ .
    (a) in orange   (b) with orange marking tape
    (c) by other effective means   (d) any of these

107. Where the service disconnecting means is power-operable, the _____ can be connected ahead of the service disconnecting means if suitable overcurrent protection and disconnecting means are provided.
    (a) control circuit   (b) panel   (c) grounding conductor   (d) none of these

108. The added ground-fault protective equipment at the wye service equipment will make it necessary to _____ the overall wiring system for proper selective overcurrent protection coordination.
    (a) redesign   (b) review   (c) inspect   (d) none of these

109. Conductor overload protection shall not be required where the interruption of the _____ would create a hazard, such as in a material handling magnet circuit or fire pump circuit. However short-circuit protection is required.
    (a) circuit   (b) line   (c) phase   (d) system

110. A fuse or an overcurrent trip unit of a circuit breaker shall be connected in series with each ungrounded _____ .
    (a) device   (b) conductor   (c) branch circuit   (d) all of these

111. Plug fuses of the Edison-base type shall be used _____ .
    (a) where overfusing is necessary
    (b) only as replacement in existing installations
    (c) as a replacement for type S fuses
    (d) only for 50 amperes and above

112. Circuit breakers shall be marked with their _____ rating in a manner that will be durable and visible after installation.
    (a) voltage   (b) ampere   (c) type   (d) all of these

113. Systems and circuit conductors are grounded to _____ .
    (a) limit voltages due to lightning, line surges, or unintentional contact with higher voltage lines
    (b) stabilize the voltage to ground during normal operation
    (c) facilitate overcurrent protection device operation in case of ground-faults
    (d) a and b

114. Where an ac system operating at less than 1,000 volts is grounded at any point, the _____ conductor shall be run to each service disconnecting means and shall be bonded to each disconnect enclosure.
    (a) ungrounded   (b) grounded   (c) grounding   (d) none of these

115. A bare No. 4 copper conductor installed near the bottom of a concrete foundation or footing that is in direct contact with the earth, may be used as a grounding electrode when at least _____ feet in length.
    (a) 25   (b) 15   (c) 10   (d) 20

116. The size of the grounding electrode conductor of a grounded ac system to a concrete-encased electrode shall not be required to be larger than No. _____ copper wire.
(a) 4   (b) 6   (c) 8   (d) 10

117. Service raceways threaded into metal service equipment such as bosses (hubs) are considered to be effectively _____ to the service metal enclosure.
(a) attached   (b) bonded   (c) grounded   (d) none of these

118. Metal raceways, enclosures, frames, and other noncurrent-carrying metal parts of electric equipment installed on a building equipped with a lightning protection system may require spacing from the lightning protection conductors typically 6 feet through air or ___ feet through dense materials such as, concrete, brick, wood, etc.
(a) 2   (b) 3   (c) 4   (d) 6

119. Where conductors are run in parallel in multiple raceways or cables, the equipment grounding conductor, where used, shall be run in parallel. The equipment grounding conductor can be smaller than required in Table 250-122 if:
(a) Equipment ground-fault protection device (GFPE) protects the circuit.
(b) The conditions of maintenance and supervision ensure that only qualified persons will service the installation.
(c) The size of each parallel equipment grounding conductor is sized based on the ampere setting of the equipment's ground-fault protection device in accordance with Table 250-122.
(d) all of these

120. Equipment bonding jumpers are not required for listed wiring devices that have mounting screws that provide the grounding continuity between the metal yoke and the flush box (self-grounding receptacles).
(a) True   (b) False

121. Line and ground connecting conductors for a surge arrester shall not be smaller than No. _____ .
(a) 14   (b) 12   (c) 10   (d) 8

Chapter 3 – Wiring Methods and Material

122. Single conductors are only permitted when installed as part of a wiring method listed in Chapter _____ .
(a) 4   (b) 3   (c) 2   (d) 9

123. UF cables used with a 24-volt landscape lighting system is permitted to have a minimum cover of _____ inches.
(a) 6   (b) 12   (c) 18   (d) 24

124. Support wires within non fire-rated assemblies are not required to be distinguishable from the suspended ceiling framing support wires.
(a) True   (b) False

125. Where equipment or devices are installed in ducts or plenum chambers used to transport environmental air and illumination is necessary to facilitate maintenance and repair, enclosed _____-type lighting fixtures shall be permitted.
(a) screw   (b) plug   (c) gasketed   (d) neon

126. Temporary electrical power and lighting is permitted for emergencies and _____ .
(a) tests   (b) experiments   (c) developmental work   (d) all of these

127. The parallel conductors in each phase or neutral shall _____ .
(a) be the same length and conductor material
(b) have the same circular mil area and insulation type
(c) be terminated in the same manner
(d) all of these

128. • Type THW insulation has a _____ rating when installed within electric discharge lighting equipment, such as through fluorescent lighting fixtures.
(a) 60°C   (b) 75°C   (c) 90°C   (d) none of these

129. Table 310-70 provides the ampacities of insulated single aluminum conductor isolated in air. If the conductor size is 8 AWG, MV-90, and the voltage range is 2,001 to 5,000, then the ampacity is _____ amperes.
(a) 64   (b) 85   (c) 115   (d) 150

130. • The intent of Article 318 is to limit the use of cable trays to industrial establishments only.
(a) True   (b) False

131. Conductors for open wiring on insulators shall be supported within _____ inches of a tap or splice.
(a) 6   (b) 8   (c) 10   (d) 12

132. An exposed wiring support system using a messenger wire to support insulated conductors is known as _____ .
(a) open wiring   (b) messenger supported wiring
(c) field wiring   (d) none of these

133. Supports for concealed knob-and-tube wiring shall be installed within _____ inches of each side of each tap or splice, and at intervals not exceeding _____ feet.
(a) 3, $2^1/_2$ (b) 2, $3^1/_2$ (c) 6, $4^1/_2$ (d) 4, $6^1/_2$

134. A field-installed wiring system for branch circuits designed for installation under carpet squares is defined as _____ .
(a) underfloor wiring   (b) undercarpet wiring
(c) flat conductor cable   (d) underfloor conductor cable

135. • Which of the following statement about MI cable is correct?
(a) It may be used in any hazardous location.
(b) A single run of cable shall not contain more than the equivalent of four quarter bends.
(c) It shall be securely supported at intervals not exceeding 10 feet.
(d) none of these

136. The use of electrical nonmetallic tubing and fittings shall be permitted: In any building not exceeding three floors above grade; for exposed work; where not subject to physical damage; or concealed within walls, floors, and ceilings.
(a) True   (b) False

137. • The use of Type AC cable is permitted for _____ .
(a) wet   (b) cable tray   (c) exposed   (d) b and c

138. • Type MC cable shall be permitted for systems _____ .
(a) of 600 volts, nominal or less only   (b) over 600 volts, only
(c) in excess of 600 volts, nominal   (d) in excess of 5,000 volts

139. Nonmetallic-sheathed cable can be installed in multifamily dwellings or other structures not exceeding _____ floors above grade.
(a) two   (b) three   (c) four   (d) none

140. Type SE service-entrance cables shall be permitted in interior wiring systems where all of the circuit conductors of the cable are of _____ .
(a) rubber   (b) thermoplastic   (c) nylon   (d) a or b

141. Type TC tray cable shall not be installed _____ .
(a) where it will be exposed to physical damage   (b) as open cable on brackets or cleats
(c) direct buried, unless identified for such use   (d) all of these

142. Nonmetallic underground conduit with conductors shall not be used in conduit larger than _____ inches
(a) 1   (b) 2   (c) 3   (d) 4

143. Intermediate metal conduit can be installed in or under cinder fill where subject to permanent moisture where protected by _____ and judged suitable for the condition.
(a) not less than 18 inches under the fill
(b) 2 inches of concrete
(c) corrosion protection judged suitable
(d) any of these

144. Rigid metal conduit can be installed in or under cinder fill where subject to permanent moisture where protected on all sides by a layer of noncinder concrete not less than _____ thick.
    (a) 2 inches   (b) 4 inches   (c) 6 inches   (d) 18 inches

145. Extreme _____ may cause some nonmetallic conduits to become brittle and therefore more susceptible to damage from physical contact.
    (a) temperatures   (b) corrosive conditions   (c) heat   (d) cold

146. • Electrical metallic tubing shall not be threaded.
    (a) True   (b) False

147. The maximum size flexible metallic tubing permitted is _____ inch(es).
    (a) $^3/_8$ (b) $^1/_2$ (c) $^3/_4$ (d) 1

148. Flexible metal conduit cannot be installed _____ .
    (a) underground
    (b) embedded in poured concrete
    (c) where subject to physical damage
    (d) all of these

149. • Liquidtight flexible metal conduit smaller than _____ inch(es) shall not be used, except as permitted in Section 350-10(a).
    (a) 3/8   (b) $^1/_2$ (c) 11   (d) $1^1/_4$

150. • The maximum number of conductors permitted in any surface raceway shall be _____ .
    (a) no more than 30 percent of the inside diameter
    (b) no greater than the number for which it was designed
    (c) no more than 75 percent of the cross-sectional area
    (d) that which is permitted in the Table

151. Underfloor raceways shall be laid so that a straight line from the center of one _____ to the center of the next _____ will coincide with the centerline of the raceway system.
    (a) termination point   (b) junction box   (c) receptacle   (d) panelboard

152. No conductor larger than a No. _____ shall be installed in a cellular metal floor raceway.
    (a) No. 1/0   (b) No. 4/0   (c) No. 1   (d) no restriction

153. The header on a cellular concrete floor raceway shall be installed _____ to the cells.
    (a) in a straight line   (b) at right angles to the cells
    (c) a and b   (d) none of these

154. Wireways are sheet metal troughs with _____ that are used to form a raceway system.
    (a) removable covers   (b) hinged covers   (c) a or b   (d) none of these

155. Flat cable assemblies shall not be installed outdoors or in wet or damp locations unless _____ for use in wet locations.
    (a) special permission is granted   (b) approved
    (c) identified   (d) none of these

156. Where a vertical busway penetrates two or more floors (in other than industrial establishments), a 4-inch high curb shall be installed around the busway floor opening to prevent liquids from entering the vertical busway. The curb shall be installed within _____ of the floor opening for the busway, and electrical equipment shall be located so that liquids retained by the four-inch curb will not damage equipment.
    (a) 12 inches   (b) 6 inches   (c) 12 feet   (d) 6 feet

157. The cablebus assembly is designed to carry _____ current and to withstand the magnetic forces of such current.
    (a) service   (b) load   (c) fault   (d) grounded

158. Metallic boxes are required to be _____ .
    (a) bonded   (b) installed   (c) grounded   (d) all of these

159.   • In combustible walls or ceilings, the front edge of an outlet box or fitting may set back of the finished surface _____ inch.
      (a) $^3/_8$ (b) $^1/_8$ (c) $^1/_2$ (d) not at all

160.   Listed boxes designed for underground installation can be directly buried where covered by _____ .
      (a) concrete   (b) gravel
      (c) noncohesive granulated soil   (d) b and c

161.   In walls constructed of wood or other _____ material, cabinets shall be flush with the finished surface or project therefrom.
      (a) wood-like   (b) thermoplastic   (c) corrosive   (d) combustible

162.   Because an auxiliary gutter is used to supplement wiring space, it is not a raceway and shall not extend a distance greater than _____ feet beyond the equipment that it supplements.
      (a) 50   (b) 30   (c) 10   (d) 25

163.   Which of the following switches are required to indicate whether they are in the "on" or "off" position?
      (a) general use switches   (b) motor circuit switches
      (c) circuit breakers   (d) all of these

164.   Each switchboard, or panelboard if used as service equipment, shall be provided with a main bonding jumper within the panelboard or one of the sections of the switchboard for connecting the grounded service conductor on its _____ side to the switchboard or panelboard frame.
      (a) load   (b) supply   (c) phase   (d) high-leg

165.   The minimum spacing of busbars of opposite polarity, held in free air in a panel is _____ inch(es) when not over 125 volts, nominal.
      (a) $^1/_2$ (b) 1   (c) 2   (d) 4

Chapter 4 – Equipment for General Use

166.   • In no case shall conductors within flexible cords and cables be associated in such a way with respect to the kind of circuit, the wiring method used, or the number of conductors such that the _____ temperature of the conductors is exceeded.
      (a) operating   (b) governing   (c) ambient   (d) limiting

167.   • A flexible cord conductor intended to be used as a _____ conductor shall have a continuous identifying marker readily distinguishing it from the other conductor or conductors.
      (a) ungrounded   (b) equipment grounding   (c) service   (d) high-leg

168.   The smallest size fixture wire permitted in the *NEC®* is No. _____ .
      (a) 22   (b) 20   (c) 18   (d) 16

169.   Lighting fixtures, lamp holders, and receptacles shall have no live parts normally exposed to contact. Cleat-type lampholders and receptacles located at least _____ feet above the floor shall be permitted to have exposed contacts.
      (a) 3   (b) 6   (c) 8   (d) none of these

170.   Metal poles that support lighting fixtures must meet the following requirements: _____ .
      (a) they must have an accessible handhole (sized 2" × 4" inch) with a raintight cover
      (b) an accessible grounding terminal must be installed accessible from the handhole
      (c) a and b
      (d) none of these

171.   Branch circuit conductors within 3 inches of a ballast within the ballast compartment shall be of a type recognized for use at temperatures not lower than 90°C such as:
      (a) THHN   (b) THW   (c) TW   (d) a or b

172. • Receptacles installed indoors or outdoors protected from the weather or other damp locations shall be in an enclosure that is _____ when the receptacle is covered.
(a) raintight   (b) weatherproof   (c) rainproof   (d) weathertight

173. Auxiliary equipment where not installed as part of electric-discharge lighting fixtures (not over 1,000 volts) shall be enclosed in accessible, permanently installed _____ .
(a) nonmetallic cabinets   (b) enclosures
(c) metal cabinets   (d) all of these

174. If a protective device rating is marked on an appliance, the branch circuit overcurrent protection device rating shall not be greater than _____ percent of the protective device rating marked on the appliance.
(a) 100   (b) 50   (c) 80   (d) 115

175. Appliances that have a unit switch with an _____ setting and that disconnect all the ungrounded conductors can serve as the disconnecting means for the appliance.
(a) on   (b) off   (c) on/off   (d) all of these

176. Baseboard heaters with factory-installed receptacle outlets can be used as the required outlets required by Section 210-52(a).
(a) True   (b) False

177. The size of the branch circuit overcurrent protective devices and conductors for an electrode-type boiler shall be calculated on the basis of _____ .
(a) 125 percent of the total load excluding motors
(b) 125 percent of the total load including motors
(c) 150 percent of the nameplate value
(d) 100 percent of the nameplate value

178. Embedded de-icing and snow-melting equipment cables, units, and panels shall not be installed where they bridge _____ unless provision is made for expansion and contraction.
(a) roads   (b) over water spans   (c) runways   (d) expansion joints

179. External surfaces of pipeline and vessel heating equipment that operate at temperatures exceeding _____°F shall be physically guarded, isolated, or thermally insulated to protect against contact by personnel in the area.
(a) 110   (b) 120   (c) 130   (d) 140

180. Branch circuit conductors supplying a single motor shall have an ampacity not less than _____ rating.
(a) 125 percent of maximum motor nameplate
(b) 125 percent of the motor full-load current
(c) 125 percent of the motor full locked rotor
(d) 80 percent of the motor full-load current

181. • A motor feeder supplying sized by Section 430-24 shall have a protective device with a rating or setting of _____ branch circuit short circuit and ground-fault protective device for any motor in the group plus the sum of the full-load currents of the other motors of the group.
(a) not greater than the largest   (b) 125 percent of the largest rating
(c) equal to the largest rating   (d) none of these

182. When a controller is not within sight of the motor location, the motor disconnecting means shall be capable of _____ .
(a) ground-fault operation
(b) opening all ungrounded conductors
(c) being locked in the open position
(d) being locked in the closed position

183. A disconnecting means serving a hermetic refrigerant motor compressor selected on the basis of the nameplate rated load current or branch circuit selection current, whichever is greater, shall have an ampere rating of _____ percent of the nameplate rated load current or branch circuit selection current.
(a) 125   (b) 80   (c) 100   (d) 115

184. A secondary tie of a transformer is a circuit operating at _____ volts or less between phases that connects two power sources or power supply points.
      (a) 600   (b) 1,000   (c) 12,000   (d) 35,000

185. Personnel doors for transformer vaults shall _____ and be equipped with panic bars, pressure plates, or other devices that are normally latched but open under simple pressure.
      (a) be clearly identified   (b) swing out
      (c) a and b   (d) a or b

186. A thermal barrier shall be required if the space between resistors and reactors and any combustible material is less than _____ inches.
      (a) 2   (b) 3   (c) 6   (d) 12

Chapter 5 – Special Occupancies

187. For Class _____ locations, Groups "A, B, C, and D", the classification involves determinations of maximum explosion pressure, maximum safe clearance between parts of a clamped joint in an enclosure, and the minimum ignition temperature of the atmospheric mixture.
      (a) I   (b) II   (c) III   (d) all of these

188. Class III locations are those that are hazardous because of the presence of _____ .
      (a) combustible dust
      (b) easily ignitable fibers or flyings
      (c) flammable gases or vapors
      (d) flammable liquids or gases

189. • Meters, instruments, and relays installed in Class I Division 2 locations can have switches, circuit breakers, and make-and-break contacts of push buttons, relays, alarm bells, and horns installed in enclosures approved for general purpose if current-interrupting contacts are _____ .
      (a) immersed in oil
      (b) enclosed within a hermetically sealed chamber
      (c) a or b
      (d) a and b

190. MC cable containing shielded cables and/or twisted pair cables in a Class I, Division 1 location shall not require the removal of the shielding material or the separation of the twisted pairs, provided the termination is by a(n) _____ means to minimize the entrance of gases or vapors and prevent propagation of flame into the cable core.
      (a) approved   (b) listed   (c) acceptable   (d) none of these

191. In Class II, Division 1 and 2 locations, an approved method of bonding is the use of _____ .
      (a) bonding jumpers with approved fittings
      (b) double lock-nut types of contacts
      (c) lock-nut bushing types of contacts
      (d) any of the above are approved methods of bonding

192. Article _____ contains the requirements for the wiring of occupancy locations used for service and repair operation in connection with self-propelled vehicles (including passenger automobiles, buses, trucks, tractors, etc.) in which volatile flammable liquids are used for fuel or power.
      (a) 500   (b) 501   (c) 511   (d) 514

193. Which of the following areas of an aircraft hanger is not classified as a Class I location?
      (a) below floor level
      (b) areas not cut off or ventilated
      (c) the vicinity of aircraft
      (d) areas suitably cut off and ventilated

194. Remote pump control wiring for gasoline dispensers must be isolated from each other to eliminate feedback.
      (a) True   (b) False

195. 15- and 20-ampere receptacles located in pediatric areas shall be _____ .
(a) tamper resistant   (b) isolated   (c) GFCI-protected   (d) specification grade

196. The maximum internal current permitted to flow through the line isolation monitor, when any point of the isolated system is grounded, shall be _____ when used in a health care facility.
(a) 15 amperes or less   (b) no more than 1 ampere
(c) 1 milliamperes   (d) 10 milliamperes

197. In places of assembly, nonmetallic raceways encased in not less than _____ of concrete shall be permitted.
(a) 1 inch   (b) 2 inches   (c) 3 inches   (d) none of these

198. Flexible cords and cables for carnival, circuses, and fairs shall be listed for wet locations and be sunlight resistant when indoors.
(a) True   (b) False

199. Conductors supplying outlets for arc and Xenon projectors of the professional type shall not be smaller than No.
_____ .
(a) 12   (b) 10   (c) 8   (d) 6

200. • A minimum of 70 percent of all recreational vehicle sites with electrical supply shall each be equipped with a _____-ampere receptacle.
(a) 15   (b) 20   (c) 30   (d) 50

CHAPTER 3
Advanced *NEC*® Calculations and *Code* Questions

Scope of Chapter 3

Unit 10

Multifamily Dwelling Unit Load Calculations

OBJECTIVES

After reading this unit, the student should be able to explain the following concepts:

Air-conditioning versus heat	Cooking equipment calculations	Dwelling-unit calculations – standard
Appliance (small) branch circuits	Dwelling-unit calculations – optional	method
Appliance demand load	method	Laundry circuit
Clothes dryer demand load		Lighting and receptacles

After reading this unit, the student should be able to explain the following terms:

General lighting	Rounding	Voltages
General use receptacles	Standard method	
Optional method	Unbalanced demand load	

10–1 MULTIFAMILY DWELLING-UNIT CALCULATIONS – STANDARD METHOD

When determining the ungrounded and grounded (neutral) conductor for multifamily dwelling units (Figure 10–1), apply the following standard method steps:

Step 1: ➼ **General Lighting and Receptacles, Small Appliance and Laundry Circuits [*Section 220-11*]**

The *NEC*® recognizes that the general lighting and receptacles, small appliance, and laundry circuits will not all be on, or loaded, at the same time. The *NEC*® permits the following demand factors to be applied to these loads [*Section 220-16*]:

(a) Total Connected Load: Determine the total connected general lighting and receptacles (3 VA per square-foot), small appliance (3,000 VA), and the laundry (1,500 VA) circuit load of all dwelling-units. The laundry load (1,500 VA) can be omitted if laundry facilities which are available to all building occupants, are provided on the premises [*Section 210-52(f)* Exception 1].

(b) Demand Factor: Apply Table 220-11 demand factors to the total connected load (Step 1a).
First 3,000 VA at 100% demand
Next 117,000 VA at 35% demand
Remainder at 25% demand

Step 2: ➼ **Air-Conditioning versus Heat [*Section 220-15, 220-21,* and *440-34*]**

Because the air-conditioning and heating loads are not on at the same time (simultaneously), it is permissible to omit the smaller of the two loads.

(a) Air-conditioning: The air-conditioning demand load shall be calculated at 125 percent of the largest air-conditioning motor VA, plus the sum of the other air-conditioning motor VAs [*Section 440-34*].

(b) Heat: Electric space-heating loads shall be computed at 100 percent of the total connected load [*Section 220-15*].

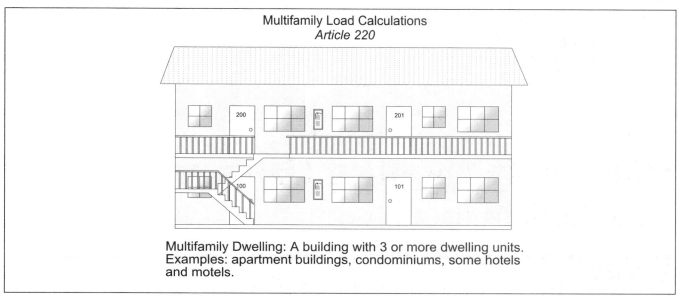

Multifamily Load Calculations
Article 220

Multifamily Dwelling: A building with 3 or more dwelling units. Examples: apartment buildings, condominiums, some hotels and motels.

Figure 10-1
Multifamily Load Calculations

Step 3: ➡ **Appliances [*Section 220-17*]**

A demand factor of 75 percent is permitted for four or more appliances fastened in place such as a dishwasher, waste disposal, trash compactor, water heater, etc. This does not apply to space-heating equipment [*Section 220-15*], clothes dryers [*Section 220-18*], cooking appliances [*Section 220-19*], or air-conditioning equipment [*Section 440-34*].

Step 4: ➡ **Clothes Dryers [*Section 220-18*]**

The feeder or service demand load for electric clothes dryers located in dwelling units shall not be less than: 5,000 watts, or the nameplate rating (whichever is greater) adjusted according to the demand factors listed in Table 220-18.

Note: A dryer load is not required if the dwelling unit does not contain an electric dryer. Laundry room dryers shall not have their loads calculated according to this method. This is covered in Unit 11.

Step 5: ➡ **Cooking Equipment [*Section 220-19*]**

Household cooking appliances rated over $1^3/_4$ kW can have their feeder and service loads calculated according to the demand factors of Section 220-19, Table and Notes.

Step 6: ➡ **Feeder and Service Conductor Size**

400 ampere and less: The ungrounded conductors are sized according to Table 310-15(b)(6) for 120/240-volt single-phase systems.

Over 400 ampere: The ungrounded conductors are sized according to Table 310-16 to the calculated unbalanced demand load.

10–2 MULTIFAMILY DWELLING-UNIT CALCULATION EXAMPLES – STANDARD METHOD

Step 1. General Lighting, Small Appliance, and Laundry Demand [Section 220-11]

➡ **General Lighting Load No. 1**

What is the demand load for an apartment building that contains 20 units? Each apartment is 840 square feet.

Note: Laundry facilities are provided on the premises for all tenants [*Section 210-52(f)*].

(a) 5,200 VA   (b) 40,590 VA   (c) 110,400 VA   (d) none of these

• Answer: (b) 40,590 VA

General lighting	(840 square feet × 3 VA)	2,520 VA
Small appliance circuits		3,000 VA
Laundry circuit		0 VA
Total connected load for one unit		5,520 VA

Demand factor [*Section 220-11*] (5,520 VA × 20 units)	110,400 VA	
First 3,000 VA at 100%	− 3,000 VA × 1.00 =	3,000 VA
Next 117,000 VA at 35%	107,400 VA × 0.35 =	37,590 VA
Total demand load		40,590 VA

➡ **General Lighting Load No. 2**

What is the general lighting net computed demand load for a 20-unit apartment building? Each unit is 990 square feet.

(a) 74,700 VA   (b) 149,400 VA   (c) 51,300 VA   (d) 105,600 VA

• Answer: (c) 51,300 VA

General lighting	(990 square feet × 3 VA)	2,970 VA
Small appliance circuits		3,000 VA
Laundry circuit		1,500 VA
Total connected load for one unit		7,470 VA

Demand factor [*Section 220-11*] (7,470 VA × 20 units)	149,400 VA	
First 3,000 VA at 100%	− 3,000 VA × 1.00 =	3,000 VA
	146,400 VA	
Next 117,000 VA at 35%	−117,000 VA × 0.35 =	40,950 VA
Remainder VA at 25%	29,400 VA × 0.25 =	+ 7,350 VA
Total demand load		51,300 VA

Step 2. Air-Conditioning versus Heat [Section 220-15 and 440-34]

➡ **Air- Conditioning versus Heat No. 1**

What is the net computed air-conditioning versus heat load for a 40-unit multifamily building that has an air conditioner (3 horsepower 230 volt) and two baseboard heaters (3 kW) in each unit?

(a) 160 kW   (b) 240 kW   (c) 60 kW   (d) 50 kW

• Answer: (b) 240 kW

Air-conditioning [*Section 440-34*] 230 volts × 17 ampere* = 3,910 VA

(3,910 VA × 1.25) + (3,910 VA × 39 units) = 157,378 VA, Omit smaller than heat [*Section 220-21*]

Heat [*Section 220-15*] 3,000 watts × 2 units = 6,000 watts × 40 units = 240,000 watts/1,000 = 240 kW

➡ **Air-Conditioning versus Heat No. 2**

What is the air-conditioning versus heat demand load for a 25-unit multifamily building that has a air-conditioning (3 horsepower 230 volt) and electric heat (5 kW)?

(a) 160 kVA   (b) 125 kVA   (c) 6 kVA   (d) 5 kVA

• Answer: (b) 125 kVA

Air-conditioning [*Section 440-34*] 230 volts × 17 ampere* = 3,910 VA

(3,910 VA × 1.25) + (3,910 × 24 units) = 98,728 VA, Omit

Heat [*Section 220-15*] 5,000 watts × 25 units = 125,000 watts/1,000 = 125 kW

* Table 430-148

Step 3. Appliance Demand Load [Section 220-17]

➡ **Appliance Load No. 1**

What is the appliance demand load for a 20-unit multifamily building that contains a 940 VA waste disposal, a 1,250 VA dishwasher, and a 4,500 VA water heater?

(a) 100 kVA   (b) 134 kVA   (c) 7 kVA   (d) 5 kVA

• Answer: (a) 100 kVA

Waste disposal	940 VA
Dishwasher	1,250 VA
Water heater	4,500 VA
Total demand load	6,690 VA × 20 units × 0.75 = 100,350 VA

➥ **Appliance Load No. 2**

What is the appliance demand load for a 35-unit multifamily building that contains a 900 VA waste disposal, a 1,200 VA dishwasher, and a 5,000 VA water heater?

(a) 71 kVA   (b) 142 kVA   (c) 107 kVA   (d) 186 kVA

• Answer: (d) 186 kVA

Waste disposal	900 VA
Dishwasher	1,200 VA
Water heater	5,000 VA
Total demand load	7,100 VA × 35 units × 0.75 = 186,375 VA/1,000 = 186 kVA

Step 4. Dryer Demand Load [Section 220-18]

➥ **Dryer Load No. 1**

A multifamily dwelling (12-unit building) contains a 4.5 kVA electric clothes dryer in each unit. What is the feeder and service dryer demand load for the building?

(a) 5 kVA   (b) 27 kVA   (c) 60 kVA   (d) none of these

• Answer: (b) 27 kVA

5 kVA × 12 units = 60 kVA × 0.45 = 27 kVA

➥ **Dryer Load No. 2**

What is the demand load for twenty 5.25 kW dryers installed in dwelling units of a multifamily building?

(a) 5 kVA   (b) 27 kVA   (c) 60 kVA   (d) 37 kVA

• Answer: (d) 37 kVA

5.25 kVA × 20 units = 105 kVA × 0.35 = 36.75 kVA

Step 5. Cooking Equipment Demand Load [Section 220-19]

❏ **Column A**

What is the feeder and service demand load for five 9 kW ranges?

(a) 9 kW   (b) 45 kW   (c) 20 kW   (d) none of these

• Answer: (c) 20 kW

❏ **Over 12 kW – Note 1**

What is the feeder and service demand load for three ranges rated 16 kW each?

(a) 15 kW   (b) 14 kW   (c) 17 kW   (d) 21 kW

• Answer: (c) 17 kW (closest answer)

Step 1: ➥ "Column A" demand load: 14 kW (3 units)

Step 2: ➥ The average range (16 kW) exceeds 12 kW by 4 kW. Increase "Column A" demand load (14 kW) by 20 percent: 14 kW × 1.2 = 16.8 kW.

❏ **Unequal Ratings over 12 kW – Note 2**

What is the feeder and service demand load for three ranges rated 9 kW and three ranges rated 14 kW?

(a) 36 kW   (b) 42 kW   (c) 78 kW   (d) 22 kW

• Answer: (d) 22 kW

Step 1: ➤ Determine the total connected load.

| 9 kW (minimum 12 kW) | 3 ranges × 12 kW = | 36 kW |
| 14 kW | 3 ranges × 14 kW = | <u>42 kW</u> |

Total connected load   78 kW

Step 2: ➤ Determine the average range rating, 78 kW/6 units = 13 kW average rating.

Step 3: ➤ Demand load Table 220-19 Column A: 6 ranges = 21 kW.

Step 4: ➤ The average range (13 kW) exceeds 12 kW by 1 kW. Increase Column A demand load (21 kW) by 5 percent:
21 kW × 1.05 = 22.05 kW.

❏ Note 3 – Less than 3¹/₂ kW – Column B

What is the feeder and service demand load for ten 3 kW ovens?

(a) 10 kW   (b) 30 kW   (c) 15 kW   (d) 20 kW

• Answer: (c) 15 kW closest answer

3 kW × 10 units = 30 kW × 0.49 = 14.70 kW

❏ Note 3 – Less than 8³/₄ kW – Column C

What is the feeder and service demand load for eight 6 kW cooktops?

(a) 10 kW   (b) 17 kW   (c) 14.7 kW   (d) 48 kW

• Answer: (b) 17 kW

6 kW × 8 units = 48 kW × 0.36 = 17.28 kW

Step 6. Service Conductor Size [Table 310-15(b)(6)]

➤ **Service Conductor Size No. 1**

What size aluminum service conductors are required for a 120/240-volt, single-phase multifamily building that has a total demand load of 93 kVA (Figure 10–2)?

(a) 300 kcmil   (b) 350 kcmil   (c) 500 kcmil   (d) 600 kcmil

• Answer: (d) 600 kcmil aluminum

I = VA/E

I = 93,000 VA/240 volts = 388 ampere,
600 kcmil aluminum

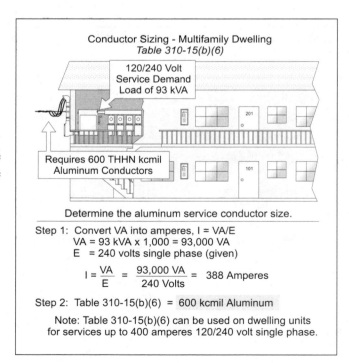

Conductor Sizing - Multifamily Dwelling
Table 310-15(b)(6)

120/240 Volt
Service Demand
Load of 93 kVA

201

Requires 600 THHN kcmil
Aluminum Conductors

101

Determine the aluminum service conductor size.

Step 1: Convert VA into amperes, I = VA/E
VA = 93 kVA x 1,000 = 93,000 VA
E = 240 volts single phase (given)

$$I = \frac{VA}{E} = \frac{93,000 \text{ VA}}{240 \text{ Volts}} = 388 \text{ Amperes}$$

Step 2: Table 310-15(b)(6) = 600 kcmil Aluminum

Note: Table 310-15(b)(6) can be used on dwelling units
for services up to 400 amperes 120/240 volt single phase.

➤ **Service Conductor Size No. 2**

What size service conductors are required for a multifamily building that has a total demand load of 270 kVA for a 120/208-volt 3-phase system?

Note: Service conductors are run in parallel.

(a) 2 – 300 kcmil per phase   (b) 2 – 350 kcmil per phase
(c) 2 – 500 kcmil per phase   (d) 2 – 600 kcmil per phase

• Answer: (c) 2 – 500 kcmil per phase

I = VA/(E × 1.732)

I = 270,000 VA/(208 volts × 1.732) = 750 ampere

Ampere per parallel set:

750 ampere/2 raceways = 375 ampere per phase.

500 kcmil conductor has an ampacity of 380 ampere [Table 310-16 at 75°C].

Two sets of 500 kcmil conductors (380 ampere × 2) can be protected by an 800-ampere protection device [*Section 240-3(b)*].

Figure 10-2
Conductor Sizing – Multifamily Dwelling

10–3 MULTIFAMILY DWELLING-UNIT CALCULATION SAMPLE – STANDARD METHOD

What is the demand load for a 25 unit apartment building? Each apartment is 1,000 square feet and contains the following: air-conditioning (28 ampere, 240 volts), heat (7.5 kW), dishwasher (1.2 kVA), waste disposal (1.5 kVA), water heater (4.5 kW), dryer (4.5 kW), and range (15.5 kW). System voltage 120/240 volt, single-phase.

Step 1. General Lighting, Small Appliance, and Laundry Demand [Section 220-11]

Each unit contains:

General lighting	(1,000 square feet × 3 VA)	3,000 VA
Small appliance circuits		3,000 VA
Laundry circuit		1,500 VA
		7,500 VA

Total demand load (7,500 VA × 25 units)   187,500 VA

First 3,000 VA at 100% =   – 3,000 VA × 1.00 =   3,000 VA

184,500 VA

Next 117,000 VA at 35%   – 117,000 VA × 0.35 = 40,950 VA

Remainder at 25%   67,500 VA × 0.25 = 16,875 VA

60,875 VA

Step 2. Air-Conditioning versus Heat [Section 220-15 and 440-34]

Air-conditioning [*Section 440-34*], 230 volts × 28 ampere = 6,440 VA

6,440 VA × 1.25 = 8,050 VA + (6,440 × 24 units) = 162,610 VA, omit [*Section 220-21*]

Heat [*Section 220-15*]

7,500 VA × 25 units = 187,500 VA

Step 3. Appliance Demand Load [Section 220-17]

Waste disposal	1,200 VA
Dishwasher	1,500 VA
Water heater	4,500 VA
Total connected load	7,200 VA × 25 units × 0.75 = 135,000 VA

Step 4. Dryer Demand Load [Section 220-18]

Total connected load = 5 kVA* × 25 units = 125 kW

Demand factor for 25 units = 32.5%

Total demand dryer load = 125 kW × 0.325 = 40.625 kW

* The minimum dryer load for standard load calculations is 5 kW

Step 5. Cooking Equipment Demand Load [Section 220-19]

Step 1: ⇒ Column A demand load for 25 units = 40 kW.

Step 2: ⇒ The average range (15.5 kW) exceeds 12 kW by 3.5 kW. Increase Column A demand load (40 kW) by 20%. Range demand load: 40 kW × 1.2 = 48 kW

Step 6. Service Conductor Size [Table 310-15(b)(6)]

Step 1: ⇒ Total general lighting, small appliance, laundry demand load	60,875 VA
Step 2: ⇒ Total heat demand load [*Section 220-15*] (7,500 VA × 25 units)	187,500 VA
Step 3: ⇒ Total appliance demand load	135,000 VA
Step 4: ⇒ Total demand dryer load	40,625 VA
Step 5: ⇒ Range demand load	48,000 VA
Total demand load	472,000 VA

Service conductor ampere = VA/E = 472,000 VA/240 volts = 1,967 ampere

Unit 10 – Multifamily Dwelling Unit Load Calculations Summary Questions

10–1 Multifamily Dwelling-Unit Load Calculations – Standard Method

1. • Each unit of a 20-unit apartment building is 840 square feet. What is the general lighting feeder demand load for the building? Note: Laundry facilities are provided on the premises for all tenants.
    (a) 5,200 VA   (b) 40,590 VA   (c) 110,400 VA   (d) none of these

2. Each unit of a 20-unit apartment building is 900 square feet. What is the general lighting net computed demand load for the building?
    (a) 74,700 VA   (b) 149,400 VA   (c) 51,300 VA   (d) 105,600 VA

3. A 40-unit multifamily building has a air-conditioning (3 horsepower 230 volt) and two baseboard heaters (3 kW) in each unit. What is the air-conditioning versus heat net computed demand load?
    (a) 160 kW   (b) 240 kW   (c) 60 kW   (d) 50 kW

4. A 25-unit multifamily building has a air-conditioning (3 horsepower 230 volt) and electric heat (5 kW). What is the air-conditioning versus heat net computed demand load?
    (a) 160 kW   (b) 125 kW   (c) 6 kW   (d) 5 kW

5. Each 16-unit multifamily building contains a waste disposal (940 VA), a dishwasher (1,250 VA), and a water heater (4,500 VA). What is the service demand load for these appliances?
    (a) 100 kVA   (b) 134 kVA   (c) 80 kVA   (d) 5 kVA

6. A 28-unit apartment building contains a waste disposal (900 VA), dishwasher (1,200 VA), and a water heater (5,000 VA) in each unit. What is the feeder and service demand load for the appliances?
    (a) 149 kVA   (b) 142 kVA   (c) 107 kVA   (d) 186 kVA

7. A multifamily dwelling (40 unit) contains a 4.5 kW electric clothes dryer in each unit. What is the feeder and service demand load for all the dryers?
    (a) 50 kW   (b) 27 kW   (c) 60 kW   (d) none of these

8. What is the demand load for ten 5.25 kW dryers installed in dwelling units of a multifamily building?
    (a) 26 kW   (b) 37 kW   (c) 60 kW   (d) 37 kW

9. What is the demand load for twelve 3.25 kW ovens?
    (a) 10 kW   (b) 18 kW   (c) 15 kW   (d) 20 kW

10. What is the demand load for eight 7 kW cooktops?
    (a) 20 kW   (b) 17 kW   (c) 14.7 kW   (d) 48 kW

11. • What is the demand load for five 12.4 kW ranges?
    (a) 9 kW   (b) 45 kW   (c) 20 kW   (d) none of these

12. What is the demand load for three ranges rated 15.5 kW?
    (a) 15 kW   (b) 14 kW   (c) 17 kW   (d) 21 kW

13. What is the feeder and service demand load for three ranges rated 11 kW and three ranges rated 14 kW?
    (a) 36 kW   (b) 42 kW   (c) 78 kW   (d) 22 kW

14. • What size aluminum service conductors are required for a 115/230-volt, single-phase multifamily building that has a total demand load of 90 kW?
     (a) 500 kcmil   (b) 600 kcmil   (c) 700 kcmil   (d) 800 kcmil

15. What size service conductors are required for a multifamily building that has a total demand load of 260 kW for a 120/208-volt, wye 3-phase system?
     (a) 2 – 300 kcmil   (b) 2 – 350 kcmil   (c) 2 – 500 kcmil   (d) 2 – 600 kcmil

For the Next Five Questions

A multifamily building has 12 units. Each is 1,500 square feet and contains the following:
System voltage 120/240 volts, single-phase.

Dryer (4.5 kW)	Washing machine (1.2 kVA)
Range (14.45 kW)	Dishwasher (1.5 kVA)
Water heater (4 kW)	Heat (5 kW)

A/C (3 horsepower with $^1/_8$ horsepower compressor fan)

16. • What is the building's net computed load for the general lighting, small appliance, and laundry in VA?
     (a) 40 kVA   (b) 108 kVA   (c) 105 kVA   (d) 90 kVA

17. What is the building's demand load in kVA for an air-conditioner (3 horsepower with a $^1/_8$ horsepower compressor fan) versus electric heat (5 kW)?
     (a) 52 kW   (b) 60 kW   (c) 30 kW   (d) 105 kW

18. What is the building's demand load for the appliances in VA?
     (a) 40 kVA   (b) 55 kVA   (c) 43 kVA   (d) 50 kVA

19. What is the building's demand load for the 4.5 kW dryer?
     (a) 60 kW   (b) 27 kW   (c) 55 kW   (d) 101 kW

20. What is the building's demand load for the 14.45 kW range?
     (a) 27 kW   (b) 35 kW   (c) 168 kW   (d) 30 kW

21. If the total demand load of a multifamily dwelling-unit is 206 kVA, what is the service and feeder conductor size? The system voltage is 120/240, single-phase.
     (a) 600 ampere   (b) 800 ampere   (c) 1,000 ampere   (d) 1,200 ampere

22. • If a service contains three sets of parallel 400 kcmil conductors, what size bonding jumper is required for each service raceway?
     (a) No. 1   (b) No. 1/0   (c) No. 2/0   (d) No. 3/0

23. • If 400 kcmil service conductors are in parallel in three raceways, what is the minimum size grounding electrode conductor required?
     (a) No. 1/0   (b) No. 2/0   (c) No. 3/0   (d) No. 4/0

10–3 Multifamily Dwelling Calculation [*Section 220-32*] – Optional Method

24. • When determining the service (using optional calculations) for a multifamily dwelling, the total connected load shall have the demand factors of Table 220-32 applied. When determining the total connected load, the largest of the _____ shall be used.
     (a) 100 percent of the air-conditioning
     (b) 125 percent of the air-conditioning
     (c) 100 percent of the heat
     (d) a or c

25. A multifamily building has 60 units. Each unit is 1,500 square feet. Using the optional dwelling-unit calculations, what is the net computed load for the building general lighting and general use receptacles, small appliance and laundry circuits?
     (a) 145 kVA   (b) 190 kVA   (c) 108 kVA   (d) 130 kVA

26. A 60-unit multifamily dwelling has an air-conditioner (3 horsepower with a 1/8 horsepower blower) and electric heat (5 kW) in each unit. Using the optional method, what is the air-conditioning versus heat demand load?
    (a) 50 kVA   (b) 61 kVA   (c) 30 kVA   (d) 72 kVA

27. Using the optional method for dwelling-unit calculations, what is a 60-unit multifamily building net demand load if each unit has a water heater (4 kW and a dishwasher (1.5 kVA)?
    (a) 80 kVA   (b) 50 kVA   (c) 30 kVA   (d) 60 kVA

28. Each unit of a 60-unit apartment building has a 4 kW dryer. Using the optional method, the demand load that would be added to the service is _____ kW.
    (a) 75   (b) 240   (c) 72   (d) 58

29. Using the optional method for dwelling-unit calculations, what is the 60-unit multifamily building net computed load for a 4.5 kW dryer?
    (a) 65 kW   (b) 25 kW   (c) 55 kW   (d) 75 kW

30. Using the optional method for dwelling-unit calculations, what is a 60-unit multifamily building demand load if each apartment has a 14 kW range?
    (a) 150 kW   (b) 50 kW   (c) 100 kW   (d) 200 kW

31. If the total demand load is 270 kVA, what is the service and feeder conductor size? The service is 208/120-volt, wye 3-phase.
    (a) 600 ampere   (b) 800 ampere   (c) 1,000 ampere   (d) 1,200 ampere

32. Each unit of a 20-unit multifamily dwelling has 900 square feet of living space and contains one of each of the following: air-conditioning (5 horsepower), heat (5 kW), water heater (5 kW), range (14 kW). The service for this apartment building is approximately _____ if the optional method calculation is used.
    (a) 200 kVA   (b) 250 kVA   (c) 280 kVA   (d) 320 kVA

33. • If the service conductors are in parallel in two raceways (500 kcmil), what size bonding jumper is required for each service raceway?
    (a) No. 2   (b) No. 1   (c) No. 1/0   (d) No. 2/0

34. • If the service conductors are in parallel in two raceways (500 kcmil), what is the minimum size grounding electrode conductor required?
    (a) No. 1/0   (b) No. 2/0   (c) No. 3/0   (d) No. 4/0

☆ Challenge Questions

General Lighting and Receptacle Calculations [Table 220- 11]

35. Each dwelling unit of a 20-unit multifamily building has 900 square feet of living space. What is the general lighting load for the multifamily dwelling-unit apartment building?
    (a) 45 kVA   (b) 60 kVA   (c) 37 kVA   (d) 54 kVA

36. An multifamily apartment building contains 20 units and each unit has 840 square feet of living space. What is the general lighting feeder demand load for the multifamily building, if laundry facilities are provided on the premises for all tenants and no laundry circuit is installed in each unit?
    (a) 35 kVA   (b) 41 kVA   (c) 45 kVA   (d) 63 kVA

Appliance Demand Factors [*Section 220-17*]

37. An multifamily apartment building contains 20 units. Each unit contains a waste disposal (900 VA), a dishwasher (1,200 VA), and a water heater (5,000 VA). What is the feeder demand load for the appliances in this building?
    (a) 106,500 VA   (b) 117,100 VA   (c) 137,000 VA   (d) 60,000 VA

Dryer Calculation [*Section 220-18*]

38. The nameplate rating for each household dryer in a 10-unit apartment building is 4 kW. This would add _____ to the service size.
    (a) 20 kW   (b) 25 kW   (c) 40 kW   (d) 50 kW

Ranges – Note 1 of Table 220-19

39. • The demand load for thirty 15.8 kW household ranges is _____ kW.
    (a) 31   (b) 47   (c) 54   (d) 33

Ranges – Note 2 of Table 220-19

40. What is the feeder demand load for five 10 kW, five 14 kW, and five 16 kW household ranges?
    (a) 210 kW   (b) 30 kW   (c) 14 kW   (d) 33 kW

41. What kW would be added to service loads for ten 12 kW, eight 14 kW, and two 9 kW household ranges?
    (a) 33 kW   (b) 35 kW   (c) 36.75 kW   (d) 29.35 kW

Ranges – Note 3 to Table 220-19

42. What is the minimum demand load for five 5 kW cooktops, two 4 kW ovens, and four 7 kW ranges?
    (a) 15.5 kW   (b) 8.8 kW   (c) 19.5 kW   (d) 18.2 kW

43. • What is the maximum dwelling-unit feeder or service demand load for fifteen 8 kW cooking units?
    (a) 38.4 kW   (b) 30 kW   (c) 120 kW   (d) none of these

Ranges – Note 5 of Table 220-19

44. A school has twenty 10 kW ranges installed in the home economics class. The minimum load this would add to the service is _____ .
    (a) 35 kW   (b) 44.8 kW   (c) 56 kW   (d) 160 kW

10–12 Neutral Calculation [*Section 220-22*]

45. The service neutral demand load for household electric clothes dryers shall be calculated at _____ of the demand load as determined by Section 220-18.
    (a) 50%   (b) 60%   (c) 70%   (d) 80%

46. The feeder neutral demand for fifteen 8 kW cooking units would be _____ .
    (a) 38.4 kW   (b) 30 kW   (c) 120 kW   (d) 21 kW

47. What is the feeder neutral load in ampere for fifteen 6 kW household dryers?
    (a) 105 ampere   (b) 125 ampere   (c) 150 ampere   (d) none of these

48. A 12-unit multifamily dwelling contains a 4 kVA electric clothes dryer in each unit. What is the feeder or service neutral demand load (in ampere)?
    (a) 115 ampere   (b) 225 ampere   (c) 80 ampere   (d) 55 ampere

49. • The feeder and service neutral demand load can be reduced 70 percent for that portion of the unbalanced load over 200 ampere. This applies to _____ systems.
    (a) 3-wire, single-phase, 120/240 volt   (b) 3-wire, single-phase, 120/208 volt
    (c) 4-wire, 3-phase, 120/208 volt   (d) a and c

50. • What is the dwelling-unit service neutral load (in ampere) for ten 9 kW ranges?
    (a) 40 ampere   (b) 55 ampere   (c) 75 ampere   (d) 105 ampere

NEC® Questions from Section 90 through Appendix C

Article 90 – Introduction

51.   Hazards often occur because of _____ .
(a) overloading of wiring systems by methods or usage not in conformity with the *Code*
(b) initial wiring not providing for increases in the use of electricity
(c) a and b
(d) none of these

52.   Equipment listed by a qualified electrical testing laboratory is not required to have the _____ wiring to be reinspected at the time of installation.
(a) external   (b) associated   (c) internal   (d) all of these

NEC® Chapter 1 – General Requirements

53.   Electrical wiring and equipment that is capable of being reached quickly for operation, renewal, or inspections without resorting to portable ladders and such is known as _____ .
(a) accessible (equipment)   (b) accessible (wiring methods)
(c) accessible, readily   (d) all of these

54.   Nonmetallic-sheathed cable, when run in the space between studs, is defined as _____ .
(a) inaccessible   (b) concealed   (c) hidden   (d) enclosed

55.   A _____ is a device, or group of devices, or other means by which the conductors of a circuit can be disconnected from their source of supply.
(a) circuit breaker   (b) fuse   (c) disconnecting means   (d) switch

56.   A building or portion of a building in which self-propelled vehicles carrying volatile flammable liquids for fuel or power are kept for use, sale, storage, and exhibition is called _____ .
(a) a garage   (b) a service station   (c) a service garage   (d) all of these

57.   _____ means that an object is not readily accessible to persons unless special means for access are used.
(a) Isolated   (b) Secluded   (c) Protected   (d) Locked

58.   Any current in excess of the rated current of equipment or the ampacity of a conductor is called _____ .
(a) overload   (b) faulted   (c) overcurrent   (d) shorted

59.   A contact device installed at an outlet for the connection of an attachment plug is known as a _____ .
(a) receptacle outlet   (b) duplex receptacle   (c) receptacle   (d) plug

60.   Equipment or materials included in a list published by an organization acceptable to the authority having jurisdiction and concerned with product evaluation are known as _____ .
(a) labeled   (b) listed   (c) approved   (d) identified

61.   Conductors and equipment supplied from a battery UPS system, a solar voltaic system, generator, transformer, or phase converter is considered a service.
(a) True   (b) False

62.   The _____ is the point of connection between the facilities of the serving utility and the premises wiring.
(a) service entrance   (b) service point
(c) overcurrent protection   (d) beginning of the wiring system

63.   Conductors shall be _____ unless otherwise provided in the *Code*.
(a) lead   (b) stranded   (c) copper   (d) aluminum

64. Some spray cleaning and lubricating compounds contain chemicals that cause severe deteriorating reactions with plastics.
    (a) True   (b) False

65. Soldered splices must be joined mechanically so as to be electrically secure before soldered.
    (a) True   (b) False

66. A minimum of _____ feet of working clearance is required to live parts operating at 300 volts, nominal, to ground where there are exposed live parts on one side and no live or grounded parts on the other side.
    (a) 2   (b) 3   (c) 4   (d) 6

67. Working space with _____ entrance(s) provided shall be so located that the edge of the entrance nearest the equipment is the minimum clear distance given in Table 110-26(a) away from such equipment.
    (a) 1   (b) 2   (c) 3   (d) 4

68. A 7-foot high fence (or wall) is considered acceptable for preventing access to an over 600 volts, nominal, installation.
    (a) True   (b) False

69. • The minimum depth of clear working space in front of electrical equipment for 5,000 volts, nominal, to ground is _____ feet when there are exposed live parts on both sides of the workspace with the operator between.
    (a) 4   (b) 5   (c) 6   (d) 9

Chapter 2 – Wiring and Protection

70. A cable containing an insulated conductor with a white outer finish can be used for 3-way or 4-way switch loops, if the white conductor is not used for the switch leg. In this application the white conductor must be permanently re-identified to indicate its use by painting or other effective means at its termination, and at each location where the conductor is visible and accessible.
    (a) True   (b) False

71. A branch circuit voltage that exceeds 277 volts to ground and does not exceed 600 volts between conductor is used to wire the auxiliary equipment of electrical discharge lamps mounted on poles. The minimum height of these lighting fixtures shall not be less than _____ feet.
    (a) 31   (b) 15   (c) 18   (d) 22

72. • A _____ receptacle without GFCI protection is permitted to be located in a dwelling unit garage for one appliance if located within the dedicated space for the appliance.
    (a) multioutlet   (b) duplex   (c) single   (d) a and b

73. There shall be a minimum of one _____-ampere branch circuit for the laundry outlet(s) in a dwelling unit.
    (a) 15   (b) 20   (c) 30   (d) b and c

74. Section 210-20(a) requires the circuit overcurrent protection device to have an ampacity of not less than 125 percent of the continuous load, plus 100 percent of the noncontinuous load.
    (a) True   (b) False

75. In a dwelling unit, each wall space of _____ feet or wider requires a receptacle.
    (a) 2   (b) 3   (c) 4   (d) 5

76. • A receptacle outlet must be installed in dwelling units for every kitchen and dining area countertop space _____, and no point along the wall line shall be more than 2 feet, measured horizontally from a receptacle outlet.
    (a) wider than 12 inches   (b) wider than 2 feet
    (c) 2 feet and wider   (d) 12 inches and wider

77. Hallways in dwelling units that are _____ feet long or longer require a receptacle outlet.
    (a) 12   (b) 10   (c) 8   (d) 15

78. For the purpose of lighting outlets in dwelling units, a vehicle door in a garage is considered as an outdoor entrance.
    (a) True   (b) False

79. Where required, drawings for feeder installations must be submitted before _____ .
    (a) completion of the installation　　　　　(b) beginning the installation
    (c) the feeders are energized or tested　　　(d) drawings are not required

80. Where fixed multi-outlet assemblies are employed, each _____ feet or fraction thereof of each separate and continuous length shall be considered as one outlet of not less than 180 volt-amperes capacity.
    (a) 5　　　　　　　(b) 5½　　　　　　　(c) 6　　　　　　　(d) 6½

81. Where it is unlikely that two or more loads will be in use simultaneously, it shall be permissible to use only the _____ loads on at any give time in computing the total load to a feeder.
    (a) smaller　　　(b) larger　　　(c) difference between the load　　　(d) none of these

82. A demand factor of _____ percent is applicable for a multifamily dwelling with ten units, if the optional calculation method is used.
    (a) 75　　　　　　　(b) 60　　　　　　　(c) 50　　　　　　　(d) 43

83. Where the farm dwelling has electric heat and the farm has electric grain drying systems, Part _____ of Article 220 shall not be used to compute the dwelling load.
    (a) A　　　　　　　(b) B　　　　　　　(c) C　　　　　　　(d) none of these

84. • Conductors on poles shall have a separation of not less than _____ where not placed on racks or brackets.
    (a) 3 inches　　　(b) 6 inches　　　(c) 12 inches　　　(d) 18 inches

85. Final spans of feeders to buildings they supply shall be kept _____ feet from windows that are designed to be opened, doors, balconies, ladders, stairs, and similar locations.
    (a) 3　　　　　　　(b) 5　　　　　　　(c) 6　　　　　　　(d) 10

86. Service conductors installed as unjacketed multiconductor cable shall have a minimum clearance of _____ feet from windows designed to be opened, doors, porches, fire escapes, and similar equipment.
    (a) 3　　　　　　　(b) 4　　　　　　　(c) 6　　　　　　　(d) 10

87. Service-drop conductors shall have a minimum of _____ feet vertical clearance from final grade over residential property and driveways, as well as those commercial areas not subject to truck traffic where the voltage is limited to 300 volts to ground.
    (a) 10　　　　　　　(b) 12　　　　　　　(c) 15　　　　　　　(d) 18

88. Service lateral conductors must have _____ .
    (a) adequate mechanical strength　　　　　(b) sufficient ampacity for the loads computed
    (c) a and b　　　　　　　　　　　　　　　(d) none of these

89. Service cables mounted in contact with a building shall be supported at intervals not exceeding _____ .
    (a) 4 feet　　　(b) 3 feet　　　(c) 30 inches　　　(d) 24 inches

90. To prevent water from entering service equipment, service entrance conductors must be _____ .
    (a) connected to service drop conductors below the level of the service head
    (b) have drip loops formed on the service entrance conductors
    (c) a or b
    (d) a and b

91. • When the service contains two to six service disconnecting means, they shall be _____ .
    (a) the same size　　　　　　　　(b) grouped at one location
    (c) in the same enclosure　　　　(d) none of these

92. In a service, overcurrent protection devices shall never be placed in the grounded service conductor except circuit breakers that simultaneously open all conductors of the circuit.
    (a) True　　　　　　(b) False

93. The ground-fault protection system shall be _____ when first installed on site.
    (a) inspected　　　(b) identified　　　(c) turned on　　　(d) performance tested

94. • A No. 12 THHN conductor can be protected by an overcurrent protection device that is greater than 20 amperes. This is permitted when the conductors are used for _____ .
(a) tap conductors
(b) motor circuit conductors
(c) Class 1 control and signaling circuit conductors
(d) all of these

95. A device which, when interrupting currents within its range, will reduce the current flowing in the faulted circuit to a magnitude substantially less than that obtainable in the same circuit, if the device were replaced with a solid conductor having comparable impedance, is defined as a _____ protection device.
(a) short-circuit   (b) overload   (c) ground-fault   (d) current-limiting overcurrent

96. A tap from a feeder that is over 10 feet but less than 25 feet is permitted without overcurrent protection at the tap point, providing the _____ .
(a) ampacity of the tap conductors is not less than $^1/_3$ the ampacity of the overcurrent protection device protecting the feeders being tapped
(b) tap conductors terminate in a single circuit breaker or set of fuses that limit the load to the ampacity of the tap conductors
(c) tap conductors are protected from physical damage
(d) all of these

97. Handles or levers of circuit breakers and similar parts that may move suddenly in such a way that persons in the vicinity are likely to be injured by being struck by them shall be _____ .
(a) guarded   (b) isolated   (c) a and b   (d) a or b

98. 300-volt cartridge fuses and fuse holders are not permitted on circuits exceeding 300 volts _____ .
(a) between conductors   (b) to ground
(c) of the circuit   (d) a or c

99. Every circuit breaker having an interrupting current rating of other than _____ amperes shall have its interrupting rating marked on the circuit breaker.
(a) 50,000   (b) 10,000   (c) 15,000   (d) 5,000

100. • Grounding electrode conductor fittings shall be protected from physical damage by being enclosed in _____ .
(a) metal   (b) wood
(c) the equivalent of a or b   (d) none of these

101. The grounding electrode for a separately derived system shall be as near as practicable to and preferably in the same area as the grounding electrode conductor connection to the system. The grounding electrode shall be:
(a) the nearest effectively grounded metal member of the building structure.
(b) the nearest effectively grounded metal water pipe, but only if it is within 5 feet from the point of entrance into the building.
(c) any metal structure that is effectively grounded.
(d) a or b

102. • The smallest diameter rod that may be used for a made electrode is _____ inch.
(a) $^1/_2$ (b) $^3/_4$ (c) 1   (d) $1^1/_4$

103. When multiple ground rods are used for a made grounding electrode, they shall be separated not less than _____ feet apart.
(a) 6   (b) 8   (c) 20   (d) 12

104. • A metal conduit used to protect the grounding electrode conductor does not need to be physically (mechanically) continuous to the grounding electrode, if the metal raceway is electrically continuous.
(a) True   (b) False

105.   The noncurrent-carrying metal parts of equipment, such as _____, shall be effectively bonded together.
(a) service raceways, cable trays, or service cable armor
(b) service equipment enclosures containing service conductors, including meter fittings, boxes, or the like, interposed in the service raceway or armor
(c) the metallic raceway or armor enclosing a grounding electrode conductor
(d) all of these

106.   A 200-ampere residential service consists of No. 2/0 THHN copper conductors. What is the minimum size bonding jumper for the service raceway?
(a) 6 aluminum   (b) 3 copper   (c) 4 aluminum   (d) 4 copper

107.   Permanently mounted electrical equipment and skids shall be grounded with an equipment bonding jumper sized as required by Section _____ .
(a) 250-50   (b) 250-66   (c) 250-122   (d) 310-15

108.   • When ungrounded conductors are increased in size to compensate for voltage drop, the equipment grounding conductor is not required to be increased, because it is not a current-carrying conductor.
(a) True   (b) False

109.   On the load side of the service disconnecting means, the _____ circuit conductor is permitted to ground meter enclosures if all meter enclosures are located near the service disconnecting means and no service ground-fault protection is installed.
(a) grounding   (b) bonding   (c) grounded   (d) phase

110.   • Where the direct current system consists of a _____, the grounding conductor shall not be smaller than the neutral conductor.
(a) 2-wire balancer set
(b) 3-wire balancer set
(c) balancer winding with overcurrent protection
(d) b or c

Chapter 3 – Wiring Methods and Material

111.   Where nails or screws are likely to penetrate nonmetallic-sheathed cable or electrical nonmetallic tubing installed through metal framing members, a steel sleeve, steel plate, or steel clip not less than _____ inch in thickness shall be used to protect the cable or tubing.
(a) $\frac{1}{16}$ (b) $\frac{1}{8}$ (c) $\frac{1}{2}$ (d) none of these

112.   • Where moisture could enter a raceway and contact energized live parts, _____ are required at one or both ends of the raceway.
(a) seals   (b) plugs   (c) bushings   (d) a and b

113.   • Raceways shall be provided with _____ where necessary to compensate for thermal expansion and contraction.
(a) expansion couplings   (b) expansion fittings
(c) bonding jumpers   (d) none of these

114.   Raceways shall be _____ between pulling points before installing conductors.
(a) mechanically completed   (b) tested for ground-faults
(c) a minimum of 80 percent completed   (d) none of these

115.   Type AC cable can be used in ducts or plenums used for environmental air.
(a) True   (b) False

116.   Receptacles for construction sites shall not be installed on the _____ or connected to _____ that supply temporary lighting.
(a) same branch circuit, feeders   (b) feeders, multiwire branch circuits
(c) same branch circuit, multiwire branch circuits   (d) all of these

117. For temporary wiring over 600 volts, _____ shall be provided to prevent access of other than authorized and qualified personnel.
      (a) fencing   (b) barriers   (c) signs   (d) a or b

118. There are four principal determinants of conductor operating temperature. One of those is _____ generated internally in the conductor as the result of load current flow.
      (a) friction   (b) magnetism   (c) heat   (d) none of these

119. Conductor derating factors shall not apply to conductors in nipples having a length not exceeding _____ inches.
      (a) 12   (b) 24   (c) 36   (d) 48

120. Thermal resistivity, as used in the *Code*, refers to the heat _____ capability through a substance by conduction.
      (a) assimilation   (b) generation   (c) transfer   (d) dissipation

121. Steel or aluminum cable tray systems can be used as equipment grounding conductors provided the cable tray sections and fittings have been _____ marked to show the cross-sectional area of metal in channel cable trays, or cable trays of one-piece construction.
      (a) legibly   (b) durably   (c) a or b   (d) a and b

122. Open conductors entering or leaving locations subject to dampness, wetness, or corrosive vapors shall have _____ formed on them and shall then pass upward and inward from the outside of the buildings, or from the damp, wet, or corrosive location, through noncombustible, nonabsorbent insulating tubes.
      (a) weather heads   (b) drip loops   (c) identification   (d) blisters

123. The maximum voltage permitted between ungrounded conductors of flat conductor cable systems is _____ volts.
      (a) 600   (b) 300   (c) 250   (d) 150

124. Bends in Type MI cable shall be so made as not to _____ the cable.
      (a) damage   (b) shorten   (c) a and b   (d) none of these

125. The *National Electrical Code*® permits ENT to be installed in _____ if used with fittings listed for the purpose.
      (a) wet locations indoors
      (b) in a concrete slab on or below grade
      (c) direct earth burial
      (d) a and b

126. Armored cable used for the connection of recessed lighting fixtures or equipment within an accessible ceiling does not need to be secured for lengths up to _____ feet.
      (a) 2   (b) 3   (c) 4   (d) 6

127. Nonmetallic-sheathed cable must closely follow the surface of the building finish or running boards when run exposed.
      (a) True   (b) False

128. • Nonmetallic-sheathed cable conductors must be rated _____ °C.
      (a) 60   (b) 75   (c) 90   (d) any of these

129. Aerial cable used for nonmetallic extensions and its tap connectors shall be provided with an approved means for _____ .
      (a) operation   (b) polarization   (c) grounding   (d) disconnect

130. 1 inch IMC raceway containing three or more conductors can be conductor filled to _____ percent.
      (a) 53   (b) 31   (c) 40   (d) 60

131. Where threadless couplings and connectors used with rigid metal conduit are installed in wet locations, they shall be of the _____ type.
      (a) raintight   (b) wet and damp location   (c) nonabsorbent   (d) weatherproof

132. Rigid nonmetallic conduit can be used to support nonmetallic conduit bodies, but the conduit bodies shall not contain devices, lighting fixtures, or other equipment.
      (a) True   (b) False

161.   The disconnecting means for a torque motor shall have an ampere rating of at least _____ percent of the motor nameplate current.
        (a) 100   (b) 115   (c) 125   (d) 175

162.   The total rating of a plug-connected room air-conditioner, where lighting units or other appliances are also supplied, shall not exceed _____ percent.
        (a) 80   (b) 70   (c) 50   (d) 40

163.   What size overcurrent protection is required for a 45 kVA transformer that has a primary current rating of 54 amperes?
        (a) 50   (b) 60   (c) 70   (d) 100

164.   The residual voltage of a capacitor shall be reduced to 50 volts or less within _____ after the capacitor is disconnected from the source of supply.
        (a) 15 seconds   (b) 45 seconds   (c) 1 minute   (d) 2 minutes

Chapter 5 – Special Occupancies

165.   All conduit referred to in hazardous locations shall be threaded with a _____ inch taper per foot.
        (a) $^1/_2$ (b) $^3/_4$ (c) 1   (d) all of these

166.   A Class I, Division 1 location is a location in which _____ .
        (a) ignitable concentrations of flammable gases or vapors can exist under normal operating conditions
        (b) ignitable concentrations of such gases or vapors may exist frequently because of repair or maintenance operations or because of leakage
        (c) breakdown or faulty operation of equipment or processes might release ignitable concentrations of flammable gases or vapors, and might also cause simultaneous failure of electric equipment.
        (d) all of these

167.   • Wiring methods permitted in Class I, Division 2 locations include _____ .
        (a) threaded rigid metal conduit
        (b) threaded steel intermediate metal conduit
        (c) general purpose boxes and fittings containing no arcing devices
        (d) all of these

168.   A standard circuit breaker mounted in a Class I, Division 2 location with make-and-break contacts and not hermetically sealed or oil immersed must be installed in a Class I, Division 1 rated enclosure.
        (a) True   (b) False

169.   The emergency controls for unattended self-service stations must be located not less than _____ feet, nor more than _____ feet from the gasoline dispensers.
        (a) 10, 25   (b) 20, 50   (c) 20, 100   (d) 50, 100

170.   In a health care facility, receptacles and attachment plugs in a hazardous (classified) location within an anesthetizing area shall be listed for use in Class I, Group _____ locations.
        (a) A   (b) B   (c) C   (d) D

171.   Wiring for temporary lighting inside of tents and concession areas for carnival, circuses, and fairs shall be securely installed, and where subject to physical damage, they shall be provided with mechanical protection.
        (a) True   (b) False

172.   In reference to mobile/manufactured homes, examples of "Appliance, Portable" could be _____, but only if these appliances are not built-in.
        (a) a refrigerator   (b) gas range equipment
        (c) a clothes washer   (d) all of these

173.   The mobile home service equipment shall be located adjacent to the mobile home and not mounted in or on the mobile home. The service equipment shall be located in sight from, but not more than ___ feet from the exterior wall of the mobile home it serves.
        (a) 15   (b) 20   (c) 30   (d) none of these

174. Where shore power accommodations provide two receptacles for an individual boat slip, and these receptacles have different voltages, only the _____ receptacle need be used for feeder and service calculations.
      (a) smaller   (b) larger   (c) all receptacles   (d) none of these

Chapter 6 – Special Equipment

175. An outdoor portable electric sign shall have a ground-fault circuit-interrupter _____ .
      (a) located on the sign
      (b) located in the power supply cord within 12 inches of the attachment plug
      (c) as an integral part of the attachment plug of supply cord
      (d) b or c

176. Where multiple driving machines are connected to a single elevator, escalator, moving walk, or pumping unit, there shall be one disconnecting means to disconnect the _____ .
      (a) motor(s)   (b) control valve operating magnets
      (c) a or b   (d) none of these

177. _____ conductor cables No. 4 and larger marked, "for use in cable trays" or "for CT use" can be within a raised floor of information technology equipment.
      (a) Green   (b) Insulated   (c) Single   (d) all of these

178. A disconnecting means for X-ray equipment shall have adequate capacity for at least _____ .
      (a) 50 percent of the input required for the momentary rating of equipment
      (b) 100 percent of the input required for the momentary rating of equipment
      (c) 50 percent of the input requirement for the long time rating
      (d) 125 percent of the input requirement

179. A center pivot irrigation machine may have hand-portable motors.
      (a) True   (b) False

180. A pool capable of holding water to a maximum depth of _____ inches is a storable pool.
      (a) 18   (b) 36   (c) 42   (d) none of these

181. Paddle fans must not be installed within _____ feet from the edge of outdoor pools, spas, and hot tubs.
      (a) 3   (b) 5   (c) 10   (d) 12

182. Forming shells for wet-niche lighting fixtures must be installed with the top level of the fixture lens at least _____ inches below the normal water level of the pool or spa.
      (a) 6   (b) 12   (c) 18   (d) 24

183. • When bonding together pool reinforcing steel and welded wire fabric (wire-mesh) with tie-wire, the tie-wires must be _____ .
      (a) stainless steel   (b) accessible   (c) made up tight   (d) none of these

184. Wet-niche lighting fixtures must be grounded by a minimum No. 12 insulated copper conductor. This grounding conductor must be installed in _____ .
      (a) rigid nonmetallic conduit
      (b) electrical metallic conduit when on or in the building
      (c) flexible metal conduit
      (d) a or b

185. Receptacle outlet(s) for _____ shall be GFCI-protected.
      (a) self-contained spa or hot tub
      (b) packaged spa or hot tub equipment assembly
      (c) field assembled spa or hot tub with a heater load of 50 amperes or less
      (d) all of these

Chapter 7 – Special Conditions

186.   Transfer equipment, including automatic transfer switches, shall be _____ .
(a) automatic
(b) identified for emergency use
(c) approved by the authority having jurisdiction
(d) all of these

187.   Where an internal combustion engines is used as the prime-mover for an emergency system, an on-site fuel supply shall be provided with an on-premise fuel supply for not less than _____ hours full demand.
(a) 2   (b) 3   (c) 4   (d) 5

188.   A legally required standby system is intended to automatically supply power to _____ .
(a) those systems classed as emergency systems
(b) selected loads
(c) a and b
(d) none of these

189.   Since Class 3 control circuits permit higher allowable levels of voltage and current, additional _____ are specified to provide protection against the electric shock hazard that could be encountered.
(a) circuits   (b) safeguards   (c) conditions   (d) requirements

190.   The maximum power source rating would be _____ amperes for a Class 2 control or signal circuit rated 25 volts (inherently limited power source). Tip: See the formula in Table 11 of Chapter 9.
(a) 3   (b) 4   (c) 6   (d) 7

191.   • Raceways shall not be used as a means of support for Class 2 or Class 3 circuit conductors.
(a) True   (b) False

192.   Splices and terminations of nonpower-limited fire alarm circuits shall be made in _____ fittings, boxes, enclosures, fire alarm devices, or utilization equipment.
(a) identified   (b) listed   (c) labeled   (d) none of these

193.   Where exposed to contact with electric light or power conductors, the noncurrent-carrying metallic members of optical fiber cables entering buildings shall be _____ .
(a) grounded at the point of emergence through an exterior wall
(b) grounded at the point of emergence through a concrete floor slab
(c) interrupted as close to the point of entrance as practicable by an insulating joint
(d) any of these

Chapter 8 – Communications Systems

194.   Communication wires and cables shall be separated at least 2 inches from conductors of _____ circuits.
(a) power   (b) lighting   (c) Class 1   (d) any of these

195.   Where practical, coaxial cables for a CATV system shall be separated by _____ from lightning conductors.
(a) 3 inches   (b) 6 inches   (c) 6 feet   (d) 2 feet

196.   • Where network-powered broadband communications system cables enter buildings, they shall _____ .
(a) where practicable, be located below the electric light or power conductors
(b) not be attached to a crossarm that carries electric light or power conductors
(c) have a vertical clearance of not less than 8 feet from all points of roofs above which they pass
(d) all of these

197.   Network-powered broadband communications system cables shall be separated at least 2 inches from conductors of _____ circuits.
(a) power   (b) lighting   (c) Class 1   (d) any of these

Chapter 9 – Tables

198. When considering the number of conductors permitted when calculating raceway conductors fill, equipment grounding or bonding conductors shall _____ .

(a) not be counted   (b) have the actual dimensions used

(c) not be counted if in a nipple   (d) not be counted if for a wye 3-phase balanced load

199. The inside diameter of a 1 inch intermediate metal conduit is _____ inch(es).

(a) 1.0   (b) 1.105   (c) 0.86   (d) 0.34

Appendix C

200. The maximum number of No. 4 RHH conductors with outer covering permitted in 1 inch electrical metallic tubing is

_____ .

(a) one   (b) two   (c) three   (d) four

Unit 11

Commercial Load Calculations

OBJECTIVES

After reading this unit, the student should be able to briefly explain the following concepts:

Part A – General
Conductor overcurrent protection
   [*Section 240–3*]
Conductor ampacity
Fraction of an ampere
General requirements
Part B – Loads
Air-conditioning
Dryers
Electric heat

Kitchen equipment
Laundry circuit
Lighting demand factors
Lighting without demand factors
Lighting miscellaneous
Multioutlet assembly [*Section 220–3(c) Exception No. 1*]
Receptacles [*Section 220–3(b)(7)*]
Banks and offices general
   lighting and receptacles

Signs [*Section 600–6(c)*]
Part C – Load Calculations
Marina [*Section 555–5*]
Mobile home park [*Section 550–22*]
Recreational vehicle park
   [*Section 551–73*]
Restaurant – Optional method
   [*Section 220–36*]
School – optional method [*Section 220–34*]

After reading this unit, the student should be able to briefly explain the following terms:

General lighting
General lighting demand factors
Nonlinear loads
VA rating

Ampacity
Continuous load
Next size up protection device
Overcurrent protection

Rounding
Standard ampere ratings for
   overcurrent protection devices
Voltages

PART A – GENERAL

11–1 GENERAL REQUIREMENTS

Article 220 provides the requirement for branch circuits, feeders, and services. In addition to this Article, other Articles are applicable such as Branch Circuits – 210, Feeders – 215, Services – 230, Overcurrent Protection – 240, Wiring Methods – 300, Conductors – 310, Appliances – 422, Electric Space-Heating Equipment – 424, Motors – 430, and Air-Conditioning – 440.

11–2 CONDUCTOR AMPACITY [ARTICLE 100]

The ampacity of a conductor is the rating, in ampere, that a conductor can carry continuously without exceeding its insulation temperature rating [*NEC*® Definition – Article 100]. The allowable ampacities listed in Table 310-16 are affected by ambient temperature, conductor insulation, and conductor bundling [*Section 310-10*] (Figure 11–1).

Continuous Loads

Conductors are sized at 125 percent of the continuous load before any derating factor and the overcurrent protection devices are sized at 125 percent of the continuous load [*Section 210-19(a), 215-2,* and *230-42*].

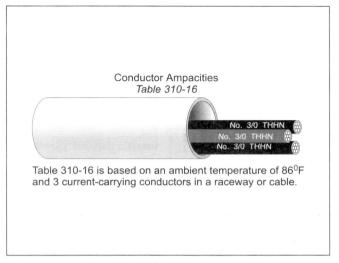

Conductor Ampacities
Table 310-16

No. 3/0 THHN
No. 3/0 THHN
No. 3/0 THHN

Table 310-16 is based on an ambient temperature of 86⁰F
and 3 current-carrying conductors in a raceway or cable.

Figure 11–1
Conductor Ampacities

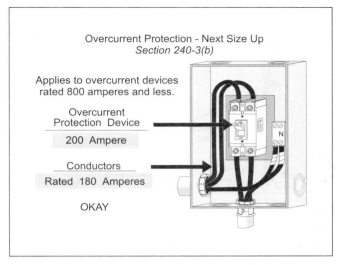

Overcurrent Protection - Next Size Up
Section 240-3(b)

Applies to overcurrent devices
rated 800 amperes and less.

Overcurrent
Protection Device

200 Ampere

Conductors

Rated 180 Amperes

OKAY

Figure 11–2
Overcurrent Protection – Next Size Up

11–3 CONDUCTOR OVERCURRENT PROTECTION [*SECTION 240-3*]

The purpose of overcurrent protection devices is to protect conductors and equipment against excessive or dangerous temperatures [*Section 240-1 FPN*]. There are many rules in the *National Electrical Code®* for conductor protection, and there are many different installation applications where the general rule of protecting the conductor at its ampacity does not apply. Examples would be motor circuits, air-conditioning, tap conductors, etc. See Section 240-3 for specific rules on conductor overcurrent protection.

Next Size Up Okay [*Section 240-3(b)*] If the ampacity of a conductor does not correspond with the standard ampere rating of a fuse or circuit breaker, as listed in Section 240-6(a), the next size up protection device is permitted. This practice only applies if the conductors do not supply multioutlet receptacles and if the next size up overcurrent protection device does not exceed 800 ampere (Figure 11–2).

Standard Size Overcurrent Devices [*Section 240-6(a)*] The following is a list of some of the standard ampere ratings for fuses and inverse time circuit breakers: 15, 20, 25, 30, 35, 40, 45, 50, 60, 70, 80, 90, 100, 110, 125, 150, 175, 200, 225, 250, 300, 350, 400, 500, 600, 800, 1,000, 1,200, and 1,600 ampere.

11–4 VOLTAGES [*SECTION* 220-2(a)]

Unless other voltages are specified, branch-circuit, feeder, and service loads shall be computed at a nominal system voltage of 120, 120/240, 208Y/120, 240, 480Y/277, 480 volts, or 600Y/347 (Figure 11–3).

11–5 ROUNDING AN AMPERE [*SECTION* 220-2(b)]

The rules for rounding an ampere require that when the calculation results in a fraction of an ampere of 0.50 and more, we "round up" to the next ampere. If the calculation is 0.49 of an ampere or less, we "round down" to the next lower ampere.

PART B – LOADS

11–6 AIR-CONDITIONING

Air-Conditioning Branch Circuit Branch circuit conductors and protection for air-conditioning equipment is marked on the equipment nameplate [*Section 110-3(b)*]. The nameplate values are determined by the use of Sections 440-32 for conductor sizing and 440-22(a) for short-circuit protection. Section 440-32 specifies that branch circuit conductors must be sized no less than 125 percent of the air-conditioner rating, and Section 440-22(a) specifies that the short-circuit protection device must be sized between 175 percent and up to 225 percent of the air-conditioning rating.

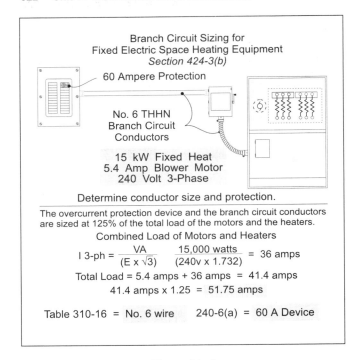

Figure 11–6
Branch Circuit Sizing for Fixed Electric Space-Heating
Equipment

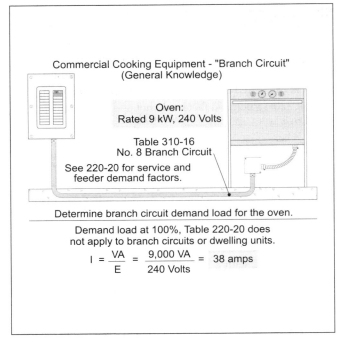

Figure 11–7
Commercial Cooking Equipment – "Branch Circuit"
(General Knowledge)

❏ Electric Heat Feeder/Service Demand Load

What is the feeder/service demand load for a building that has seven 3-phase, 208-volt, 10 kW heat strips with a 5.4-ampere blower motor (1,945 VA) for each unit?

(a) 84 kVA   (b) 53 kVA   (c) 129 kVA   (d) 154 kVA

  • Answer: (a) 84 kVA

  10,000 VA + 1,945 VA = 11,945 VA $\times$ 7 units = 83,615 VA/1,000 = 83.615 kVA

11–9 KITCHEN EQUIPMENT

Kitchen Equipment Branch Circuit

Branch circuit conductors and overcurrent protection for commercial kitchen equipment are sized according to the appliance nameplate rating.

❏ Kitchen Equipment Branch Circuit No. 1

What is the branch circuit demand load (in ampere) for one 9 kW oven rated 240 volts (Figure 11–7)?

(a) 38 ampere   (b) 27 ampere   (c) 32 ampere   (d) 33 ampere

  • Answer: (a) 38 ampere

  I = P/E

  I = 9,000 watts/240 volts

  I = 38 ampere

❏ Kitchen Equipment Branch Circuit No. 2

What is the branch circuit load for one 14.47 kW range rated 208 volts, 3-phase?

(a) 60 ampere   (b) 40 ampere   (c) 50 ampere   (d) 30 ampere

  • Answer: (b) 40 ampere

  I = P/(E $\times$ 1.732)

  I = 14,470 watts/(208 volts $\times$ 1.732)

  I = 40 ampere

Kitchen Equipment Feeder/Service Demand Load [*Section 220-20*]

The service demand load for thermostatic control or intermittent use for commercial kitchen equipment is determined by applying the demand factors from Table 220-20, to the total connected kitchen equipment load. *The feeder or service demand load cannot be less than the two largest appliances.*

Note: The demand factors of Table 220-20 do not apply to space heating, ventilating, or air-conditioning equipment.

❏ **Kitchen Equipment Feeder/Service No. 1**

What is the demand load for the following kitchen equipment loads (Figure 11–8)?

Water heater	5 kW	Booster heater	7.5 kW
Mixer	3 kW	Oven	5 kW
Dishwasher	1.5 kW	Waste disposal	1 kW

(a) 15 kW   (b) 23 kW   (c) 12.5 kW   (d) none of these

• Answer: (a) 15 kW

Water heater	5 kW
Booster heater	7.5 kW
Mixer	3 kW
Oven	5 kW
Dishwasher	1.5 kW
Waste disposal	1 kW
Total connected	23 kW × 0.65 = 14.95 kW

❏ **Kitchen Equipment Feeder/Service No. 2**

What is the demand load for the following kitchen equipment loads?

Water heater	10 kW*	Booster heater	15 kW*
Mixer	4 kW	Oven	6 kW
Dishwasher	1.5 kW	Waste disposal	1 kW

(a) 24.4 kW   (b) 38.2 kW   (c) 25.0 kW   (d) 18.9 kW

• Answer: (c) 25 kW*

Water heater	10 kW
Booster heater	15 kW
Mixer	4 kW
Oven	6 kW
Dishwasher	1.5 kW
Waste disposal	1 kW
Total connected	37.5 kW × 0.65 = 24.4 kW

* The demand load cannot be less than the two largest appliances 10 kW + 15 kW = 25 kW.

11–10 LAUNDRY EQUIPMENT

Laundry equipment circuits are sized to the appliance nameplate rating. For exam purposes, it is generally accepted that a laundry circuit is not considered a continuous load and assume all commercial laundry circuits to be rated 1,500 VA unless noted otherwise in the question.

❏ **Laundry Equipment**

What is the demand load for 10 washing machines located in a laundry room (Figure 11–9)?

(a) 1,500 VA   (b) 15,000 VA
(c) 1,125 VA   (d) none of these

• Answer: (b) 15,000 VA

1,500 VA × 10 units = 15,000 VA

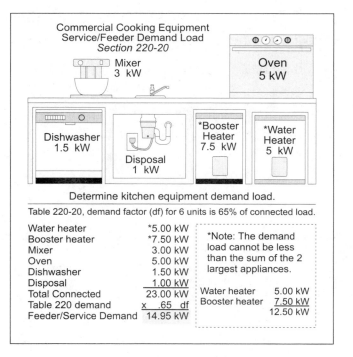

Figure 11–8
Commercial Cooking Equipment Service/Feeder
Demand Load

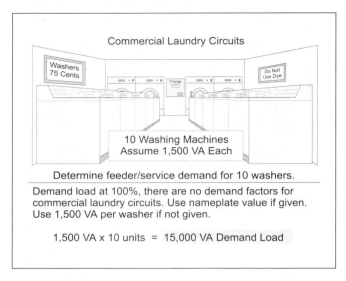

Figure 11–9
Commercial Laundry Circuits

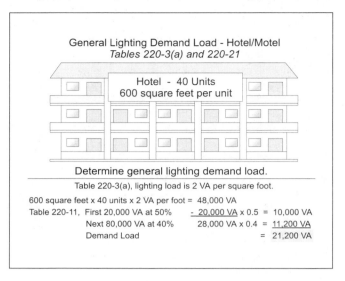

Figure 11–10
General Lighting Demand Load – Hotel/Motel

11–11 LIGHTING – DEMAND FACTORS [Table 220-3(a) and 220-11]

The *NEC*® requires a minimum load per square foot for general lighting depending on the type of occupancy [Table 220-3(a)]. For the guest rooms of hotels and motels, hospitals, and storage warehouses, the general lighting demand factors of Table 220-11 can be applied to the general lighting load.

Hotel or Motel Guest Rooms – General Lighting [Table 220-3(a) and Table 220-11]

The general lighting demand load of 2 VA each square foot [Table 220-3(a)] for the guest rooms of hotels and motels is permitted to be reduced according to the demand factors listed in Table 220-11.

General Lighting Demand Factors

First 20,000 VA at 50 percent demand factor
Next 80,000 VA at 40 percent demand factor
Remainder VA at 30 percent demand factor

❏ Hotel General Lighting Demand

What is the general lighting demand load for a 40-room hotel? Each unit contains 600 square feet of living area (Figure 11–10).

  (a) 48 kVA   (b) 24 kVA   (c) 20 kVA   (d) 21 kVA

    • Answer: (d) 21 kVA

40 units × 600 square feet (24,000 square feet × 2 VA) 48,000 VA

First 20,000 VA at 50 %	$-20{,}000\ \text{VA} \times 0.5 = 10{,}000\ \text{VA}$
Next 80,000 VA at 40 %	$28{,}000\ \text{VA} \times 0.4 = \underline{11{,}200\ \text{VA}}$
	21,200 VA

11–12 LIGHTING WITHOUT DEMAND FACTORS [Table 220-3(a), 215-2 and 230-42]

The general lighting load for commercial occupancies other than guest rooms of motels and hotels, hospitals, and storage warehouses is assumed continuous and shall be calculated at 125 percent of the general lighting load as listed in Table 220-3(a) [Table 220-11].

❏ Store Lighting

What is the general lighting load load for a 21,000 square-foot store (Figure 11–11)?

  (a) 40 kVA   (b) 63 kVA   (c) 79 kVA   (d) 81 kVA

    • Answer: (c) 79 kVA

21,000 square feet × 3 VA × 1.25 = 78,750 VA

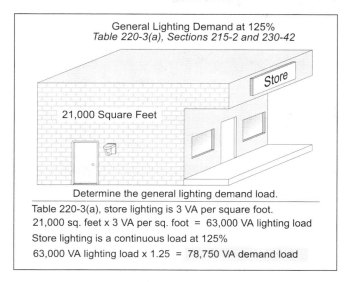

Figure 11–11
General Lighting Demand at 125%

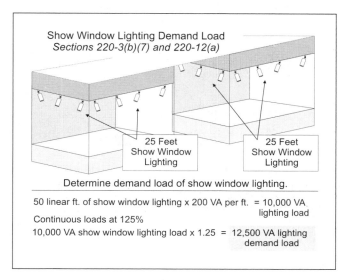

Figure 11–12
Show-Window Lighting Demand Load

❏ Club Lighting

What is the general lighting load for a 4,700 square-foot dance club?

(a) 4,700 VA   (b) 9,400 VA   (c) 11,750 VA   (d) 250 kVA

   • Answer: (c) 11,750 VA

   4,700 square feet × 2 VA × 1.25 = 11,750 VA

❏ School Lighting

What is the general lighting load for a 125,000 square-foot school?

(a) 125 kVA   (b) 375 kVA   (c) 475 kVA   (d) 550 kVA

   • Answer: (c) 475 kVA

   125,000 square feet × 3 VA × 1.25 = 468,750 VA

11–13 LIGHTING – MISCELLANEOUS

Show-Window Lighting [*Section 220-3(b)(7)* and *220-12(a)*]

The demand load for each linear foot of show-window lighting shall be calculated at 200 VA per foot. Show-window lighting is assumed to be a continuous load; See Example D3 in Appendix D for the requirements for show-window branch circuits.

❏ Show-Window Load

What is the demand load in kVA for 50 feet of show-window lighting (Figure 11–12)?

(a) 6 kVA   (b) 7.5 kVA   (c) 9 kVA   (d) 12.5 kVA

   • Answer: (d) 12.5 kVA

   50 feet × 200 VA per foot = 10,000 VA × 1.25 = 12,500 VA

11–14 MULTI-OUTLET RECEPTACLE ASSEMBLY [*SECTION 220-3(b)(8)*]

Each 5 feet, or fraction of that, of multioutlet receptacle assembly shall be considered to be 180 VA for service calculations. When a multi-outlet receptacle assembly is expected to have a number of appliances used simultaneously, each foot or fraction of a foot shall be considered as 180 VA for service calculations. A multi-outlet receptacle assembly is not generally considered to be a continuous load.

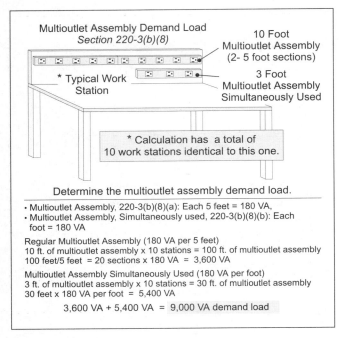

Figure 11–13
Multi-outlet Assembly Demand Load

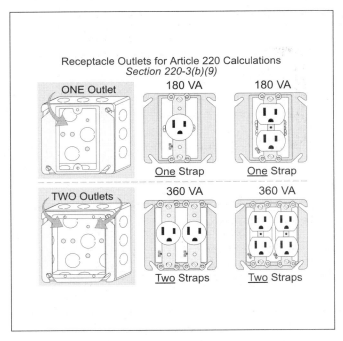

Figure 11–14
Receptacle Outlet Load Calculations

❏ Multi-outlet Receptacle Assembly

What is the demand load for 10 work stations that have 10 feet of multi-outlet receptacle assembly and 3 feet of multi-outlet receptacle assembly simultaneously used (Figure 11–13)?

   (a) 5 kVA   (b) 6 kVA   (c) 7 kVA   (d) 9 kVA

    • Answer: (d) 9 kVA

    100 feet/5 feet per section (20 sections × 180 VA)   3,600 VA

    30 feet/1 foot per section (30 sections × 180 VA)   <u>5,400 VA</u>

                                          9,000 VA/1,000 = 9 kVA

11–15 RECEPTACLE VA LOAD [*SECTION 220-13*]

Receptacle VA Load

The minimum load for each commercial or industrial general use receptacle outlet shall be 180 VA per strap [*Section 220–3(b)(9)*]. Receptacles are generally not considered to be a continuous load (Figure 11–14).

Number of Receptacles Permitted on a Circuit

The maximum number of receptacle outlets permitted on a commercial or industrial circuit is dependent on the circuit ampacity. The number of receptacles per circuit is calculated by dividing the VA rating of the circuit by 180 VA for each receptacle strap.

❏ Receptacles Per Circuit [*Section 220-3(b)(9)*]

How many receptacle outlets are permitted on a 15-ampere, 120-volt circuit (Figure 11–15)?

   (a) 10   (b) 13   (c) 15   (d) 20

    • Answer: (a) 10

The total circuit VA load for a 15-ampere circuit is 120 volts × 15 ampere = 1,800 VA.

The number of receptacle outlets per circuit = 1,800 VA/180 VA = 10 receptacles.

Note: 15-ampere circuits are permitted for commercial and industrial occupancies according to the *National Electrical Code*®, but some local codes require a minimum 20-ampere rating for commercial and industrial circuits [*Section 310-5*].

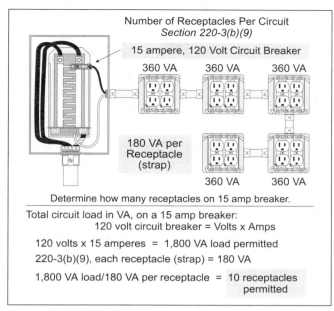

Figure 11–15
Number of Receptacles Per Circuit

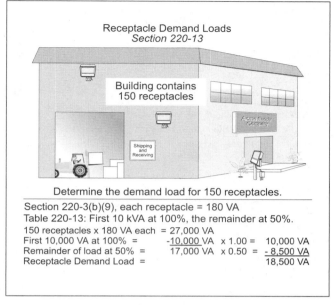

Figure 11–16
Receptacle Demand Loads

Receptacle Service Demand Load [*Section 220-13*]

The feeder and service demand load for commercial receptacles is calculated at 180 volt-ampere per receptacle strap [*Section 220-3(b)(9)*]. The demand factors of Table 220-13 can be used for that portion of the receptacle load in excess of 10 kVA, including the multi-outlet receptacle load [*Section 220–3(b)(8)*]. Receptacle loads are generally not considered to be a continuous load.

Table 220-13 Receptacle Demand Factors:

First 10 kVA at 100 percent demand factor
Remainder kVA at 50 percent demand factor

❏ Receptacle Service Demand Load

What is the service demand load for one hundred and fifty, 20-ampere, 120-volt general use receptacles in a commercial building (Figure 11–16)?

(a) 27 kVA   (b) 14 kVA   (c) 4 kVA   (d) none of these

• Answer: (d) none of these

Total receptacle load   (150 receptacles × 180 VA)   27,000 VA
First 10 kVA at 100%   − 10,000 VA × 1.00 = 10,000 VA
Remainder at 50%   17,000 VA × 0.50 =   8,500 VA
Total receptacle demand load   18,500 VA/1,000 = 18.5 kVA

11–16 BANKS AND OFFICES GENERAL LIGHTING AND RECEPTACLES

Some testing agencies include the receptacle demand load as part of the general lighting load for banks and offices. If that is the case, the general lighting demand load for banks and offices would be calculated as 3.5 VA each square foot times 125 percent for continuous lighting load, plus the receptacle demand load after applying Table 220-13 demand factors.

Receptacle Demand [Table 220-13] The receptacle demand load is calculated at 180 volt-ampere for each receptacle outlet [*Section 220-3(b)(9)*] if the number of receptacles are known, or 1 VA for each square foot if the number of receptacles are unknown [Table 220–3 (a) Note b].

❏ Bank General Lighting and Receptacle

What is the general lighting demand load (including receptacles) for an 18,000 square-foot bank? The number of receptacles is unknown (Figure 11–17).

(a) 68 kVA   (b) 110 kVA   (c) 84 kVA   (d) 93 kVA

• Answer: (d) 93 kVA

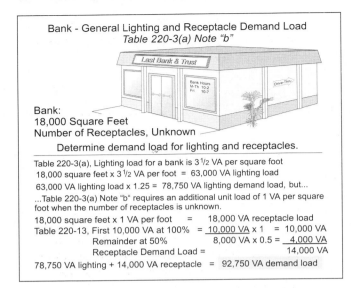

Figure 11–17
Bank – General Lighting and Receptacle Demand Load

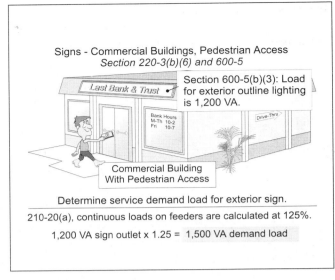

Figure 11–18
Signs – Commercial Buildings, Pedestrian Access

(a) General Lighting (18,000 square feet × 3.5 VA × 1.25)   78,750 VA

(b) Receptacle Load (18,000 square feet × 1 VA)   18,000 VA

First 10 kVA at 100%   10,000 VA × 1.00 =   10,000 VA

Remainder at 50%   8,000 VA × 0.50 =   4,000 VA

Total demand load   92,750 VA/1,000 = 92.75 kVA

❏ **Office General Lighting and Receptacle**

What is the general lighting demand load for a 28,000 square-foot office building that has 160 receptacles?

(a) 111 kVA   (b) 110 kVA   (c) 128 kVA   (d) 142 kVA

   • Answer: (d) 142 kVA

(a) General Lighting (28,000 square feet 3.5 VA × 1.25)   122,500 VA.

(b) Receptacle Load (160 receptacles × 180 VA)   28,800 VA

First 10,000 VA at 100%   10,000 VA × 1.00 =   10,000 VA

Remainder at 50%   18,800 VA × 0.50 =   9,400 VA

                                                                            141,900 VA/1,000 = 142 kVA

11–17 SIGNS [*SECTION 220-3(b)(6)* and *600-5*]

The *NEC®* requires each commercial occupancy that is accessible to pedestrians to be provided with at least one 20 ampere branch circuit for a sign [*Section 600-5(a)*]. The load for the required exterior signs or outline lighting shall be a minimum of 1,200 VA [*Section 220-3(b)(6)* and *600-5(b)(3)*]. A sign outlet is considered to be a continuous load, and the feeder load must be sized at 125 percent of the continuous load [*Section 210-20(a), 215-2(a), 230-42* and *384-16(d)*].

❏ **Sign Demand Load**

What is the demand load for one electric sign (Figure 11–18)?

(a) 1,200 VA   (b) 1,500 VA   (c) 1,920 VA   (d) 2,400 VA

   • Answer: (b) 1,500 VA

   1,200 VA × 1.25 = 1,500 VA

11–18 NEUTRAL CALCULATIONS [*SECTION 220-22*]

The neutral load is considered the maximum unbalanced demand load between the grounded (neutral) conductor and any one ungrounded (hot) conductor as determined by the calculations in Article 220, Part B. This means that line-to-line loads are not considered when sizing the neutral conductor.

Reduction over 200 Ampere

For balanced 3-wire, single-phase and 4-wire, 3-phase wye systems, the neutral demand load can be reduced 70 percent for that portion of the unbalanced load over 200 ampere.

❏ **Reduction over 200 Ampere**

What is the neutral current for a balanced 400-ampere 3-wire 120/240-volt feeder?

(a) 400 ampere   (b) 340 ampere   (c) 300 ampere   (d) none of these

• Answer: (b) 340 ampere

Total neutral load	400 ampere	
First 200 ampere at 100%	200 ampere × 1.00 =	200 ampere
Remainder at 70%	200 ampere × 0.70 =	140 ampere
		340 ampere

Reduction Not Permitted

The neutral demand load shall not be permitted to be reduced for 3-wire, single-phase 208Y/120-, or 480Y/277-volt circuits consisting of two line wires and the common conductor (neutral) of a 4-wire, 3-phase wye system. This is because the common (neutral) conductor of a 3-wire circuit connected to a 4-wire, 3-phase wye system carries approximately the same current as the phase conductors; see Section 310–15 (b)(4)(b). This can be proven with the following formula:

$$I_{Neutral} = \sqrt{(L_1^2 + L_2^2) - (L_1 \times L_2)}$$

$Line_1$ = Neutral Current of Line 1
$Line_2$ = Neutral Current of Line 2

❏ **Three-Wire Wye Neutral Current**

What is the neutral current for a balanced 300-ampere 3-wire 208Y/120-volt feeder, supplied from a 4-wire, 3-phase wye system (Figure 11–19)?

(a) 100 ampere   (b) 200 ampere   (c) 300 ampere   (d) none of these

• Answer: (c) 300 ampere

$$I_{Neutral} = \sqrt{L_1^2 + L_2^2 - (L_1 \times L_2)} = \sqrt{(300^2 + 300^2) - (300 \times 300)} = \sqrt{180{,}000 - 90{,}000} = \sqrt{90{,}000} = 300 \text{ ampere}$$

Nonlinear Loads

The neutral demand load cannot be reduced for nonlinear loads supplied from a 4-wire, wye-connected, 3-phase system because they produce *triplen harmonic currents* that add on the neutral conductor, which can require the neutral conductor to be larger than the ungrounded conductor load; see Section 220-22 FPN No. 2.

PART C – LOAD CALCULATION EXAMPLES

MARINA [SECTION 555-6]

The *National Electrical Code*® permits a demand factor to apply to the receptacle outlets for boat slips at a marina. The demand factors of Section 555-6 are based on the number of receptacles on the feeder. The receptacles must also be balanced between the lines to determine the number of receptacles on any given line.

❏ **Marina Receptacle Outlet Demand**

What size 120/240-volt, single-phase service is required for a marina that has twenty 20-ampere, 120-volt receptacles and twenty 30-ampere, 240-volt receptacles (Figure 11–20)?

(a) 200 ampere   (b) 400 ampere   (c) 600 ampere   (d) 800 ampere

• Answer: (c) 600 ampere

	Line 1	Line 2
Ten 20 ampere, 120 volt	200 ampere	200 ampere
Twenty 30 ampere, 240 volt	600 ampere	600 ampere
	800 ampere	800 ampere × 0.7 = 560 ampere per line

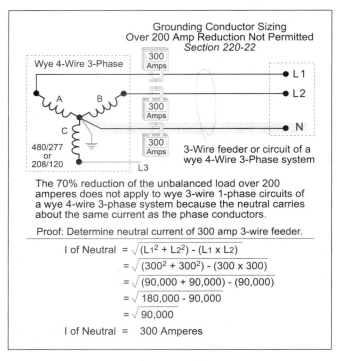

Figure 11–19
Grounding Conductor Sizing

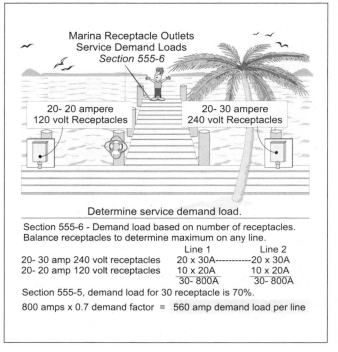

Figure 11–20
Marina Receptacle Outlet – Service Demand Loads

MOBILE/MANUFACTURED HOME PARK [SECTION 550-22]

The service demand load for a mobile/manufactured home park is sized according to the demand factors of Table 550-22 to the larger of 16,000 VA for each mobile/manufactured home lot or the calculated load for each mobile/manufactured home site according to Section 550-13.

❏ Mobile/Manufactured Home Park

What is the demand load for a mobile/manufactured home park that has facilities for 35 sites? The system is 120/240 volts, single-phase (Figure 11–21).

    (a) 400 ampere   (b) 600 ampere
    (c) 800 ampere   (d) 1,000 ampere
       • Answer: (b) 600 ampere nearest answer
       16,000 VA $\times$ 35 sites $\times$ 0.24 = 134,400 VA
       I = VA/E = 134,400 VA/240 volts = 560 ampere

RECREATIONAL VEHICLE PARK [SECTION 551-73]

Recreational vehicle parks are calculated according to the demand factor of Table 551-73. The total calculated load is based on:

    • 2,400 VA for each 20-ampere supply facilities site,

    • 3,600 VA for each 20- and 30-ampere supply facilities site, and

    • 9,600 VA for each 50-ampere, 120/240-volt supply facilities site.

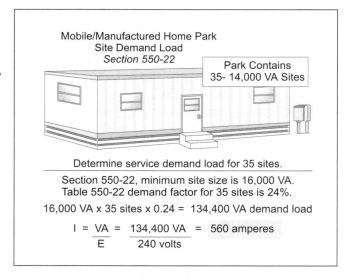

Figure 11–21
Mobile/Manufactured Home Park Site Demand Load

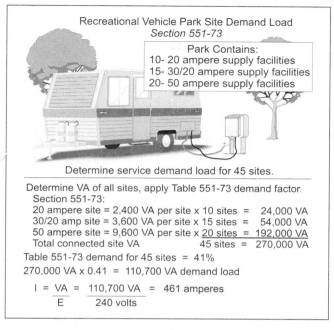

Figure 11–22
Recreational Vehicle Park Site Demand Load

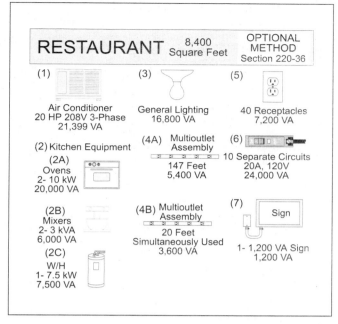

Figure 11–23
Restaurant Demand Load Optional Method

❏ Recreational Vehicle Park

What is the demand load for a recreational vehicle park that has ten 20-ampere supply facilities, fifteen 20/30-ampere supply facilities, and twenty 50-ampere supply facilities? The system is 120/240 volts, single-phase (Figure 11–22).

    (a) 400 ampere   (b) 500 ampere   (c) 600 ampere   (d) 700 ampere

    • Answer: (b) 500 ampere

The service for the recreational vehicle park is sized according to the demand factors of Table 551-73

Ten 20-ampere sites (2,400 VA × 10 sites)	24,000 VA
Fifteen 20/30-ampere sites (3,600 VA × 15 sites)	54,000 VA
Twenty 50-ampere sites (9,600 VA × 20 sites)	192,000 VA
Demand factor [Table 551-73]	270,000 VA × 0.41 = 110,700 VA

I = VA/E = 110,700 VA/240 volts = 461 ampere

RESTAURANT – OPTIONAL METHOD [SECTION 220-36]

An optional method of calculating the demand service load for a restaurant is permitted. The following steps can be used to determine the service size:

Step 1: ➧   Determine the total connected load. Add the nameplate rating of all loads at 100 percent and include both the air conditioner and heat load.

Step 2: ➧   Apply the demand factors from from Table 220-36 to the total connected load (Step 1).

All-Electric Restaurant Demand Factors	
0 – 250 kVA	80%
251 – 280 kVA	70%
281 – 325 kVA	60%
Over 326 kVA	50%

Not All-Electric Restaurant Demand Factors	
0 – 250 kVA	100%
251 – 280 kVA	90%
281 – 325 kVA	80%
326 – 375 kVA	70%
376 – 800 kVA	65%
Over 800 kVA	50%

❏ Restaurant Optional Method

Using the optional method, what is the service size for the loads (Figure 11–23)?

Step 1: ➡   Determine the total connected load.

(1) Air-conditioning demand [Table 430-150 and 440-34]
VA = 208 volts $\times$ 59.4 ampere $\times$ 1.732   21,399 VA

(2) Kitchen equipment

Ovens	10 kVA $\times$ 2 units =	20,000 VA
Mixers	3 kVA $\times$ 2 units =	6,000 VA
Water heaters	7.5 kVA $\times$ 1 unit =	7,500 VA

(3) Lighting [Table 220-3(a)] (8,400 square feet $\times$ 2 VA)   16,800 VA

(4) Multi-outlet assembly [*Section 220-3(b)(8)*]
147 feet/5 feet = thirty 5-foot sections (180 VA $\times$ 30)   5,400 VA
Simultaneously used (180 VA $\times$ 20)   3,600 VA

(5) Receptacles [*Section 220-3(b)(9)*] (180 VA $\times$ 40 receptacles)   7,200 VA
(6) Separate circuits (120 volts $\times$ 20 ampere $\times$ 10 circuits)   24,000 VA
(7) Sign [*Section 220-3(b)(6)* and *600-5*]   1,200 VA

Totals	
(1) Air-conditioning	21,399 VA
(2) Kitchen equipment	33,500 VA
(3) Lighting	16,800 VA
(4) Multi-outlet assembly	9,000 VA
(5) Receptacles	7,200 VA
(6) Separate circuits	24,000 VA
(7) Sign	1,200 VA
Total connected load	113,099 VA

Step 2: ➡   Apply the demand factors from Table 220-36 to the total connected load.

<u>All-Electric</u> If the restaurant is all electric, a demand factor of 80 percent is permitted to apply to the first 250 kVA.

113,099 VA $\times$ 0.8 = 90,479 VA

I = VA/(E $\times$ $\sqrt{3}$) = 90,479 VA/(208 volts $\times$ 1.732) = 251 ampere

<u>Not All-Electric</u> If the restaurant is not all electric, then a demand factor of 100 percent applies to the first 250 kVA.

113,099 VA at 100% = 113,099 VA

I = VA/(E $\times$ $\sqrt{3}$) = 113,099 VA/(208 volts $\times$ 1.732) = 314 ampere

SCHOOL – OPTIONAL METHOD [SECTION 220-34]

An optional method of calculating the demand service load for a school is permitted. The following steps can be used to determine the service size:

Step 1: ➡   Determine the total connected load. Add the nameplate rating of all loads at 100 percent and select the larger of the air-conditioning versus heat load.

Step 2: ➡   (a) Determine the average VA each square foot by dividing the total connected load (Step 1) by the square feet of the building.

(b) Determine the demand VA each square foot by applying the demand factors of Table 220-34 to the average VA each square foot (Step 2).

(c) Determine the school net VA by multiplying the demand VA each square foot (Step 3) by the square footage of the school building.

❏ **School Optional Method**

What is the service size for the following loads? The system voltage is 208Y/120 volt, 3-phase (Figure 11–24).

(1) Air-conditioning	50,000 VA
(2) Cooking equipment	40,000 VA
(3) Lighting	100,000 VA
(4) Multi-outlet assembly	10,000 VA
(5) Receptacles	40,000 VA
(6) Separate circuits	40,000 VA
	280,000 VA

(a) 300 ampere   (b) 400 ampere
(c) 500 ampere   (d) 600 ampere

• Answer: (c) 500 ampere

Determine the average VA per square foot. Total connected load/square feet area:

280,000 VA/10,000 square feet = 28 VA per square foot.

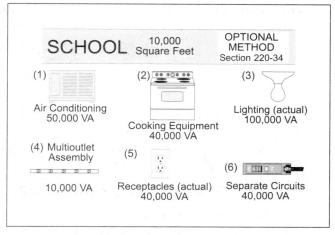

Figure 11–24
School – Optional Service Demand Load

Apply the demand factors from Table 220-34 to the VA per square foot

Average VA per square foot	28 VA
First 3 VA at 100%	– 3 VA × 1.00 = 3.00 VA
	25 VA
Next 17 VA at 75%	– 17 VA × 0.75 = 12.75 VA
Remainder at 25%	8 VA × 0.25 = 2.00 VA
	17.75 VA

Net VA per square foot = 17.75 VA × 10,000 square feet = 177,500 VA

Service Size – I = VA/(E × √3) = 177,500 VA/(208 volts × 1.732) = 493 ampere

---

Service Demand Load Using the Standard Method

For exam preparation purposes, you are not expected to calculate the total demand load for a commercial building. However, you are expected to know how to determine the demand load for the individual loads (Steps 1 through 10 below). For your own personal knowledge, you can use the following steps to determine the total demand load for a commercial building for the purpose of sizing the service.

Part A – Determine the Demand Load for Each Type of Load

Step 1: ➟ Determine the general lighting load: Table 220-3(a) and Table 220-11.

Step 2: ➟ Determine the receptacle demand load, [*Section 220-13*].

Step 3: ➟ Determine the appliance demand load at 100 percent.

Step 4: ➟ Determine the demand load for show windows at 125 percent, [*Section 220-12*].

Step 5: ➟ Determine the demand load for multi-outlet assembly, [*Section 220-3(b)(8)*].

Step 6: ➟ Determine the larger of:

Air-conditioning, largest air-conditioning load at 125 percent plus all other loads at 100 percent, [*Section 440-33*]

Heat at 100 percent, [*Section 220-15*].

Step 7: ➟ Determine the demand load for educational cooking equipment, [*Section 220-19* Note 5].

Step 8: ➟ Determine the kitchen equipment demand load, [Table 220-20].

Step 9: ➟ All other noncontinuous loads at 100 percent, [*Section 220-10(b)*].

Step 10: ➟ All other continuous loads at 125 percent, [*Section 215-2, 215-3, 230-42*].

Part B – Determine the Total Demand Load

Add up the individual demand loads of Steps 1 through 10.

Part C – Determine the Service Size

Divide the total demand load (Part B) by the system voltage.

Single-Phase = I = VA/E   Three-Phase = I = VA/(E × 1.732)

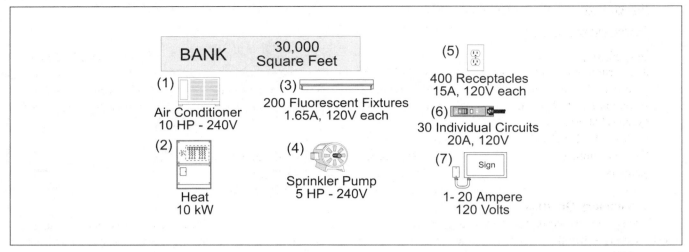

Figure 11–25
Bank

PART D – LOAD CALCULATION EXAMPLES

BANK (120/240 Volt, Single-Phase) (Figure 11–25)

(1) Air-conditioning [Article 440, 430-24, and Table 430-148]
   10 horsepower single-phase (230 volts × 50 ampere × 1.25)   14,375 VA

(2) Heat [*Section 220-15* and *220-21*] 10 kW omit

(3) Lighting [Table 220-12(b)]
   General Lighting (30,000 square feet × 3.5 VA = 105,000 VA* × 1.25)   131,250 VA
   Actual Lighting (200 units × 1.65 ampere × 120 volts × 1.25 = 49,500 VA, omit)

(4) Motor [Table 430-148] 5 horsepower single-phase (230 volts × 28 ampere)   6,440 VA
   (Some exams use 240 volts × 28 ampere = 6,720 VA)

(5) Receptacles [*Section 220-13*] (400 receptacles × 180 VA)   72,000 VA
                                                          −10,000 VA   × 1.00 =   10,000 VA*
                                                           62,000 VA   × 0.50 =   31,000 VA*

(6) Separate Circuits (30 circuits × 20 ampere × 120 volts)   72,000 VA*
(7) Sign [*Section 220-3(b)(6), and 600-5(b)(3)*] (1,200 VA* × 1.25)   1,500 VA
*Neutral load

SUMMARY	Overcurrent Protection	Conductor	Neutral
(1) Air conditioner	14,375 VA	14,375 VA *1	0 VA
(3) Lighting	131,250 VA	131,250 VA	105,000 VA *2
(4) Motor	16,100 VA *3	6,440 VA	0 VA
(5) Receptacles	41,000 VA	41,000 VA	41,000 VA
(6) Separate Circuits	72,000 VA	72,000 VA	72,000 VA
(7) Sign	+1,500 VA	+1,500 VA	+1,200 VA *2
	276,225 VA	266,565 VA	219,200 VA

*1 Includes 25 percent for the largest motors (14,375 × 1.25)

*2 The neutral reflects the feeder conductor load at 100 percent, not 125 percent [*Section 220-22*]!

*3 Feeder protection is sized according to the largest motor short-circuit ground-fault protection device [*Section 430-63*].
28 ampere × 2.5 = 70 ampere, 70 ampere × 230 volts = 16,100 VA

<u>Feeder/service protection device</u> I = VA/E = (276,225 VA /240 volts) = 1,151 ampere

<u>Feeder and service conductor</u> I = VA/E = 266,565 VA/240 volts = 1,111 ampere

<u>Neutral Ampere</u> The neutral conductor is permitted to be reduced according to the requirements of Section 220-22 for that portion of the unbalanced load that exceed 200 ampere. There shall be no reduction of the neutral capacity for that portion of the load that consists of nonlinear loads supplied from a 4-wire, wye-connected, 3-phase system. Since this system is a 120/240-volt, single-phase system, we are permitted to reduce the neutral by 70 percent for that portion that exceeds 200 ampere.

I = VA/E I = 219,200/240 volts	913 ampere	
First 200 ampere	− 200 ampere × 1.00 =	200 ampere
Remainder	713 ampere × 0.7 =	+ 499 ampere
		699 ampere

❏ Summary Questions

The service overcurrent protection device must be sized no less than 1,151 ampere. The service conductors must have an ampacity no less than the overcurrent protection device rating [*Section 240-3(c)*]. The grounded (neutral) conductor must not be less than 699 ampere.

➤ **Overcurrent Protection: What size service overcurrent protection device is required?**

(a) 800 ampere   (b) 1,000 ampere   (c) 1,200 ampere   (d) 1,600 ampere

   • Answer: (c) 1,200 ampere [*Section 240-6(a)*].

➤ **Service Conductor Size: What size service THHN conductors are required in each raceway if the service is parallel in four raceways?**

(a) 250 kcmil   (b) 300 kcmil   (c) 350 kcmil   (d) 400 kcmil

   • Answer: (c) 350 kcmil

   1,200 ampere/4 raceways = 300 ampere per raceway, sized based on 75°C terminal rating [*Section 110-14(c)(2)*]
   350 kcmil rated 310 ampere × 4 = 1,240 ampere [*Section 240-3(c)* and Table 310-16]

Note: 300 kcmil THHN has an ampacity of 320 ampere at 90°C, but we must size the conductors at 75°C, not 90°C.

➤ **Neutral Size: What size service grounded (neutral) conductor is required in each of the four raceways?**

(a) No. 1/0   (b) No. 2/0   (c) 250 kcmil   (d) 350 kcmil

   • Answer: (b) No. 2/0

   The grounded (neutral) conductor must be sized no less than:

   (1) 12¹/₂ percent area of the line conductor, 350,000 × 4 = 1,400,000 × 0.125 = 175,000 [*Section 250-24(b)*]
      175,000/4 = 43,750 kcmil, No. 3 per raceway [Chapter 9, Table 8].

   (2) When paralleling conductors, no grounded (neutral) conductor can be smaller than No. 1/0 [*Section 310-4*].

   (3) The service neutral conductor must have an ampacity of at least 699 ampere.
      699 ampere/4 = 175 ampere, Table 310-16, No. 2/0 is required at 75°C [*Section 110-14(c)*]

➤ **Grounding Electrode [*Section 250-66(b)*]: What size grounding electrode conductor is required to a concrete encased electrode?**

(a) No. 4   (b) No. 1/0   (c) No. 2/0   (d) No. 3/0

   • Answer: (a) No. 4

OFFICE BUILDING (480Y/277 Volt, Three-Phase) (Figure 11–26)

(1) Air-conditioning (Table 430-150 and 440-34)

460 volts × 7.6 ampere × 1.732 = (6,055 VA × 15 units) + (6,055 VA × 0.25)   92,339 VA

(2) Lighting (*NEC®*) [Table 220-3(a) and 220-10(b)]

28,000 square feet × 3.5 VA × 1.25 = 122,500 VA (omit)

Lighting Electric Discharge (Actual) [*Section 215-2 and 230-42*]

500 lights × 0.75 ampere × 277 volts = 103,875 VA* × 1.25   129,844 VA

(3) Lighting, Track (Actual) [*Section 220-12(b)*] 150 VA per 2 feet

200 feet/2 feet = 100 sections × 150 VA = 15,000 VA* × 1.25   18,750 VA

(4) Receptacles [*Section 220-13*](Actual) (200 receptacles × 180 VA) 36,000 VA

$$\underline{-10,000\ VA}\ \times 100\% =\quad 10,000\ VA^*$$

26,000 VA ×   50% =   13,000 VA*

(5) Multi-outlet Assembly [*Section 220-3(b)(8)*]

147 feet/5 feet = thirty 5-foot sections (30 sections × 180 VA)   5,400 VA*

Simultaneously used (20 feet × 180 VA)   3,600 VA*

(6) Separate Circuits (noncontinuous loads) (120 volts × 20 ampere × 30 circuits)   72,000 VA*

(7) Sign [*Section 220-3(b)(6)* and *600-5 (b)(3)*] (1,200 VA* × 1.25)   1,500 VA

*Neutral load

SUMMARY	Overcurrent Protection	Conductor	Neutral
(1) Air-conditioning	92,339 VA	92,339 VA	0 VA
(2) Lighting (actual)	129,844 VA	129,844 VA	103,875 VA *[1]
(3) Lighting (track)	18,750 VA	18,750 VA	15,000 VA *[1]
(4) Receptacles (actual)	23,000 VA	23,000 VA	23,000 VA
(5) Multi-outlet Assembly	9,000 VA	9,000 VA	9,000 VA
(6) Separate Circuits	72,000 VA	72,000 VA	72,000 VA
(7) Sign	1,500 VA	1,500 VA	1,200 VA *[1]
	346,433 VA	346,433 VA	224,075 VA

*[1] The neutral reflects the feeder load at 100 percent, not 125 percent.

Feeder/service protection device I = VA/(E × $\sqrt{3}$), I = 346,433 VA/(480 × 1.732) = 416 ampere

Feeder and service conductor ampere I = VA/(E × $\sqrt{3}$), I = 346,433 VA/(480 × 1.732) = 416 ampere

Neutral conductor I = VA/(E × $\sqrt{3}$), I = 224,075 VA/(480 volts × 1.732) = 270 ampere*

❏ **Summary**

The service overcurrent protection device and conductor must be sized no less than 416 ampere, and the grounded (neutral) conductor must not be less than 283 ampere.

➥ **Overcurrent Protection: What size service overcurrent protection device is required?**

(a) 300 ampere   (b) 350 ampere

(c) 450 ampere   (d) 500 ampere

• Answer: (c) 450 ampere [*Section 240-6(a)*]

Since the total demand load for the overcurrent device is 416 ampere, the minimum service permitted is 450 ampere.

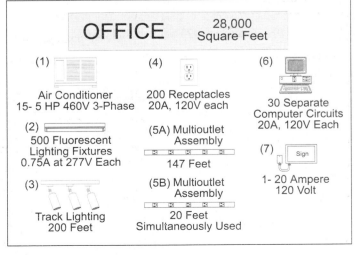

Figure 11–26
Office

➤ **Service Conductor Size: What size service conductors are required if total calculated demand load is 416 ampere?**

(a) 300 kcmil   (b) 400 kcmil   (c) 500 kcmil   (d) 600 kcmil

• Answer: (d) 600 kcmil rated 420 ampere [*Section 240-3(b)* and *Table 310-16*]

The service conductors must have an ampacity of at least 416 ampere, protected by a 450 ampere overcurrent protection device [*Section 240-3(b)*], and the conductors must be selected based on 75°C insulation rating [*Section 110-14(c)(2)*].

➤ **Neutral Size: What size service grounded (neutral) conductor is required for this service?**

(a) No. 1/0   (b) 3/0   (c) 250 kcmil   (d) 300 kcmil

• Answer: (d) 300 kcmil

The grounded (neutral) conductor must be sized no less than:

(1) The required grounding electrode conductor [*Section 250-24(b)* and *Table 250-66*], No. 2.

(2) An ampacity of at least 270 ampere, Table 310-16, 300 kcmil is rated 280 ampere at 75°C [*Section 110-14(c)*].

* Note: Because there is 125 ampere of electric discharge lighting [103,875 VA/(480 × 1.732)], the neutral conductor is not permitted to be reduced for that portion of the load in excess of 200 ampere [*Section 220-22*].

RESTAURANT – Standard Load Calculation (208Y/120 Volt, Three-Phase) (Figure 11–27)

(1) Air-conditioning [Table 430-150, and 440-34] (208 volts × 59.4 ampere × 1.732 × 1.25) = 26,749 VA

(2) Kitchen equipment [*Section 220-20*]

Ovens	(10 kW × 2 units)	20.0 kW
Mixer	(3 kW × 2 units)	6.0 kW
Water heaters	(7.5 kW × 1 unit)	7.5 kW
Total connected		33.5 kW × 0.7 × 1,000 = 23,450 kW

(3) Lighting *NEC*® [Table 220-3(a) and 220-10(b)]

8,400 square feet 3 2 VA 3 1.25 = 21,000 VA, omit

(4) Lighting, Track 150 VA per two feet [*Section 220-12(b)*]

50 feet/2 = 25 sections 3 150 VA 3 1.25	4,688 VA
100 lights 3 1.65 ampere 3 120 volts 3 1.25	24,750 VA

(5) Multi-outlet Assembly [*Section 220-3(b)(8)*]

50/5 = ten 5-foot sections (10 sections 3 180 VA)	1,800 VA
Simultaneous used (20 feet 3 180 VA)	3,600 VA

(6) Receptacles (Actual) [*Section 220-13*] (40 receptacles 3 180 VA)   7,200 VA

(7) Separate Circuits [*Section 215-2(a)* and *230–42*](120 volts 3 20 ampere 3 10 circuits)   24,000 VA

(8) Sign [*Section 220-3(b)(6)* and *600-5 (b)(3)*] (1,200 VA 3 1.25)   1,500 VA

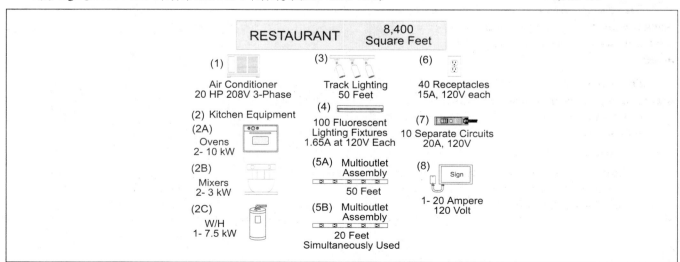

Figure 11–27 Restaurant

SUMMARY	Overcurrent Protection	Conductor
(1) Air-conditioning	26,749 VA	26,749 VA
(2) Kitchen equipment	23,450 VA	23,450 VA
(3) Lighting (track)	4,688 VA	4,688 VA
(4) Lighting (actual)	24,750 VA	24,750 VA
(5) Multi-outlet Assembly	5,400 VA	5,400 VA
(6) Receptacles (actual)	7,200 VA	7,200 VA
(7) Separate Circuits	24,000 VA	24,000 VA
(8) Sign	1,500 VA	1,500 VA
	117,737 VA	117,737 VA

Feeder and service protection device $I = VA/(E \times \sqrt{3}) = 117{,}737\ VA/(208 \times 1.732) = 327$ ampere

Feeder and service conductor $I = VA/(E \times \sqrt{3}) = 117{,}737\ VA/(208 \times 1.732) = 327$ ampere

❑ **Summary Questions**

The service overcurrent protection device must be sized no less than 327 ampere. The service conductors must be no less than 327 ampere.

➡ **Overcurrent Protection: What size service overcurrent protection device is required?**

(a) 300 ampere   (b) 350 ampere   (c) 450 ampere   (d) 500 ampere

• Answer: (b) 350 ampere [*Section 240-6(a)*]

Since the total demand load for the overcurrent device is 327 ampere, the minimum service permitted is 350 ampere [*Section 240-6(a)*].

➡ **Service Conductors Size: What size service conductors are required?**

(a) No. 4/0   (b) 250 kcmil   (c) 300 kcmil   (d) 400 kcmil

• Answer: (d) 400 kcmil

The service conductors must have an ampacity of at least 327 ampere and be protected by a 350-ampere protection device [*Section 240-3(b)*] next size up is permitted, and the conductors must be selected according to Table 310-16, 400 kcmil based on 75°C insulation rating [*Section 110-14(c)(2)*] is rated 335 ampere.

Unit 11 – Commercial Load Calculations Summary Questions

Part A – General

11–2 Conductor Ampacity [Article 100]

1. The _____ of a conductor is the rating in ampere a conductor can carry continuously without exceeding its insulation temperature rating. The allowable ampacities as listed in Table 310-16 are affected by ambient temperature, current flow, conductor insulation, and conductor bundling [*Section 310-10*].
    (a) load rating   (b) ampacity   (c) demand load   (d) continuous factor

11–3 Conductor Overcurrent Protection [*Section 240-3*]

2. The purpose of conductor _____ is to protect the conductors against excessive or dangerous temperatures. If the ampacity of a conductor does not correspond with the standard ampere rating of a fuse or circuit breaker, the next size up protection device is permitted. This applies only if the conductors do not supply multi-outlet receptacles and if the next size overcurrent protection device does not exceed 800 ampere.
    (a) short-circuit protection   (b) ground-fault protection
    (c) overload protection   (d) overcurrent protection

3. The following is a list of some of the standard ampere ratings for overcurrent protection devices (fuses and inverse time circuit breakers): 15, 25, 35, 45, 80, 90, 110, 175, 250, and 350.
    (a) True   (b) False

11–4 Voltages [*Section 220-2(a)*]

4. Unless other voltages are specified, branch-circuit, feeder, and service loads shall be computed at a nominal system voltage of _____ .
    (a) 600Y/347   (b) 240/120   (c) 208Y/120   (d) any of these

11–5 Rounding an Ampere [*Section 220-2(b)*]

5. There are no *NEC*® rules for rounding when a calculation results in a fraction of an ampere, but it does contain the following note: "except where the computations result in a _____ of an ampere or larger, such fractions may be dropped."
    (a) 0.05   (b) 0.5   (c) 0.49   (d) 0.51

Part B – Loads

11–6 Air-Conditioning

6. What is the feeder or service demand load required for the air-conditioning of a 6-unit office building? Each unit contains one air conditioner rated 3 horsepower 230 volts.
    (a) 13 kVA   (b) 24 kVA   (c) 45 kVA   (d) 51 kVA

11–7 Dryers

7. What size branch circuit conductor and overcurrent protection is required for a 6 kW dryer rated 240 volts, located in the laundry room of a multifamily dwelling?
    (a) No. 12 with a 20-ampere OCPD   (b) No. 10 with a 20-ampere OCPD
    (c) No. 12 with a 30-ampere OCPD   (d) No. 10 with a 30-ampere OCPD

8. What is the feeder and service demand for eight 6.75 kW dryers?
    (a) 70 kW   (b) 54 kW   (c) 35 kW   (d) 27 kW

11–8 Electric Heat

9. • What size conductor and protection is required for a single-phase, 15 kW, 480-volt heat strip that has a 1.6-ampere blower motor?
    (a) No. 10 THHN with 30-ampere protection   (b) No. 6 THHN with 45-ampere protection
    (c) No. 10 THHN with 40-ampere protection   (d) No. 4 THHN with 70-ampere protection

10. What is the feeder and service demand load in VA for a building that has four 20 kW, 230-volt, single-phase heat strips that have a 5.4-ampere blower motor for each unit?
    (a) 100 kVA   (b) 50 kVA   (c) 125 kVA   (d) 85 kVA

11–9 Kitchen Equipment

11. What is the branch circuit demand load in ampere for one 11.4 kW oven rated 240 volts?
    (a) 60 ampere   (b) 27 ampere   (c) 48 ampere   (d) 33 ampere

12. • What is the feeder and service demand load for the following?
    Water heater   9 kW
    Booster heater   12 kW
    Mixer   3 kW
    Oven   2 kW
    Dishwasher   1.5kW
    Waste disposal   1 kW
    (a) 19 kW   (b) 21 kW   (c) 29 kW   (d) 12 kW

13. What is the feeder and service demand load for the following?
    Water heater   14 kW
    Booster heater   11 kW
    Mixer   7 kW
    Oven   9 kW
    Dishwasher   1.5 kW
    Waste disposal   3 kW
    (a) 25 kW   (b) 30 kW   (c) 20 kW   (d) 45 kW

11–10 Laundry Equipment

14. What is the feeder and service demand load for 7 washing machines at 1,500 VA each?
    (a) 10,500 VA   (b) 15,000 VA   (c) 1,125 VA   (d) none of these

11–11 Lighting – Demand Factors [Table 220-3(a) and 220-11]

15. What is the general lighting feeder or service demand load for a 24-room motel, 685 square feet each?
    (a) 10 kVA   (b) 15 kVA   (c) 20 kVA   (d) 25 kVA

16. What is the general lighting and receptacle feeder or service demand load for a 250,000 square-foot storage warehouse that has 200 receptacles?
    (a) 25 kVA   (b) 55 kVA   (c) 95 kVA   (d) 105 kVA

11–12 Lighting without Demand Factors [Table 220-3(a), 215-2, and 230-42]

17. What is the feeder and service (overcurrent protection device) general lighting demand load for a 3,200 square-foot dance club?

(a) 3,200 VA   (b) 6,400 VA   (c) 8,000 VA   (d) 12,000 VA

18. What is the feeder and service general lighting load for a 90,000 square-foot school?

(a) 238 kVA   (b) 338 kVA   (c) 90 kVA   (d) 270 kVA

11–13 Lighting – Miscellaneous

19. How many 2 × 4 fluorescent fixtures, each rated 277 volts, 0.8 ampere, can be connected to a 20-ampere circuit? The four lamps are rated 40 watts each and the fixture is to be on for more than 3 hours.

(a) 5   (b) 7   (c) 9   (d) 20

20. How many 360-watt, 120-volt incandescent fixtures can be connected to a 20-ampere, 120-volt circuit continuously?

(a) 5   (b) 7   (c) 9   (d) 12

21. • What is the VA demand feeder and service load for 130 feet of show-window lighting?

(a) 26 kVA   (b) 33 kVA   (c) 9 kVA   (d) 11 kVA

11–14 Multi-outlet Receptacle Assembly [*Section 220-3(b)(8)*]

22. What is the feeder demand load for 50 feet of multi-outlet assembly and 10 feet of multi-outlet assembly simultaneously used?

(a) 3,600 VA   (b) 12,000 VA   (c) 7,200 VA   (d) 5,500 VA

11–15 Receptacle VA Load [*Section 220-13*]

23. How many receptacle outlets are permitted on a 20-ampere, 120-volt circuit?

(a) 10   (b) 13   (c) 15   (d) 20

24. What is the service demand load for 110 receptacles (15 or 20 ampere, 125 volts) in a commercial building?

(a) 5 kVA   (b) 10 kVA   (c) 15 kVA   (d) 20 kVA

25. What is the service demand load for the general lighting and receptacles for a 30,000 square-foot bank?

(a) 111 kVA   (b) 152 kVA   (c) 123 kVA   (d) 175 kVA

26. • What is the service general lighting demand and receptacle load for a 10,000 square-foot office building with 75 receptacles?

(a) 25 kVA   (b) 35 kVA   (c) 45 kVA   (d) 55 kVA

11–17 Signs [*Section 220-3(b)(b) and 600-5*]

27. What is the feeder demand load for sizing the overcurrent protection device for one electric sign?

(a) 1,400 VA   (b) 1,500 VA   (c) 1,920 VA   (d) 2,400 VA

11–18 Neutral Calculations [*Section 220-22*]

28. What is the neutral current for a balanced 150-ampere, 3-wire, 208Y/120-volt feeder?

(a) 0 ampere   (b) 150 ampere   (c) 250 ampere   (d) 300 ampere

Part C – Load Calculations

Marina [*Section 555-6*]

29. A marina has the following: 24 slips (20-ampere, 240-volt receptacles) and 30 slips designed for boats over 20 feet (30-ampere, 240-volt receptacles). What size service is required for the marina?

(a) 400 ampere   (b) 550 ampere   (c) 900 ampere   (d) 1,000 ampere

Restaurant Calculations

50.   • The branch circuit rating required for a 12 kW range in a restaurant would be _____ .
       (a) 50 ampere   (b) 45 ampere   (c) 35 ampere   (d) 30 ampere

51.   • A new restaurant has a total connected lighting load of 30 kVA. The kitchen equipment includes two gas stoves, one gas grill, three gas ovens, one 75-gallon gas water heater, one 5 kW dishwasher, two 2 kW coffee makers, five 2 kW kitchen appliances on their own circuit, and ten 1.5 kVA small appliance circuits. Using the optional method, the service demand load would be closest to _____ .
       (a) 64 kVA   (b) 50 kVA   (c) 45 kVA   (d) 38 kVA

School – Optional Method [*Section 220-34*]

52.   • Using the optional method, what is the demand load (VA per square foot) for a 10,000 square foot school that has a total connected load of 320 kVA?
       (a) 15.75 VA per square foot   (b) 18.75 VA per square foot
       (c) 29.00 VA per square foot   (d) 12.75 VA per square foot

53.   Using the optional method, what is the demand load (kVA) for a 20,000 square-foot school that has a total connected load of 160 kVA?
       (a) 135 kVA   (b) 106 kVA   (c) 120 kVA   (d) 112 kVA

NEC® Questions Random Order Section 90 through Chapter 9

54.   The area of square inches for aluminum No. 1/0 bare conductor is _____ .
       (a) 0.087   (b) 0.109   (c) 0.137   (d) 0.173

55.   Interconnecting cables under information technology equipment raised floors provide a significant fire and smoke load and must be listed as Type _____ cable when used under raised floors of a computer room.
       (a) RF   (b) UF   (c) LS   (d) DP

56.   Chapters 1 through 4 of the *NEC*® apply _____ .
       (a) generally to all electrical installations   (b) to special installations and conditions
       (c) to special equipment and material   (d) all of these

57.   • For antenna conductors for an amateur transmitting and receiving station where of hard-drawn copper, the maximum open span length is 200 feet and requires a minimum of No. _____ .
       (a) 14   (b) 12   (c) 10   (d) 8

58.   The repair of hard-service cord No. _____ and larger shall be permitted if conductors are spliced in accordance with Section 110-14(b) and the completed splice retains the insulation, outer sheath properties, and usage characteristics of the cord being spliced.
       (a) 16   (b) 15   (c) 14   (d) 12

59.   GFCI protection is required for all 125-volt, single-phase receptacles located within 5 feet measured _____ from the inside walls of a hydromassage bathtub.
       (a) vertically   (b) horizontally   (c) across   (d) none of these

60.   All _____-ampere, 125-volt receptacle outlets that are not a part of the permanent wiring of the building or structure and are in use by personnel for temporary power shall have ground-fault circuit-interrupter protection for personnel.
       (a) 15   (b) 20   (c) 30   (d) all of these

61.   Nonmetallic cable trays shall be made of _____ material.
       (a) fire-resistant   (b) waterproof   (c) corrosive   (d) flame-retardant

62.   A maximum of _____ No. 8 TW conductors can be installed in $1^1/_2$-inch electrical metallic tubing.
       (a) 11   (b) 22   (c) 18   (d) 8

63.   Each unit of a data processing system for use on a branch circuit shall have a nameplate that includes the _____ .
       (a) rating in volts   (b) operating frequency   (c) total load in amperes   (d) all of these

64. • All metal raceways, boxes, and enclosures from a hazardous (classified) location to the service or source of a separately derived system must be _____ according to the requirements of Section 250-94.
(a) grounded   (b) continuous   (c) rigid metal conduit   (d) bonded

65. X-ray equipment mounted on a permanent base with wheels for moving while completely assembled is defined as _____ .
(a) portable   (b) mobile   (c) movable   (d) a room

66. An isolating switch is one that is _____ .
(a) not readily accessible to persons unless special means for access are used
(b) capable of interrupting the maximum operating overload current of a motor
(c) intended for use in general distribution and branch circuits
(d) intended for isolating an electrical circuit from the source of power

67. Coaxial cables used in power-limited fire alarm systems shall have an overall insulation rated at _____ volts.
(a) 100   (b) 300   (c) 600   (d) 1,000

68. The scope of Article 680 includes _____ .
(a) wading and decorative pools   (b) fountains
(c) hydromassage bathtubs   (d) all of these

69. Unless an individual switch is provided for each lighting fixture located over combustible material, lampholders shall be located at least _____ feet above the floor, or shall be so located or guarded that the lamps cannot be readily removed or damaged.
(a) 3   (b) 6   (c) 8   (d) 10

70. Single, locking, and grounding-type receptacle for water-pump motor or other loads directly related to the circulation system of a permanently installed pool or fountain can be located between _____ feet from the inside walls of the pool or fountain, if GFCI-protected.
(a) 3 to 6   (b) 5 to 10   (c) 10 to 15   (d) 10 to 20

71. Fire alarm systems include _____ .
(a) fire alarm   (b) guard tour   (c) sprinkler water flow   (d) all of these

72. Fire alarm systems, multiconductor control, or power cable listed for use can be installed in the space over a hung ceiling used for environmental air.
(a) True   (b) False

73. Incandescent lamp lighting fixtures shall be marked to indicate the maximum allowable _____ of lamps.
(a) voltage   (b) amperage   (c) rating   (d) wattage

74. Lighting fixtures or ceiling fans mounted less than 12 feet above the water level cannot be installed within _____ feet of an outdoor pool, fountain, or spa.
(a) 3   (b) 5   (c) 10   (d) 8

75. Electric space heating cables shall be furnished complete with factory-assembled non-heating leads at least _____ in length.
(a) 6 inches   (b) 18 inches   (c) 3 feet   (d) 7 feet

76. Overhead branch circuit, feeder, or utility service conductors that operate at 480 volts line-to-line can be installed over pools and other structures. The clearance from the water must be at least _____ feet.
(a) 14   (b) 16   (c) 18   (d) 20

77. Type UF cable shall not be used _____ .
(a) in any hazardous (classified) location
(b) embedded in poured cement, concrete, or aggregate
(c) where exposed to direct rays of the sun, unless identified as sunlight-resistant
(d) all of these

78. Two conductors can fill the percent area of a conduit _____ percent.
    (a) 40   (b) 31   (c) 53   (d) 30

79. Underground rigid nonmetallic wiring when located less than 5 feet from the inside wall of the pool or spa must be buried not less than _____ inches.
    (a) 6   (b) 10   (c) 12   (d) 18

80. The pool light junction box connected to a conduit that extends directly to a forming shell or mounting bracket of a no-niche fixture shall be _____ for this use.
    (a) listed   (b) labeled   (c) marked   (d) identified

81. Motor overload protection shall not be shunted or cut out during the starting period if the motor is _____ .
    (a) not automatically started   (b) automatically started
    (c) manually started   (d) all of these

82. A lighting fixture located within 5 feet from the inside wall of an indoor spa or hot tub must be _____ .
    (a) a minimum of 7 feet 6 inches above maximum water level
    (b) protected by a ground-fault circuit-interrupter
    (c) a or b
    (d) a and b

83. For industrial machinery, straight exposed vertical risers of intermediate metal conduit with threaded couplings are permitted, if supported at the top and bottom no more than _____ feet apart.
    (a) 10   (b) 12   (c) 15   (d) 20

84. All areas designated as hazardous shall be properly _____ and the documentation shall be available to those authorized to design, install, inspect, maintain, or operate electrical equipment at these locations.
    (a) cleaned   (b) documented   (c) maintained   (d) all of these

85. The junction box that extends to the forming shell must be listed as a pool light junction box. In addition, the junction box shall be located not less than _____ inches above the ground level, or pool deck, or not less than _____ inches above the maximum water level.
    (a) 8, 4   (b) 4, 8   (c) 6, 12   (d) 12, 6

86. The size and number of conductors shall be that for which the cablebus is designed, and in no case smaller than No. _____ .
    (a) 1/0   (b) 2/0   (c) 3/0   (d) 4/0

87. A pool light junction box that has a raceway that extends directly to underwater pool light forming shells shall be located not less than _____ feet from the outdoor pool or spa.
    (a) 2   (b) 3   (c) 4   (d) 6

88. Audible and visual signal devices for an emergency system shall be provided, where practicable, for the purpose(s) of indicating _____ .
    (a) that the battery is carrying load   (b) of derangement of the emergency source
    (c) that the battery charger is not functioning (d) all of these

89. Emergency circuit wiring must be designed and located in such a manner so as to minimize the hazards that might cause failure because of _____ .
    (a) flooding   (b) fire   (c) icing   (d) all of these

90. The alternate source for emergency systems _____ be required to have ground-fault protection of equipment.
    (a) shall   (b) shall not

91. Where overcurrent protection is provided as part of the industrial machine, the machine shall be marked _____ .
    (a) overcurrent protection provided at machine supply terminals
    (b) this unit contains overcurrent protection
    (c) fuses or circuit breaker enclosed
    (d) with the overcurrent protection category and type

92. A ground-fault circuit-interrupter (GFCI) shall be installed in the branch circuit supplying swimming pool lighting fixtures that operates at more than _____, so that there is no shock hazard during relamping.
    (a) 6 volts   (b) 14 volts   (c) 15 volts   (d) 18 volts

93. Optional standby systems are typically installed to provide an alternate source of power for _____ .
    (a) data processing and communication systems   (b) emergency systems for health care facilities
    (c) emergency systems for hospitals   (d) none of these

94. Circuits and equipment operating at less than 50 volts shall use receptacles that are not rated less than _____ amperes.
    (a) 10   (b) 15   (c) 20   (d) 30

95. Access to equipment shall not be denied by an accumulation of wires and cables that prevent removal of panels, including suspending ceiling tiles. The intention of this Section is to limit the quantity of wires and cables that can be installed (laid) directly upon lay-in type suspended ceilings. This applies to _____ .
    (a) Class 2 and 3 cables
    (b) communication cables
    (c) CATV coaxial cables
    (d) all of these

96. A Class 1 signaling circuit shall not exceed _____ volts.
    (a) 130   (b) 150   (c) 220   (d) 600

97. Class 1 control circuits using No. 18 shall use insulation types _____ .
    (a) RFH-2, RFHH-2, or RFHH-3   (b) TF, TFF, TFN, or TFFN
    (c) RHH, RHW, THWN, or THHN   (d) a and b

98. The power for Class 2 or Class 3 circuits shall be _____ .
    (a) inherently limited requiring no overcurrent protection
    (b) limited by a combination of a power source and overcurrent protection
    (c) inherently limited and requires overcurrent protection
    (d) a and b

99. A storable swimming pool is capable of holding water to a maximum depth of _____ inches.
    (a) 24   (b) 36   (c) 42   (d) 48

100. Ground-fault circuit-interrupters (GFCI) for a pool or spa 120-volt wet-niche light shall only be of the circuit breaker type.
    (a) True   (b) False

101. All snap switches and dimmers must be effectively grounded so that they can effectively ground a metal faceplate whether or not a metal faceplate is installed.
    (a) True   (b) False

102. Where the service overcurrent protection devices are locked or sealed, or otherwise are not readily accessible, _____ overcurrent protection devices shall be installed in a readily accessible location, and shall be of lower ampere rating than the service overcurrent protection device.
    (a) sub-service   (b) branch circuit   (c) additional   (d) protected

103. When sizing a junction box, a straight pull shall not be less than _____ for a junction box containing conductors not over 600 volts.
    (a) 8 times the diameter of the largest raceway
    (b) 6 times the diameter of the largest raceway
    (c) 48 times the outside diameter of the largest shielded conductor
    (d) 36 times the largest conductor

104. Each service disconnect shall be permanently _____ to identify it as part of the service disconnecting means.
    (a) identified   (b) positioned   (c) marked   (d) none of these

105.   When determining a Class I, Division 2 location, _____ are factors that merit consideration in determining the classification and extent of each location.
(a) the quantity of flammable material that might escape in case of accident
(b) the adequacy of ventilating equipment
(c) the record of the industry or business with respect to explosions or fires
(d) all of these

106.   Transformers of more than _____ kVA rating shall be installed in a transformer room of fire-resistant construction.
(a) 35,000   (b) $87^1/_2$ (c) $112^1/_2$ (d) 75

107.   • Conduit bodies with conductors larger than No. 6 shall have a cross-sectional area at least twice that of the largest conduit to which they are connected.
(a) True   (b) False

108.   If the disconnect is not within sight of the fixed electric space heater (without supplementary overcurrent protection devices), it must be capable of being _____ .
(a) locked   (b) locked closed   (c) locked open   (d) within sight

109.   Overcurrent protection devices shall be readily accessible except for _____ as provided in Section 364-12.
(a) feeder taps   (b) cable bus   (c) busways   (d) wireways

110.   The ampacities as listed in the Tables of Article 310 are based on temperature alone and do not take _____ into consideration.
(a) continuous loads   (b) voltage drop   (c) insulation   (d) wet locations

111.   • _____ identified for use on lighting track shall be designed specifically for the track on which they are to be installed.
(a) Fittings   (b) Receptacles   (c) Devices   (d) all of these

112.   Interlocks on access doors or panels shall not be required if the applicator is an induction heating coil at dc ground potential or operating at less than _____ volts ac.
(a) 50   (b) 120   (c) 150   (d) 240

113.   Type TFN or other approved insulation used on a nonpower-limited fire alarm circuit is required for _____ .
(a) conductor sizes Nos. 16 and 18   (b) all conductors up to 600 volts
(c) conductors sizes Nos. 18, 16, and 14   (d) conductor sizes No. 14 and larger

114.   The circular mil area of a No. 12 conductor is _____ .
(a) 10,380   (b) 26,240   (c) 6,530   (d) 6,350

115.   Exposed power-limited fire alarm circuit cables must be installed in _____ when passing through a floor or wall to a height of 7 feet above the floor unless adequate protection can be afforded by the building construction or unless an equivalent solid guard is provided.
(a) rigid metal conduit   (b) rigid nonmetallic conduit
(c) EMT   (d) any of these

116.   Cables and conductors of two or more power-limited fire alarm circuits can be installed in the same cable, enclosure, or raceway.
(a) True   (b) False

117.   Five conductors are fill-limited to _____ percent of the conduit or tubing area.
(a) 35   (b) 60   (c) 40   (d) 55

118.   A conductive general-use type of optical fiber cable installed as wiring within buildings is Type _____ .
(a) FPLP   (b) CL2P   (c) OFPN   (d) OFC

119.   • Where communications cables enter buildings, they shall _____ .
(a) where practicable, be located below the electric light or power conductors
(b) not be attached to a crossarm that carries electric light or power conductors.
(c) have a vertical clearance of not less than 8 feet from all points of roofs above which they pass.
(d) all of these

120. In communication circuits, the bonding together of all separate electrodes can be bonded together with a minimum size jumper of No. _____ copper.
(a) 10   (b) 8   (c) 6   (d) 4

121. Rigid metal conduit that is directly buried outdoors must have at least _____ inches of cover.
(a) 6   (b) 12   (c) 18   (d) 24

122. _____ communication raceways shall be listed as having adequate fire-resistant characteristics capable of preventing the carrying of fire from floor to floor.
(a) Plenum   (b) Riser   (c) General purpose   (d) none of these

123. • The minimum size conductor permitted in parallel for elevator lighting is No. _____ provided the ampacity is equivalent to a No. 14 wire.
(a) 14   (b) 20   (c) 16   (d) 1/0

124. Where individual open conductors enter a building or other structure, they shall enter through roof bushings or through the wall in an upward slant through individual, noncombustible, nonabsorbent insulating _____ .
(a) tubes   (b) raceways   (c) sleeves   (d) bushels

125. Each conduit leaving a Class I, Division 1 location require a seal located on either side of the hazardous location boundary. Unions, couplings, boxes, or fittings are permitted between the seal and the point where the conduit leaves the Division 1 location.
(a) True   (b) False

126. The _____ branch loads served by the generating equipment cannot be automatically shed upon generating equipment overloading.
(a) emergency   (b) equipment   (c) life safety   (d) none of these

127. The minimum point of attachment of the service drop conductors to a building shall in no case be less than _____ feet.
(a) 8   (b) 10   (c) 12   (d) 15

128. Type ITC cables with a gas/vaportight continuous corrugated aluminum sheath, an overall jacket of suitable polymeric material, and provided with termination fittings listed for the application can be installed in Class I, Division 1 _____ establishments with restricted public access.
(a) commercial   (b) industrial   (c) institutional   (d) all of these

129. In a multiple-occupancy building where electric service and electrical maintenance are provided by the building management under continuous building management supervision, the service disconnecting means supplying more than one occupancy can be accessible to authorized _____ only.
(a) inspectors   (b) tenants
(c) management personnel   (d) none of these

130. Conductors of Class 2 and Class 3 circuits shall not be placed in any enclosure, raceway, cable, or similar fittings with conductors of Class 1 or electric light or power conductors, except when they are _____ .
(a) insulated for the maximum voltage   (b) totally comprised of aluminum conductors
(c) separated by a barrier   (d) all of these

131. A vertical run of 4/0 copper must be supported at intervals not exceeding _____ feet.
(a) 80   (b) 100   (c) 120   (d) 40

132. Class 2 and 3 control conductors shall be installed in _____ when used in hoistways.
(a) rigid metal conduit   (b) electrical metallic tubing
(c) intermediate metal conduit   (d) all of these

133. • The ampacity of supply branch circuit conductors and the overcurrent protection devices for X-ray equipment shall not be less than _____ .
(a) 50 percent of the momentary rating   (b) 100 percent of the long-time rating
(c) the larger of a or b   (d) the smaller of a or b

134. Grounding electrodes that consist of rods, require a minimum of _____ feet in contact with the soil.
     (a) 10   (b) 8   (c) 6   (d) 12

135. • Which of the following must be bonded?
     (a) electric equipment within 5 feet of the inside pool
     (b) pool steel structure steel
     (c) metal fittings greater than 4 inches within or attached to the pool
     (d) all of these

136. The ultimate trip current of a thermally protected motor (full-load current not exceeding 9 amperes) shall not exceed _____ percent of motor full-load current given in Tables 430-148, 430-149, and 430-150.
     (a) 140   (b) 156   (c) 170   (d) none of these

137. The pool structure, including the reinforcing metal of the pool shell and deck to be bonded.
     (a) True   (b) False

138. Service cables shall be equipped with a raintight _____ .
     (a) raceway   (b) service head   (c) cover   (d) all of these

139. The items required to be connected to the common bonding grid of a pool shall be done with a minimum No. 8 _____ conductor.
     (a) solid insulated   (b) solid bare   (c) solid covered   (d) any of these

140. • Service conductors shall not be spliced.
     (a) True   (b) False

141. Wet niche lighting fixtures shall be connected to an equipment grounding conductor not smaller than No. _____ .
     (a) 10   (b) 6   (c) 8   (d) 12

142. To use the optional method for calculating a service to a school, the school must be equipped with _____ .
     (a) cooking facilities   (b) electric space heating   (c) air-conditioning   (d) b or c

143. To qualify as a lighting and appliance branch circuit panelboard, the number of circuits rated 30 ampere or less with neutrals must be _____ of the total.
     (a) more than 10 percent   (b) 10 percent
     (c) 24   (d) 42

144. Overhead spans of open conductors not over 600 volts, nominal, shall be a minimum of _____ feet above finished grade, sidewalks, or from any platform or projection from which they might be reached where the supply conductors are limited to 150 volts to ground and accessible to pedestrians only.
     (a) 10   (b) 12   (c) 15   (d) 18

145. A recessed incandescent fixture shall be so installed that adjacent combustible material will not be subjected to temperatures in excess of _____°C.
     (a) 75   (b) 90   (c) 125   (d) 150

146. Surface-mounted fluorescent lighting fixtures in clothes closets can be installed on the wall above the door or on the ceiling, provided there is a minimum clearance of _____ inches between the fixture and the nearest point of a storage space.
     (a) 3   (b) 6   (c) 9   (d) 12

147. Feeder conductors supplying fire pump motors and accessory equipment shall be sized no less than _____ percent of the sum of the motor full-load currents as listed in Table 430-148 and/or 430-150, plus 100 percent of the ampere rating of the fire pump accessory equipment.
     (a) 100   (b) 125   (c) 250   (d) 600

148. A fixture that weighs more than 50 pounds is permitted to be supported to an outlet box or fitting that is designed and listed for the weight of the fixture.
     (a) True   (b) False

149. Wiring from emergency source or emergency distribution overcurrent protection to emergency loads shall _____ other wiring.
    (a) not enter the same raceway or box with   (b) not enter the same cable or cabinet with
    (c) be kept entirely independent of all   (d) all of these

150. Lighting fixtures attached to the suspended ceiling framing shall be secured to the framing member with screws, bolts, rivets, or clips _____ and identified for use with the type of ceiling framing member(s) and fixture(s) involved.
    (a) marked   (b) labeled   (c) identified   (d) listed

151. Class 1 power limited circuits shall be supplied from a source having a rated output of not more than 30 volts. If the voltage rating were 25 volts, the maximum volt-amperes of the circuit would be _____ VA.
    (a) 750   (b) 700   (c) 1,000   (d) 1,200

152. Splices and taps shall not be located within lighting fixture _____ .
    (a) arms or stems   (b) bases or screw shells   (c) a and b   (d) a or b

153. Where practical, a separation of at least _____ feet shall be maintained between open conductors of communication systems on buildings and lightning conductors.
    (a) 6   (b) 8   (c) 10   (d) 12

154. • An 800-ampere fuse rated 600 volt _____ on a 250-volt system.
    (a) shall not be used   (b) must be used   (c) can be installed   (d) none of these

155. Conductors shall be _____ to provide ready and safe access in underground and subsurface enclosures into which persons enter for installation and maintenance.
    (a) bundled   (b) tied together   (c) color coded   (d) racked

156. A night club lighting dimmer installed in an ungrounded conductor shall have overcurrent protection rated at no more than _____ percent of its own rating.
    (a) 50   (b) 70   (c) 80   (d) 125

157. Any pit or depression below a garage floor level shall be considered to be a Class I, Division _____ location up to floor level.
    (a) 1   (b) 2   (c) 3   (d) not classified

158. _____ conductors shall be used for wiring on fixture chains and other movable parts.
    (a) Solid   (b) Covered   (c) Insulated   (d) Stranded

159. Equipment termination provisions can be used with the higher rated conductors at the ampacity of the higher rated conductors, provided the equipment is _____ and identified for use with the higher rated conductors.
    (a) listed   (b) tested   (c) installed   (d) rated

160. Where communications wire and cables are installed in a raceway, the raceway shall be of a type permitted in Chapter 3 and the raceway shall be installed in accordance with Chapter 3 requirements.
    (a) True   (b) False

161. • Working space shall not be used for _____ .
    (a) storage   (b) raceways   (c) lighting   (d) accessibility

162. A bare No. 10 solid copper wire has a cross-section area of _____ square inches.
    (a) 0.008   (b) 0.101   (c) 0.012   (d) 0.106

163. Conductors of light or power may occupy the same enclosure or raceway with conductors of power-limited fire alarm circuits.
    (a) True   (b) False

164. The dedicated equipment space for electrical equipment is required for panelboards measured from the floor to a height of _____ feet above the equipment or to the structural ceiling, whichever is lower.
    (a) 3   (b) 6   (c) 12   (d) 30

165.   The feeder to a swimming pool panelboard at a separate building or structure is permitted to be supplied with any Chapter 3 wiring method, however the feeder must have a separate insulated copper equipment grounding conductor.
(a) True   (b) False

166.   Listed outlet boxes or outlet box systems that are identified for the purpose shall be permitted to support suspended ceiling fans that weigh no more than _____ pounds.
(a) 50   (b) 60   (c) 70   (d) all of these

167.   • Pool-associated motors must be grounded by a minimum No. 12 insulated copper conductor. This grounding conductor must be installed in _____ .
(a) rigid nonmetallic conduit
(b) electrical metallic tubing, where installed on or within buildings
(c) flexible metal conduit
(d) a or b

168.   The top shield installed over all floor-mounted Type FCC cable shall completely _____ all cable runs, corners, connectors, and ends.
(a) cover   (b) encase   (c) protect   (d) none of these

169.   All electric equipment, including power supply cords, used with storable pools shall be protected by _____ .
(a) fuses   (b) circuit breakers   (c) double-insulation   (d) GFCI

170.   Hydromassage bathtub equipment shall be _____ without damaging the building structure or building finish.
(a) readily accessible   (b) accessible   (c) within sight   (d) none of these

171.   • For installations to supply only limited loads of a single branch circuit, the service disconnecting means shall have a rating of not less than _____ amperes.
(a) 15   (b) 20   (c) 25   (d) 30

172.   Interior metal water piping systems must be bonded to the _____ .
(a) grounded conductor at the service
(b) service equipment enclosure
(c) equipment grounding bar or bus at any panelboard within the building
(d) a and b

173.   In general, areas where _____ are handled and stored may present severe corrosive conditions, particularly when wet or damp.
(a) laboratory chemicals and acids   (b) acids and alkali chemicals
(c) acids and water   (d) chemicals and water

174.   A _____ shall be permitted in lieu of a box or terminal fitting at the end of a conduit where the raceway terminates behind an unenclosed switchboard or similar equipment.
(a) bushing   (b) bonding bushing   (c) coupling   (d) connector

175.   The maximum number of bends between pull points cannot exceed _____ degrees, including any offsets.
(a) 320   (b) 270   (c) 360   (d) unlimited

176.   Soft-drawn or medium-drawn copper lead-in conductors for television equipment antenna systems shall be permitted where the maximum span between points of support is less than _____ feet.
(a) 35   (b) 30   (c) 20   (d) 10

177.   Ground-fault protection of equipment shall be provided for solidly grounded wye electrical services of more than 150 volts to ground, but not exceeding 600 volts phase-to-phase for each service disconnecting means rated _____ amperes or more.
(a) 1,000   (b) 1,500   (c) 2,000   (d) 2,500

178.   The outer sheath of copper sheath MI cable provides _____ .
(a) an adequate path for grounding purposes   (b) a moisture seal
(c) mechanical protection   (d) all of these

179. The maximum length of an unprotected feeder tap conductor (indoors) is _____ feet.
(a) 15 (b) 20 (c) 50 (d) 100

180. • Insulated conductors used in wet locations shall be _____ .
(a) moisture-impervious metal-sheathed
(b) RHW, TW, THW, THHW, THWN, XHHW
(c) listed for wet locations
(d) any of these

181. • A disconnecting means is required to disconnect the _____ from all ungrounded supply conductors.
(a) motor (b) motor or controller (c) controller (d) motor and controller

182. The receiving station outdoor antenna conductor (where the span is 75 feet) shall be at least a No. _____ if a copper-clad steel conductor is used.
(a) 10 (b) 12 (c) 14 (d) 17

183. In network-powered broadband communications systems, all separate electrodes can be bonded together with a minimum size jumper of No. _____ copper.
(a) 10 (b) 8 (c) 6 (d) 4

184. When conduit nipples having a maximum length not exceeding 24 inches are installed between boxes:
I. The nipple can be filled to 75 percent.   II. Section 310-15(b)(2) adjustment factors do not apply
III. Section 310-15(b)(2) adjustment factors apply   IV. The nipple can be filled to 60 percent.
(a) I and II (b) II and IV (c) III and IV (d) I and II

185. Where the calculated number of conductors, all of the same size, includes a decimal fraction, the next higher whole number shall be used where this decimal is _____ or larger.
(a) 0.5 (b) 0.08 (c) 0.008 (d) 0.8

186. A 250 kcmil bare cable has a conductor diameter of _____ inch.
(a) 0.557 (b) 0.755 (c) 0.575 (d) 0.690

187. Service laterals that are not encased in concrete and buried 18 inches or more below grade shall have their location identified by a warning ribbon placed in the trench at least _____ inches above the underground installation.
(a) 6 (b) 12 (c) 18 (d) none of these

188. The alternating current resistance of 1,000 feet of No. 4/0 aluminum would have the same resistance as 1,000 feet of _____ copper when installed in a steel raceway.
(a) No. 1 (b) No. 1/0 (c) No. 2/0 (d) 250 kcmil

189. Metal raceways, cable armor, boxes, cable sheathing, cabinets, elbows, couplings, fittings, supports, and support hardware shall be of materials suitable for _____ .
(a) corrosive locations
(b) wet locations
(c) the environment in which they are to be installed
(d) none of these

190. Where the opening to an outlet, junction, or switch point is less than 8 inches in any dimension, each conductor shall be long enough to extend at least ___ inches outside the opening of the enclosure.
(a) 0 (b) 3 (c) 6 (d) 12

191. The alternating current resistance for 1,000 feet of No. 4/0 aluminum conductor in a steel raceway is _____ ohm.
(a) 0.063 (b) 0.10 (c) 10.0 (d) 0.11

192. The voltage at the motor terminals, when the motor is operating at 115 percent of the full-load current rating of the motor shall not drop more than ___ percent below the voltage rating of the motor.
(a) 5 (b) 10 (c) 15 (d) any of these

193.   The transmitter enclosure of a radio or TV station shall have interlocks to disconnect the power to the transmitter in the enclosure access door if the voltage between conductors is over _____ volts.

(a) 150   (b) 250   (c) 350   (d) 480

194.   Coaxial cable is permitted to be placed in a raceway, compartment, outlet box, or junction box with the conductors of light or power circuits, or Class 1 circuits when _____ .

(a) installed in rigid metal conduit

(b) separated by a permanent barrier

(c) insulated

(d) none of these

195.   Motors shall be located so that adequate _____ is provided and so that maintenance, such as lubrication of bearings and replacing of brushes, can be readily accomplished.

(a) space   (b) ventilation   (c) protection   (d) all of these

196.   Motor control circuit transformers rated less than 2 amperes can have the primary protection device set no more than _____ percent of the rated primary current rating.

(a) 150   (b) 200   (c) 400   (d) 500

197.   A Group G atmosphere contains combustible dusts such as flour, grain, wood, plastic, and other chemicals.

(a) True   (b) False

198.   The maximum load permitted on a heating element in a pool heater shall not exceed _____ amperes.

(a) 20   (b) 35   (c) 48   (d) 60

199.   A $1\frac{1}{2}$ inch rigid metal conduit nipple with three conductors can be filled to an area of _____ square inches.

(a) 0.88   (b) 1.07   (c) 1.24   (d) 1.34

200.   Where nipples 24 inches or less are used, the ampacity adjustment factor for more than three current-carrying conductors _____ to this condition.

(a) does not apply   (b) shall be applied

Unit 12

Delta/Delta and Delta/Wye Transformer Calculations

OBJECTIVES

After reading this unit, the student should be able to briefly explain the following concepts:

Part A – Delta/Delta Transformers	Delta transformer sizing	Phase current
Current flow	Delta panel schedule in kVA	Wye phase versus line wye
Delta transformer voltage	Delta panelboard and conductor sizing	transformer loading and balancing
Delta high-leg	Delta neutral current	Wye transformer sizing
Delta primary and secondary	Delta maximum unbalanced load	Wye panel schedule in kVA
line currents	Delta/delta example	Wye panelboard and conductor sizing
Delta primary or secondary phase	**Part B – Delta/Wye Transformers**	Wye neutral current
currents	Wye transformer voltage	Wye maximum unbalanced load
Delta phase versus line	Wye voltage triangle	Delta/wye example
Delta current triangle	Wye transformer current	Delta versus wye
Delta transformer balancing	line current	

After reading this unit, the student should be able to briefly explain the following terms:

Delta connected	Phase load – delta	Wye secondary
kVA rating	Phase load – wye	Winding
Line	Phase voltage	Wye connected
Line current	Ratio	Primary/secondary voltage – ampere
Line voltage	Balanced load	Three-phase load
Phase	Unbalanced load	Single-phase load
Phase current	Delta secondary	

INTRODUCTION

This unit deals with 3-phase delta/delta and delta/wye transformer sizing and balancing. The system voltages used in this Unit were selected because of their common industry configuration. Delta/delta and delta/wye concepts are opposite of each other, and it is almost impossible to remember the differences between the systems. But only a few of the concepts contained in this Unit are ever on an electrician's exam.

DEFINITIONS

Delta Connected

Delta connected means the windings of three single-phase transformers are connected in series to form a closed circuit. A line can be traced from one point of the delta system, through all transformers, then back to the original starting point (series). Many call it a *delta high-leg* system because the voltage from Line 2 to to ground is 208 volts (Figure 12–1).

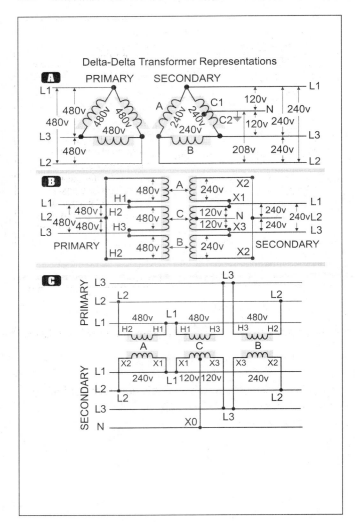

Figure 12–1
Delta-Delta Transformer Representations

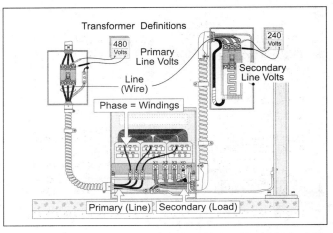

Figure 12–2
Transformer Definitions

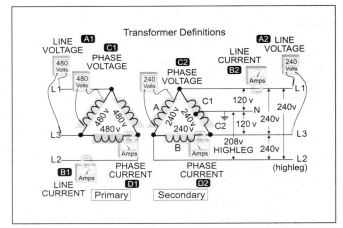

Figure 12–3
Transformer Definitions

kVA Rating

Transformers are rated in kilovolt-ampere (kVA) and are sized according to the VA rating of the loads they supply.

Line

The line is considered the electrical system supply (hot, ungrounded) conductors (Figure 12–2 and Figure 12– 3).

Line Current

The line current for both delta and wye systems is the current on the line (hot, ungrounded) conductors, calculated according to the following formulas (Figure 12–3, B1 and B2):

$$\text{Line Current (single-phase)} = \frac{\text{Line Power}}{\text{Line Voltage, } I_{\text{Line}}} = \frac{VA_{\text{Line}}}{E_{\text{Line}}}$$

$$\text{Line Current (three-phase)} = \frac{\text{Line Power}}{(\text{Line Voltage} \times \sqrt{3}), I_{\text{Line}}} = \frac{VA_{\text{Line}}}{(E_{\text{Line}} \times \sqrt{3})}$$

Line Voltage

The line voltage is the voltage that is measured between any two-line (hot ungrounded) conductors. It is also called line-to-line voltage. Line voltage is greater than phase voltage for wye systems, and line voltage is the same as *phase voltage* for delta systems (Figure 12–3, A1 and A2).

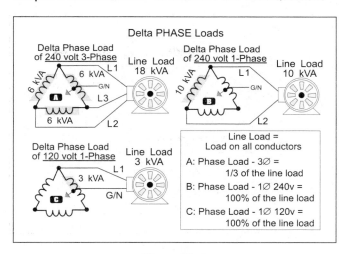

Figure 12–4
Delta Phase Loads

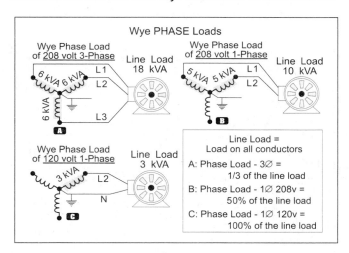

Figure 12–5
Wye Phase Loads

Phase

The *phase* winding is the coil shaped conductors that serve as the primary or secondary of the transformer (Figure 12–3, C1 and C2).

Phase Current

The phase current is the current of the transformer winding. For delta systems, the phase current is less than the line current. For wye systems, the phase current is the same as the line current (Figure 12–3, D1 and D2).

Phase Load Delta

The phase load is the load on the transformer winding (Figure 12–4).

The phase load of a three-phase, 240-volt load = 1/3 of the line load.

The phase load of a single-phase, 240-volt load = line load.

The phase load of a single-phase, 120-volt load = line load.

Phase Load Wye

The phase load is the load on the transformer winding (Figure 12–5).

The phase load of a three-phase, 208-volt load = $^1/_3$ of the line load.

The phase load of a single-phase, 208-volt load = $^1/_2$ of the line load.

The phase load of a single-phase, 120-volt load = line load.

Phase Voltage

The phase voltage is the internal transformer voltage generated across any one winding of a transformer. For a delta secondary, the phase voltage is equal to the line voltage. For wye secondaries, the phase voltage is less than the line voltage (Figure 12–6).

Ratio (Voltage)

The ratio is the relationship between the number of primary winding turns as compared to the number of secondary winding turns. The ratio is a comparison between the primary phase voltage to the secondary phase voltage. For typical delta/delta systems, the ratio is 2:1. For typical wye systems, the ratio is 4:1 (Figure 12–7).

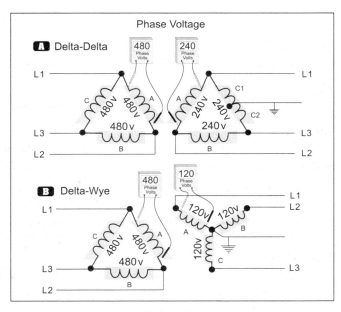

Figure 12–6
Phase Voltage

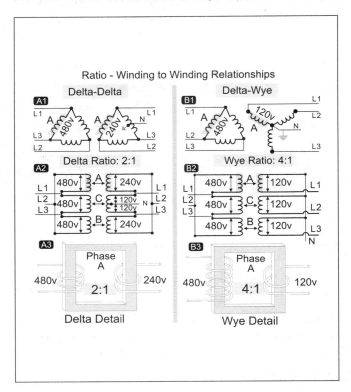

Figure 12–7
Ratio – Winding to Winding Relationships

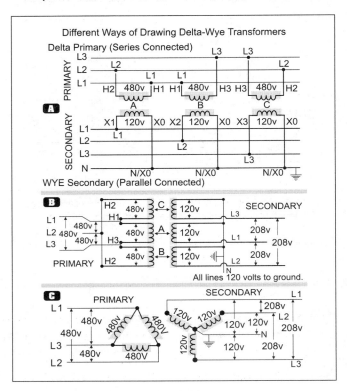

Figure 12–8
Different Ways of Drawing Delta-Wye Transformers

Unbalanced Load (Neutral Current)

The unbalanced load is the load on the secondary grounded (neutral) conductors.

Delta Secondary – Unbalanced Current

The unbalanced load is calculated using:

Line 1 Neutral Current less Line 3 Neutral Current, $I_{Neutral} = I_{Line1} - I_{Line3}$

Note: Line 2 (high-leg) and its voltage to ground is approximately 208 volts, therefore no neutral loads are connected to this line.

Wye Secondary – Unbalanced Current

The unbalanced load is calculated using:

$$I_{Neutral} = \sqrt{(L_1^2 + L_2^2 + L_3^2) - (L_1 \times L_2 + L_2 \times L_3 + L_1 \times L_3)}$$

Winding

The *primary winding* is the winding(s) on the input side of a transformer and the *secondary winding(s)* is the output side of a transformer.

Wye Connected

Wye connected means a connection of three single-phase transformers of the same rated voltage to a common point (neutral) and the other ends are connected to the line conductors (Figure 12–8).

12–1 CURRENT FLOW

When a load is connected to the secondary of a transformer, current will flow through the secondary conductor windings. The current flow in the secondary creates an electromagnetic field that opposes the primary electromagnetic field. The secondary flux lines effectively reduce the strength of the primary flux lines. As a result, less *counter-electromotive force* is generated in the primary winding conductors. With less CEMF to oppose the primary applied voltage, the primary current automatically increases in direct proportion to the secondary current (Figure 12–9).

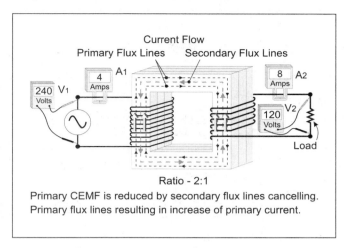

Figure 12–9
Current Flow

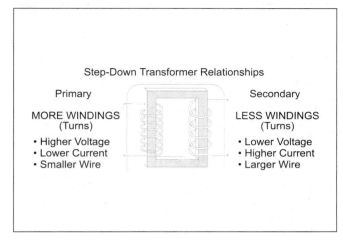

Figure 12–10
Step-Down Transformer Relationships

Note: The primary and secondary line currents are inversely proportional to the voltage ratio of the transformer. This means that the winding with the most number of turns will have a higher voltage and lower current as compared to the winding with the least number of turns, which will have a lower voltage and higher current (Figure 12–10).

The following Tables show the current relationship between kVA and voltage for common size transformers:

Single-Phase Transformers I = VA/E			
kVA Rating	Current at 208 Volts	Current at 240 Volts	Current at 480 Volts
7.5	36 ampere	31 ampere	16 ampere
10	48 ampere	42 ampere	21 ampere
15	72 ampere	63 ampere	31 ampere
25	120 ampere	104 ampere	52 ampere
37.5	180 ampere	156 ampere	78 ampere

Three-Phase Transformers I = VA/(E × $\sqrt{3}$)			
kVA Rating	Current at 208 Volts	Current at 240 Volts	Current at 480 Volts
15	42 ampere	36 ampere	18 ampere
22.5	63 ampere	54 ampere	27 ampere
30	83 ampere	72 ampere	36 ampere
37.5	104 ampere	90 ampere	45 ampere
45	125 ampere	108 ampere	54 ampere
50	139 ampere	120 ampere	60 ampere
75	208 ampere	180 ampere	90 ampere
112.5	313 ampere	271 ampere	135 ampere

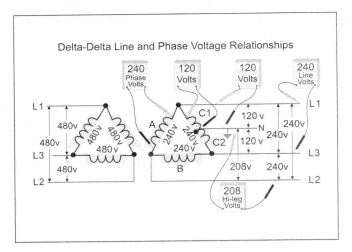

Figure 12–11
Delta-Delta Line and Phase Voltage Relationships

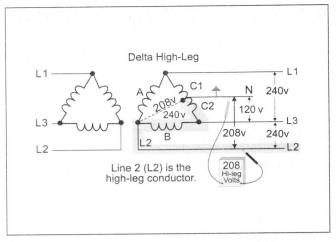

Figure 12–12
Delta High-Leg

PART A – DELTA/DELTA TRANSFORMERS

12–2 DELTA TRANSFORMER VOLTAGE

In a delta configured transformer, the line voltage equals the phase voltage ($E_{Line} = E_{Phase}$) (Figure 12–11).

Primary Delta Voltage

LINE Voltage	PHASE Voltage
L_1 to L_2 = 480 volts	Phase winding A = 480 volts
L_2 to L_3 = 480 volts	Phase winding B = 480 volts
L_3 to L_1 = 480 volts	Phase winding C = 480 volts

Secondary Delta Voltage

LINE Voltage	PHASE Voltage	NEUTRAL Voltage
L_1 to L_2 = 240 volts	Phase winding A = 240 volts	Neutral to L_1 = 120 volts
L_2 to L_3 = 240 volts	Phase winding B = 240 volts	Neutral to L_2 = 208 volts
L_3 to L_1 = 240 volts	Phase winding C = 240 volts	Neutral to L_3 = 120 volts

12–3 DELTA HIGH-LEG

The term high-leg (bastard leg, identified with an orange cover [*Seciton 384-3(e)*]) is used to identify the conductor that has a higher voltage to ground. The high-leg voltage is the vector sum of the voltage of transformer "A" and "C_1", or transformers "B" and "C_2," which equals 120 volts × 1.732 = 208 volts for a 120/240 volt secondary.

Note: The actual voltage is often less than the nominal system voltage because of voltage drop (Figure 12–12).

❏ **High-Leg Voltage**

What is the high-leg voltage if the secondary voltage is 230/115 volts?

(a) **115 volts**   (b) **230 volts**   (c) **199 volts**   (d) **none of these**

   • Answer: (c) 199 volts
   115 volts × 1.732 = 199.18 volts

12–4 DELTA PRIMARY AND SECONDARY LINE CURRENTS

In a delta configured transformer, the line current does not equal the phase current.

The primary or secondary line current of a transformer can be calculated by the formula:

$$I_{Line} = \frac{VA_{Line}}{E_{Line} \times \sqrt{3}}$$

❏ Primary Line Current

What is the primary line current for a 150 kVA, 480- to 240/120-volt, 3-phase transformer (Figure 12–13, Part A)?

(a) 416 ampere   (b) 360 ampere

(c) 180 ampere   (d) 144 ampere

• Answer: (c) 180 ampere

$I_{Line} = VA_{Line}/(E_{Line} \times \sqrt{3})$

$I_{Line} = 150,000 \text{ VA}/(480 \text{ volts} \times 1.732) = 180 \text{ ampere}$

❏ Secondary Line Current

What is the secondary line current for a 150 kVA, 480- to 240-volt, 3-phase transformer (Figure 12–13, Part B)?

(a) 416 ampere   (b) 360 ampere   (c) 180 ampere   (d) 144 ampere

• Answer: (b) 360 ampere

$I_{Line} = VA_{Line}/(E_{Line} \times \sqrt{3})$

$I_{Line} = 150,000 \text{ VA}/(240 \text{ volts} \times 1.732) = 360 \text{ ampere}$

12–5 DELTA PRIMARY OR SECONDARY PHASE CURRENTS

The phase current of a transformer winding is calculated by dividing the phase VA by the phase volts:

$$I_{Phase} = VA_{Phase}/E_{Phase}$$

The phase load of a three-phase, 240-volt load = line load/3.

The phase load of a single-phase, 240-volt load = line load.

The phase load of a single-phase, 120-volt load = line load.

❏ Primary Phase Current

What is the primary phase current for a 150 kVA, 480 to 240/120 volt, 3-phase transformer (Figure 12–14, Part A)?

(a) 416 ampere   (b) 360 ampere   (c) 180 ampere   (d) 104 ampere

• Answer: (d) 104 ampere

$$VA_{Phase} = \frac{150,000 \text{ VA}}{3 \text{ Phases}} = 50,000 \text{ VA}$$

$$I_{Phase} = \frac{VA_{Phase}}{E_{Phase}}$$

$I_{Phase} = 50,000 \text{ VA}/480 \text{ volts} = 104 \text{ ampere}$

❏ Secondary Phase Current

What is the secondary phase current for a 150 kVA, 480- to 240/120-volt, 3-phase transformer (Figure 12–14, Part B)?

(a) 416 ampere   (b) 360 ampere

(c) 208 ampere   (d) 104 ampere

• Answer: (c) 208 ampere

Phase power = 150,000 VA/3 = 50,000 VA

$$I_{Phase} = \frac{VA_{Phase}}{E_{Phase}} = 50,000 \text{ VA}/240 \text{ volts}$$

$I_{Phase} = 208 \text{ ampere}$

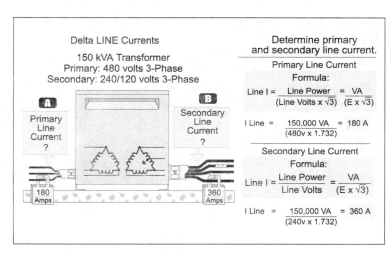

Figure 12–13
Delta Line Currents

12–6 DELTA PHASE VERSUS LINE

Since each line conductor from a delta transformer is actually connected to two transformer windings (phases), the effects of loading on the line (conductors) can be different than on the phase (winding).

❏ **Phase VA (three-phase)**

A 36 kVA, 240-volt, 3-phase load has the following effect on a delta system (Figure 12–15):

LINE: Total line power = 36 kVA
$I_{Line} = VA_{Line}/(E_{Line} \times \sqrt{3})$
$I_{Line} = 36{,}000 \text{ VA}/(240 \text{ volts} \times \sqrt{3})$
$I_{Line} = 87$ ampere, or

$I_{Line} = I_{Phase} \times \sqrt{3}$
$I_{Line} = 50$ ampere $\times 1.732 = 87$ ampere

PHASE: Phase power = 12 kVA (winding)
$I_{Phase} = VA_{Phase}/E_{Phase}$
$I_{Phase} = 12{,}000 \text{ VA}/240 \text{ volts} = 50$ ampere, or

$I_{Phase} = I_{Line}/\sqrt{3}$
$I_{Phase} = 87$ ampere$/1.732 = 50$ ampere

❏ **Phase VA (single-phase) 240 Volt**

A 10 kVA, 240-volt, single-phase load has the following effect on a delta/delta system (Figure 12–16):
LINE: Total line power = 10 kVA
$I_{Line} = VA_{Line}/E_{Line}$
$I_{Line} = 10{,}000 \text{ VA}/240 \text{ volts} = 42$ ampere

PHASE: Phase power = 10 kVA (winding)
$I_{Phase} = VA_{Phase}/E_{Phase}$
$I_{Phase} = 10{,}000 \text{ VA}/240 \text{ volts} = 42$ ampere

❏ **Phase VA (single-phase) 120 Volt**

A 3 kVA, 120-volt, single-phase load has the following effect on the system (Figure 12–17):
LINE: Line power = 3 kVA
$I_{Line} = VA_{Line}/E_{Line}$
$I_{Line} = 3{,}000 \text{ VA}/120 \text{ volts} = 25$ ampere

PHASE: Phase power = 3 kVA
(C_1 or C_2 winding)
$I_{Phase} = VA_{Phase}/E_{Phase}$
$I_{Phase} = 3{,}000 \text{ VA}/120 \text{ volts} = 25$ ampere

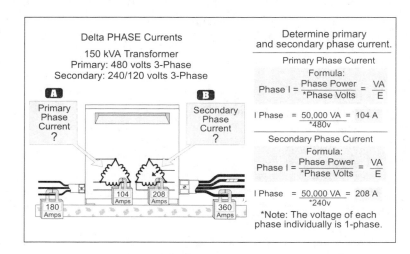

Figure 12–14
Delta Phase Currents

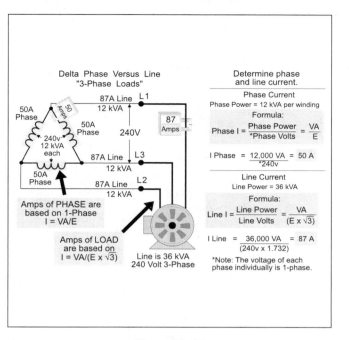

Figure 12–15
Delta Phase versus Line Loads (3-Phase)

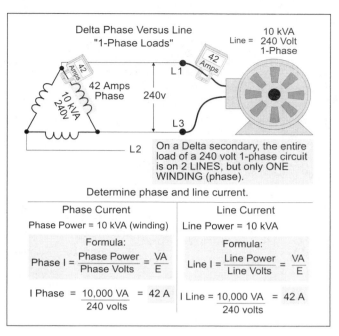

Figure 12–16
Delta Phase versus Line Load (1-Phase)

12–7 DELTA CURRENT TRIANGLE

The 3-phase line and phase current of a delta system are not equal; the difference is the square root of 3 ($\sqrt{3}$).

$$I_{Line} = I_{Phase} \times \sqrt{3} \qquad I_{Phase} = I_{Line} / \sqrt{3}$$

The delta triangle can be used to calculate delta 3-phase line and phase currents. Place your finger over the desired item and the remaining items show the formula to use (Figure 12–18).

12–8 DELTA TRANSFORMER BALANCING

To properly size a delta/delta transformer, the transformer phases (windings) must be balanced. The following steps are used to balance the transformer:

Step 1: ➤ Determine the VA rating of all loads.

Step 2: ➤ Balance 3-phase loads: $1/3$ on Phase A, $1/3$ on Phase B, and $1/3$ on Phase C.

Step 3: ➤ Balance single-phase, 240-volt loads (largest to smallest): 100 percent on Phase A or 100 percent on Phase B. It is permissible to place some 240-volt, single-phase load on Phase C when necessary for balance.

Step 4: ➤ Balance the 120-volt loads (largest to smallest): 100 percent on C1 or 100 percent on C2.

❏ **Delta Transformer Balancing**

Balance and size a 480 to 240/120 volt 3-phase transformer for the following loads: One 36 kVA 3-phase heat strip, two 10 kVA single-phase 240-volt loads, and three 3 kVA 120-volt loads (Figure 12–19).

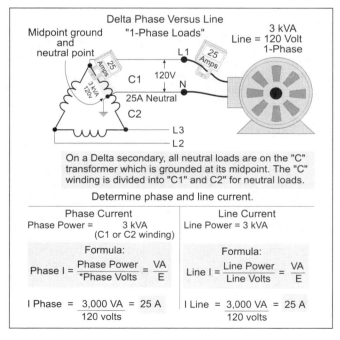

Figure 12–17
Delta Phase versus Line Load (1-Phase)

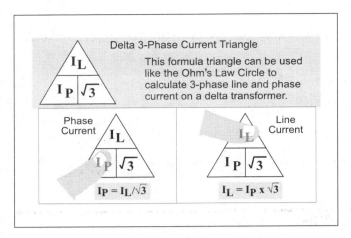

Figure 12–18
Delta 3-Phase Current Triangle

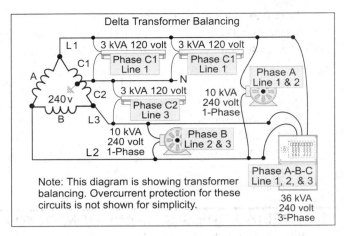

Figure 12–19
Delta Transformer Balancing

Note: For convenience of formatting, the symbol "Ø" is used to represent the word "phase."

	Phase A (L$_1$ and L$_2$)	Phase B (L$_2$ and L$_3$)	C$_1$ (L$_1$)	C$_2$ (L$_3$)	Line Total
One 36 kVA, 240 volt, 3Ø	12 kVA	12 kVA	6 kVA	6 kVA	36,000 VA
Two 10 kVA, 240 volt, 1Ø	10 kVA	10 kVA			20,000 VA
Three 3 kVA, 120 volt, 1Ø			6 kVA *	3 kVA *	9,000 VA
	22 kVA	22 kVA	12 kVA	9 kVA	65,000 VA

*Indicates neutral (120-volt) loads.

12–9 DELTA TRANSFORMER SIZING

Once you balance the transformer, size the transformer according to the load of each phase. The "C" transformer must be sized using two times the highest of "C$_1$" or "C$_2$." The "C" transformer is actually a single unit. If one side has a larger load, that side determines the transformer size.

❏ Delta Transformer Sizing

What size 480- to 240/120-volt, 3-phase transformer is required for the following loads: one 36 kVA 3-phase heat strip, two 10 kVA single-phase, 240-volt loads, and three 3 kVA 120-volt loads?

(a) three single-phase 25 kVA transformers   (b) one 3-phase 75 kVA transformer

(c) a or b   (d) none of these

• Answer: (c) a or b

Phase winding A = 22 kVA

Phase winding B = 22 kVA

Phase winding C = (12 kVA of C$_1$ × 2) = 24 kVA

12–10 DELTA PANEL SCHEDULE IN kVA

When balancing a panelboard in kVA, be sure that 3-phase loads are balanced $1/3$ on each line and 240-volt, single-phase loads are balanced on each line.

❏ **Delta Panel Schedule in kVA**

Balance a 240/120-volt, 3-phase panelboard for the following loads in kVA: one 36 kVA 3-phase heat strip, two 10 kVA single-phase, 240-volt loads, and three 3 kVA 120-volt loads.

	Line 1	Line 2	Line 3	Line Total
36 kVA, 240 volt, 3Ø	12 kVA	12 kVA	12 kVA	36,000 VA
10 kVA, 240 volt, 1Ø	5 kVA	5 kVA		10,000 VA
10 kVA, 240 volt, 1Ø		5 kVA	5 kVA	10,000 VA
Three 3 kVA, 120 volt, 1Ø	6 kVA *		3 kVA *	9,000 VA
	23 kVA	22 kVA	20 kVA	65,000 VA

*Indicates neutral (120-volt) loads.

12–11 DELTA PANELBOARD AND CONDUCTOR SIZING

When selecting the sizing of *panelboards and conductors*, you must balance the line loads in ampere.

❏ **Delta Conductors Sizing**

Balance a 240/120-volt, 3-phase panelboard for the following loads in ampere: one 36 kVA 3-phase heat strip, two 10 kVA single-phase, 240-volt loads, and three 3 kVA 120-volt loads.

	Line 1	Line 2	Line 3	Line Ampere
36 kVA, 240 volt, 3Ø	87 Amperes	87 Amperes	87 Amperes	36,000 VA/(240 volts × 1.732)
10 kVA, 240 volt, 1Ø	42 Amperes	42 Amperes		10,000 VA/240 volts
10 kVA, 240 volt, 1Ø		42 Amperes	42 Amperes	10,000 VA/240 volts
Three 3 kVA, 120 volt, 1Ø	50 Amperes *		25 Amperes *	3,000 VA/120 volts
	179 Amperes	171 Amperes	154 Amperes	

*Indicates neutral (120-volt) loads.

Why balance the panel in ampere? Why not take the VA per phase and divide by phase voltage?

   • Answer: Line current of a 3-phase load is calculated by the formula

   $I_L = VA/(E_{Line} \times \sqrt{3})$

   $I_L = 36,000 \text{ VA}/(240 \text{ volts} \times 1.732) = 87$ ampere per line.

   Note: If you took the per line power of 12,000 VA and divided by one line voltage of 120 volts, you come up with an incorrect line current of 12,000 VA/120 volts = 100 ampere.

12–12 DELTA NEUTRAL CURRENT

The neutral current for a delta secondary is calculated as Line$_1$ neutral current less Line$_3$ neutral current.

❏ **Delta Neutral Current**

What is the neutral current for the following loads: one 36 kVA 3-phase heat strip, two 10 kVA, single-phase 240-volt loads, and three 3 kVA 120-volt loads?

In our continuous example, Line 1 neutral current = 50 ampere and Line 3 neutral current = 25 ampere.

(a) 0 ampere   (b) 25 ampere   (c) 50 ampere   (d) none of these

   • Answer: (b) 25 ampere
   Neutral current = 50 ampere less 25 ampere
   Neutral current = 25 ampere

	Line 1	Line 2	Line 3	Ampere Calculation
36 kVA, 240 volts, 3Ø	87 Amperes	87 Amperes	87 Amperes	36,000 VA/(240 volts × 1.732)
10 kVA, 240 volts, 1Ø	42 Amperes	42 Amperes		10,000 VA/240 volts
10 kVA, 240 volts, 1Ø		42 Amperes	42 Amperes	10,000 VA/240 volts
Three 3 kVA, 120 volts, 1Ø	50 Amperes *		25 Amperes *	3,000 VA/120 volts

*Indicates neutral (120-volt) loads.

12–13 DELTA MAXIMUM UNBALANCED LOAD

The *maximum unbalanced load* (neutral) is the largest phase 120-volt neutral current. The neutral current is the actual current on the neutral conductor.

❏ Maximum Unbalanced Load

What is the maximum unbalanced load for the following loads: one 36 kVA 3-phase heat strip, two 10 kVA single-phase, 240-volt loads, and three 3 kVA 120-volt loads?

| (a) 0 ampere | (b) 25 ampere | (c) 50 ampere | (d) none of these |

   • Answer: (c) 50 ampere

   Maximum unbalanced current equals the line with the largest neutral current.

	Line 1	Line 2	Line 3	Ampere Calculation
36 kVA 240 volt, 3Ø	87 Amperes	87 Amperes	87 Amperes	36,000 VA/(240 volts × 1.732)
10 kVA 240 volt, 1Ø	42 Amperes	42 Amperes		10,000 VA/240 volts
10 kVA 240 volt, 1Ø		42 Amperes	42 Amperes	10,000 VA/240 volts
Three 3 kVA 120 volt, 1Ø	50 Amperes*		25 Amperes*	3,000 VA/120 volts

*Indicates neutral (120-volt) loads.

12–14 DELTA/DELTA EXAMPLE

One – Dishwasher 4.5 kW, 120v

One – A/C 10 horsepower 240 volt, 3Ø

One – Heaters 18 kW 240 volt 3Ø

Two – Ranges 14 kW, 240 volt 1Ø

Two – 3 horsepower motors, 240 volt, 1Ø

One – Water heater 10 kW, 240 volt, 1Ø

Eight – Lighting circuit 1.5 kW 120 volt

Motor VA

VA (single-phase) = Table Volts × Table Ampere, Table 430-148

3 horsepower VA = 230 volts × 17 ampere = 3,910 VA, Table 430-150

VA (3-phase) = Table Volts × Table Ampere × $\sqrt{3}$, Table 430-150

10 horsepower VA = 230 volts × 28 ampere × 1.732 = 11,154

Note: Some exam testing agencies use 240 volts instead of the table volts.

	Phase A (L₁ and L₂)	Phase B (L₂ and L₃)	C₁ (L₁)	C₂ (L₃)	Line Total
Heat 18 kW (omit A/C)	6,000 VA	6,000 VA	3,000 VA	3,000 VA	18,000 VA
Ranges 14 kW, 240 volt, 1Ø	14,000 VA	14,000 VA			28,000 VA
Water heater, 10 kW 240 volt, 1Ø	10,000 VA				10,000 VA
3 horsepower 240 volt, 1Ø		3,910 VA			3,910 VA
3 horsepower 240 volt, 1Ø		3,910 VA			3,910 VA
Dishwasher 4.5 kW, 120 volt			4,500 VA *		4,500 VA
Lighting (8 – 1.5 kW), 120 volt	_____	_____	4,500 VA *	7,500 VA*	12,000 VA
	30,000 VA	27,820 VA	12,000 VA	10,500 VA	80,320 VA

*Indicates neutral (120-volt) loads.

Note: Phase totals (30,000 VA, 27,820 VA, 22,500 VA) should add up to the Line total (80,320 VA). This is done as a check to make sure all items have been accounted for and added correctly.

10 horsepower VA = Table Volts × Table Ampere × 1.732

VA = 230 volts × 28 ampere × 1.732 = 11,154 VA (omit)

➡ Transformer Size

What size transformers are required?

| (a) three 30 kVA single-phase transformers | (b) one 90 kVA 3-phase transformer |
| (c) a or b | (d) none of these |

   • Answer: (c) a or b

Phase A = 30 kVA
Phase B = 28 kVA
Phase C = 24 kVA (12 kVA × 2)

➤ High-Leg Voltage

What is the high-leg voltage to the ground?

(a) 120 volts   (b) 208 volts   (c) 240 volts   (d) none of these

• Answer: (b) 208 volts

The vector sum of "A" and "C1" or transformers "B" and "C2," = 120 volts × 1.732 = 208 volts.

➤ Neutral kVA

What is the maximum kVA on the neutral?

(a) 3 kVA   (b) 6 kVA   (c) 7.5 kVA   (d) 9 kVA

• Answer: (d) 9 kVA

If Phase C2 loads are not on, then phase C1 neutral loads would impose a 9 kVA load to neutral. (4.5 kW + 4.5 kW). This does not include the 6 kVA on transformer "C" from the heat load since there are no neutrals involved.

➤ Maximum Unbalanced Current

What is the maximum unbalanced load on the neutral?

(a) 25 ampere   (b) 50 ampere   (c) 75 ampere   (d) 100 ampere

• Answer: (c) 75 ampere

I = P/E

I = 9,000 VA/120 volts = 75 ampere

Note: The neutral must be sized to the carry maximum unbalanced neutral current, which in this case is 75 ampere.

➤ Neutral Current

What is the current on the neutral?

(a) 0 ampere   (b) 13 ampere   (c) 25 ampere   (d) none of these

• Answer: (b) 13 ampere

C1 = 9,000 VA/120 volts = 75 ampere
C2 = 7,500 VA/120 volts = 62.5 ampere
Neutral current = 75 ampere less 62.5 ampere = 12.5 ampere

➤ Phase VA Load (single-phase)

What is the phase load of each 3 horsepower 240-volt, single-phase motor?

(a) 1,855 VA   (b) 3,910 VA   (c) 978 VA   (d) none of these

• Answer: (b) 3,910 VA

VA = 230 volts × 28 ampere = 3,910 VA

On a delta system, a 240-volt, single-phase load is 100 percent on one transformer.

➤ Phase VA Load (three-phase)

What is the phase load for the 3-phase, 10 horsepower motor?

(a) 11,154 VA   (b) 3,718 VA   (c) 5,577 VA   (d) none of these

• Answer: (b) 3,718 VA

VA = Volts × Ampere × 1.732
VA = 230 volts × 28 ampere × 1.732, VA = 11,154 VA
11,154 VA/3 phases = 3,718 VA per phase

➤ High-Leg Conductor Size

If only the 3-phase 18 kW load is on the high-leg, what is the current on the high-leg conductor?

(a) 25 ampere   (b) 43 ampere   (c) 73 ampere   (d) 97 ampere

• Answer: (b) 43 ampere

To calculate current on any conductor:
$$I = P/(E \times \sqrt{3})$$
I = 18,000 VA/(240 volts × 1.732) = 43 ampere

➤ **Voltage Ratio**

What is the phase voltage ratio of the transformer?

(a) 4:1   (b) 1:4   (c) 1:2   (d) 2:1

    • Answer: (d) 2:1

    480 primary phase volts to 240 secondary phase volts

❏ **Panel Loading – VA**

Balance the loads on the panelboard in VA.

	Line 1	Line 2	Line 3	Line Total
Heat 18 kW 240 volts, 3Ø	6,000 VA	6,000 VA	6,000 VA	18,000 VA
Ranges 14 kW 240 volts, 1Ø	7,000 VA	7,000 VA		14,000 VA
Ranges 14 kW 240 volts, 1Ø		7,000 VA	7,000 VA	14,000 VA
Water heater 240 volts, 1Ø	5,000 VA	5,000 VA		10,000 VA
3 horsepower 240 volts, 1Ø Motor		1,955 VA	1,955 VA	3,910 VA
3 horsepower 240 volts, 1Ø Motor		1,955 VA	1,955 VA	3,910 VA
Dishwasher 1Ø 120V	4,500 VA*			4,500 VA
Lighting (8 circuits) 1Ø 120V	4,500 VA*		7,500 VA *	4,500 VA
	27,000 VA	28,910 VA	24,410 VA	80,320 VA

* Indicates neutral (120-volt) loads.

❏ **Panel Sizing – Ampere**

Balance the loads from the previous example on the panelboard in ampere.

	Line 1	Line 2	Line 3	Ampere Calculation
Heat 18 kW 240 volts, 3Ø	43 amp	43 amp	43 amp	18,000 VA/(240 volts × 1.732)
Ranges 14 kW 240 volts, 1Ø	58 amp	58 amp		14,000 VA/240 volts
Ranges 14 kW 240 volts, 1Ø		58 amp	58 amp	14,000 VA/240 volts
Water heater 240 volts, 1Ø	42 amp	42 amp		10,000 VA/240 volts
3 horsepower 240 volts, 1Ø Motor		17 amp	17 amp	FLC
3 horsepower 240 volts, 1Ø Motor		17 amp	17 amp	FLC
Dishwasher 1Ø 120 volts	38 amp*			4,500 VA/120 volts
Lighting (8 circuits) 1Ø 120 volts	38 amp*		63 amp*	7,500 VA/120 volts
	219 amp	235 ampere	198 amp	

*Indicates neutral (120-volt) loads.

PART B – DELTA/WYE TRANSFORMERS

12–15 WYE TRANSFORMER VOLTAGE

In a wye configured transformer, the line voltage does not equal the phase voltage (Figure 12–20).

Reminder: In a delta configured transformer, line current does not equal phase current.

Primary Voltage (Delta)

LINE Voltage	**PHASE Voltage**
L_1 to L_2 = 480 volts	Phase Winding A = 480 volts
L_2 to L_3 = 480 volts	Phase Winding B = 480 volts
L_3 to L_1 = 480 volts	Phase Winding C = 480 volts

Secondary Voltage (Wye)

LINE Voltage	**PHASE Voltage**	**NEUTRAL Voltage**
L_1 to L_2 = 208 volts	Phase Winding A = 120 volts	Neutral to L_1 = 120 volts
L_2 to L_3 = 208 volts	Phase Winding B = 120 volts	Neutral to L_2 = 120 volts
L_3 to L_1 = 208 volts	Phase Winding C = 120 volts	Neutral to L_3 = 120 volts

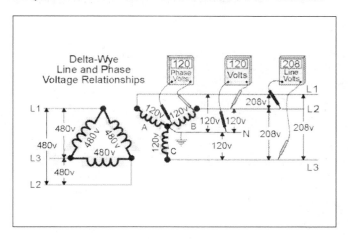

Figure 12–20
Delta-Wye Line and Phase Voltage Relationships

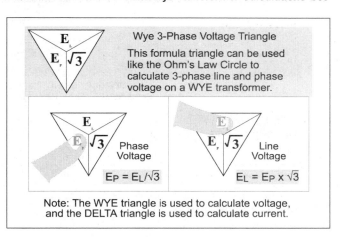

Figure 12–21
Wye 3-Phase Voltage Triangle

12–16 WYE LINE AND PHASE VOLTAGE TRIANGLE

The line voltage and phase voltage of a wye system are not the same. They differ by a factor of $\sqrt{3}$ (1.732).

$$\text{Line Voltage } I_{Line} = E_{Phase} \times \sqrt{3} \qquad\qquad \text{Phase Voltage } I_{Phase} = E_{Line}/\sqrt{3}$$

The wye voltage triangle (Figure 12–21) can be used to calculate wye 3-phase line and phase voltage. Place your finger over the desired item, and the remaining items show the formula to use. These formulas can be used when either one of the phase or line voltages is known.

12–17 WYE TRANSFORMERS CURRENT

In a wye configured transformer, the 3-phase and single-phase 120-volt line current equals the phase current ($I_{Phase} = I_{Line}$) (Figure 12–22).

12–18 WYE LINE CURRENT

The line current of both delta and wye transformers can be calculated by the following formula:

$$I_{Line} = VA_{Line}/(E_{Line} \times \sqrt{3})$$

❏ **Primary Line Current Question**

What is the primary line current for a 150 kVA, 480- to 208Y/120-volt, 3-phase transformer (Figure 12–23, Part A)?

(a) 416 ampere   (b) 360 ampere
(c) 180 ampere   (d) 144 ampere
   • Answer: (c) 180 ampere
   $I_{Line} = VA_{Line}/(E_{Line} \times \sqrt{3}) = 150,000 \text{ VA}/(480 \text{ volts} \times 1.732) = 180$ ampere

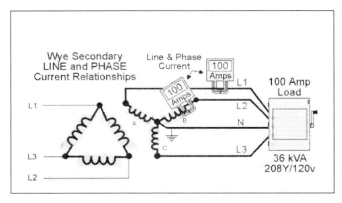

Figure 12–22
Wye Secondary Line and Phase Current Relationships

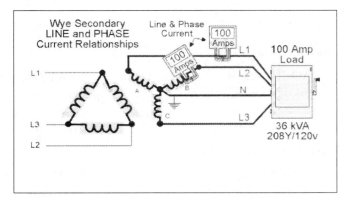

Figure 12–23
Delta-Wye Line Currents

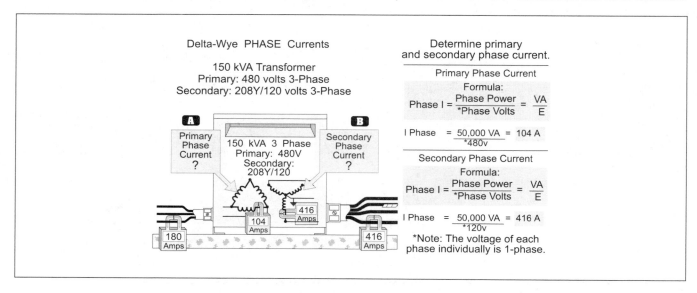

Figure 12–24
Delta-Wye Phase Currents

❏ Secondary Line Current Question

What is the secondary line current for a 150 kVA, 480- to 208Y/120-volt, 3-phase transformer (Figure 12–23, Part B)?

(a) 416 ampere   (b) 360 ampere   (c) 180 ampere   (d) 144 ampere

• Answer: (a) 416 ampere

I_{Line} = VA_{Line}/(E_{Line} × $\sqrt{3}$) = 150,000 VA/(208 volts × 1.732) = 416 ampere

12–19 WYE PHASE CURRENT

The phase current of a wye configured transformer winding is the same as the line current. The phase current of a delta configured transformer winding is less than the line current by $\sqrt{3}$ (1.732) .

❏ Primary Phase Current Question

What is the primary phase current for a 150 kVA, 480- to 208Y/120-volt, 3-phase transformer (Figure 12–24, Part A)?

(a) 416 ampere   (b) 360 ampere   (c) 180 ampere   (d) 104 ampere

• Answer: (d) 104 ampere

$$I_{Line} = \frac{VA_{Line}}{E_{Line} \times \sqrt{3}} = 150,000 \text{ VA}/(480 \text{ volts} \times 1.732) = 180 \text{ ampere}$$

$$I_{Phase} = \frac{VA_{Phase}}{E_{Phase}} = 50,000 \text{ VA}/480 \text{ volts} = 104 \text{ ampere, or } I_{Phase} = I_{Line}/\sqrt{3} = 180 \text{ ampere}/1.732 = 104 \text{ ampere}$$

❏ Secondary Phase Current Question

What is the secondary phase current for a 150 kVA, 480- to 208Y/120-volt, 3-phase transformer (Figure 13-24, Part B)?

(a) 416 ampere   (b) 360 ampere   (c) 208 ampere   (d) 104 ampere

• Answer: (a) 416 ampere

$$I_{Line} = \frac{VA_{Line}}{E_{Line} \times \sqrt{3}} = 150,000 \text{ VA}/(208 \text{ volts} \times 1.732) = 416 \text{ ampere, or}$$

$$I_{Phase} = \frac{VA_{Phase}}{E_{Phase}} = 50,000 \text{ VA}/120 \text{ volts} = 416 \text{ ampere}$$

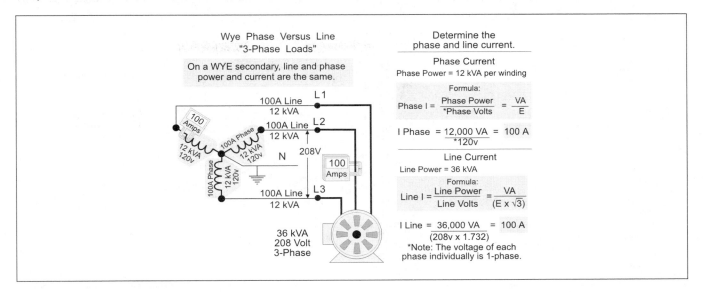

Figure 12–25
Wye Phase versus Line Load (3-Phase)

12–20 WYE PHASE VERSUS LINE CURRENT

Since each line conductor from a wye transformer is connected to a different transformer winding (phase), the effects of 3-phase loading on the line are the same as the phase (Figure 12–25).

❏ **Phase VA Load (three-phase)**

A 36 kVA, 208-volt, 3-phase load has the following effect on the system:

LINE: Line power = 36 kVA

$I_{Line} = VA_{Line}/(E_{Line} \times \sqrt{3})$

$I_{Line} = 36,000 \ VA/(208 \ volts \times \sqrt{3}) = 100 \ ampere$

PHASE: Phase power = 12 kVA (any winding)

$I_{Phase} = VA_{Phase}/E_{Phase}$

$I_{Phase} = 12,000 \ VA/120 \ volts = 100 \ ampere$

❏ **Phase VA Load (single-phase) 208 Volt**

A 10 kVA, 208-volt, single-phase load has the following effect on the system (Figure 12–26):

LINE: Line power = 10 kVA

$I_{Line} = VA_{Line}/E_{Line}$

$I_{Line} = 10,000 \ VA/208 \ volts = 48 \ ampere$

PHASE: Phase power = 10 kVA (winding)

$I_{Phase} = VA_{Phase}/E_{Phase}$

$I_{Phase} = 5,000 \ VA/120 \ volts = 42 \ ampere$

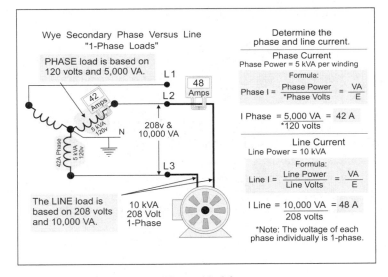

Figure 12–26
Wye Phase versus Line Load (1-Phase)

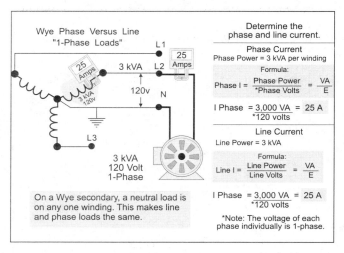

Figure 12–27
Wye Phase versus Line Load (1-Phase)

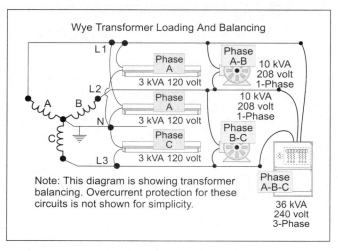

Figure 12–28
Wye Transformer Loading and Balancing

❏ **Phase VA Load (single-phase) 120 Volt**

A 3 kVA, 120-volt, single-phase load has the following effect on the system (Figure 12–27):

LINE: Line power = 3 kVA

$I_{Line} = VA_{Line}/E_{Line}$

$I_{Line} = 3,000$ VA/120 volts = 25 ampere

PHASE: Phase power = 3 kVA (any winding)

$I_{Phase} = VA_{Phase}/E_{Phase}$

$I_{Phase} = 3,000$ VA/120 volts = 25 ampere

12–21 WYE TRANSFORMER LOADING AND BALANCING

To properly size a delta/wye transformer, the secondary transformer phases (windings) or the line conductors must be balanced. Balancing the panel (line conductors) is identical to balancing the transformer for wye configured transformers. The following steps should be helpful to balance the transformer:

Step 1: ➥   Determine the VA rating of all loads.

Step 2: ➥   Balance 3-phase loads: $^1/_3$ on Phase A, $^1/_3$ on Phase B, and $^1/_3$ on Phase C.

Step 3: ➥   Balance single-phase, 208-volt loads (largest to smallest): 50 percent on each phase (A to B, B to C, or A to C).

Step 4: ➥   Balance the 120-volt loads (largest to smallest): 100 percent on any phase.

❏ **Transformer Loading and Balancing**

Balance and size a 480- to 208Y/120-volt, 3-phase phase transformer for the following loads: One 36 kVA, 3-phase heat strip, two 10 kVA, single-phase, 208-volt loads, and three 3 kVA, 120-volt loads (Figure 12–28).

	Phase A (L₁)	Phase B (L₂)	Phase C (L₃)	Line Total
36 kVA, 208 volt, 3Ø	12 kVA	12 kVA	12 kVA	36 kVA
10 kVA, 208 volt, 1Ø	5 kVA	5 kVA		10 kVA
10 kVA, 208 volt, 1Ø		5 kVA	5 kVA	10 kVA
Three 3 kVA, 120 volt, 1Ø	6 kVA *		3 kVA *	9 kVA
	23 kVA	22 kVA	20 kVA	65 kVA

* Indicates neutral (120-volt) loads.

12–22 WYE TRANSFORMER SIZING

Once you balance the transformer, you can size the transformer according to the load of each phase.

❏ **Transformer Sizing Example**

What size 480- to 208Y/120-volt, 3-phase, phase transformer is required for the following loads: one 36 kVA, 3-phase heat strip, two 10 kVA single-phase, 208-volt loads, and three 3 kVA 120-volt loads?

(a) three single-phase 25 kVA transformers   (b) one 3-phase 75 kVA transformer

(c) a or b   (d) none of these

• Answer: (c) a or b

Phase A = 23 kVA

Phase B = 22 kVA

Phase C = 20 kVA

	Phase A (L₁)	Phase B (L₂)	Phase C (L₃)	Line Total
36 kVA, 208 volt, 3Ø	12 kVA	12 kVA	12 kVA	36 kVA
10 kVA, 208 volt, 1Ø	5 kVA	5 kVA		10 kVA
10 kVA, 208 volt, 1Ø		5 kVA	5 kVA	10 kVA
Three 3 kVA, 120 volt, 1Ø	6 kVA *		3 kVA *	9 kVA
	23 kVA	22 kVA	20 kVA	65 kVA

* Indicates neutral (120-volt) loads.

12–23 WYE PANEL SCHEDULE IN kVA

When balancing a panelboard in kVA, be sure that 3-phase loads are balanced $1/3$ on each line and 208-volt single-phase loads are balanced $1/2$ on each line.

❏ **Panel Schedule, kVA**

Balance and size a 208Y/120-volt, 3-phase phase panelboard in kVA for the following loads: one 36 kVA, 3-phase heat strip, two 10 kVA, single-phase, 208-volt loads, and three 3 kVA, 120-volt loads.

	Line 1	Line 2	Line 3	Line Total
36 kVA, 208 volt, 3Ø	12 kVA	12 kVA	12 kVA	36,000 VA
10 kVA, 208 volt, 1Ø	5 kVA	5 kVA		10,000 VA
10 kVA, 208 volt, 1Ø		5 kVA	5 kVA	10,000 VA
Two 3 kVA, 120 volt, 1Ø	6 kVA *			6,000 VA
One 3 kVA, 120 volt, 1Ø			3 kVA *	3,000 VA
	23 kVA	22 kVA	20 kVA	65,000 VA

* Indicates neutral (120-volt) loads.

12–24 WYE PANELBOARD AND CONDUCTOR SIZING

When selecting and sizing the panelboard and conductors, we must balance the line loads in ampere.

❏ **Panelboard Conductors Sizing**

Balance and size a 208Y/120-volt, 3-phase panelboard in ampere for the following loads: one 36 kVA, 3-phase heat strip, two 10 kVA, single-phase, 208-volt loads, and three 3 kVA, 120-volt loads.

	Line 1	Line 2	Line 3	Ampere Calculation
36 kVA, 208 volt, 3Ø	100 amp	100 amp	100 amp	36,000/(208 volts × 1.732)
10 kVA, 208 volt, 1Ø	48 amp	48 amp		10,000 VA/208 volts
10 kVA, 208 volt, 1Ø		48 amp	48 amp	10,000 VA/208 volts
Two 3 kVA, 120 volt, 1Ø	50 amp*			6,000 VA/120 volts
One 3 kVA, 120 volt, 1Ø	_____	_____	25 amp*	3,000 VA/120 volts
	198 amp	196 amp	173 amp	

*Indicates neutral (120-volt) loads.

12–25 WYE NEUTRAL CURRENT

To determine the neutral current of a wye system, we use the following formula:

$$I_{Neutral} = \sqrt{(L_1{}^2 + L_2{}^2 + L_3{}^2) - (L_1 \times L_2 + L_2 \times L_3 + L_1 \times L_3)}$$

❏ Neutral Current

Balance and size the neutral current for the following loads: one 36 kVA, 3-phase heat strip, two 10 kVA, single-phase, 208-volt loads, and three 3 kVA, 120-volt loads. L_1 = 50 ampere, L_2 = 0 ampere, L_3 = 25 ampere.

(a) 0 ampere　　　　　(b) 25 ampere　　　　　(c) 43 ampere　　　　　(d) 50 ampere

• Answer: (c) 43 ampere

Based on the previous example.

$$I_{Neutral} = \sqrt{(L_1{}^2 + L_2{}^2 + L_3{}^2) - (L_1 \times L_2 + L_2 \times L_3 + L_1 \times L_3)}$$

$$I_{Neutral} = \sqrt{(50^2 + 0^2 + 25^2) - [(50 \times 0) + 0 \times 25)(50 \times 25)]} = \sqrt{(2,500 + 625) - 1,250} = \sqrt{1,875} = 43.3 \text{ ampere}$$

12–26 WYE MAXIMUM UNBALANCED LOAD

The maximum unbalanced load (neutral) is the largest phase 120-volt neutral current with other lines off.

❏ Maximum Unbalanced Load

Balance and size the maximum unbalanced load for the following loads: one 36 kVA, 3-phase heat strip, two 10 kVA, single-phase, 208-volt loads, and three 3 kVA, 120-volt loads. L1 = 50 ampere, L2 = 0 ampere, L3 = 25 ampere.

(a) 0 ampere　　　　　(b) 25 ampere　　　　　(c) 50 ampere　　　　　(d) none of these

• Answer: (c) 50 ampere

The maximum unbalanced current equals the largest line neutral current.

	Line 1	Line 2	Line 3	Ampere Calculation
36 kVA, 208 volt, 3Ø	100 amp	100 amp	100 amp	36,000/(208 volts × 1.732)
10 kVA, 208 volt, 1Ø	48 amp	48 amp		10,000 VA/208 volts
10 kVA, 208 volt, 1Ø		48 amp	48 amp	10,000 VA/208 volts
Two 3 kVA, 120 volt, 1Ø	50 amp*			6,000 VA/120 volts
One 3 kVA, 120 volt, 1Ø	_____	_____	25 amp*	3,000 VA/120 volts
	198 amp	196 amp	173 amp	

* Indicates neutral (120-volt) loads.

12–27 DELTA/WYE EXAMPLE

Dishwasher (4.5 kW) 120 volts　　　　　　　　Two – 3 horsepower motors, 208 volts, 1Ø
A/C motor (10 horsepower) 208 volts, 3Ø　　　Water heater (10 kW) 208 volts, 1Ø
Heat (18 kW) 208 volts, 3Ø　　Eight – Lighting circuits (1.5 kW), 120 volts
Two – Ranges (14 kW) 208 volts, 1Ø

	Phase A (L₁)	Phase B (L₂)	Phase C (L₃)	Line Total
Heat (A/C omitted *)	6,000 VA	6,000 VA	6,000 VA	18.0 kVA
Ranges 14 kW 208 volt, 1Ø	7,000 VA	7,000 VA		14.0 kVA
Ranges 14 kW 208 volt, 1Ø		7,000 VA	7,000 VA	14.0 kVA
Water Heater 10 kW 208 volt, 1Ø	5,000 VA		5,000 VA	10.0 kVA
3 horsepower 208 volt, 1Ø Motor	1,945 VA	1,945 VA		3.89 kVA
3 horsepower 208 volt, 1Ø Motor		1,945 VA	1,945 VA	3.89 kVA
Dishwasher 4.5 kW 120 volt			4,500 VA *	4.5 kVA
Lighting (8 – 1.5 kW circuits)	6,000 VA *	3,000 VA *	3,000 VA *	12.0 kVA
	25,945 VA	26,890 VA	27,445 VA	80.28 kVA

Note: The phase totals (26 kVA, 27 kVA and 27.5 kVA) should add up to the line total (80.5 kVA). This method is used as a check to make sure all items have been accounted for and added correctly.

Motor VA

VA Single-phase = Table Volts × Table Ampere

3 horsepower VA = E × I = 208 volts × 18.7 ampere = 3,890 VA

VA Per Phase 3,890/2 = 1,945 VA per phase

VA Three-phase = Table Volts × Table Ampere × $\sqrt{3}$

10 horsepower A/C, VA = Table Volts × Table Ampere × 1.732

10 horsepower A/C, VA = 208 volts × 30.8 ampere × 1.732 = 11,096 VA (omit, smaller than 18 kW heat)

➥ Transformer Sizing

What size transformers are required?

(a) three 30 kVA, single-phase transformers   (b) one 90 kVA, 3-phase transformer

(c) a or b   (d) none of these

• Answer: (c) a or b

Phase A = 26 kVA

Phase B = 27 kVA

Phase C = 28 kVA

➥ Maximum kVA on Neutral

What is the maximum kVA on the neutral?

(a) 3 kVA   (b) 6 kVA   (c) 7.5 kVA   (d) 9 kVA

• Answer: (c) 7.5 kVA

120-volt loads only, do not count phase to phase (208-volt) loads.

➥ Maximum Neutral Current

What is the maximum current on the neutral?

(a) 50 ampere   (b) 63 ampere   (c) 75 ampere   (d) 100 ampere

• Answer: (b) 63 ampere

I = P/E = 7,500 VA/120 volts = 63 ampere.

The neutral must be sized to carry the maximum unbalanced neutral current, which in this case is 63 ampere.

➥ Neutral Current Question

What is the current on the neutral?

(a) 25 ampere   (b) 33 ampere   (c) 50 ampere   (d) 63 ampere

• Answer: (b) 33 ampere

$$I_{Neutral} = \sqrt{L_1{}^2 + L_2{}^2 + L_3{}^2 - (L_1 \times L_2 + L_2 \times L_3 + L_1 \times L_3)}$$

L_1 = 6,000 VA/120 volts = 50 ampere

L_2 = 3,000 VA/120 volts = 25 ampere

L_3 = 7,500 VA/120 volts = 63 ampere

$$I_{Neutral} = \sqrt{(50^2 + 25^2 + 63^2) - [(50 \times 25) + (25 \times 63) + 50 \times 63]}$$

$$I_{Neutral} = \sqrt{(2,500 + 625 + 3,969) - (1,250 + 1,575 + 3,150)} = \sqrt{7,049 - 5,975} = \sqrt{1,119} = 33 \text{ ampere}$$

➥ Phase VA (three-phase)

What is the per phase load of each 3 horsepower 208-volt, single-phase motor?

(a) 1,945 VA   (b) 3,910 VA   (c) 978 VA   (d) none of these

• Answer: (a) 1,945 VA

VA (single-phase) = Table Volts × Table Ampere

3 horsepower VA = 208 volts × 18.7 ampere = 3,890 VA

208-volt, single-phase load is on two transformer phases = 3,890 VA/2 phases = 1,945 VA

➥ Phase VA (single-phase)

What is the per phase load for the 3-phase, 10 horsepower A/C motor?

(a) 11,154 VA   (b) 3,698 VA   (c) 5,577 VA   (d) none of these

• Answer: (b) 3,698 VA

10 horsepower VA = Table volts × Table ampere × 1.732

VA = 208 volts × 30.8 ampere × 1.732 = 11,095 VA

Three-phase load is on three phases = 11,095 VA/3 phases = 3,698 VA per phase

➤ **Voltage Ratio**

What is the phase voltage ratio of the transformer?

(a) 4:1   (b) 1:4   (c) 1:2   (d) 2:1

• Answer: (a) 4:1

480 primary phase volts to 120 secondary phase volts

❏ **Balance Panel – VA**

Balance the loads on the panelboard in VA.

	Line 1	Line 2	Line 3	Line Total
Heat 18 kW(A/C 11 kW/ omitted)	6,000 VA	6,000 VA	6,000 VA	18.0 kVA
Ranges 14 kW, 208 volts, 1Ø	7,000 VA	7,000 VA		14.0 kVA
Ranges 14 kW, 208 volts, 1Ø		7,000 VA	7,000 VA	14.0 kVA
Water heater 10 kW, 208 volts, 1Ø	5,000 VA		5,000 VA	10.0 kVA
3 horsepower 208 volts, 1Ø	1,945 VA	1,945 VA		3.9 kVA
3 horsepower 208 volts, 1Ø		1,945 VA	`,945 VA	3.9 kVA
Dishwasher 4.5 kW, 120 volt			4,500 VA *	4.5 kVA
Lighting (8 – 1.5 kW circuits)	<u>6,000 VA *</u>	<u>3,000 VA *</u>	<u>3,000 VA *</u>	<u>12.0 kVA</u>
	25,945 VA	26,890 VA	27,445 VA	80.3 kVA

*Indicates neutral (120-volt) loads.

❏ **Panelboard Balancing and Sizing**

Balance the previous example loads on the panelboard in ampere.

	Line 1	Line 2	Line 3	Ampere Calculations
Heat 18 kW Heat 3Ø	50.0 amp	50.0 amp	50.0 amp	18,000 VA/(208 volts × 1.732)
Range 14 kW, 208 volts, 1Ø	67.0 amp	67.0 amp		14,000 VA/208 volts
Range 14 kW, 208 volts, 1Ø		67.0 amp	67.0 amp	14,000 VA/208 volts
Water heater 10 kW, 208 volt,1Ø	48.0 amp		48.0 amp	10,000 VA/208 volts
3 horsepower 208 volts, 1Ø Motor	18.7 amp	18.7 amp		Motor FLC
3 horsepower 208 volts, 1Ø Motor		18.7 amp	18.7 amp	Motor FLC
Dishwasher 1Ø 120 volt			38.0 amp	4,500 VA/120 volts
Lighting (8 – 1.5 kW circuits)	<u>50.0 amp</u>	<u>25.0 amp</u>	<u>25.0 amp</u>	1,500 VA/120 volts
	233.7 amp	246.4 amp	246.7 amp	

12–28 DELTA VERSUS WYE

Wye

Phase CURRENT is the SAME as line current.

$I_{Phase} = I_{Line}$ or $I_{Line} = I_{Phase}$

Wye Phase VOLTAGE is DIFFERENT than line voltage.

Line Voltage is greater than Phase Voltage by the square root of 3 ($\sqrt{3}$).

$E_{Line} = E_{Phas} \times \sqrt{3}$ or $E_{Phase} = E_{Line} / \sqrt{3}$

Delta

Delta phase VOLTAGE is the SAME as the line voltage.

$E_{Phase} = E_{Line}$ or $E_{Line} = E_{Phase}$

Delta phase CURRENT is DIFFERENT than the line current.

Line current is greater than Phase Current by the square root of 3 (1.732).

$I_{Line} = I_{Phase} \times \sqrt{3}$ or $I_{Phase} = I_{Line} / \sqrt{3}$

Unit 12 – Delta/Delta and Delta/Wye Transformers Summary Questions

Definitions

1. Delta connected means the windings of three single-phase transformers (same rated voltage) are connected in _____ . A delta-connected transformer is represented by the greek letter delta Δ.
    (a) series   (b) parallel   (c) series-parallel   (d) a and b

2. Which system is called the high-leg system because the voltage from one conductor to ground is between 190 and 208 volts-to-ground?
    (a) delta   (b) wye

3. The _____ is the electrical system ungrounded conductors.
    (a) load   (b) line   (c) phase   (d) system

4. The _____ voltage is the voltage measured between any ungrounded conductors.
    (a) load   (b) line   (c) phase   (d) system

5. • Line voltage of a _____ system is greater than phase voltage and line voltage of a _____ system is the same as phase voltage.
    (a) delta, delta   (b) wye, wye   (c) delta, wye   (d) wye, delta

6. The _____ is the coil shape conductor that serve as the primary or secondary of the transformer.
    (a) line   (b) load   (c) phase   (d) none of these

7. The phase current is the current of the transformer winding. For _____ systems, the phase current is less than the line current. For _____ systems, the phase current is the same as the line current.
    (a) wye, wye   (b) delta, delta   (c) wye, delta   (d) delta, wye

8. The phase load is the load on the transformer winding. For delta systems, the phase load is _____ .
    (a) three-phase, 240-volt load = line load/3   (b) single-phase, 240-volt load = line load
    (c) Single-phase, 120-volt load = line load   (d) all of these

9. The phase load is the load on the transformer winding. For wye systems, the phase load is _____ .
    (a) three-phase, 208-volt load = line load/3   (b) single-phase, 208-volt load = line load/2
    (c) single-phase, 120-volt load = line load   (d) all of these

10. The phase voltage is the internal transformer voltage generated across any one "winding" of a transformer. For _____ systems, the phase voltage is equal to the line voltage. For _____ systems, the phase voltage is less than the line voltage.
    (a) wye, wye   (b) delta, delta   (c) wye, delta   (d) delta, wye

11. The ratio is the relationship between the number of primary winding turns compared to the number of secondary winding turns. The ratio is a comparison between the primary phase voltage and the secondary phase voltage. For typical delta/delta systems, the ratio is _____, and for typical wye systems, the ratio is _____ .
    (a) 1:2, 1:4   (b) 2:1, 4:1   (c) 4:1, 2:1   (d) none of these

12. What is the turns ratio of a 480- to 208Y/120-volt transformer?
    (a) 4:1   (b) 1:4   (c) 1:2   (d) 2:1

13. • The unbalanced load is the load on the secondary grounded (neutral) conductors. For _____ systems, the formula is: $I_{Neutral} = \sqrt{(L_1^2 + L_2^2 + L_3^2) - (L_1 \times L_2 + L_2 \times L_3 + L_1 \times L_3)}$
    (a) delta   (b) wye   (c) a or b   (d) none of these

14. The unbalanced (neutral) current for a _____ system can be calculated by: $I_{Neutral} = I_{Line\ 1} - I_{Line\ 3}$.
    (a) delta   (b) wye   (c) a or b   (d) none of these

15. _____ connected means a connection of three single-phase transformers to a common point (neutral), and the other end is connected to the line conductors.
    (a) Delta   (b) Wye   (c) a or b   (d) none of these

12–1 Current Flow

16. When a load is connected to the secondary of a transformer, current flows through the secondary conductor winding. The current flow in the secondary creates an electromagnetic field that opposes the primary electromagnetic field.
    (a) True   (b) False

17. The primary and secondary line currents are directly proportional to the ratio of the transformer.
    (a) True   (b) False

Part A – Delta/Delta Transformer

12–2 Delta Transformer Voltage

18. In a delta configured transformer, the line voltage equals the phase voltage.
    (a) True   (b) False

12–3 Delta High-Leg

19. If the secondary voltage of a delta/delta transformer is 220/110 volts, the high-leg voltage to ground (or neutral) is _____ volts.
    (a) 190   (b) 196   (c) 202   (d) 208

12–4 Delta Primary and Secondary Line Currents

20. In a delta configured transformer, the line current equals the phase current.
    (a) True   (b) False

21. What is the primary line current for a 45 kVA, 480- to 240/120-volt, 3-phase transformer?
    (a) 124 ampere   (b) 108 ampere   (c) 54 ampere   (d) 43 ampere

22. What is the secondary line current for a 45 kVA, 480- to 240-volt, 3-phase transformer?
    (a) 124 ampere   (b) 108 ampere   (c) 54 ampere   (d) 43 ampere

12–5 Delta Primary or Secondary Phase Currents

23. The phase current of a transformer winding is calculated by dividing the phase load by the phase volts: I Phase = VAPhase/EPhase.
    (a) True   (b) False

24. What is the primary phase current for a 15 kVA, 480- to 240/120-volt, single-phase transformer?
    (a) 124 ampere   (b) 62 ampere   (c) 45 ampere   (d) 31 ampere

25. What is the secondary phase current for a 15 kVA, 480- to 240/120-volt, single-phase transformer?
    (a) 124 ampere   (b) 62 ampere   (c) 45 ampere   (d) 31 ampere

12–6 Delta Phase versus Line

26. A 15 kVA, 240-volt, 3-phase load has the following effect on a delta system:
(a) I_{Line} = 15,000 VA/(240 volts $\times \sqrt{3}$) = 36 ampere
(b) I_{Phase} = 5,000 VA/240 volts = 21 ampere
(c) $I_{Line} = I_{Phase} \times \sqrt{3}$ = 21ampere $\times$ 1.732 = 36 ampere
(d) all of the above

27. • A 5 kVA, 240-volt, single-phase load has the following effect on a delta/delta system:
(a) Line Power = 5 kVA   (b) Phase Power = 5 kVA
(c) I_{Phase} = 5,000 VA/240 volts = 21 ampere   (d) all of the above

28. • A 2 kVA, 120-volt, single-phase load has the following effect on a delta system:
(a) Line Power = 2 kVA   (b) I_{Line} = 2,000 VA/120 volts = 17 ampere
(c) Phase Power = 3 kVA (C1 or C2 winding)   (d) a and b

12–7 Delta Current Triangle

29. The line and phase current of a delta system are not equal. The difference between line and phase current can be described by which of the following equations?
(a) $I_{Line} = I_{Phase} \times \sqrt{3}$ (b) $I_{Phase} = I_{Line}/\sqrt{3}$
(c) a and b   (d) none of these

12–8 and 12–9 Delta Transformer Balancing and Sizing

30. To properly size a delta/delta transformer, the transformer must be balanced in kVA. Which of the following steps should be used to balance the transformer?
(a) Balance 3-phase loads, $^1/_3$ on Phase A, $^1/_3$ on Phase B, and $^1/_3$ on Phase C.
(b) Balance single-phase, 240-volt loads, 100 percent on Phase A or Phase B.
(c) Balance the 120-volt loads, 100 percent on Phase C1 or C2.
(d) all of these

31. • Balance and size a 480- to 240/120-volt, 3-phase delta transformer for the following loads: One 18 kVA 3-phase heat strip, two 5 kVA, single-phase, 240-volt loads, and three 2 kVA, 120-volt loads. What size transformer is required?
(a) three single-phase 12.5 kVA transformers
(b) one 3-phase 37.5 kVA transformer
(c) a or b
(d) none of these

12–10 Delta Panel Schedule in kVA

32. When balancing a panelboard in kVA be sure that 3-phase loads are balanced $^1/_3$ on each line and 240-volt, single-phase loads are balanced on each line.
(a) True   (b) False

33. • Balance a 120/240-volt delta 3-phase panelboard in kVA for the following loads: one 18 kVA 3-phase heat strip, two 5-kVA, single-phase, 240-volt loads, and three 2-kVA, 120-volt loads.
(a) The largest line load is 13 kVA.   (b) The smallest line load is 10 kVA.
(c) Total line load is equal to 34 kVA.   (d) all of these

12–11 Delta Panelboard and Conductor Sizing

34. When selecting and sizing panelboards and conductors, we must balance the line loads in ampere. Balance a 120/240-volt, 3-phase panelboard in ampere for the following loads: one 18 kVA, 3-phase heat strip, two 5 kVA, single-phase 240-volt loads, and three 2 kVA, 120-volt loads.
(a) The largest line is 98 ampere.   (b) The smallest line is 81 ampere.
(c) Each line equals to 82 ampere.   (d) a and b

12–12 Delta Neutral Current

35.　What is the neutral current for: one 18 kVA, 3-phase heat strip, two 5 kVA, single-phase, 240-volt loads, and three 2 kVA, 120-volt loads?

(a) 17 ampere　　　　(b) 34 ampere　　　　(c) 0 ampere　　　　(d) a and b

12–13 Delta Maximum Unbalanced Load

36.　• What is the maximum unbalanced load in ampere for: One 18 kVA, 3-phase heat strip, two 5 kVA, single-phase, 240-volt loads, and three 2 kVA, 120-volt loads?

(a) 25 ampere　　　　(b) 34 ampere　　　　(c) 0 ampere　　　　(d) none of these

37.　What is the phase load for a 18 kVA, 3-phase heat strip?

(a) 18 kVA　　　　(b) 9 kVA　　　　(c) 6 kVA　　　　(d) 3 kVA

38.　• What is the transformer winding phase load for a 5 kVA, single-phase, 240-volt load?

(a) 5 kVA　　　　(b) 2.5 kVA　　　　(c) 1.25 kVA　　　　(d) none of these

39.　If only a 3-phase, 12 kW load is on the high-leg, what is the current on the high-leg conductor?

(a) 29 ampere　　　　(b) 43 ampere　　　　(c) 73 ampere　　　　(d) 97 ampere

40.　What is the ratio of a 480- to 240/120-volt transformer?

(a) 4:1　　　　(b) 1:4　　　　(c) 1:2　　　　(d) 2:1

Part B Delta/Wye Transformers

12–15 Wye Transformer Voltage

41.　In a wye configured transformer, the line voltage equals the phase voltage.

(a) True　　　　(b) False

12–17 Wye Transformers Current

42.　In a wye configured transformer, the line current equals the phase current.

(a) True　　　　(b) False

12–18 Wye Line Current

43.　The line current of both delta and wye 3-phase transformers can be calculated by the formula:
$I_{Line} = VA_{Line}/(E_{Line} \times \sqrt{3})$.

(a) True　　　　(b) False

44.　What is the primary line current for a 22 kVA, 480- to 208Y/120-volt, 3-phase transformer?

(a) 61 ampere　　　　(b) 54 ampere　　　　(c) 26 ampere　　　　(d) 22 ampere

45.　What is the secondary line current for a 22 kVA, 480- to 208Y/120-volt, 3-phase transformer?

(a) 61 ampere　　　　(b) 54 ampere　　　　(c) 27 ampere　　　　(d) 22 ampere

12–19 Wye Phase Current

46.　The phase current of a wye configured transformer winding is the same as the line current. The phase current of a delta configured transformer winding is less than the line current by $\sqrt{3}$.

(a) True　　　　(b) False

47.　• What is the primary phase current for a 22 kVA, 480- to 208Y/120-volt, 3-phase transformer?

(a) 15 ampere　　　　(b) 22 ampere　　　　(c) 36 ampere　　　　(d) 61 ampere

48.　• What is the secondary phase current for a 22 kVA, 480- to 208Y/120-volt, 3-phase transformer?

(a) 15 ampere　　　　(b) 22 ampere　　　　(c) 36 ampere　　　　(d) 61 ampere

12–20 Wye Phase versus Line Current

49. Since each line conductor from a wye transformer is connected to a different transformer winding (phase), the effects of loading on the line are the same as the phase. A 12 kVA, 208-volt, 3-phase load has the following effect on the system:
(a) $I_{Line} = VA_{Line}/(E_{Line} \times \sqrt{3}) = 12,000 \text{ VA}/(208 \text{ volts} \times \sqrt{3}) = 33$ ampere
(b) Phase Power = 4 kVA
(c) $I_{Phase} = VA_{Phase}/E_{Phase} = 4,000 \text{ VA}/120 \text{ volts} = 30$ ampere
(d) all of these

50. A 5 kVA, 208-volt, single-phase load has the following effect on the system:
(a) Line Power = 5 kVA
(b) $I_{Line} = VA/E = 5,000 \text{ VA}/208 \text{ volts} = 24$ ampere
(c) $I_{Phase} = VA_{Phase}/E_{Phase} = 2,500 \text{ VA}/120 \text{ volts} = 21$ ampere
(d) all of these

51. A 2 kVA, 120-volt, single-phase load has the following effect on the system:
(a) Line Power = 2 kVA
(b) $I_{Line} = VA/E = 2,000 \text{ VA}/120 \text{ volts} = 17$ ampere
(c) $I_{Phase} = VA_{Phase}/E_{Phase} = 2,000 \text{ VA}/120 \text{ volts} = 17$ ampere
(d) all of these

52. • What is the phase load of a 3 horsepower 208-volt, single-phase motor?
(a) 1,945 VA   (b) 3,890 VA   (c) 978 VA   (d) none of these

53. • What is the per phase load for a 3-phase, 7.5 horsepower 208-volt motor?
(a) 2,906 VA   (b) 8,718 VA   (c) 4.359 VA   (d) none of these

12–21 and 12–22 Wye Transformer Balancing and Sizing

54. To properly size a delta/wye transformer, the secondary transformer phases (windings) or the line conductors must be balanced. Balancing the panel (line conductors) is identical to balancing the transformer for wye configured transformers. Which of the following steps should be used to balance the transformer?
(a) balance 3-phase loads, 1/3 on each phase
(b) balance single-phase, 208-volt loads 50 percent on each phase
(c) balance the 120-volt loads 100 percent on any phase
(d) all of these

55. • Balance and size a 480- to 208Y/120-volt, 3-phase transformer for the following loads: one 18 kVA, 3-phase heat strip, two 5 kVA, single-phase, 208-volt loads, and three 2 kVA, 120-volt loads. What size transformer is required?
(a) three single-phase 12.5 kVA transformers   (b) one 3-phase 37.5 kVA transformer
(c) a or b   (d) none of these

12–23 Wye Panel Schedule in kVA

56. When balancing a panelboard in kVA be sure that 3-phase loads are balanced $1/3$ on each line and 208-volt, single-phase loads are balanced $1/2$ on each line. Balance a 208Y/120-volt, 3-phase panelboard in kVA for the following loads: one 18 kVA, 3-phase heat strip, two 5 kVA, single-phase, 208-volt loads, and three 2 kVA, 120-volt loads.
(a) the largest line is 12.5 kVA   (b) the smallest line is 10.5 kVA
(c) a and b   (d) none of these

12–24 Wye Panelboard and Conductor Sizing

57. • When selecting and sizing panelboards and conductors, we must balance the line loads in ampere. Balance a 208Y/120-volt, 3-phase panelboard in ampere for the following loads: one 18 kVA, 208-volt, 3-phase heat strip, two 5 kVA, single-phase, 208-volt loads, and three 2 kVA, 120-volt loads.
(a) The largest line current is 108 ampere.   (b) The smallest line is 91 ampere.
(c) a and b   (d) none of these

12–25 Wye Neutral Current

58. To determine the neutral current for a 3-wire wye system we must use the following formula:

$$I_{Neutral} = \sqrt{(L_1^2 + L_2^2 + L_3^2) - (L_1 \times L_2 + L_2 \times L_3 + L_1 \times L_3)}$$

    (a) True   (b) False

59. • What is the neutral current for: one 18 kVA, 3-phase heat strip, two 5 kVA, single-phase, 208-volt loads and three 2 kVA, 120-volt loads?

    (a) 47 ampere   (b) 29 ampere   (c) 34 ampere   (d) a and b

12–26 Wye Maximum Unbalanced Load

60. • What is the maximum unbalanced load in ampere for: one 18 kVA, 3-phase heat strip, two 5 kVA, single-phase, 208-volt loads, and three 2 kVA, 120-volt loads.

    (a) 17 ampere   (b) 34 ampere   (c) 0 ampere   (d) a and b

12–27 Delta versus Wye

61. Which of the following statements are true for wye systems?

    (a) Phase current is the same as line current: $I_{Phase} = I_{Line}$.
    (b) Phase voltage is the same as the line voltage.
    (c) $E_{Line} = E_{Phase} \times \sqrt{3}$
    (d) a and c

62. Which of the following statements are true for delta systems?

    (a) Phase voltage is the same as line voltage: $E_{Phase} = E_{Line}$.
    (b) Phase current is the same as the line current.
    (c) Line current is greater than phase current: $I_{Line} = I_{Phase} \times \sqrt{3}$.
    (d) a and c

☆ Challenge Questions

Part A – Delta/Delta Transformers

The following statement applies to the next seven questions:

The following loads are to be connected to a 3-phase delta/delta, 480-volt to 240-volt transformer.
Heat 18 kW, 240-volt, 3-phase heat
A/C 25 horsepower 240-volt, 3-phase synchronous motor
Two motors 1 horsepower 240-volt, single-phase motors
Two 2 horsepower 120-volt motors
Three 1,900 watt, 120-volt lighting loads

Balance the above loads as closely as possible, then answer the next seven questions.

63. • Three single-phase transformers are connected in series. What size transformers are required for the above loads?

    (a) 10/10/10 kVA   (b) 15/15/15 kVA   (c) 10/10/20 kVA   (d) 15/15/10 kVA

64. • The line VA load of the 25 horsepower 3-phase synchronous motor would be _____ VA.

    (a) 9,700   (b) 21,113   (c) 32,031   (d) 28,090

65. • After balancing the loads, the maximum unbalanced neutral current is _____ ampere.

    (a) 27   (b) 1.5   (c) 48   (d) 148

66. • The transformer phase load of one 1 horsepower motor would be _____ VA.

    (a) 1,380   (b) 2,760   (c) 2,300   (d) 1,150

67. • The line load of one of the 1 horsepower motors would be _____ VA.

    (a) 1,380   (b) 2,760   (c) 2,300   (d) 1,150

68. • The transformer phase VA load of the 25 horsepower 3-phase synchronous motor is closest to _____ VA.
    (a) 3,520   (b) 7,038   (c) 21,100   (d) 32,100

69. • If the only load on the high-leg is the 3-phase 25 horsepower motor, the high-leg conductor would have to be sized a minimum of _____ ampere.
    (a) 13   (b) 66   (c) 18   (d) 22

The following statement applies to the next two questions:
A 3-phase 37.5 kVA delta/delta transformer has a primary of 480 volts and a secondary of 240/120 volts.

70. • The primary line current is closest to _____ ampere.
    (a) 45   (b) 90   (c) 50   (d) 65

71. • The maximum overcurrent device size for the 37.5 kVA delta/delta transformer is _____ ampere.
    (a) 40   (b) 45   (c) 50   (d) 60

The following statement applies to the next question:
Seven heaters 1,500 watt, 120 volt
Five 3 kW, 240-volt, single-phase heaters
Equipment circuits 30 kW, 120 volt
Lighting circuits 40 kW, 120

72. • If the only loads on the center tapped transformer (C1 and C2) are the 120-volt loads, what is the total kW load on this transformer?
    (a) 30–40 kW   (b) 50–60 kW   (c) 80–90 kW   (d) over 100 kW

73. • The total load on the center tapped 240/120-volt transformer is 100 kW, which consists of 80 kW of 120 volt balanced loads, and 20 kW of 240-volt loads. What is the maximum unbalanced neutral current?
    (a) 167 ampere   (b) 0 ampere   (c) 192 ampere   (d) 333 ampere

74. • On a delta/delta 3-phase transformer, a 30,000 VA, 240-volt, 3-phase load would have a phase current of _____ ampere on the secondary winding.
    (a) 125   (b) 72   (c) 42   (d) none of these

75. • What is the secondary line current of a 3-phase, 45 kVA, delta/delta transformer that has a turn ratio of 2:1 and a primary voltage of 460?
    (a) 113 ampere   (b) 55 ampere   (c) 36 ampere   (d) cannot be determined

76. • A 2,200/220-volt, 3-phase transformer is connected delta/delta. The secondary phase current is 300 ampere and the secondary line current is 240 volts. What is the primary line current?
    (a) 40 ampere   (b) 20 ampere   (c) 50 ampere   (d) 30 ampere

77. • For a given 3-phase transformer, the connection giving the highest secondary voltage is _____ .
    (a) wye primary, delta secondary   (b) delta primary, delta secondary
    (c) wye primary, wye secondary   (d) delta primary, wye secondary

Part B – Delta/Wye Transformers

78. • A delta/wye, 480/208- 120-volt, 3-phase transformer has a secondary line current of 200 ampere. What is the primary line current?
    (a) 31 ampere   (b) 72 ampere   (c) 87 ampere   (d) 53 ampere

79. • If the secondary phase voltage of a delta/wye connected transformer is 277 volts, and a 10 kVA, 3-phase load was connected to that secondary, the line current would be _____ ampere.
    (a) 7   (b) 12   (c) 32   (d) none of these

The following statement applies to the next two questions:

Motor 15 horsepower 208-volt, 3-phase   Two dryers 1.5 kW, 208-volt dryers
Four lighting circuits 3 kW, 120 volt   Cooktop 8 kW, 208 volt
Five heaters 2.5 kW, 208 volt

80.   • Each 2.5 kW heat strip has a line load of _____ .
         (a) 1,250 watts   (b) 12,500 watts   (c) 2,500 watts   (d) no load

81.   • The load per line of the 3-phase motor can be determined by which of these formulas?
         (a) $\sqrt{3} \times$ Volts $\times$ Ampere   (b) (Volts $\times$ Ampere)/3
         (c) Volts $\times$ Ampere   (d) none of these

82.   • If a 3-phase motor has a total load of 12,000 VA, and is connected to a delta/wye transformer, the load on one phase of
         the secondary will be _____ VA.
         (a) 12,000   (b) 4,000   (c) 3,000   (d) 1,500

83.   • A 3-phase 480Y/208-120-volt transformer provides power to nine 2,000 VA, 120-volt lights and nine 2,000 watt,
         208-volt heaters. The maximum neutral current is _____ ampere.
         (a) 50   (b) 75   (c) 87   (d) 95

84.   • There are five 2,500 watt, 208-volt heat strips on the secondary of a 3-phase 208Y/120-volt transformer. What is the
         load per phase for each heat strip?
         (a) 1,403 watts   (b) 1,250 watts   (c) 2,500 watts   (d) 12,500 watts

85.   • The phase load of the 10 horsepower 3-phase 208Y/120-volt motor is closest to _____ VA.
         (a) 3,700   (b) 5,500   (c) 4,300   (d) 11,100

The following statement applies to the next question:

The system is 3-phase 208Y/120-volt and the loads are:
Nine – 2 kVA, 120-volt lighting loads
Five – 3 kVA, 120-volt loads

86.   • The neutral conductor must be size to carry _____ ampere.
         (a) 10   (b) 25   (c) 90   (d) 100

NEC® Questions Random Order Section 90 through Chapter 9

87.   Outlet boxes can be secured to independent support wires, which are taut and secured at both ends if the box is
         supported to the independent support wires, with fitting and methods identified for the purpose.
         (a) True   (b) False

88.   Steel or aluminum cable tray systems can be used as an equipment grounding conductor provided the cable tray
         sections and fittings are identified for _____ purposes.
         (a) grounding   (b) special   (c) industrial   (d) all

89.   In a balanced, 208Y/120 volt, 3-phase, 4-wire system, the grounded (neutral) conductor will carry _____ amperes, if the
         loads supplied are linear loads and no harmonic currents are present.
         (a) full load   (b) zero   (c) fault current   (d) none of these

90.   All boxes and conduit bodies, covers, extension rings, plaster rings, and the like shall be durably and legibly marked
         with the manufacturer's name or trademark.
         (a) True   (b) False

91.   Lampholders installed in wet or damp locations shall be of the _____ type.
         (a) waterproof   (b) weatherproof   (c) moistureproof   (d) moisture resistant

92. • A 277-volt fluorescent lighting fixture (in an office, continuously loaded) draws 1.34 amperes. The number of such lighting fixtures that may be connected on a 2-wire 277-volt circuit protected by a 20-ampere circuit breaker is _____ .
(a) 8   (b) 9   (c) 10   (d) 12

93. For nonpower limited fire alarm circuits, a No. 18 conductor is considered protected if the overcurrent device protecting the system is not over _____ amperes.
(a) 15   (b) 10   (c) 20   (d) 7

94. The _____ pool bonding conductor must terminated either by exothermic welding or by pressure connectors that are labeled as being suitable for the purpose.
(a) No. 8   (b) insulated   (c) copper   (d) all of these

95. Where more than one nominal voltage system exists in a building, each ungrounded conductor of a multiwire branch circuit where accessible, shall be identified by phase and system. The identification can be _____ permanently posted at each branch circuit panelboard.
(a) color coding   (b) phase tape   (c) tagging   (d) any of these

96. The enclosure for a transformer, or ground-fault circuit-interrupter that is connected to a conduit that extends directly to a pool light forming shell shall be listed for this purpose.
(a) labeled   (c) marked   (c) identified   (d) none of these

97. Isolated ground receptacles installed in nonmetallic boxes must be covered with a nonmetallic faceplate because a metal faceplate cannot be grounded to the circuit equipment grounding conductor.
(a) True   (b) False

98. The maximum overcurrent protection in amperes, when required, for a Class 2 remote-control circuit is _____ amperes.
(a) 1.0   (b) 2.0   (c) 5.0   (d) 7.5

99. The conductors on the load side of a transformer or any GFCI protection device that supplies power to a pool light, shall not be in the same raceway or enclosure with other conductors.
(a) True   (b) False

100. Where switches, cutouts, or similar equipment operating at 600 volts, nominal, or less are installed in a room where there are exposed energized parts at over 600 volts, nominal, the high-voltage equipment shall be effectively separated from the space occupied by the low-voltage equipment by a suitable _____ .
(a) partition   (b) fence   (c) screen   (d) any of these

101. The disconnecting means for each circuit leading to a sign located within a fountain must be located in accordance with Section _____ .
(a) 430-102   (b) 440-14   (c) 680-12   (d) all or these

102. The maximum spacing of receptacle outlets on countertops in a mobile home is _____ feet.
(a) 6   (b) 12   (c) 3   (d) none of these

103. • Class 1 control circuit conductors shall have overcurrent protection _____ .
(a) in accordance with the values specified in Table 310-16 through 310-31 for No. 14 and larger
(b) not exceeding 7 amperes for No. 18, and 10 amperes for No. 16
(c) and derating factors do not apply
(d) all of these

104. • A No. 12 THHN conductor can be protected by an overcurrent protection device that is greater than 20 amperes. This is permitted when the conductors are used for _____ .
(a) tap conductors   (b) motor circuit conductors
(c) Class 1 control and signaling circuit conductors   (d) all of these

105. Flexible cords and cables used for temporary wiring shall be protected _____ .
(a) from accidental damage   (b) where passing through doors
(c) from sharp corners and projections (d) all of these

106. In locations where _____ would be exposed to physical damage, enclosures or guards shall be so arranged and of such strength as to prevent physical damage.
    (a) motor control panels   (b) electrical equipment
    (c) generators   (d) circuit breakers

107. GFCI protection is required for all 125-volt, 15- and 20-ampere, single-phase, receptacles installed _____ of commercial, industrial, and all other nondwelling occupancies.
    (a) outside   (b) in wet locations   (c) within 6 feet of sinks   (d) in bathrooms

108. • For the purpose of determining the placement of receptacles in a dwelling-unit kitchen, a(n) _____ countertop is measured from the connecting edge.
    (a) island   (b) usable   (c) peninsular   (d) cooking

109. _____ is not a magnetic metal and there will be no heating due to inductive hysteresis heating.
    (a) Steel   (b) Iron   (c) Aluminum   (d) all of these

110. A receptacle outlet for the laundry is not required in a dwelling unit in a multifamily building where laundry facilities are provided on the premises that are available to all building occupants.
    (a) True   (b) False

111. Lighting track shall not be installed less than _____ feet above the finished floor except where protected from physical damage or track operating at less than 30 volts RMS open-circuit voltage.
    (a) 4   (b) 5   (c) 5$^1/_2$ (d) 6

112. Loads computed for dwelling unit small-appliance circuits can be included with the _____ load and subjected to the demand factors permitted in Table 220-11 for the general lighting load.
    (a) general lighting   (b) feeder   (c) appliance   (d) receptacle

113. • The *Code* applies to the installation _____ .
    (a) of electrical conductors and equipment within or on public and private buildings
    (b) of outside conductors and equipment on the premises
    (c) of optical fiber cable
    (d) all these

114. • For antenna conductors for an amateur transmitting and receiving station where of hard-drawn copper, the maximum open span length is 200 feet and requires a minimum of No. _____ .
    (a) 14   (b) 12   (c) 10   (d) 8

115. • Admitting close approach, not guarded by locked doors, elevation, or other effective means, is commonly referred to as _____ .
    (a) accessible (equipment)   (b) accessible (wiring methods)
    (c) accessible, readily   (d) all of these

116. • Where communications cables enter buildings, they shall _____ .
    (a) where practicable, be located below the electric light or power conductors
    (b) not be attached to a crossarm that carries electric light or power conductors
    (c) have a vertical clearance of not less than 8 feet from all points of roofs above which they pass
    (d) all of these

117. • The connection between the grounded (neutral) conductor and the equipment grounding conductor at the service is accomplished with the use of a _____ jumper.
    (a) main bonding   (b) bonding
    (c) equipment bonding   (d) circuit bonding

118. • Where network-powered broadband communications system cables enter buildings, they shall _____ .
    (a) where practicable, be located below the electric light or power conductors
    (b) not be attached to a crossarm that carries electric light or power conductors
    (c) have a vertical clearance of not less than 8 feet from all points of roofs above which they pass
    (d) all of these

119. • Conductors for an appliance circuit supplying more than one appliance or appliance receptacle in an installation operating at less than 50 volts shall not be smaller than No. _____ copper or equivalent.
  (a) 18　　　　　　　(b) 14　　　　　　　(c) 12　　　　　　　(d) 10

120. • Electrical nonmetallic tubing is permitted for direct earth burial when used with fittings listed for this purpose.
  (a) True　　　　　　(b) False

121. • Pool-associated motors must be grounded by a minimum No. 12 insulated copper conductor. This grounding conductor must be installed in _____ .
  (a) rigid nonmetallic conduit
  (b) electrical metallic tubing, where installed on or within buildings
  (c) flexible metal conduit
  (d) a or b

122. • A device that, by insertion in a receptacle, establishes connection between the conductors of the attached flexible cord and the conductors connected permanently to the receptacle is called an _____ .
  (a) attachment plug　　(b) plug cap　　　　(c) cap　　　　　　(d) all of these

123. • Auxiliary gutters shall not contain more than _____ conductors at any cross-section.
  (a) 30　　　　　　　(b) 40 current-carrying　　(c) 20　　　　　　(d) 30 current-carrying

124. • In no case shall conductors within flexible cords and cables be associated in such a way with respect to the kind of circuit, the wiring method used, or the number of conductors such that the _____ temperature of the conductors is exceeded.
  (a) operating　　　　(b) governing　　　　(c) ambient　　　　(d) limiting

125. • When bonding together pool reinforcing steel and welded wire fabric (wire-mesh) with tie-wire, the tie-wires must be _____ .
  (a) stainless steel　　(b) accessible　　　　(c) made up tight　　(d) none of these

126. • Where devices or plug-in connections for tapping off feeder or branch circuits from busways consist of an externally operable fusible switch, _____ shall be provided for operation of the disconnecting means from the floor.
  (a) ropes　　　　　　(b) chains　　　　　　(c) hook sticks　　　(d) any of these

127. • Ground-fault protection is required for the feeder disconnect if _____ .
  (a) the feeder is rated 1,000 amperes or more　　　(b) it is a solidly grounded wye system
  (c) the system voltage is 480Y/277 volts　　　　　(d) all of these

128. • Meters, instruments, and relays installed in Class I, Division 2 locations can have switches, circuit breakers, and make-and-break contacts of push buttons, relays, alarm bells, and horns installed in enclosures approved for general purpose if current-interrupting contacts are _____ .
  (a) immersed in oil
  (b) enclosed within a hermetically sealed chamber
  (c) a or b
  (d) a and b

129. • Which of the following must be bonded?
  (a) electric equipment within 5 feet of the inside pool
  (b) pool steel structure steel.
  (c) metal fittings greater than 4 inches within or attached to the pool
  (d) all of these

130. • The connected load to which the demand factors of Table 220-32 apply shall include the _____ rating of all appliances that are fastened in place, permanently connected or located to be on a specific circuit: ranges, wall-mounted ovens, counter-mounted cooking units, clothes dryers, water heaters, and space heaters.
  (a) calculated　　　　(b) nameplate　　　　(c) circuit　　　　　(d) overcurrent protection

131.   • The essential electrical systems in a health care facility shall have sources of power from:
(a) a normal source generally supplying the entire electrical system.
(b) one or more alternate sources for use when the normal source is interrupted.
(c) a or b
(d) a and b

132.   • The cross sectional area of 1 inch IMC is approximately _____ square inch(es).
(a) 1.22   (b) 0.62   (c) 0.96   (d) 2.13

133.   • A receptacle outlet must be installed in dwelling units for every kitchen and dining area countertop space _____, and no point along the wall line shall be more than 2 feet, measured horizontally from a receptacle outlet.
(a) wider than 12 inches   (b) wider than 2 feet
(c) 2 feet and wider   (d) 12 inches and wider

134.   • For a circuit to be considered a multiwire branch circuit, it must have _____ .
(a) two or more ungrounded conductors with a potential difference between them
(b) a grounded conductor having equal potential difference between it and each ungrounded conductor of the circuit
(c) a grounded conductor connected to the neutral (grounded) conductor of the system
(d) all of these

135.   • Electrical nonmetallic tubing is not permitted in places of assembly, unless it is encased in at least _____ of concrete.
(a) 1 inch   (b) 2 inches   (c) 3 inches   (d) 4 inches

136.   • A conductor encased within material of composition or thickness that is not recognized by the *Code* is called a _____ conductor.
(a) noninsulating   (b) bare   (c) covered   (d) protected

137.   • For the purpose of determining the placement of receptacles in a dwelling-unit kitchen, a(n) _____ countertop is measured from the connecting edge.
(a) island   (b) usable   (c) peninsular   (d) cooking

138.   • Underground outdoor pool or spa equipment rooms or pits must have adequate drainage to prevent water accumulation during abnormal operation.
(a) True   (b) False

139.   • The case or housing of apparatus or the fence surrounding an installation to prevent accidental contact of persons to energized parts is called _____ .
(a) guarded   (b) covered   (c) protection   (d) an enclosure

140.   • 3/8 inch flexible metal conduit used for the supply to recessed lighting fixtures must have the raceway secured every _____ inches.
(a) 18   (b) 36
(c) 48   (d) is not required to be secured

141.   • In a grounded system, the conductor that connects the circuit grounded (neutral) conductor at the service to the grounding electrode is called the _____ conductor.
(a) main grounding   (b) common main
(c) equipment grounding   (d) grounding electrode

142.   • An underground rigid nonmetallic raceway must be not less than _____ feet from the inside wall of the pool or spa, unless space limitations prevent otherwise.
(a) 8   (b) 10   (c) 5   (d) 25

143.   • Open wiring on insulators within _____ feet from the floor shall be considered exposed to physical damage.
(a) 4   (b) 2   (c) 7   (d) none of these

144.   • What size THHN conductor is required for a 50-ampere circuit if the equipment terminals are listed for 75°C conductor sizing? Tip: Table 310-16 lists conductor ampacities.
(a) No. 10   (b) No. 8   (c) No. 6   (d) all of these

145. • What size copper grounding electrode conductor is required for a service that has three 500 kcmil copper conductors per phase?

(a) No. 1   (b) No. 1/0   (c) No. 2/0   (d) No. 3/0

146. • Wireways shall be permitted for _____ .

(a) exposed work   (b) concealed work

(c) wet locations if of raintight construction   (d) a and c

147. • At dwelling units, a 125-volt receptacle is required to be installed within 20 feet of the inside wall of the pool.

(a) True   (b) False

148. • The necessary equipment, usually a circuit breaker or switch and fuse and their accessories, is referred to as _____ .

(a) service equipment   (b) service

(c) service disconnect   (d) service overcurrent protection device

149. • Wiring methods permitted in Class I, Division 2 locations include _____ .

(a) threaded rigid metal conduit

(b) threaded steel intermediate metal conduit

(c) general purpose boxes and fittings containing no arcing devices

(d) all of these

150. • Ventilated heating or cooling equipment (including ducts) that service the electrical room or space cannot be installed in the dedicated space above a panelboard or switchboard.

(a) True   (b) False

151. • Where a cable containing a conductor with white or natural gray outer finish is used for 3-way or 4-way switch loops, the white or natural gray conductor can be used for the supply to the switch and reidentification of the white or natural gray conductor is not required.

(a) True   (b) False

152. • Working space shall not be used for _____ .

(a) storage   (b) raceways   (c) lighting   (d) accessibility

153. • A Class III, Division_____ location is where easily ignitable fibers or combustible flying material are stored or handled but not manufactured.

(a) 1   (b) 2   (c) 3   (d) all of these

154. • The rating of the branch circuit short circuit ground-fault protection device for an individual motor-compressor shall not exceed _____ percent of the rated-load current or branch circuit selection current, whichever is greater, if the protection device will carry the starting current of the motor.

(a) 100   (b) 125   (c) 175   (d) none of these

155. • A _____ receptacle without GFCI protection is permitted to be located in a dwelling-unit garage for one appliance if located within the dedicated space for the appliance.

(a) multi-outlet   (b) duplex   (c) single   (d) a and b

156. • A 6-foot-wide control panel operating at over 600 volts requires a minimum of _____ entrances to the required working space about electrical equipment.

(a) one   (b) two   (c) three   (d) four

157. • Heating, air-conditioning, and refrigeration equipment on rooftops or in attics and crawl spaces, shall have receptacle outlets for use in servicing these units. The receptacle shall be _____ .

(a) GFCI-protected   (b) at an accessible location

(c) within 25 feet of the equipment   (d) all of the above

158. • A mobile home that is factory-equipped with gas or oil-fired heating and cooking appliances can be supplied with a listed mobile home power-supply cord rated _____ amperes.

(a) 30   (b) 35   (c) 40   (d) 50

159.   • The minimum depth of clear working space in front of electrical equipment for 5,000 volts, nominal, to ground is
_____ feet when there are exposed live parts on both sides of the workspace with the operator between.

(a) 4   (b) 5   (c) 6   (d) 9

160.   • Dwelling-unit GFCI protection is required for all 125-volt 15- and 20-ampere, receptacles installed to supply
countertop appliances in kitchens.

(a) True   (b) False

161.   • Service conductors between the street main and the first point of connection to the service entrance run underground
are known as the _____ .

(a) utility service   (b) service lateral

(c) service drop   (d) main service conductors

162.   • The coupler of a mobile home is included when determining the number of branch circuits for a mobile home.

(a) True   (b) False

163.   • Motor disconnecting means shall be a motor-circuit switch rated in _____, be a molded case switch, or a circuit
breaker.

(a) horsepower   (b) watts   (c) amperes   (d) locked-rotor current

164.   • In dwelling units, outdoor receptacles can be connected to the 20 ampere small-appliance branch circuit.

(a) True   (b) False

165.   • A small mobile home park has six mobile homes. What is the total park electrical wiring system load after applying
the demand factors permitted in Article 550?

(a) 4,640 VA   (b) 27,840 VA   (c) 96,000 VA   (d) none of these

166.   • The 3 volt-ampere per square foot for dwelling unit general lighting includes all 15- and 20-ampere general use
receptacles. The floor area shall _____ open porches, garages, or unused or unfinished spaces, not adaptable for future
use.

(a) include   (b) not include

167.   • A motel conference room is designed for the assembly of more than 100 persons. The fixed wiring methods require
_____ .

(a) rigid nonmetallic conduit   (b) Type MC cable

(c) nonmetallic sheathed cable   (d) Type AC cable

168.   • Which rooms in a dwelling unit must have a switch-controlled lighting outlet?

(a) every habitable room   (b) bathrooms

(c) hallways and stairways   (d) all of these

169.   • Exposed live parts of motors and controllers operating at _____ volts or more between terminals shall be guarded
against accidental contact by enclosure or by location.

(a) 24   (b) 50   (c) 150   (d) 60

170.   • A branch circuit rated 20 amperes serves four receptacles. The rating of the receptacles must not be less than _____
amperes.

(a) 10   (b) 15   (c) 25   (d) 30

171.   • In a commercial garage, the pit shall be classified as a _____ location, unless provisions are made for six air changes
per hour.

(a) Class I, Division 2   (b) Class II, Division 2   (c) Class II, Division 1   (d) Class I, Division 1

172.   • Transformers with nonflammable dielectric fluid rated over 35,000 volts installed in a vault shall be furnished with a
_____ when installed indoors.

(a) fluid confinement area

(b) pressure relief vent

(c) means for absorbing or venting any gasses generated by arcing

(d) all of these

173.   • Multiwire branch circuits shall _____ .
(a) supply only line-to-neutral loads
(b) provide a means to disconnect simultaneously all ungrounded conductors if the multiwire branch circuit supplies more than one device or equipment on the same yoke for commercial installations
(c) have their conductors originate from different panelboards
(d) none of these

174.   • ³/₈ inch listed liquidtight flexible metal conduit with fittings listed for grounding can be used as a grounding path when _____ .
(a) a 20-ampere overcurrent protection device protects circuit conductors and the total length of the liquidtight is 6 feet or less
(b) a 60-ampere overcurrent protection device protects conductors and the total length of the liquidtight is 6 feet or less
(c) a and b
(d) none of these

175.   • A single receptacle installed on an individual branch circuit must be rated at least _____ percent of the rating of the circuit.
(a) 50   (b) 60   (c) 90   (d) 100

176.   • When bonding pool and spa equipment, a solid No. 8 copper conductor must be run back to the service equipment. This conductor must be unbroken.
(a) True   (b) False

177.   • When replacing an ungrounded receptacle in a bedroom of a dwelling unit, if a grounding means does not exist in the receptacle enclosure, you must use a _____ .
(a) nongrounding receptacle   (b) grounding receptacle
(c) GFCI-type receptacle   (d) a or b

178.   • A two-family dwelling unit, such as a duplex apartment, is considered a single-family dwelling.
(a) True   (b) False

179.   • Working space distances shall be measured from the _____ of equipment or apparatus if such are enclosed.
(a) front   (b) opening   (c) a or b   (d) none of these

180.   • An overload may be caused by a short-circuit or ground-fault.
(a) True   (b) False

181.   • Where liquidtight flexible conduit is used to connect to equipment and flexibility where required, a separate _____ conductor must be installed.
(a) bond jumper   (b) bonding
(c) equipment grounding conductor   (d) none of these

182.   • Warning signs for over 600 volts shall read "Warning – High Voltage – Keep Out".
(a) True   (b) False

183.   • When one electrical circuit controls another circuit through a relay, that first circuit is called a _____ .
(a) control circuit   (b) remote-control circuit   (c) signal circuit   (d) controller

184.   • Liquidtight flexible metal conduit smaller than _____ inch(es) shall not be used, except as permitted in Section 350-10(a).
(a) ³/₈   (b) ¹/₂   (c) 11   (d) 1¹/₂

185.   • The _____ of any system is the ratio of the maximum demand for a system, or part of a system, to the total connected load of a system under consideration.
(a) load   (b) demand factor   (c) minimum load   (d) computed factor

186.   • Type MV is defined as a single or multiconductor solid dielectric insulated cable rated _____ volt or higher.
(a) 601   (b) 1,001   (c) 2,001   (d) 6,001

187.   • Which of the following statement about MI cable is correct?
(a) It may be used in any hazardous location.
(b) A single run of cable shall not contain more than the equivalent of four quarter bends.
(c) It shall be securely supported at intervals not exceeding 10 feet.
(d) none of these

188.   • Doorbell wiring, rated as a Class 2 control circuit in a dwelling unit _____ run in the same raceway or enclosure with light and power conductors.
(a) is permitted with 600 volt insulation to be
(b) shall not be
(c) shall be
(d) is permitted, if the insulation is equal to the highest voltage conductor installed, to be

189.   • Locations where flammable paints are dried but in which the ventilating equipment is interlocked with the electrical equipment may be designated as a _____ location.
(a) Class I, Division 2   (b) Unclassified   (c) Class II, Division 2   (d) Class II, Division 1

190.   • A disconnecting means is required to disconnect the _____ from all ungrounded supply conductors.
(a) motor   (b) motor or controller   (c) controller   (d) motor and controller

191.   • A feeder to a panelboard is run with No. 6 THHN conductor. The maximum size overcurrent protection for these conductors are _____ amperes. Note: Terminal rating of 60°C.
(a) 60   (b) 40   (c) 90   (d) 100

192.   • The ampacity of supply branch circuit conductors and the overcurrent protection devices for X-ray equipment shall not be less than _____ .
(a) 50 percent of the momentary rating
(b) 100 percent of the long-time rating
(c) the larger of a or b
(d) the smaller of a or b

193.   • Underground gasoline dispenser wiring shall be installed in _____ .
(a) threaded rigid metal conduit
(b) threaded intermediate metal conduit
(c) rigid nonmetallic conduit when buried not less than two feet of earth
(d) any of these

194.   • Conductors that supply two or more motors shall have an ampacity _____ .
(a) not less than the total sum of the full-load current rating plus 125 percent of the highest motor in the group
(b) equal to the sum of the full-load current rating of all the motors plus 125 percent of the highest motor in the group
(c) equal to the sum of the full-load current rating of all the motors plus 25 percent of the highest motor in the group
(d) not less than 125 percent of the sum of all the motors in the group

195.   • In Class 2 alternating current circuit installations, the maximum voltage for wet contact that is most likely to occur is _____ volts peaks.
(a) 10.5   (b) 21.2   (c) 100   (d) none of these

196.   • It shall be permissible to extend busways vertically through dry floors if totally enclosed (unventilated) where passing through and for a minimum distance of _____ feet above the floor to provide adequate protection from physical damage.
(a) 6   (b) 6½   (c) 8   (d) 10

197.   • A synthetic nonflammable insulating medium which, when decomposed by electric arcs, produces predominantly nonflammable gaseous mixtures is known as _____ .
(a) oil   (b) geritol   (c) askarel   (d) phenol

198.   • The *Code* covers all of the following electrical installations except _____ .
(a) floating dwellings
(b) in or on private and public buildings
(c) industrial substations
(d) public utilities generation and transmission facilities

199.   • Raceways shall not be used as a means of support for Class 2 or Class 3 circuit conductors.
(a) True   (b) False

200.   • The purpose of the *Code* is the practical safeguarding of persons and property from hazards arising from the use of electricity. The *Code* contains provisions considered necessary for safety regardless of _____ .
(a) efficient use
(b) convenience
(c) good service or future expansion
(d) all of these

ANSWERS
Units 1 through 12

The *Code* references to each question apply to the 1999 *National Electrical Code*®. In addition one of the unique feature of this Instructor's Guide is the percentages 70%, 85%, and 90% included after most of the answers. These numbers indicate the percentages of my students who answered the questions correctly (rounded off to the nearest 5%). This information will help you to determine which concept the students grasped, and which questions are particularly difficult. With this information, you can evaluate which subjects need more teaching or review.

Note: Some answers do not have a percentage because they reflect questions based on the 1999 *NEC*®, for which I don't have any statistical data at the time of this printing.

Errors: Although I have taken great care in insuring the each question is valid and that the answers are correct, I am not perfect. There were many changes to the 1999 *NEC*® and I'm sure I missed some. If you feel that I have made an error, please let me know by contacting Delmar Publishers by mail addresssed to Electrical Editor, Delmar Publishers, 3 Columbia Circle, Box 15015, Albany, NY 12212. You can also send e-mail to info@delmar.com.

God Bless, Mike Holt

1. (a) 0.5   95%
$^1/_2$ = 1 divided by 2 = .5

2. (d) .2   85%
4/18 = 4 divided by 18 = .22

3. (d) .075   85%
75 watts/1,000 = .075 kW

4. (b) 110 is greater than 100 watts   70%
100 watts/.9 = 111 watts

5. (d) multiplier

6. (b) 23 amperes   80%
115% = 1.15 20 × 1.15 = 23 amperes

7. (a) 72   95%
90 amperes × .8 = 72 amperes

8. (d) all of the above

9. (c) 100 feet   90%
41,000/421.8 = 99.3 feet

10. (b) 50 ampere   80%
$I = P/(E \times 1.732)$
$P = 18$ kW × 1,000 = 18,000 watts
E = 208 volts
I = 18,000 watts/(208 volts × 1.732)
I = 50 ampere

11. (b) left   90%

12. (b) .75   95%

13. (c) 2.25   90%

14. (c) 3   90%

15. (b) 9.6 kVA   90%
change % to decimal and add 1
20% = .20 = 1.2, 8 kVA × 1.2 = 9.6 kVA

16. (a) .8   85%
1/1.25 = .8

17. (c) 80 ampere   65%
reciprocal of 125% = 80% = .8
100 × .8 = 80 ampere

18. (c) 30   65%
5 + 7 + 8 + 9 = 29 answers rounded to nearest 5

19. (b) 50 watts   95%
$P = 16^2$ ampere × .2 ohm
P = 51.2 watts

20. (c) 3 square inches   70%
Area = πr^2 = 3.14 × 1^2
Area = 3.14 square inches

21. (c) 16   90%
4 × 4 = 16

22. (c) 144   95%
12 × 12 = 144

23. (b) 32   85%

24. (a) 1.732   95%

25. (d) $D = \dfrac{CM \times VD}{2 \times L \times I}$ 60%

26. (c) $P = I \times E$ 95%

27. (d) all of these   95%

28. (a) True   90%

29. (b) direct   85%

30. (c) Alternating   95%

31. (d) silver, copper, gold, aluminum   95%

32. (c) volt   95%

33. (a) silver, gold.   85%

34. (d) ohm   95%

35. (b) False   85%

36. (a) intensity   80%

37. (c) power   90%

38. (d) directly, inversely   95%

39. (d) all of these   90%

40. (d) impedance   85%

41. (a) 6.4 volts   75%
$E_{VD} = I \times R$, I = 16 ampere
$R = \dfrac{2 \text{ ohms}}{1,000} \times 200 = .4$ ohm
E_{VD} = 16 ampere × .4 ohm
E_{VD} = 6.4 VD

42. (a) .14 ohm   95%
Formula: R = E/I
R = 7.3 volts/50 amperes
R = .144 ohm

43. (a) 175 watts   90%
$P = I \times E$
$P = 24 \times 7.2$
$P = 173$ watts

44. (c) 42 ampere
$I = P/E$
$P = 10,000$ VA, $E = 240$ volts
$I = 10,000$ VA/240 volts
$I = 41.67$ ampere

45. (c) a and b   50%

46. (a) True   90%

47. (b) 150 watts   95%
$P = I^2 \times R$
$I = 16$ ampere, $R = .6$ ohm
$P = 16$ ampere $^2 \times .6$ ohm
$P = 153.6$ watts

48. (a) 43 watts   75%
$P = I \times E$
$P = 1^2$ ampere $\times 3.6$ volts
$P = 43$ watts

49. (a) $33.55   70%
$P = I^2 \times R$
$P = 12$ ampere
$P = 12$ ampere $^2 \times .30$ ohm
$P = 43.2$ watts per hour
Power
Day = 43.2 watts $\times$ 24 hours = 1,037 watts
Year = 1,037 watts $\times$ 365 days = 378,505 watts
Year kWH = 378,505/1,000 = 378.51 kWH
$ = 378.51 kWH $\times$ .086 = $32.55

50. (b) 2.5 kW   55%
$R = E^2/PR = 230$ volts
$R = 230$ volts^2/10,000 watts
$R = 5.29$ ohm

$P = E^2/R = 115$ volts
$P = 115^2/5.29$ ohm
$P = 2,500$ watts/1,000 = 2.5 kW

51. (d) ammeter   95%

52. (a) direct current   70%

53. (d) at right angles (perpendicular)   95%

54. (b) coil and power supply   80%

☆ Challenge Answers

55. (d) 1,000 VA   65%

56. (d) Salt water   70%

57. (d) Current   65%

58. (a) material   85%

59. (a) increase, $E = I \times R$ 70%

60. (b) Reduced by half   75%
Current is inversely proportion to resistance.
Example: $I = E/R$
$I = 100$ volts/*10 ohm*
$I = 10$ ampere
$I = 100$ volts/*5 ohm*
$I = 20$ ampere

61. (c) a shorted coil   55%
If the reading is less than 30 ohm, the length of the coils conductor must be less than required (shorted).

62. (d) any of these   65%

63. (d) 20 watts   70%
The formula I^2R in the question has nothing to do with the actual calculation. The two knowns given are Volts(E) and Resistance(R).
$P = E^2/R$
$P = 10$ volts^2/5 ohm
$P = 20$ watts

64. (c) $P = I^2 \times R$ 75%

65. (a) higher   60%
Power increases with increased resistance.
Example: $P = I^2 \times R$
$P = 10$ ampere $^2 \times$ *10 ohm* = 1000 watts
$P = 10$ ampere $^2 \times$ *20 ohm* = 2000 watts

66. (b) 4.5 ampere   85%
Formula $I = E/R$
$E = 120$ volts, $R = 26.45$ ohm
$I = 120$ volts/26.45 ohm
$I = 4.54$ ampere

67. (b) less
Resistance does not change with changing voltage. Power (watts) decreases with reduced voltage.
$P = E^2/R$
$P = $ *120 volt* ^{2}/100 ohm = 140 watts
$P = $ *100 volt* ^{2}/100 ohm = 100 watts

68. (c) use more power   65%
Power (watts) increases with increased voltage.
$P = E^2/R$
$P = $ *120 volt* ^{2}/100 ohm = 140 watts
$P = $ *100 volt* ^{2}/100 ohm = 100 watts

69. (d) 1,227 watts   70%
When the voltage is reduced the power will also be reduced.
$P = E^2/R$
$P = $ *230 volt* ^{2}/35.27 ohm = 1,500 watts
$P = $ *208 volt* ^{2}/35.27 ohm = 1,227 watts

70. (b) 18 watts   60%
$P = I^2 \times R$
$P = 3$ ampere $^2 \times 2$ ohm
$P = 18$ watts

Unit 2 – Electrical Circuits

1. (a) True   95%

2. (a) True   85%

3. (a) True   85%

4. (c) voltage drop   85%

5. (d) constant   85%

6. (b) False   70%
   voltage is additive

7. (b) False   90%

8. (b) parallel   95%

9. (d) all of these   70%

10. (b) False   95%

11. (a) True   95%

12. (b) False   80%
    power is additive

13. (d) all of these   80%

14. (d) none of these   85%

$$R_T = \frac{\text{Resistance of equal resistors}}{\text{Number of resistors}}$$

$$R_T = \frac{6\ \text{ohm}}{3\ \text{resistors}}$$

$R_T = 2$ ohm

15. (d) 9 ohm   85%
    $R = E^2/P$
    Coffee pot resistance = 120 volts^2/500 watts
    Coffee pot resistance = 24 ohm
    Skillet resistance =120 volts^2/1,000 watts
    Skillet resistance = 14.4 ohm

$$R_T = \frac{(R_1 \times R_{2)}\ \text{Product}}{(R_1 \times R_{2)}\ \text{Sum}}$$

$$R_T = \frac{(24\ \text{ohm} \times 14.4\ \text{ohm})}{(24\ \text{ohm} + 14.4\ \text{ohm})}$$

$R_T = 9$ ohm

16. (a) 5 ohm   85%

$$R_T = \frac{1}{{}^1/_{R1} + {}^1/_{R2} + {}^1/_{R3}}$$

$$R_T = \frac{1}{{}^1/_{20} + {}^1/_{20} + {}^1/_{10}}$$

$$R_T = \frac{1}{.05 + .05 + .10} = \frac{1}{.2}$$

$R_T = 5$ ohm

17. (b) 1, 2, and 3   70%

18. (c) series-parallel circuit   85%

19. (a) True, Article 100   85%

20. (d) 100%   70%

21. (a) 0%   75%

22. (a) True   95%

23. (c) 20 ampere   70%

24. (b) False   80%

25. (a) 0 ampere   95%

26. (b) False   70%

27. (d) 40 ampere   70%
    Depends on the harmonic content of the
    nonlinear load.

28. (d) b and c   85%

29. (b) grounded   70%

☆ Challenge Answers

30. (a) one ampere   60%
    The current through any resistor of a circuit is
    equal to the current of the circuit. The current
    of this circuit is equal to:
    I = E/R
    E = 12 volts, R = 12 ohm
    I = 12 volts/12 ohm
    I = 1 ampere

31. (d) all of these   70%
     Voltage in a series circuit is distributed among all
     equal resistors equally according to Kirchoff's Law on
     voltage. This means that since there are four resistors,
     each resistor will have one-quarter of the source
     voltage. 120 volts/4 = 30 volts.

32. (d) all of these   70%

33. (c) 10 volts

     $E = I \times R$ 50%
     E = 30 volts, R_T = 22.5 ohm
     I = 30 volts/22.5 ohm = 1.33 ampere

     $E = I \times R_2$
     I = 1.33 ampere, R_2 = 7.5 ohm
     E = 1.33 ampere $\times$ 7.5 ohm
     E = 9.98 volts

34. (c) 6 volts. This is tricky. By placing   85%
     the voltage meter across the switch, the circuit con-
     ductors and the load are used as part of the voltage
     meter leads.

35. (c) parallel   70%

36. (b) parallel   70%

37. (d) b and c   40%

38. (b) all in parallel   50%
     The greater the operating voltage, the greater the
     power.
     Series Example: 120-volt circuit, four 10 ohm resis-
     tors: If connected in series, each resistor would oper-
     ate at one-quarter of the voltage.
     $P = E^2/R$
     P = 30 volts²/10 ohm
     P = 90 watts
     Parallel Example: 120-volt circuit, four 10 ohm resis-
     tors: If connected in parallel, each resistor would oper-
     ate at 120 volts.
     $P = E^2/R$
     P = 120 volts²/10 ohm
     P = 1,440 watts

39. (b) 1 ohm   75%
     Rule: The total resistance of a parallel circuit is always
     less than the smallest resistor:
     Formula:
     $$R_T = \frac{1}{\frac{1}{R} + \frac{1}{R_2} + \frac{1}{R_3}}$$

     $$R_T = \frac{1}{\frac{1}{2} + \frac{1}{3} + \frac{1}{7}}$$

     $$R_T = \frac{1}{(0.5 + .333 + .1429)}$$

     $$R_T = \frac{1}{9759} = 1.02 \text{ ohm}$$

40. (c) ammeter 3 only   75%

41. (d) 48 ohm   80%
     $R = E^2/P$
     E = 24 volt, P = 12 watts
     $R = E^2/P$ = 24 volts²/12 watts
     R = 576/12 watts
     R = 48 ohm

42. (a) 22 ohm   85%
     Formula: $R_T = \dfrac{R_1 \times R_2 \text{ (Product)}}{R_1 + R_2 \text{ (Sum)}}$

     R_1 (Bell 1) = E/I
     E = 30 volts, I = .75 ampere
     R_1 = 30 volts/.75 ampere
     R_1 = 40 ohm
     R_2 (Bell 2) = 48 ohm
     $RT = \dfrac{R_1 \times R_2}{R_1 + R_2} = \dfrac{40 \text{ ohm} \times 48 \text{ ohm}}{40 \text{ ohm} + 48 \text{ ohm}}$
     R_T = 1920/88
     R_T = 21.82 ohm

43. (d) all of these   70%

44. (a) 1.5 volts   60%
     $E = I \times R$
     I = The current through R3 is one-half of .75 ampere
     (resistors in parallel)
     I = .75 ampere/2, = .375 ampere
     R = 4 ohm
     E = .375 ampere $\times$ 4 ohm
     E = 1.5 volts

45. (b) 3 volts   65%
     $E = I \times R$
     I = .75 ampere, R = 4 ohm (resistance total)
     E = .75 ampere $\times$ 4 ohm
     E = 3 volts

46. (d) 10 ohm 85%

Calculate the parallel resistance and then add the resistance of resistor R1.

$$\frac{\text{Resistance of one resistor}}{\text{Number of resistors}} = \frac{15 \text{ ohm}}{3}$$

R = 5 ohm

Note: The 3 parallel resistor can be thought of as a single 5 ohm resistor in series with resistor R1.

Resistance total = $R_1 + (R_{2, 3, 4})$

R_T = 5 ohm (R_1) + 5 ohm ($R_{2, 3, 4}$)

R_T = 10 ohm

47. (a) 60 volts 70%

$E_{VD} = I \times R$

I = 12 ampere, R = 5 ohm

E_{VD} = 12 ampere $\times$ 5 ohm

E_{VD} = 60 volts

3–12 Multiwire Circuits

48. (b) .58 ampere 60%

If the neutral is opened, the multiwire circuit becomes one 240-volt series circuit.

$I = E/R_T$

E = 240 volts, $R_T = R_1 + R_2$

$R_1 = E_2/P = 130$ volts^2/75 watts = 225 ohm

$R_2 = E_2/P = 120$ volts^2/75 watts = 192 ohm

Resistance total 417 ohm

$I = E/R_T$

I = 240 volts/417 ohm

I = .575 ampere

NEC® Answers

49. (c) 230-52 90%

50. (c) 230-54(b) Exception 95%

51. (c) 230-54(g) 90%

52. (b) 230-70(a) 95%

53. (d) 230-70(c) 90%

54. (c) 230-72(b) 75%

55. (d) 230-76 80%

56. (d) 230-79(b) 95%

57. (b) 230-80 85%

58. (a) 230-72(c) 90%

59. (c) 230-93 75%

60. (c) 230-95(a) 75%

61. (b) 230-95(c) FPN No. 1 80%

62. (a) 230-202(b) 90%

63. (c) 240-3 90%

64. (a) 240-3(b), 310-16, 240-6(a), 110-14(c)(1) 70%

65. (c) 240-6(a) 80%

66. (c) 240-10 80%

67. (a) 240-13 95%

68. (a) 240-20(b) 60%

69. (a) 240-21

70. (b) 240-21(c)(3) 85%

71. (c) 240-24(a) 90%

72. (a) 240-33 90%

73. (b) 240-51(a) 95%

74. (d) 240-54(b) and 250-51(a) 55%

75. (b) 240-54(e) 90%

76. (d) 240-60(c) 85%

77. (c) 240-82 85%

78. (a) 240-83(b) 90%

79. (a) 240-85 90%

80. (d) 250-2(d) 90%

81. (b) 250-8 95%

82. (d) 250-24(a)(1) 70%

The grounding electrode conductor can be connected to the grounded service conductor at any accessible point between the load end of the service drop (or service lateral) and the service disconnecting means.

83. (c) 250-24(b)(1) 95%

84. (d) 250-28(b) 95%

85. (b) 250-32(e) 95%

86. (c) 250-50(a)(2) 75%

87. (a) 250-50(b) 90%

88. (b) 250-52(c)(1) 90%

89. (d) 250-52(c)(3) 85%

90. (b) 250-56 60%

91. (a) 250-58, 250-52 95%

92. (d) 250-64 95%

93. (c) Table 250-66 70%

94. (b) 250-68(a) Exception 80%

95. (d) 250-70 and 250-8 95%

96. (c) 250-86 Exception No. 2 90%

97. (d) 250-94

98. (d) 250-96(a) 95%

100. (a) 250-102(c)   80%

When determining the size of bonding jumpers to parallel service raceways, and the bonding jumper is run in parallel, the service entrance conductors in each raceway determine the size. Table 250-66 requires a No. 1/0 bonding jumper for each service raceway.

101. (d) 250-102(e)   90%

102. (a) 250-106

103. (c) 250-112(b)   95%

104. (b) 250-118(6)   90%

105. (d) Table 250-122   90%

106. (a) 250-122(c)   70%

Multiple circuits in one raceway require one equipment grounding conductor. Its size is based on the largest overcurrent protection device of the multiple circuits.

107. (d) 250-126   80%

108. (d) 250-142(b)   90%

109. (b) 250-146(a)

110. (a) 250-146(b)

111. (a) 250-148 Exception   65%

112. (a) 250-170   95%

113. (b) 280-2   90%

114. (b) 280-23   85%

115. (d) 300-3(b), see 300-5(i)   80%

116. (d) 300-4(b)(1)   90%

117. (a) 300-4(c)   95%

118. (c) Table 300-5 Column 3   65%

119. (b) 300-5 Table in the heading   95%

120. (a) 300-5(e)   90%

121. (d) 300-5(h)   60%

122. (d) 300-6(a)   95%

123. (a) 300-7(a)   90%

124. (d) 300-11(a)(1)

125. (b) 300-13(b)   85%

126. (a) 300-15(f)   95%

127. (d) 300-17   95%

128. (d) 300-18(b)

129. (c) 300-20(a)   85%

130. (d) 300-22(a)   85%

131. (a) 300-22(b)   85%

132. (d) 300-22(i)   65%

133. (d) 305-3(a)   90%

134. (d) 305-4(c)   95%

135. (c) 305-4(d)   75%

136. (c) 305-4(g)   95%

137. (d) 305-6(a)

138. (d) 305-6(b)

139. (d) 310-4 Exception No. 2   95%

140. (d) 310-4

141. (a) 310-6   80%

142. (b) 310-10 FPN   85%

143. (a) Table 310-13   90%

144. (c) 310-13 Table   95%

145. (c) 310-15(a)(2) Exception   90%

146. (d) 310-15(b)(3)   70%

147. (c) 310-15(b)(5)   80%

148. (a) 310-17, 310-69   80%

149. (b) 310-60(c)(2)(b)   90%

150. (a) 318-2

151. (d) 318-5   90%

152. (d) 318-6(j)

153. (a) 318-7(b)(2) Note B   75%

154. (c) 318-8(c)   85%

155. (c) 310-15(b)(2)(a)   90%

156. (c) 320-7   95%

157. (a) 320-11   90%

158. (c) 320-14   60%

159. (d) 321-3(a)   90%

160. (c) 324-3   95%

161. (a) 324-7   90%

162. (c) 326-1   70%

163. (b) 328-4   85%

164. (d) 328-5   95%

165. (b) 328-10   80%

166. (c) 328-13   90%

167. (d) 328-31   80%

168. (d) 330-12(1)   85%

169. (b) 330-13(1)   95%

170. (a) 330-20   95%

171. (d) 330-22   85%

172. (c) 331-3(2)   95%

173. (d) 331-3(8)

174. (b) 331-4(8) and 518-4   75%

175. (d) 331-11   95%

176. (d) 333-3 and 333-4(1) and (8)   95%

177. (c) 333-8   95%

178. (b) 333-12(a)   90%

179. (d) 334-4   80%
Direct burial only where identified for such use, but
not always.

180. (a) 334-20   90%

181. (d) 336-4   65%

182. (d) 336-5(8)   90%

183. (c) 336-18   95%

184. (a) 336-18, Exception No. 3

185. (a) 336-30(a)(2)   70%

186. (b) 338-1(c)   70%

187. (d) 339-1(a)   75%

188. (d) 339-3(b)   90%

189. (d) 340-4   95%

190. (a) 342-3(c) refers to 336-5(a)(1)   80%

191. (d) 342-6   85%

192. (d) 342-7(b)(3)   95%

193. (d) 343-4(1)(2)(3)   90%

194. (b) 345-3(a) and 346-3(a)   90%

195. (c) 345-7, Chapter 9, Table 4   75%

196. (a) 345-9(b)   95%

197. (b) 345-12(a)(1)   95%

198. (d) 346-3(a)   65%

199. (c) 346-5   90%

200. (b) 346-10 and Table 346-10   75%
The general rule applies. Since the question did not
specify (one shot bender), don't use Table 346-10
Exception Table. Note the term field bend is in the
general rule.

Notes:

Unit 3 – Understanding Alternating Current

1. (a) True — 80%

2. (a) True — 100%

3. (b) False — 90%

4. (d) all of these — 95%

5. (b) alternating current — 60%

6. (d) none of these — 80%

7. (d) hertz — 80%

8. (a) peak — 80%

9. (d) degrees — 95%

10. (d) lead and lag — 80%

11. (d) Instantaneous — 75%

12. (a) Peak — 90%

13. (b) Root-mean-square — 60%

14. (c) induction — 95%

15. (c) self-induced voltage — 80%

16. (d) all of these — 80%

17. (b) increasing — 85%

18. (a) True — 95%

19. (d) none of these — 80%

20. (a) True — 80%

21. (a) True — 80%

22. (d) impedance — 95%

23. (b) False — 70%
   caused by ac power

24. (d) skin effect — 95%

25. (a) True — 100%

26. (d) all of these — 60%

27. (a) Capacitance — 100%

28. (b) charged — 90%

29. (d) shorted — 80%

30. (a) conducting — 95%

31. (b) condenser — 100%

32. (a) leads the applied voltage by 90° — 80%

33. (a) True — 90%

34. (c) apparent power — 80%

35. (a) True — 95%

36. (a) True — 90%

37. (b) R/Z — 90%

38. (d) a and b — 75%

39. (d) none of these — 100%

40. (c) power factor — 80%

41. (a) True — 80%

42. (b) $296 — 80%
   Power per day = 10 fixtures × 150 watts × 6 hours
   Power per day = 9,000 watts
   kW per year =

   $$\frac{(9{,}000 \text{ watts power day} \times 365 \text{ days})}{1{,}000}$$

   kW per year = 3,285 kW

   Cost per year = 3,285 kW × .09 = $295.65

43. (d) 75% — 80%
   Power Factor = Watts/ VA

   $$\text{Power factor} = \frac{(4 \text{ fixtures} \times 34 \text{ watts})}{(120 \text{ volts} \times 1.5 \text{ ampere})}$$

   $$\text{Power factor} = \frac{136 \text{ watts}}{180 \text{ volt-ampere}}$$

   Power factor = .7555 or 75.55%

44. (d) 2,400 VA — 75%
   Note: (a) is not the correct answer because apparent power is expressed in VA not watts.

45. (d) none of these — 70%
   Note: True power is always less than apparent power.

46. (a) True — 70%

47. (c) 1,200 watts — 70%
   Note: (a) is not the correct answer because true power is expressed in watts not VA.

48. (b) 30 kVA   90%

$$kVA = \frac{(Volts \times Ampere)}{1,000}$$

$$kVA = \frac{(240 \text{ volts} \times 125 \text{ ampere})}{1,000}$$

$$kVA = 30 \text{ kVA}$$

49. (b) 35 kVA   90%

$$kVA = \frac{Watts}{Power\ Factor}$$

$$kVA = \frac{30,000 \text{ watts}}{0.85 \text{ Power Factor}}$$

$$kVA = 35,294 \text{ watts}/1,000$$
$$kVA = 35$$

50. (d) 6 circuits   60%
Circuit = 120 volts × 20 ampere = 2,400 VA

$$Lights\ per\ circuit = \frac{2,400 \text{ VA}}{300 \text{ watts}} = 8 \text{ lights}$$

$$No.\ of\ circuits = \frac{42 \text{ fixtures}}{8 \text{ fixture per circuit}}$$

No. of circuits = 5.25 circuits

51. (c) 7 circuits   70%
Circuit capacity = 120 volts × 20 ampere
Circuit capacity = 2,400 VA

$$Load\ VA = \frac{Watts}{PF}$$

$$Load\ VA = \frac{300 \text{ watts}}{.85 \text{ power factor}} = 353 \text{ VA}$$

$$Lights\ per\ circuit = \frac{2,400 \text{ VA}}{353 \text{ VA}} = 6 \text{ lights}$$

$$No.\ of\ circuits = \frac{42 \text{ fixtures}}{6 \text{ fixtures per circuit}}$$

No. of circuits = 7

52. (a) True   100%
53. (b) 73 percent   95%

$$Efficiency = \frac{Output\ Watts}{Input\ Watts}$$

$$Efficiency = \frac{1,320 \text{ watts}}{1,800 \text{ watts}}$$

Efficiency = .7333 or 73.33%

54. (b) 1.5 ampere   100%

$$Input\ Watts = \frac{Output\ Watts}{Efficiency}$$

$$Input\ Watts = \frac{160 \text{ watts}}{.88 \text{ EFF}}$$

Input Watts = 182 watts

$$Input\ Ampere = \frac{Watts}{Volts} = \frac{182 \text{ watts}}{120 \text{ volts}}$$

Input Ampere =1.5 amperes

55. (a) 970 watts   90%
Output Watts = Input Watts × Efficiency
Output Watts = 1,000 watts × .97 EFF
Output Watts = 970 watts

☆ Challenge Answers

56. (d) all of these   90%
57. (a) reduced voltage drop   85%
When the system voltage is increased, the current will
decrease and voltage drop is determined by E = I × R.
58. (b) ease of voltage variation   90%
59. (d) all of these   80%
60. (c) one second   80%
61. (a) 1/120th of a second   90%
62. (a) the same   70%
63. (c) 35 ampere   80%
Effective (RMS) Current = Peak Current × .707
RMS Current = 50 ampere × .707
RMS Current = 35 ampere
64. (b) False   50%
The peak value of alternating current is equal to:
E_{Peak} = 120 volts × 1.41
E_{Peak} = 169 volts
65. (c) effective   60%
66. (c) Lenz's Law   90%
67. (c) XL   90%
68. (a) True   90%
69. (c) ohm   90%
Inductance is measured in Henrys, but inductive reactance is measured in ohm.
70. (a) remain constant   75%
Inductive reactance changes with frequency, not with
voltage or current.

71. (a) impedance 80%

72. (c) alternating current 80%
Impedance is the opposition to current flow because of resistance, capacitive reactance (X_C), and inductive reactance (X_L)

73. (a) higher than 95%

74. (b) shorted 80%
A capacitor is constructed of two layers of foils separated by dielectric insulation. If the lamp lights up, that means that the capacitor must be shorted.

75. (c) dielectric 95%

76. (c) farads 90%

77. (a) ohm 90%

78. (b) lags the current by 90° 75%

79. (c) $X_L = X_C$ are in series 80%

80. (c) apparent power 80%

81. (b) 2.3 kVA 60%
Apparent Power = Volts × Ampere
Volts = 120, Ampere = 19.2
Apparent Power = 120 volts × 19.2 ampere
Apparent Power = 2,304 VA/1000, = 2.304 kVA

82. (d) resistive loads 80%
Power factor is unity (100%) for resistive loads.

83. (c) 65 ampere 75%
Current = Watts/(Volts × 1.732 × Power Factor)
Watts = 24,000, Volts = 230, Power Factor = .92
Current = 24,000 watts/(230 volts × 1.732 × .92 PF)
Current = 65.48 ampere

84. (c) series-parallel 70%
Series to measure the current and parallel to measure the voltage.

85. (d) b and c 70%

86. (c) $I^2 \times R$ 80%

87. (c) 24,367 watts 60%
True Power (watts) = Apparent Power × Power Factor
Apparent Power = Volts × Ampere × 1.732
Volts = 208, Ampere = 76
Apparent Power = 208 volts × 76 ampere × 1.732
Apparent Power = 27,379 volt-ampere

True Power = Apparent Power × Power Factor
Apparent Power = 27,379, PF = .89
True Power = 27,379 VA × .89 PF
True Power = 24,367 watts

88. (b) 1.91 kW 70%
True Power = Apparent Power × Power Factor
Apparent Power = 2,100, Power Factor = .91
True Power = 2,100 VA × .91 PF
True Power = 1,911/1000 = 1.911 kW

89. (c) Horsepower × 746/Watts Input 60%
Motor efficiency is equal to motor output watts (Horsepower × 746 watts) divided by motor input watts.

90. (c) 90% 80%
Efficiency = Output/Input
Output = 4,000 VA
Input Watts = 3,600 VA
Efficiency = 3,600 VA output/4,000 VA input
Efficiency = .9 or 90%
Note: Efficiency is never greater than 100%

NEC® Answers

91. (d) 346-12(b)(1) 95%

92. (b) 346-16(c) 90%

93. (b) 347-2(d) 85%

94. (d) 347-3 85%

95. (d) 347-8 95%

96. (d) 347-12 and FPN 70%

97. (c) 348-10 80%

98. (d) 348-16 90%

99. (c) 349-20(b) 90%

100. (a) 350-5(7) 95%

101. (d) 350-12 Table 80%

102. (b) 350-20 95%

103. (a) 351-6(b) and Table 350-12 90%

104. (a) 351-9 Exception 75%

105. (c) 351-23(b)(3) 90%

106. (d) 352-4 90%

107. (a) 352-7 95%

108. (a) 352-22(a) 70%

109. (b) 354-6 85%

110. (c) 354-14 90%

111. (b) 356-6 75%

112. (b) 358-2 90%

113. (c) 358-5 80%

114. (c) 358-9 90%

115. (d) 362-2 75%

116. (d) 362-7 90%

117. (b) 362-11 90%

118. (d) 363-4 90%

119. (b) 363-10   80%
120. (d) 364-4(b)   90%
121. (a) 364-6(b)(1)   75%
122. (c) 364-28   90%
123. (d) 365-2(a)   90%
124. (c) 365-3(d)   95%
125. (a) 370-2   80%
126. (b) 370-15(a)   90%
127. (a) 370-16(b)(1)   90%
128. (b) 370-16(2) and (3)   60%
129. (b) 370-17(c) Exception   75%
130. (a) 370-22   85%
131. (d) 370-23(c)   90%
132. (c) 370-23(e)   90%
133. (a) 370-24   60%
134. (d) 370-27(a) Exception
135. (b) 370-28(b)(2)   90%
136. (a) 370-40(b)   95%
137. (c) 370-71(a)   90%
138. (d) 370-71(b)(1)   60%
139. (a) 373-4   80%
140. (c) 373-10(b)   80%
141. (a) 374-3(b)   95%
142. (d) 374-6(a)   75%
      $4'' \times {}^{1}/_{2}'' \times 1{,}000$ ampere = 2,000 ampere
143. (d) 374-8(a)   75%
144. (a) 380-3(b)
145. (c) 380-8(b)   95%
146. (a) 380-10(b)   95%
147. (d) 380-14(b)(2)   75%
148. (b) 384-3(a)(1)   90%
149. (b) 384-3(f)   90%
150. (d) 384-8(a)   95%
151. (c) 384-10   95%
152. (a) 384-15   90%
153. (d) 384-16(d)   80%
154. (b) 384-32   80%
155. (a) 400-4 Note 2   95%
156. (d) 400-7(a)(b)   90%
157. (d) 400-8   95%
158. (d) 400-11   80%
159. (b) 400-22(b)   80%
160. (d) 402-5   90%

161. (b) 402-10   95%
162. (b) 410-4(a)   80%
      Do you really think a damp location fixture can be used in a wet location?
163. (c) 410-6   80%
164. (b) 410-8(c)   75%
165. (a) 410-9   95%
166. (c) 410-15(b)(1) Exception No. 2   95%
167. (b) 410-16(a)   95%
168. (a) 410-18(a)   95%
169. (b) 410-27(b), see 402-6   95%
170. (b) 410-28(e)   90%
171. (b) 410-30(c)(1)   80%
172. (b) 410-35(a)   80%
173. (b) 410-47   90%
174. (b) 410-52   95%
175. (a) 410-56(d)   80%
176. (a) 410-56(i)   90%
177. (c) 410-57(b)
178. (b) 410-57(f)   65%
179. (c) 410-65(c)   95%
180. (d) 410-67(c)
181. (c) 410-73(e)(1)   85%
182. (b) 410-78   95%
183. (b) 410-100   95%
184. (c) 410-101(c)   80%
185. (b) 422-10(a)   90%
186. (c) 422-11(e)(3)   80%
      16.3 ampere $\times$ 150% = 24.45 ampere
187. (c) 422-14   80%
188. (b) 422-16(b)(2)
189. (a) 422-31(a)   95%
190. (c) 422-44   75%
191. (b) 422-60(a)   95%
192. (b) 424-11   95%
193. (c) 424-22(b) Be careful.   60%
194. (c) 424-35   85%
195. (a) 424-38(b)(4)   95%
196. (a) 424-66   80%
197. (b) 424-83   80%
198. (d) 426-20(e)   90%
199. (a) 426-34   95%
200. (d) 427-2 FPN   95%

1. (a) True   90%

2. (a) magnetic field   90%

3. (b) rotor   85%

4. (c) commutator   85%

5. (b) False   50%
   results in higher CEMF

6. (b) False   50%
   decreases the rate

7. (c) a or b   80%

8. (d) synchronous   95%

9. (c) Wound rotor

10. (b) Universal   95%

11. (c) CBA   50%
    CBA is a rotation opposite of *ABC*

12. (a) 277, 480   80%

13. (c) 40 horsepower   95%
    Horsepower = Watts/746 watts
    Horsepower = 30,000 watts/746 watts

14. (a) 11 kW   95%
    Output = Horsepower × 746 watts
    Output = 15 horsepower × 746 watts
    Output = 11,190 watts

15. (a) 3.75 kW   85%
    Motor output is not
    effected by phase, voltage, efficiency or
    power factor.
    Output Watts = Horsepower × 746 watts
    Output Watts = 5 horsepower × 746 watts
    Output Watts = 3,750 watts

16. (a) True   75%

17. (b) 21.6 ampere   95%
    Step 1: Determine motor output watts
    Output Watts = Horsepower × 746 watts
    Output Watts = 5 horsepower × 746 watts
    Output Watts = 3,730 watts
    Step 2: Determine motor input watts
    Input = Output/Efficiency
    Input = 3,730 watts/.90 efficiency
    Input = 4,144 watts

Step 3: Determine motor input VA
VA = Watts/Power Factor
VA = 4,144 watts/.80 power factor
VA = 4,069 VA
Step 4: Determine the motor ampere
I = VA/E
I = 5,180 VA/240 volts

$$I = \frac{\text{Horsepower} \times 746 \text{ watts}}{\text{Volts} \times \text{Efficiency} \times \text{Power Factor}}$$

$$I = \frac{5 \text{ horsepower} \times 746 \text{ watts}}{240 \text{ volts} \times .90 \text{ EFF} \times .80 \text{ PF}}$$

I = 21.6 ampere

18. (b) 58 ampere   90%
    Step 1: Determine motor output watts
    Output Watts = Horsepower × 746 watts
    Output Watts = 20 horsepower × 746 watts
    Output Watts = 14,920 watts
    Step 2: Determine motor input watts
    Input = Output/Efficiency
    Input = 14,920 watts/.80 efficiency
    Input = 18,650 watts
    Step 3: Determine motor input VA
    VA = Watts/Power Factor
    VA = 18,650 watts/.90 power factor
    VA = 20,722 VA
    Step 4: Determine the motor ampere
    I = VA/(E $\sqrt{3}$)
    I = 20,722 VA/(208 volts × 1.732)

$$I = \frac{\text{Horsepower} \times 746 \text{ watts}}{\text{Volts} \sqrt{3} \times \text{Efficiency} \times \text{Power Factor}}$$

$$I = \frac{20 \text{ horsepower} \times 746 \text{ watts}}{208 \text{ volts} \times 1.732 \times .8 \text{ EFF} \times .90 \text{ PF}}$$

I = 58 ampere

19. (d) transformer   90%

20. (d) mutual induction   95%

21. (b) primary, secondary   95%

22. (a) True   90%

23. (c) less, lower   80%

24. (d) more, higher   80%

25. (c) isolation   95%

26. (d) all of these   80%

27. (a) True   95%

28. (a) eddy currents   75%

29. (d) small   80%

30. (d) hysteresis loss   95%

31. (a) ratio   95%

32. (a) 1:2   75%
     240:480 = 1:2

33. (d) kVA   95%

34. (a) True   55%

35. (d) most, least   65%

☆ Challenge Answers

36. (a) direct current series motor   80%

37. (b) direct current motor   75%

38. (c) run the same as before   60%
     We must reverse the field or armature leads, not the
     line (supply) leads.

39. (b) False   60%
     We must reverse the field or armature leads, not the
     line (supply) leads.

40. (d) wound rotor motor   80%

41. (d) rotor or armature   65%

42. (b) 6,100 VA   80%
     VA_{Input} = Volts × Motor Ampere × 1.732
     VA_{Input} = 230 volt × 15.2 ampere × 1.732
     VA_{Input} = 6,055 VA

43. (b) 1,840 VA   95%
     VA_{Input} = Volts × Motor Ampere
     VA_{Input} = 115 volts × 16 ampere
     VA_{Input} = 1,840 VA

44. (b) saturated   70%

45. (d) all of these   70%

46. (a) higher   55%
     A ratio of 2:1 means that the secondary voltage is less;
     the power remains the same. Therefore, the current on
     the secondary will be greater, I = P/E.

47. (d) .06 kVA   85%
     Primary VA = Secondary VA
     Secondary VA = Volt × Ampere
     volts = 12, ampere = 5
     Secondary VA = 12 volts × 5 ampere
     Secondary VA = 60 VA or .06 kVA
     Primary VA = .06 kVA

48. (d) 0.5 ampere   75%
     The primary voltage is 10 times greater, the power
     remains the same; therefore, the current on the
     primary will be is 10 times lower.
     5 ampere × 0.1 =.5 ampere

49. (a) primary   50%
     A current transformer is a clamp-ammeter. The con-
     ductor that carries current is the primary and the meter
     is the secondary.

50. (b) secondary   95%

51. (b) secondary   85%
     Secondary has lower voltage and the power remains
     the same; the current will be greater than the primary.

52. (c) 2 ampere   80%
     If the primary voltage is 5 times more than the sec-
     ondary, then the primary current will be 5 times less
     than the secondary.
     I Primary = 10 ampere/5
     I Primary = 2 ampere

53. (b) 17 ampere   80%
     I = P/E
     P = 200 watts, E = 12 volts
     I = 200 watts/12 volts
     I = 16.67 ampere

54. (a) 6.8 ampere   95%
     I = P/E
     P = 1,630 VA (1,500 watts/.92 efficiency)
     E = 240 volts
     I = 1,630 VA/240 volts
     I = 6.79 ampere

55. (b) 42 kVA   70%
     The output is 36,000 VA
     (100 ampere × 208 volts × 1.732). The input must be
     greater than the output because of efficiency.
     Input = Output/Efficiency or
     $VA_{Primary}$ = Secondary VA/Efficiency
     $VA_{Primary}$ = 36,000 VA/.86 efficiency
     $VA_{Primary}$ = 41,860 VA

56. (b) 50 ampere   70%
     $I_{Primary}$ 3-phase = Primary VA/(E_{PRI} × $\sqrt{3}$)
     $I_{Primary}$ = 41,891 VA/(480 volt × 1,732)
     $I_{Primary}$ = 50.38 ampere

57. (a) 6 volts   75%
     The turns (voltage) ratio is 200 volts/10 volts or 20/1,
     which means that the secondary voltage will be 20
     times less than the primary.
     Secondary Volts = 120 volts/20 times
     Secondary Volts = 6 volts

<u>58.</u> (b) 4.38 ampere   90%
  I = P/E
  P = 500 VA Output/.95 Efficiency
  P = 526 VA, E = 120 volts
  I = 526 VA/120 volts
  I = 4.38 ampere

<u>59.</u> (a) 526 VA   90%
  Primary Power = Secondary Power/Efficiency
  Primary Power = 500 watts/.95 EFF
  Primary Power = 526 watts

Transformer Miscellaneous

<u>60.</u> (a) 200 VA   85%
  The load (output) is given as two 100-watt lamps, the efficiency only affects the input, not the output.

<u>61.</u> (d) a and b   60%
  At 100% efficiency the primary VA will be the same as the secondary VA.
  $VA_{Secondary}$ = Volts × Ampere
  $VA_{Secondary}$ = 24 volts × 5 ampere
  $VA_{Secondary}$ = 120 VA
  $VA_{Secondary}$ = 120 VA

<u>62.</u> (b) 217 VA   85%
  Input = Output/Efficiency
  Input = 200 watts/.92 EFF
  Input = 217 watts

<u>63.</u> (c) 31 kW   95%
  Input = Output/Efficiency
  Input = 20 kW/.65 Eff
  Input = 30.8 kW

NEC® Answers

64. (a) 427-22
65. (d) 430-7(a)(6)   90%
66. (c) 430-12(a)   75%
67. (a) 430-14(b) Exception   90%
68. (c) 430-24   75%
69. (c) 430-31   80%
70. (a) 430-32(b)(1)   80%
71. (d) 430-36   75%
72. (d) 430-43
73. (a) 430-53(a)(1)   80%
74. (a) 430-72(b) Exc. 1, Column B, Note 2   85%
75. (a) 430-74(a)   95%
76. (a) 430-83(a)(1)   95%
77. (c) 430-102(b)   70%
78. (a) 430-103   95%
79. (d) 430-109   95%

80. (b) 430-110(a) and 430-150 Table   90%
81. (d) 430-142   90%
82. (d) 440-14   95%
83. (d) 440-41(a)   80%
84. (d) 440-64   80%
85. (c) 455-6   90%
86. (a) 445-10
87. (c) 450-3(b)(2) Note 1   80%
  If the primary protection is set up to 250% of the primary current rating, then secondary overcurrent protection is required. The secondary protection device must be sized no greater than 125% of secondary current. 42 ampere × 1.25 = 53 ampere, the next size up (60 ampere) is permitted 240-6(a).

88. (c) 450-13
89. (d) 450-24   75%
90. (d) 450-43(a)   90%
91. (d) 460-2(b)   90%
92. (d) 460-28(a)   95%
93. (c) 470-18(b)   95%
94. (b) 480-8(a)   95%
95. (d) 500-3(a)   95%
96. (b) 500-5(a)(3)   85%
97. (a) 500-5(b)(1)   95%
98. (b) 500-5(b)(3) FPN No. 2   90%
99. (d) 500-5(d)   85%
100. (a) 500-7(b) FPN No. 1   80%
101. (a) 500-8   95%
102. (c) 500-9(b)   80%
103. (d) 501-4(a)(1)   95%
104. (a) 501-4(b)   80%
105. (b) 501-5   90%
106. (c) 501-5(b)(2)
107. (d) 501-5(c)(3)   85%
108. (d) 501-5(e)(1) Exception No. 2
109. (a) 501-6(b)(2)   85%
110. (d) 502-4(a)(2)   80%
111. (c) 503-9(b)   80%
112. (a) 511-2   95%
113. (d) 511-3(b)   60%
114. (c) 511-7(a)   95%
115. (d) 513-3(d)   95%
116. (b) 514-5(a)   90%
117. (d) 514-5(b)   85%

118.	(d) 514-8 Exception No. 2	75%
119.	(b) 516-2(e)	75%
120.	(d) 517-3 Patient care area	95%
121.	(d) 517-18(b)	90%
122.	(b) 517-30(a) and (b)	
123.	(c) 517-60(a)(1)	90%
124.	(d) 517-64(a)(1)	95%
125.	(b) 518-1	95%
126.	(b) 518-4(a)	75%
127.	(a) 520-24	95%
128.	(b) 520-53(g)	95%
129.	(d) 525-12(a) and 225-18	
130.	(a) 525-13(e)	
131.	(a) 525-40	
132.	(b) 540-11(b)	80%
133.	(a) 545-1	80%
134.	(d) 547-9(c)	
135.	(a) 550-5(a)	80%
136.	(b) 550-8(g)	
137.	(c) 550-13(b)(5), see Table	95%
138.	(b) 550-23(f)	
139.	(a) 551-72	
140.	(b) 555-3	80%
141.	(c) 555-5	80%
142.	(a) 555-11	95%
143.	(a) 600-5(b)(2)	75%
144.	(c) 600-6(a)	80%
145.	(b) 600-9(a)	95%
146.	(d) 604-6(a)(1)	75%
147.	(b) 610-61	95%
148.	(a) 620-12(a)(2)	85%
149.	(b) 620-61(b)(2)	90%
150.	(b) 625-22	
151.	(a) 645-5(d)(2)	75%
152.	(d) 645-10	75%
153.	(c) 660-5	80%
154.	(d) 665-24	90%
155.	(b) 668-2 Definition	85%
156.	(a) 675-2 Definition	90%
157.	(a) 680-4 definition, spa or hot tub	70%
158.	(d) 680-5(a)	95%
159.	(d) 680-6(a)(1), 680-41(a)	80%
160.	(d) 680-6(a)(3)	80%

161.	(b) 680-6(d)	
162.	(c) 680-10	75%
163.	(a) 680-12	
164.	(d) 680-20(b)(1)	85%
165.	(b) 680-21(a)(5)	95%
166.	(b) 680-22	70%
167.	(b) 680-22	
168.	(b) 680-22(a)(3)	95%
169.	(d) 680-25(b)(1)	
170.	(c) 680-27(c)	80%
171.	(a) 680-40	80%
172.	(d) 680-51(e)	90%
173.	(a) 680-70 and 680-71	85%
174.	(c) 695-7	
175.	(a) 700-6	
176.	(d) 700-9(b)	
177.	(b) 700-12(a)	95%
178.	(b) 700-12(e)	75%

We want the battery to operate if the power to the lighting circuit is interrupted.

179.	(d) 701-11(b)(1)(2)(3)	95%
180.	(d) 720-4	75%
181.	(a) 725-8(a)	90%
182.	(d) 725-23	75%
183.	(d) 725-41(a)	
184.	(b) 725-41(a) FPN, Table 11in Chapter 9	75%
185.	(b) 725-54(a)(1)	75%
186.	(a) 725-54(b)(3)	95%
187.	(c) 760-1	80%
188.	(b) 760-3(b) and 300-22(a)	75%
189.	(d) 760-30(a)(1)	95%
190.	(b) 760-42	85%
191.	(d) 760-61(a)	90%
192.	(a) 770-6	
193.	(a) 800-4	
194.	(d) 800-33	90%
195.	(d) 800-51(a)	
196.	(c) 800-53(b) Exception No. 1	80%
197.	(d) 810-16(a) and Table 810-16(a)	90%
198.	(c) 810-54	90%
199.	(d) 820-40(a)(1, 3, 5)	95%
200.	(a) 830-10(i)(3)	

Unit 5 – Raceway, Outlet Box, and Junction Boxes Calculations

1. (c) Appendix C   85%

2. (a) True, Note 3 of Table 1, Chapter 9   90%

3. (d) 60%, Note 4 of Table 1, Chapter 9   90%

4. (d) 29, Appendix C, Table C1   85%

5. (d) 7, Appendix C, Table C2   85%

6. (a) 7, Appendix C, Table C3   95%

7. (b) 11, Appendix C, Table C4   85%

8. (c) 300 kcmil, Appendix C, Table 8A   80%

9. (c) 40%, Table 1 of Chapter 9   95%

10. (a) .0333 square inches, Table 5 of Chapter 9   75%

11. (a) .0209 square inches, Table 5 of Chapter 9   80%

12. (d) .0211 square inches, Table 5 of Chapter 9   95%

13. (b) .0353 square inches, Table 5 of Chapter 9   95%

14. (a) .013 [Chapter 9, Table 8]   75%
No. 8 Solid = .013 square inches
No. Stranded = .017 square inches

15. (d) all of these   85%

16. (a) 2 inch   90%
Step 1: Area of the conductors [Table 5 of Chapter 9]
3– No. 3/0 THHN:
.2679 square inches $\times$ 3 = .8037 square inches
1– No. 2 THHN:
.1158 square inches $\times$ 1 = .1158 square inches
1– No. 6 THHN:
.0507 square inches $\times$ 1 = .0507 square inches
Step 2: Total square inch area of the conductors:
.9702 square inches
Step 3: Permitted conductor fill at 40 percent fill
[Table 1 of Chapter 9 and Table 4 of Chapter 9]
2 inch schedule 40 PVC area = 1.15 square inches

17. (a) 1$^1/_2$ inch   85%
Step 1: Find the square inch area of the conductors, Chapter 9, Table 5.
3– No. 4/0 THHN:
.3237 square inches $\times$ 3 = .9711 square inches
1– No. 1/0 THHN:
.1855 square inches $\times$ 1 = .1855 square inches
1– No. 4 THHN:
.0824 square inches $\times$ 1 = .0824 square inches
Step 2: Total square inch area of the conductors:
1.239 square inches
Step 3: Size the conduit at 60 percent fill
[Chapter 9, Note 3] using Table 4
1 1/4 inch:
1.526 square inches $\times$ .6 = .9156 square inches
– Too small
1$^1/_2$ inch:
2.071 square inches $\times$ .6 = 1.2426 square inches
– Just Right
2 inch:
3.408 square inches $\times$ .6 = 2.0448 square inches
– Too Big

18. (d) 11   65%
    Step 1: Area of conductor fill permitted for a $^3/_4$ inch nipple Chapter 9, Table 4:
    .549 square inches $\times$ .6 = .3294 square inches
    Step 2: Square inch area of the existing conductors
    4– No. 10 THHN:
    .0211 square inches $\times$ 4 = .0844 square inches

    1– No. 10 bare:
    .0110 square inches $\times$ 1 = .0110 square inches
    (Note 2 of Chapter 9)

    Total area of existing conductors = .0954 square inches

    Step 3: Subtracting the area of the existing conductors from the area of permitted conductor fill
    Spare Space Area :
    Permitted Area Fill less Existing Conductors Area
    Spare space area = .3294 square inches – .0954
    Spare space area = .2340 square inches

    Step 4: Determining the number of conductors that can be added to the available spare space
    Number of conductors permitted:
    Spare Space Area/Area of Conductors
    Number of conductor permitted:
    .2340 square inches/.0211
    Number of conductors permitted = 11 conductors

19. (a) 4 $\times$ 1 inch square [Table 370-16(a)]   90%
    Insulation is not a factor, 9 – No. 14

20. (a) 8, Table 370-16(a)   90%

21. (d) all of these, Table 370-16(b)   95%

22. (d) all of these, Table 370-16(b)   80%

23. (d) all of these, Table 370-16(b)   75%

24. (d) b and c, Table 370-16(a)(1) Exception   80%

25. (a) Yes   95%
    The exception to 370-16(a) permits us to omit fixture wires that enter the outlet box from a *fixture canopy*.
    Step 1: Determine the number and size of conductors

14/2 romex	1 – No. 14's
Ground wire	1 – No. 14
One cable clamp	1 – No. 14

    Step 2: Volume of the conductors [Table 370-16(b)]:
    4 conductors $\times$ 2 cubic inches = 8.00 cubic inches

26. (c) 4 $\times$ 2 $^1/_8$ inch square   75%
    Step 1: Determine the number and size of conductors

12/2 romex	2 – No. 12 conductors
12/3 romex	3 – No. 12 conductors
Cable clamps	1 – No. 12 conductors
Switch	2 – No. 12 conductors
Receptacles	2 – No. 12 conductors
Ground wire	1 – No. 12 conductors
Total Number	11 – No. 12 conductors

    Step 2: Determine the volume (cubic inch) of the above conductors [Table 370-16(b)]
    11 conductors $\times$ 2.25 = 24.75 cubic inches
    Step 3: Select the outlet box from Table 370-16(a)
    4 $\times$ 2$^1/_8$ inch square = 30.3 cubic inches

27. (b) 2   70%
    Step 1: Determine the number and size of the existing conductors

Two receptacles	4 – No. 14 conductors
Five No. 12	5 – No. 14 conductors
Two ground wires	1 – No. 14 conductor
Total conductors	10 – No. 14 conductors

    Step 2: Determine volume of the existing conductors [Table 370-16(b)]
    10 conductors $\times$ 2 cubic inches = 20 cubic inches
    Step 3: Determine the space space
    4 $\times$ 1$^1/_2$ square box = 21 cubic inches
    Volume of box and ring:
    21 cubic inches + 3.6 (ring) = 24.6 cubic inches

    Spare Space:
    Total Volume less Conductor Volume:
    24.6 cubic inches – 20 = 4.6 cubic inches
    Step 4: No. 14 conductors permitted in spare space:
    Spare Space/Conductor Volume:
    4.6 cubic inches/2 cubic inches = 2 conductors

28. (d) all of the above are correct, 370-28(a)(1)(2)   85%
    370-28(a)(1)(2)

29. (d) none of these   80%
    370-28(a)(2) Exception and Table 373-6(a)

30. (d) 20 inches, Section 370-28 90%
Left wall to the right wall angle pull
(6 × 2.5 inches) + 2.5 = 17.5 inches
Left wall to the right wall straight pull
8 × 2.5 inches = 20 inches
Right wall to left wall angle pull
No calculation
Right wall to the left wall straight pull
8 × 2.5 inches = 20 inches

31. (d) 15 inches, Section 370-28 70%
Bottom wall to the top wall angle pull
(6 × 2.5 inches) = 15 inches
Bottom wall to the top wall straight pull
No calculation
Top wall to the bottom wall angle pull
No calculation
Top wall to the bottom wall straight pull
No calculation

32. (d) 15 inches, Section 370-28 90%
(6 × 2.5 inches) = 15 inches

33. (d) 14 inches, Section 370-28 85%
Left wall to the right wall angle pull
(6 × 2 inches) + 2 inches = 14 inches
Left wall to the right wall straight pull
No calculation
Right wall to left wall angle pull
No calculation
Right wall to the left wall straight pull
No calculation

34. (d) 14 inches, Section 370-28 95%
Bottom wall to the top wall angle pull
No calculation
Bottom wall to the to wall straight pull
No calculation
Top wall to bottom wall angle pull
(6 × 2 inches) + 2 inches = 14 inches
Top wall to the bottom wall straight pull
No calculation

35. (d) 12 inches 95%

☆ Challenge Answers

36. (a) 11 No. 2 THW conductors 75%
Area conductor fill for a 3 inch schedule 40 PVC
2.907 square inches [Chapter 9, Table 4]
Area No. 1 RHW without cover
.1901 square inches [Chapter 9, Table 5]
Area of seven No. 10 RHW conductors
.1901 × 7 conductors = 1.3307 square inches
Spare space
2.907 square inches − 1.3307 = 1.5763

Quanity of No. 2 THW permitted in spare space
1.5763 square inches/.1333 = 11.8 conductors

Note: The answer is 11 conductors, we only round up
when all of the conductors are the same size (total
cross-sectional area including insulation), See Note 7
of Chapter 9, Table 1.

37. (b) 22 cubic inches 75%

2– No. 10 passing through	2 – No. 10
1– 1 yoke (receptacle)	2 – No. 12
4– No. 14 spliced and buried	4 – No. 14
2– No. 12 for terminating	2 – No. 12
1– No. 12 bonding jumper	0 – No. 12

Total – two No. 10 conductors, four No. 12 conduc-
tors, and four No. 14 conductors
Volume of the conductors: [Table 370-16(b)]
No. 10
2.5 cubic inches × 2 conductors = 5.0 cubic inches
No. 12
2.25 cubic inches × 4 conductors = 9.0 cubic inches
No. 14
2 cubic inches × 4 conductors = 8.0 cubic inches
Total 22.0 cubic inches

38. (a) 9 conductors 50%
370-16(a)(1 and 2), Table 370-16(a)
Volume of the conductors:

2 – No. 18	2 – No. 18 conductors
1 – 14/3	3 – No. 14 conductors
1 – ground	1 – No. 14 conductor
1 – switch	2 – No. 14 conductors
2 – clamps	1 – No. 14 conductor
	9 – No. 14 conductors

39. (c) 24 inches, Section 370-28   80%
    Straight: Left to Right:
    8 × 3 inches = 24 inches
    Right to Left:
    8 × 3 inches = 24 inches
    Angle: Left to Right:
    (6 × 3 inches) + 2.5 inches + 2 = 22.5 inches
    Right to Left:
    (6 × 3 inches) + 2.5 inches = 20.5 inches

40. (a) 16 inches, Section 370-28(a)(2)   70%
    Straight: None possible
    Angle: Top to Bottom: None possible
    Bottom to Top:
     (6 × 2 inches) + 2 inches + 2 inches = 16 inches

41. (a) 12 inches, Section 370-28(a)(2)   80%
    6 × 2 inches = 12 inches

42. (b) $4^1/_2$ inches   70%
    370-28(a)(2) Exceptions, Table 373-6(a)

NEC® Answers

43. (c) 90-1(b)   90%

44. (d) 90-2(a)   70%

45. (c) 90-2(b)(4)   90%

46. (b) 90-3   85%

47. (b) 90-5(c)

48. (a) 90-8(a)   90%

49. (b) 100 Accessible, Wiring Methods   80%

50. (c) 100 Approved   90%

51. (d) 100 Automatic   95%

52. (d) 100 Branch Circuit, Appliance   90%

53. (d) 100 Branch Circuit, Multiwire   70%
    If you missed this question, you probably answered the question quickly without checking the responses in (b) and (c). Slow down and read the complete paragraph of the *NEC*®, not just the first line. Read all of the answer choices before you pick (a).

54. (b) 100 Circuit Breaker, Inverse Time   85%

55. (c) 100 Conduit Body   90%

56. (d) 100 Controller   90%

57. (c) 100 Device   90%

58. (a) 100 Dusttight   90%

59. (b) 100 Dwelling Unit, Two-Family   65%

60. (c) 100 Explosionproof Apparatus   95%

61. (a) 100 Feeder

62. (c) 100 Grounded Conductor   90%

63. (a) 100 Ground-fault Protection Equipment   90%

64. (a) 100 Interrupting Rating   95%

65. (c) 100 Lighting Outlet   85%

66. (a) 100 Location, Dry   80%

67. (c) 100 Outlet   85%

68. (c) 100 Panelboard   90%

69. (b) 100 Raintight   80%

70. (c) 100 Sealable Equipment   90%

71. (a) 100 Service and Feeder

72. (a) 100 Service Equipment   75%

73. (a) 100 Signaling Circuit   90%

74. (d) 100 Switch, Bypass Isolation   90%

75. (a) 100 Thermal Protector, FPN   90%

76. (c) 100 Voltage, Nominal   90%

77. (d) 110-3(a)   80%

78. (b) 110-6, Table 310-16   85%

79. (a) 110-9   95%

80. (d) 110-10   90%

81. (b) 110-11   70%

82. (d) 110-12(c)   80%

83. (d) 110-14(a)   85%

84. (b) 110-14(c)   90%

85. (b) 110-14(c)(1)(c)

86. (a) 110-22

87. (c) 110-26(a)(1)   75%

88. (c) 110-26(a)(1) Condition 2

89. (b) 110-26(a)(2)

90. (b) 110-26(b)   85%

91. (d) 110-26(e)   85%

92. (a) 110-26(f)(1)(a)   70%

93. (d) 110-27(a)(1, 3, 4)   90%

94. (a) 110-27(c)   80%

95. (d) 110-31(a)(1)   95%

96. (a) 110-33(a)(1)   85%

97. (b) 110-34(b)   95%

98. (b) 110-34(c)   65%
    The *NEC*® requires the word "Danger" in warning signs.

99. (d) 200-1   95%

100. (c) 200-6(a)   95%

101. (b) 200-6(d)   80%

102. (b) 200-10(b)   90%

103. (a) 210-4(c)   60%

104. (b) 210-4(d) FPN   80%

105. (a) 210-7(b)   90%
106. (d) 210-8(a)(1) and (b)(1)   80%
107. (c) 210-8(a)(2)
108. (a) 210-8(a)(6)   75%
109. (b) 210-8(b)(2) Exception
110. (a) 210-11(c)(3) Exception
111. (c) 210-19(a) FPN 4 and 215-2(d) FPN 2   80%
112. (d) 210-21(b)(1)   70%
113. (b) Table 210-21(b)(2) and Table 210-24   70%
      I know you want to say 20 ampere, but read these
      Code Sections carefully.
114. (a) 210-23(a)
115. (a) 210-50(a)   80%
116. (c) 210-52(a)(2)   80%
117. (b) 210-52(b)(1)   70%
118. (c) 210-52(c)(1)   90%
119. (a) 210-52(c)(5)   90%
120. (d) 210-52(d)
121. (d) 210-52(g)   85%
122. (a) 210-60(b)
123. (a) 210-70(a)(1)
124. (d) 210-70(a)(1 and 2)   75%
125. (d) 210-70(a)(3)
126. (c) 215-2(b)   90%
127. (c) 215-6   80%
128. (b) 220-2(b)
129. (a) 220-3(b)(10)   95%
130. (c) 220-11   75%
131. (c) 220-12(b)
132. (c) 220-17   85%
133. (c) 220-19 Note 3   80%
      $6 \text{ kW} \times 4 \text{ units} = 24 \text{ kW} \times 0.50 = 12 \text{ kW}$
134. (b) 220-22   60%
135. (c) 220-30(a)   85%
136. (d) 220-32(c)(2)   85%
137. (a) 220-36   90%
138. (b) 220-40(b)   75%
139. (b) 225-4   90%
140. (c) 225-6(b)   90%
141. (d) 225-14(d)(2)   65%
142. (c) 225-19(a)   85%
143. (b) 225-19(d)   85%
144. (c) 225-19(e)   85%

145. (d) 225-25   95%
146. (a) 230-2   95%
147. (d) 230-7   65%
148. (a) 230-9
149. (c) 230-24   95%
150. (b) 230-24(a) Exception No. 4   95%
151. (b) 230-24(d) 680-8   70%
152. (d) 230-30 Exception   90%
153. (c) 230-31(b) Exception   80%
154. (d) 230-41 Exception   85%
155. (d) 230-46
156. (b) 230-50(b)   80%
157. (c) 230-51(c)   90%
158. (a) 230-54(a)   90%
159. (a) 230-54(c)   95%
160. (c) 230-55 Exception   70%
161. (a) 230-70(b)   90%
162. (c) 230-71(a)
163. (b) 230-72(c), 230-91(b)   95%
164. (a) 230-77   65%
165. (d) 230-81   90%
166. (c) 230-90   85%
167. (c) 230-92
168. (c) 230-94 Exception No. 4   90%
169. (a) 230-95(b)   90%
170. (b) 230-95(c) FPN No. 3   75%
171. (a) 240-1 FPN   80%
172. (d) 240-3(d)
173. (c) 240-4(b)(1)   85%
174. (c) 240-8   80%
175. (c) 240-12   90%
176. (c) 240-20(a)   85%
177. (a) 240-20(b)(2),(3)   80%
178. (b) 240-21(3)(b)(3)   90%
179. (c) 240-22(a)   90%
180. (d) 240-24(c, d, and e)   95%
181. (c) 240-50(c)   95%
182. (d) 240-52   90%
183. (d) 240-54(d)   90%
184. (d) 240-60(b)   95%
185. (d) 240-61   85%
186. (c) 240-83(a)   90%
187. (b) 240-83(d)   95%

188. (a) 250-2(b)   85%   195. (b) 250-50(a)(2)

189. (b) 250-6(d)   70%   196. (d) 250-52   85%

190. (b) 250-12   95%   197. (a) 250-52(c)(2)   75%

191. (d) 250-24(a)(5)   198. (a) 250-54   90%

192. (c) 250-24(b)(2)   199. (a) 250-58   85%

193. (a) 250-32(a)   200. (d) 250-62   90%

194. (d) 250-50 Table   55%

   $500,000 \times 3 = 1,500,000$, which exceed 1,100 kcmil

Notes:

Unit 6 – Conductor Sizing and Protection Calculations

1. (d) a and c, Table 310-13   85%

2. (a) True, Table 310-16, 240-3   85%

3. (d) No. 4/0, 110-6   95%

4. (d) a and b, Table 310-5   90%

5. (a) True, 110-14(c)(1)   95%

6. (c) No. 4, 110-14(c)(1)   85%
   Conductors must be selected according to the temperature rating of the equipment (60°C). Conductors must be selected according to the lowest temperature rating of the equipment (60°C). THHN conductors can be used, but the conductor must be sized according to the 60°C terminal rating of the equipment, this requires a No. 4 conductor.

7. (c) No. 6, 110-14(c)(2)   70%
   Conductors must be selected according to the temperature rating of the equipment (75°C). THHN conductors can be used but the conductors must be sized according to the 75°C terminal rating. This requires a No. 6 THHN (rated 65 ampere at 75°C).

8. (a) True, 110-14(c)(2)   95%

9. (b) No. 10, 110-14(c)(2)   65%

10. (a) No. 1/0   85%

11. (a) True   85%

12. (c) No. 3/0   90%

13. (b) False, 310-4   70%

14. (a) True, 310-4   90%

15. (a) True, 310-4 FPN   50%

16. (a) True, 310-4 Exception   60%

17. (a) True, 310-4, 250-122   90%

18. (b) False, 310-4   70%

19. (d) No. 1/0, 310-4 and Table 250-122   90%

20. (c) No. 500, Table 310-16   80%
    Rated 380 ampere each $\times$ 2 = 760 ampere, next size up protection (800 ampere) permitted.

21. (d) No. 1/0, 310-4   65%
    250 ampere/2 = 125 ampere = No. 1 Table 310-16
    Note: The smallest size parallel conductor is No. 1/0

22. (a) True   80%
    210-19(a) FPN 4, 215-2(b) FPN 2, 310-15 FPN

23. (a) True   85%

24. (c) 5,000, 10,000   95%
    Circuit breakers, 5,000 ampere, 240-83(c)
    Fuses, 10,000 ampere, 240-60(c)

25. (d) any of these, 240-6(a)   90%

26. (a) 80%   80%
    210-20(a), 220-3(a), 215-2(a), 230-42(a), 384-16(d)

27. (b) False, 240-3(a)   80%

28. (c) No. 4, Table 310-16   85%

29. (d) all of these, 310-10 FPN   80%

30. (b) False   70%
    Table 310-16 heading, 30°C and 3 current-carrying conductors.

31. (d) 57 ampere   70%
    Table 310-16, 55 ampere $\times$ 1.04 = 57.2 ampere.
    Ampacity increased if the ambient temperature is less than 78°F.

32. (b) No. 12 THHN, 240-3(d)   80%
    The conductor must have an ampacity of 16 ampere after applying the ambient temperature derating factor and must be protected by a 20-ampere protection device. Section 240-3(d) requires the conductor to be No. 12.
    Ampacity = Table Ampere $\times$ Temperature Adjustment
    No. 14 THHN 25 ampere $\times$ 0.91 = 23 ampere
    No. 12 THHN 30 ampere $\times$ 0.91 = 27 ampere
    No. 10 THHN 40 ampere $\times$ 0.91 = 36 ampere

33. (b) 24 inches, 310-15(b)(2)(a)   95%

34. (b) 136 ampere, 310-15(b)(2)(a)   70%
No. 1/0 THHN at 90°C = 170 ampere
Ampacity = Table Ampacity × Bundle Adjustment
Ampacity = 170 ampere × 0.8
Ampacity = 136 ampere

35. (c) 10 THHN   80%
The conductors must have an ampacity of 21 ampere
and must be protected by the 30-ampere circuit pro-
tection device. Table 310-16 note on the bottom of the
Table requires a No. 10 wire for a 30-ampere protec-
tion device.
Ampacity = Table Ampacity × Bundle Adjustment
No. 14 THHN – 25 ampere × 0.7 = 17.5 ampere
No. 12 THW – 30 ampere × 0.7 = 21 ampere
No. 10 THW – 40 ampere × 0.7 = 28 ampere

36. (b) 26 ampere   85%
Ampacity = Ampacity × Temperature × Bundle
Temperature adjustment = 0.91
Bundle adjustment = 0.7
Ampacity = 40 ampere × 0.91 × 0.7
Ampacity = 25.5 ampere

37. (b) False, 310-15(b)(4)(a)   75%

38. (b) False, 310-15(b)(4)(c)   80%

39. (b) False, 310-15(b)(4)(b)   75%

40. (a) True, 310-15(a)(2)   65%

41. (a) True, 110-14(c)   80%

✯ Challenge Answers

42. (d) 35 ampere, 240-6(a), 384-16(d)   60%
Overcurrent devices must be sized not less than 125%
of the continuous load. The overcurrent protection
device must be sized not less than:
27 ampere × 1.25 = 33.75 ampere.

43. (c) 60 ampere, 240-6(a), 384-16(d)   65%
The overcurrent protection device must be sized not
less than 125% of the continuous load.
45 ampere × 1.25 = 56.25 ampere

44. (d) 90 ampere, 240-6(a), 384-16(d)   65%
Overcurrent devices must be sized not less than 125%
of the continuous load. The overcurrent device must
be sized not less than:
65 ampere × 1.25 = 81.25 ampere.

45. (c) 150 ampere   80%
240-6, 215-2(a), 230-42(a), and 384-16(d)
Overcurrent devices must be sized not less than 125%
of the continuous load. The overcurrent protection
device must be sized not less than:
103 ampere × 1.25 = 128.75 ampere.

46. (c) .82   90%
Table 310-16. Bottom, 60°C wire at 102°F. Be sure to
use a straight edge when using a table.

47. (c) 90°C, Table 310-16   70%
No correction factors listed for 60°C or 75°C.

48. (c) 144 ampere   85%
Table 310-16 and 310-15(b)(2)(a)
Ampacity = Table Ampacity × Bundle Adjustment
Table 310-16 Ampacity:
No. 4/0 XHHW Al Wet location = 180 ampere*
Six current-carrying conductors factor = 0.8
Ampacity = Table Ampacity × Bundle Adjustment
Ampacity = 180 ampere × 0.8
Ampacity = 144 ampere

*Note: Table 310-13 requires that when XHHW is
used in a wet location we must use the 75°C ampacity
column of Table 310-16. See definition of "Location,
Wet".

49. (c) 16 ampere   85%
Table 310-16 and 310-15(b)(2)(a)
Ampacity = Ampere × Temperature × Bundle
Table 310-16 Ampacity
No. 10 RHW Aluminum = 30 ampere at 75°C
Temperature Adjustment
75°C wire at 75°F = 1.05
15 current-carrying conductors factor = 0.5
Ampacity = 30 ampere × 1.05 × 0.5
Ampacity = 15.75 ampere

50. (b) No. 12 THHN,   65%
Table 310-16 and 310-15(b)(2)(a)
Ampacity = Ampere × Temperature × Bundle
Table ampacity:
No. 14 THHN – 25 ampere at 90°C
No. 12 THHN – 30 ampere at 90°C
No. 10 THHN – 40 ampere at 90°C
No. 8 THHN – 55 ampere at 90°C
Ambient Temperature Correction
90°C wire at 35°C = 1.04
9 current-carrying conductors adjustment = 0.7
No. 14 THHN
Ampacity = 25 ampere × 1.04 × 0.7
Ampacity = 18.2 ampere
No. 12 THHN
Ampacity = 30 ampere × 1.04 × 0.7
Ampacity = 21.84 ampere

51. (c) 35 ampere   70%
Bundle factors do not apply to raceways that are 24
inches or less (nipples) [*Section 310-15(b)(2)(a) Exc. 3*]
Table 310-16 Ampacity
No. 10 THW at 75°C = 35 ampere

52. (b) 68 ampere   70%
Table 310-16, 310-15(b)(2)(a) Exc. 3
Ampacity = Ampere × Temperature × Bundle
Table Ampacity 310-16
6 THHW = 75 ampere at 90°C

Note: THHW is listed in both the 75°C as well as 90°C column of Table 310-16. Use the 90°C column for dry locations or if the question does not specify a wet location.

Temperature Adjustment 90°C wire at 39°C = 0.91
Bundle adjustment = Does not apply to nipples
Ampacity = 75 ampere × 0.91 = 68 ampere

53. (c) 9 conductors   70%
310-15(b)(4)(c), 310-15(b)(5)

4 wire incandescent lighting	= 3 conductors
4 wire fluorescent lighting	= 4 conductors
2 wire receptacles	= 2 conductors
1 ground wire	= 0 conductor
Total current-carrying	= 9 conductors

54. (b) 7 conductors, 310-15(b)(2)(a)   70%
Current-Carrying Conductors

4 wire incandescent lighting	= 3 conductors
4 wire fluorescent lighting	= 4 conductors
1 ground wire	= 0 conductors
Total	= 7 conductors

NEC® Answers

55. (c) Table 250-66   65%
There is no exception that permits the grounding electrode conductor to be smaller than No. 8.

56. (b) 250-68(a)   90%

57. (a) 250-68(a) and 250-70   95%

58. (d) 250-86 Exception No. 1   75%

59. (d) 250-92(b)

60. (b) 250-96(a)   70%

61. (a) 250-96(b), 300-10 Exception No. 3   85%

62. (b) 250-102(c) and 250-66 Table   85%

63. (d) 250-102(d) and 250-122 Table   85%

64. (b) 250-104(3)

65. (d) 250-110(1, 2, and 3)   90%

66. (a) 250-114(3)   85%

67. (d) 250-119   95%

68. (b) 250-122(c)   90%

69. (b) Table 250-122 Note

70. (b) 250-136(a)   85%

71. (a) 250-146   75%

72. (a) 250-146(a)   85%

73. (a) 250-148   90%

74. (b) 250-164   90%

75. (d) 250-172 and Exception   55%
Read the exception carefully.

76. (a) 280-12   85%

77. (c) 300-2(a)   85%

78. (a) 300-3(c)(1)   65%
This Section requires all conductors to have the insulation rating to be no less than the maximum circuit voltage, in this case 277 volts.

79. (c) 300-4(b)(1)

80. (d) 300-5 Table Column 1   85%
To be buried less than 24 inches, GFCI protection is required.

81. (a) 300-5 Table Column 5   70%

82. (c) 300-5(d)   90%

83. (d) 300-5(f)   95%

84. (d) 300-5(i), see 300-3(b)   85%

85. (a) 300-6(c)   75%

86. (c) 300-10   70%

87. (a) 300-13(a)   90%

88. (b) 300-14

89. (a) 300-16(b)

90. (a) 300-18(a)

91. (c) 300-19(a) Table   75%
This Section requires a support at the top and a support not greater than 80 feet.

92. (a) 300-21   90%

93. (b) 300-22(b)   95%

94. (b) 300-22(c) FPN   90%

95. (a) 300-23   95%

96. (b) 305-3(d)   95%

97. (d) 305-4(c)

98. (c) 305-4(e)   80%

99. (d) 305-4(j)

100. (c) 305-6(a) Exception No. 1   85%

101. (c) 310-3   95%

102. (c) 310-4   95%

103. (a) 310-5 Table   80%

104. (a) 310-10   75%

105. (d) 310-11(c)   90%

106. (d) Table 310-13   95%

107. (d) Table 310-15(b)(2)   80%

108. (a) 310-15(b)(4)(c)   95%
109. (b) 310-16   85%
110. (d) 310-60(a)   80%
111. (c) 310-71   90%
112. (d) 318-4   95%
113. (c) 318-6(b)   90%
114. (c) 318-7(b)(2) Table Note B   80%
115. (c) 318-8(a)   90%
116. (b) 318-8(d)   90%
117. (d) 320-1   90%
118. (d) 320-7   95%
119. (a) 320-12   90%
120. (d) 320-14(2)   85%
121. (c) 321-6   90%
122. (b) 324-4   90%
123. (b) 324-9   90%
124. (c) 325-1   90%
125. (c) 328-2 Definition of FCC cable   90%
126. (b) 328-4   90%
127. (b) 328-6   95%
128. (d) 328-11   95%
129. (c) 328-14   90%
130. (d) 328-37   80%
131. (d) 330-13   90%
132. (d) 330-15   90%
133. (b) 330-21   95%
134. (b) 331-1   90%
135. (c) 331-3(6)
136. (a) 331-4(1)   85%
137. (a) 331-7   95%
138. (a) 331-14   90%
139. (d) 333-7(a)
140. (a) 333-11   95%
141. (b) 333-12(b)   90%
142. (b) 334-10(a)   95%
143. (c) 334-22   85%
144. (a) 336-4
145. (b) 336-5(a)(1)   95%
146. (c) 336-6(c)   100%
147. (a) 336-18   95%
148. (b) 336-26
149. (d) 338-1(b)   90%
150. (c) 338-4

151. (d) 339-3(a)(4)   90%
152. (a) 339-3(b)(10)
153. (d) 342-1   90%
154. (c) 342-5   95%
155. (b) 342-7(a)(2)   95%
156. (d) 343-3(1)(2)(3)   90%
157. (c) 343-6   95%
158. (c) 345-5   90%
159. (b) 345-9(a)   95%
160. (d) 345-12(a)   95%
161. (a) 345-12(b)(4)   80%
162. (b) 346-3(b)   95%
163. (d) 346-9(a)   90%
164. (c) Table 346-10 Exception   90%
165. (c) 346-12(b)(2)   85%
166. (d) 347-1   95%
167. (d) 347-2(f)   80%
168. (a) 347-4 Exception No. 2
169. (b) 347-8 Table   95%
170. (d) 348-7   95%
171. (b) 348-12   90%
172. (d) 349-3 and 349-4   75%
     Read the *NEC*®.
173. (d) 350-5(1) Table 310-13 wet location   95%
174. (d) 350-10   80%
     The *Code* doesn't limit the length of 3/8″ or larger
     flex.
175. (d) 350-18 Exception No. 3   70%
     Read the exception again.
176. (d) 351-4(a) and (a)(3)   80%
177. (a) 351-8   85%
178. (c) 351-9   75%
179. (a) 352-1(b)(2)   90%
180. (d) 352-5   80%
181. (d) 352-8   95%
182. (c) 354-3(b)   90%
183. (d) 354-7   90%
184. (b) 356-1   90%
185. (d) 356-9   80%
186. (c) 358-2   95%
187. (d) 358-6   80%
188. (d) 358-13   90%
189. (b) 362-4   90%

190. (b) 362-8	80%	
191. (d) 363-3	95%	
192. (d) 363-6	80%	
193. (b) 363-14	90%	
194. (b) 364-5	95%	
195. (d) 364-12 and 380-8	65%	

196. (a) 365-1	95%
197. (a) 365-2(a)	95%
198. (a) 365-7(2)	55%
199. (c) 370-3	60%
200. (c) 370-16(a)	80%

Notes:

Unit 7 – Motor Calculations

1. (a) True, 240-3(g)   90%

2. (c) No. 10   80%
   Table 430-148 lists the FLC for a 5 horsepower motor
   as 28 ampere 430-22(a).
   28 ampere × 1.25 = 35 ampere
   Table 310-16 No. 10 THHN at 75°C rated 35 ampere.

3. (a) True, 430-32, 430-52   90%

4. (c) a and b   55%

5. (d) all of these, 430-31   75%

6. (d) a and b, 430-51   95%

7. (b) False, 430-32(a)(1)   50%

8. (a) True, 430-34   80%

9. (c) 125%, 430-32   90%

10. (c) 125%, 430-32   90%

11. (b) 115%, 430-32   80%

12. (a) 20 ampere   90%
    Overloads are sized according to the nameplate cur-
    rent rating, not the motor FLC 430-32(a)(1)
    16 ampere × 1.25 = 20 ampere

13. (d) 40 ampere   95%
    Table 430-150 FLC = 32 ampere
    32 ampere × 1.25 = 40 ampere
    430-32(a)(1) and 240-6(a)

14. (d) all of these, 430-52, Table 430-152   95%

15. (b) larger, 430-52, Table 430-152   80%

16. (d) all of these, Table 430-152   90%

17. (d) 225%,   90%
    430-52(c)(1) Exception No. 2b

18. (a) 125%, 115%, 250%   80%
    430-22, 430-32, Table 430-152

19. (d) all of these, 430-22(a) and Table 310-16   80%
    30.8 ampere × 1.25 = 38.5 ampere = No. 8 THHN
    Overload protection, 430-32(a)(1)
    29 ampere (nameplate) × 1.15 = 33 ampere
    Short-circuit and ground-fault protection:
    30.8 ampere (FLC) × 2.50 = 77 ampere
    Next size up circuit breaker = 80 ampere
    430-52(c)(1) Exc. 1, Table 430-152 and 240-6(a)

20. (d) I and II, 430-24   90%

21. (d) I and II, 430-62(a)   80%

22. (d) all of these   65%
    Step 1: FLC Table 430-150
    30 horsepower motor = 32 ampere
    10 horsepower motor = 14 ampere
    Step 2: Branch circuit conductors 240-22(a)
    30 horsepower: 32 ampere × 1.25 = 40 ampere
    No. 8, Table 310-16 rated 50 amperes

    10 horsepower: 14 ampere × 1.25 = 18 ampere
    No. 14, rated 20 amperes

    Step 3: Branch circuit protection size 240-6(a)
    430-52(c)(1) Exception No. 1 and Table 430-152:
    30 horsepower: 32 ampere × 2.5 = 80 ampere
    10 horsepower: 14 ampere × 2.5 ampere = 35 ampere

    Step 4: Feeder conductor 430-24:
    (32 ampere × 1.25) + 14 ampere = 54 ampere,
    No. 6 THHN conductor

    Step 5: Feeder protection 240-6(a) and 430-62:
    Not be greater than the largest branch circuit device
    plus the sum of the the FLC's on the same line,
    80 ampere + 14 ampere = 94 ampere,
    Next size down = 90 ampere.

23. (b) full-load current   85%

24. (d) IV only   75%
    25 horsepower = 26 ampere Table 430-150
    20 horsepower = 27 ampere Table 430-150
    15 horsepower = 21 ampere Table 430-150
    3 horsepower, 1-phase, 120 volt, FLC of 34 ampere

25. (d) b and c   80%
    Table 430-150
    VA of 5 horsepower 460 volts =
    460 volt × 7.6 ampere × $\sqrt{3}$ = 6,055 VA
    VA of 5 horsepower 115 volts =
    230 volts × 15.2 ampere × $\sqrt{3}$ = 6,055 VA

26. (a) 3,890 VA   80%
    VA of 3 horsepower, 208 volt 1-phase
    208 volts × 18.7 ampere = 3,890 VA

✫ Challenge Answers

27. (c) 21 ampere   55%
430-22 Exception 1 and Table 430-22(a).
Branch circuit conductor ampacity shall not be less than 85% of the motor nameplate ampere. Table 430-22(a) Intermittent and 5 minute rated motor
25 ampere × .85 = 21.25 ampere

28. (b) 24 ampere   90%
430-32(a)(1). The motor overload for this motor must be sized *no more than* 115% of the motor nameplate current rating.
21.5 ampere × 1.15 = 24.73 ampere

29. (a) 30.8 ampere   75%
430-34. Because the service factor is greater than 1.15, the maximum overload must *not be greater* than 140% of the motor nameplate current rating (not the motor FLC).
22 ampere × 1.4 = 30.8 ampere

30. (a) 31.2 ampere   905%
430-32(a)(2), Table 430-148.
The ultimate trip device must be sized not more than 156% of the motor FLC rating.
1 horsepower, 120 volts
20 ampere × 1.56 = 31.2 ampere

31. (d) 40 ampere   80%
240-6(a), 430-6(a), 430-52(c)(1) Exception No. 1, Table 430-148, and Table 430-152. The branch circuit protection device shall not be greater than 150% of the motor FLC. Table 430-152.
Branch protection must not exceed:
24 ampere × 1.5 = 36 ampere, Exception No. 1 of 430-52(c)(1) permits the next size up, 40 ampere 240-6(a)
Note: Be careful, do not use Table 430-149 to determine the motor full load ampere.

32. (a) 125 ampere   60%
240-6(a), 430-6(a), 430-52(c)(1) Exception No. 1, Table 430-148 and Table 430-152.
The motor branch circuit protection device must not be greater than 250% of the motor FLC.
Branch protection must not exceed:
50 ampere × 2.5 = 125 ampere

33. (c) 700 ampere   80%
240-6(a), 430-6(a), Table 430-147, and Table 430-152.
The branch circuit protection device shall not be greater than 150% of the motor FLC. Table 430-152.
Branch protection must not exceed:
425 ampere × 1.5 = 637.5 ampere, but
Exception No. 1 of 430-52(c)(1) permits the next size up, 700 ampere.

34. (c) No. 4/0   50%
430-24(a) and Table 310-16. The feeder conductor must be sized not less than 125% of the largest motor FLC, plus 100% of the FLC's of all other motors on the same line. Motor FLC's – Tables 430-148 and 150.
15 horsepower 208 volt, 3-phase = 46.2 ampere
3 horsepower 208 volt, 1-phase = 18.7 ampere
1 horsepower 120 volt = 16 ampere

	L1	L2	L3
3 – 15 horsepower, 3-phase	46.2	46.2	46.2
	46.2	46.2	46.2
	46.2	46.2	46.2
3 – 3 horsepower, 1-phase	18.7	18.7	
		18.7	18.7
	18.7		18.7
3 – 1 horsepower, 120v	16.0	16.0	16.0
	192.0	192.0	192.0

Feeder conductor must not be less than:
(46.2 ampere × 1.25) + 46.2 ampere + 46.2 ampere + 18.7 ampere + 18.7 ampere + 16 ampere = 203.55 ampere
Note: No. 3/0 THHN is rated for 200 ampere at 75°C

35. (d) 90 ampere   50%
240-6(a), 430-62.The feeder protection device shall not be greater than; the largest feeder protection device plus the sum of the motor FLC's on the same line. Largest branch circuit protection device:
5 horsepower motor – 430-52(c)(1), Table 430-152.
The branch circuit protection device shall not exceed 250% of the motor FLC.
28 ampere × 2.5 = 70 ampere
Feeder protection not to exceed:
70 ampere device + 20 ampere = 90 ampere

36. (d) 175 ampere   65%
240-6(a), 430-62 and Table 430-152. Feeder protection must be sized *not greater than* the largest branch circuit protection device plus the FLC's of all other motors on the same line. Branch circuit protection devices, shall not be greater than the values of Table 430-152, except at permitted by Exception No. 1 of 430-52(c)(1).

Motor 1 = 40 horsepower,
52 ampere × 250% = 130 ampere,
Next size up, 150 ampere.

Motor 2 = 20 horsepower,
27 ampere × 250% = 67.5 ampere,
Next size up, 70 ampere.

Motor 3 = 10 horsepower,
14 ampere × 250% = 35 ampere,

Motor 4 = 5 horsepower,
7.6 ampere × 250% = 19 ampere, next size up, 20 ampere.

Feeder protection device must not be greater than:
150 ampere branch overcurrent device + 27 ampere + 14 ampere + 7.6 ampere = 199 ampere; the next size down 175 ampere, 240-6(a).

37. (d) 110 ampere   50%
430-62 and Table 430-152. Feeder protection must be sized *not greater than* the largest branch circuit protection device, plus the FLC's of other motors on the same line. Branch circuit protection devices, shall not be greater than the values of Table 430-152, except as permitted in Exception No. 1 of Section 430-52(c)(1).

$1/2$ horsepower motor
9.8 ampere × 1.75 = 17 ampere, next size up, 20 ampere.

To size the feeder protection and conductor, we must balance the twenty-two, 115-volt motors on two lines (Line 1 and Line 2). This results in only eleven motors per line, therefor the feeder protection device *must not be greater than*, 20 ampere (largest branch branch overcurrent device) plus the ten other 115 volt motors at 9.8 ampere, 20 ampere + 98 ampere = 118 ampere, next size down, 110 ampere.

38. (a) 225 ampere   55%
240-6(a), 430-62. The feeder short-circuit protection device *shall not be greater* than the largest branch circuit protection device, plus the sum of the FLC's of all other motors on the same line.

Branch circuit protection for 25 horsepower 208 volt 3-phase motor must be sized not more than 250% of the motor full load current [*Section 430-53(c)(1)* and Table 430-152].

74.8 ampere × 2.5 = 187 ampere,
Next size up, 200 ampere

Feeder protection shall not be greater than:
200 ampere + 34 ampere = 234 ampere,
Next size down, 225 ampere.

NEC® Answers

39. (c) 370-16(b)(1), Exception
40. (c) 370-16(b)(5)   90%
41. (c) 370-18   75%
42. (b) 370-22 Exception   50%
43. (d) 370-23(d)(1)
44. (b) 370-23(e)   60%
45. (d) 370-24 The answer is $15/16$ inch   50%
46. (a) 370-27(c) Exception   90%
47. (d) 370-29   90%
48. (c) 370-40(c)   90%
49. (d) 370-71(a)   95%
50. (c) 373-2(a)   90%
51. (a) 373-5(c)   75%
The question specified cables, not open wiring or knob-and-tube wiring.
52. (a) 374-1   85%
53. (d) 374-5(a)(1) Current-carrying conductors   55%
54. (d) 374-6(a)   75%
Copper (assumed) is 1,000 ampere per square inch. 1,000 ampere × 1.5 inches = 1,500 ampere.
55. (d) 374-9(a)   75%
56. (b) 380-2(b) Exception No. 1   95%
57. (a) 380-8(a)   95%
58. (b) 380-9(c)   90%
59. (d) 380-14(a)   90%
60. (b) 380-17   95%
61. (a) 384-3(a)(2)   95%
62. (a) 384-5   90%
63. (d) 384-9   90%
64. (d) 384-13   95%
65. (a) 384-15   95%
66. (c) 384-20   80%
67. (a) 400-4 Table   95%
68. (a) 400-5(a) Table Note   80%
Be careful, be sure to read the notes on the bottom of the table and use Column "B."
69. (b) 400-8   95%
70. (d) 400-10   95%
71. (a) 400-22   95%
72. (b) 402-3   85%

73. (a) 402-10   75%
74. (a) 410-4(a)   95%
75. (c) 410-4(d)   90%
76. (b) 410-8(c)   90%
77. (b) 410-8(d)   90%
78. (b) 410-15(a)   95%
79. (c) 410-15(b)(5)   90%
80. (c) 410-16(a)   95%
81. (a) 410-16(h)   80%
82. (d) 410-24   95%
83. (d) 410-28(d)   95%
84. (a) 410-30(b)   75%
85. (c) 410-31 Ex. 2   95%
86. (b) 410-42(a)   95%
87. (a) 410-53   80%
88. (b) 410-56(f)(2)   75%
89. (c) 410-57(a)   95%
90. (d) 410-57(b)(2)   95%
91. (a) 410-58(e)   95%
92. (a) 410-67(c)   90%
93. (b) 410-68   80%
94. (c) 410-76(b)   80%
95. (d) 410-80(b)   75%
   You know it is not 120 volts. Don't you have 240 volt
   ranges and dryers? 250 and 600 volts would be good
   guesses, but the answer is 1,000 volts.
96. (b) 410-101(a)   90%
97. (d) 410-101(c)(9)
98. (c) 422-10(a)   75%
99. (b) 422-12   95%
100. (b) 422-16(b)(1)   95%
101. (d) 422-18(a)   95%
102. (b) 422-31(b)   90%
103. (b) 422-46   95%
104. (c) 424-3(b)   95%
105. (a) 424-13   90%
106. (b) 424-22(b)   75%
107. (c) 424-36   80%
108. (a) 424-39   75%
109. (c) 424-72(a)   80%
110. (c) 424-91(a)   80%
111. (d) 426-2 FPN   90%
112. (c) 426-30   75%
113. (c) 426-50(a)   90%
114. (c) 427-4   95%
115. (c) 430-6(a)(1)   75%
116. (a) 430-9(b)   95%
117. (a) Table 430-12(c)(1)   95%
118. (a) 430-17   95%
119. (d) 430-28(2)   80%
120. (d) 430-32(a)   85%
121. (d) 430-34   80%
122. (c) 430-37 Table 430-37   75%
123. (c) 430-44   75%
124. (b) 430-61   55%
125. (c) 430-72(b) Exc. 2, Note 3 to Column C   85%
126. (b) 430-74(b)   80%
127. (a) 430-87   90%
128. (a) 430-102(b) Exception
129. (a) 430-107   90%
130. (a) 430-109(b)   80%
131. (c) 430-111(b)(3)   80%
132. (a) 440-1   95%
133. (d) 440-14   95%
134. (a) 440-55(b)   80%
135. (c) 445-3   95%
136. (b) 455-8(a)   95%
137. (c) 450-3(b)
138. (b) 450-11   90%
139. (d) 450-21(c)   80%
140. (c) 450-42   80%
141. (a) 450-46   95%
142. (d) 460-8(a)   95%
143. (c) 480-2   90%
144. (b) 480-9(a)   90%
145. (a) 500-3(a) FPN   95%
146. (a) 500-5(a)(1)   95%
147. (a) 500-5(b)(2)   95%
148. (d) 500-5(c)   85%
149. (c) 500-7   95%
150. (a) 500-9(a)   85%
151. (b) 500-8(a)(1)   75%
152. (a) 501-3(a)   85%
153. (d) 501-4(a)(1)   85%
154. (a) 501-4(b)   95%
155. (a) 501-5(a)(1) Exception   90%

156.	(b) 501-5(b)(2) Exception	95%
157.	(a) 501-5(c)(6)	
158.	(a) 501-6(a)	95%
159.	(c) 501-16(b)	85%
160.	(a) 502-6(a)(3)	85%
161.	(d) 503-13(a)	90%
162.	(c) 511-3(a)	80%
163.	(b) 511-3(c)	90%
164.	(b) 511-7(b)	80%
165.	(d) 513-3(b)	85%
166.	(b) 514-5(a)	85%
167.	(a) 514-5(a)	95%
168.	(d) 514-7(a)	95%
169.	(d) 515-2 Table	90%
170.	(a) 516-2(e)	85%
171.	(b) 517-3 FPN Patient care area	80%
172.	(a) 517-18(a), Exception No. 3	
173.	(b) 517-19(b)(1)	80%
	6 single receptacles or 3 duplex receptacles	
174.	(d) 517-35(a)	75%
175.	(d) 517-64(a)(1, 2 and 3)	90%
176.	(d) 518-2	90%
177.	(d) 518-4(a)	
178.	(c) 520-42	80%

179.	(d) 520-73	95%
180.	(a) 525-13(d)	
181.	(d) 525-18	
182.	(c) 540-2	80%
183.	(d) 545-3	80%
184.	(a) 550-2	
185.	(c) 550-5(a) Exception No. 1	70%
186.	(d) 550-8(b)	95%
187.	(b) 550-22 Table	65%
	316,000 VA × 0.29 = 27,840 VA	
188.	(d) 551-45(b)	75%
189.	(b) 551-73(a)	50%
190.	(d) 555-3 FPN No. 3	95%
191.	(d) 555-6	95%
192.	(b) 600-5(a)	95%
193.	(a) 600-6(a)	
194.	(d) 600-8(c)	75%
195.	(a) 600-21(d)	75%
196.	(b) 604-6(a)(3)	80%
197.	(a) 620-11(a)	80%
198.	(a) 620-22(a)	
199.	(d) 620-85	
200.	(d) 630-12(a)	95%

Notes:

Unit 8 – Voltage Drop Calculations

1. (a) greater   90%

2. (d) all of these   90%

3. (a) Silver   90%

4. (c) Aluminum   90%

5. (c) circular mils, 110-6   75%

6. (a) True   95%

7. (a) positive   90%

8. (d) Table 8, Table 9   85%

9. (a) .2 ohm   90%
   (400 feet/1,000 feet) $\times$ .491 ohm = .1964 ohm

10. (a) .023 ohm   90%
    (150 feet/1,000 feet) $\times$ .154 ohm = .0231 ohm

11. (d) .56 ohm   90%
    (1,400 feet/1,000 feet) $\times$ .403 ohm = .5642 ohm

12. (b) .16 ohm   90%
    (800 feet/1,000 feet) $\times$ .201 ohm = .1608 ohm

13. (c) .02 ohm, Chapter 9, Table 8   95%
    R_{DC} = (120 feet/1,000 feet) $\times$ .154 ohm
    R_{DC} = .02 ohm

14. (d) .5 ohm   85%
    R = (1249 feet/1000 feet) $\times$ .403 ohm
    R = .503 ohm

15. (a) True   90%

16. (a) True   75%

17. (c) eddy currents   90%

18. (b) skin effect   90%

19. (d) all of these   90%

20. (a) True   95%

21. (a) .03 ohm, Chapter 9, Table 9   85%
    Resistance of 1,000 feet of No. 2/0 = .1 ohm
    R at 320 feet = (320 feet/1,000 feet) $\times$ .10
    R = .032 ohm

22. (a) .032 ohm, Chapter 9, Table 9   90%

23. (b) .022 ohm, Chapter 9, Table 9   95%
    R_{AC} = (220 feet/1,000 feet) $\times$ .1 ohm
    R_{AC} = .022 ohm

24. (a) 0.01 ohm, Chapter 9, Table 9   85%
    R_{AC} = (369 feet/1,000 feet) $\times$ .027 ohm
    R_{AC} = .01 ohm

25. (a) .04 ohm, Chapter 9, Table 9   85%
    R = (160 feet/1,000 feet) $\times$ .25 ohm
    R = .04 ohm

26. (b) No. 2   90%
    Chapter 9 Table 9, No. 1/0 Aluminum in a steel race-
    way has a resistance of .2 ohm, a No. 2 copper con-
    ductor can be used (.2 ohm).
    Note: No. 2 THHN has an ampacity of 130 ampere at
    75° C, Table 310-16.

27. (a) .014 ohm   90%
    Note: Assume 1,000 feet

    $$R_{Total} = \frac{\text{Resistance of One Conductor}}{\text{\# of Parallel Conductors}}$$

    $$R_{Total} = \frac{.0429 \text{ ohm}}{3 \text{ resistors}}$$

    $$R_{Total} = .0143 \text{ ohm}$$

28. (d) .067 ohm   75%

    $$R = \frac{\text{Resistance of One Conductor}}{\text{Number of Parallel Conductors}}$$

    $$R = \frac{.2 \text{ ohm}}{3 \text{ resistors}}$$

    R_{Total} = .0666 ohm, round to .067

29. (a) True   85%

30. (b) suggest   85%

31. (a) Inductive   75%

32. (b) Electronic   75%

33. (a) True   85%

34. (b) 3,400 watts   75%

Power = E^2/R. Since the applied voltage (200 volts) is less the the rated volts (230 volts), the actual power will be less than rated power (4,500 watts). The resistance of a 4,500 watt water heater rated 230 volts is:

R=E^2/P

R = 230 volts^2/4,500 watt

R = 11.76 ohm.

The power consumed by a 11.76 ohm resistor connected to a 200 volt power supply is:

P = E^2/R

P = 200 volts^2/11.76 ohm

P = 3,400 watts

35. (d) all of these   90%

36. (b) 6.24 volts   95%

VD = 208 volts × .03 = 6.24 volts

37. (d) 7.2 volts   95%

VD = 240 volts × .03 = 7.2 volts

38. (b) 4.75 volts   90%

$$VD = \frac{2 \times K \times I \times D}{CM}$$

K = 12.9 ohm, copper
I = 12 ampere
D = 100 feet
CM = 6,530

$$VD = \frac{2 \times 12.9 \text{ ohm} \times 2 \text{ ampere} \times 100 \text{ feet}}{6,530 \text{ circular mils}}$$

VD = 4.74 volts dropped

39. (b) 9.5 volts   85%

$$VD = \frac{2 \times K \times I \times D}{CM}$$

K = 12.9 ohm, Copper
I = 24 ampere
D = 160 feet
CM = 10,380

$$VD = \frac{2 \times 12.9 \text{ ohm} \times 24 \text{ ampere} \times 160 \text{ feet}}{10,380 \text{ circular mils}}$$

VD = 9.5 volts

40. (d) 4.4 volts   85%

$$VD = \frac{\sqrt{3} \times K \times I \times D}{CM}$$

$\sqrt{3}$ = 1.732
K = 21.2 ohm, aluminum
I = P/(E × $\sqrt{3}$)
I = 36,000 VA/(208 volts × 1.732) = 100 ampere

D = 100 feet
CM = 83,690

$$VD = \frac{1.732 \times 21.2 \text{ ohm} \times 100 \text{ ampere} \times 100 \text{ feet}}{83,690 \text{ circular mils}}$$

VD = 4.39 volts dropped

41. (d) No. 3 THHN   90%

$$CM = \frac{2 \times K \times I \times D}{VD}$$

K = 12.9 ohm, copper
I = 52 ampere, not FLC
D = 110 feet
VD = 3.45 volts (115 volts × .03)

$$CM = \frac{2 \times 12.9 \text{ ohm} \times 52 \text{ ampere} \times 110 \text{ feet}}{3.45 \text{ volts drop}}$$

CM = 42,776, Chapter 9, Table 8 = No. 3

42. (b) No. 8 THHN   75%

$$CM = \frac{2 \times K \times I \times D}{VD}$$

K = 12.9 ohm, copper
I = 26 ampere at 230 volts
D = 110 feet
VD = 6.9 volts (230 volts × .03)

$$CM = \frac{2 \times 12.9 \text{ ohm} \times 26 \text{ ampere} \times 110 \text{ feet}}{6.9 \text{ volts drop}}$$

CM = 10,694, Chapter 9, Table 8 = No. 8

43. (a) No. 10 THHN   85%

$$CM = \frac{\sqrt{3} \times K \times I \times D}{VD}$$

K = 12.9 ohm, copper
I = 18 ampere
D = 300 feet
VD = 14.4 volts (480 volts × .03)

$$CM = \frac{1.732 \times 12.9 \text{ ohm} \times 18 \text{ ampere} \times 300 \text{ feet}}{14.4 \text{ volts drop}}$$

CM = 8,379, Chapter 9, Table 8, No. 10

44. (c) 145 feet   85%

$$D = \frac{CM \times VD}{2 \times K \times Q \times I}$$

CM = 16,510
VD = 7.2 volts (240 volts × .03)
K = 12.9 ohm
Q = Less than No. 2/0, does not apply
I = 31.25 ampere (7,500/240)
Note. Do not confuse distance (D) with length (L).

This formula gives the distance between two points, not the length of conductors in the run.

45. (c) 345 feet   85%

$$D = \frac{CM \times VD}{\sqrt{3} \times K \times I}$$

CM = 26,240
VD = 13.8 volts (460 volts × .03)
K = 12.9

$$I = 47 \text{ ampere}, = \frac{P}{E\sqrt{3}}, = \frac{37,500}{(460 \times 1.732)}$$

46. (d) 147 ampere   85%
CM = No. 1/0 = 105,600
VD = 7.2 volts (240 × .03)
K = 12.9 ohm, Copper
D = 200 feet

$$I = \frac{105,600 \text{ circular mils} \times 7.2 \text{ volts drop}}{2 \text{ conductors} \times 12.9 \text{ ohm} \times 190 \text{ feet}}$$

I = 147 ampere
Note. The maximum load permitted on No. 1/0 THHN is 150 ampere at 75°C, Table 310-16 and 110-14(c)(2).

47. (a) 91 ampere   85%

$$I = \frac{CM \times VD}{\sqrt{3} \times K \times D}$$

CM = No. 2, 66,360
VD = 13.8 volts (460 volts × .03)
K = 21.2 ohm, Aluminum
D = 300 feet

$$I = \frac{66,360 \text{ CM} \times 13.8 \text{ volts}}{\sqrt{1.732} \times 21.2 \text{ ohms} \times 300 \text{ feet}}$$

I = 83 ampere.
Note. The maximum load permitted on No. 2 THHN Aluminum is 90 ampere at 75°C, Table 310-16 and 110-14(c)(2).

48. (c) No. 4 THHN   85%
Step 1. Determine the voltage drop of the existing conductors:

$$VD = \frac{2 \times K \times I \times D}{CM}$$

K = 21.2 ohm, Aluminum
I = 50 ampere
D = 65 feet
CM = 41,740 (Chapter 9 Table 8)

$$VD = \frac{2 \times 21.2 \text{ ohm} \times 50 \text{ ampere} \times 65 \text{ feet}}{41,740 \text{ circular mils}}$$

VD = 3.3 volts
Step 2. Determine the voltage drop permitted for the extension by subtracting the voltage drop of the existing conductors from the permitted voltage drop. The voltage drop permitted for the extension would be 2.94 volts (6.24 volts – 3.3 volts).
Step 3. Determine the extended conductor size:

$$CM = \frac{2 \times K \times I \times D}{VD}$$

K = 12.9 ohm, Copper
I = 50 ampere
D = 85 feet
VD = 2.94 volts

$$CM = \frac{2 \times 12.9 \text{ ohm} \times 50 \text{ ampere} \times 85 \text{ feet}}{2.94 \text{ volts dropped}}$$

CM = 37,296, Chapter 9 Table 8 = No. 4
Note. No. 4 THHN is rated for 70 ampere at 60°C, Table 310-16 and 110-14(c)(1) which is sufficient to carry the load.

49. (a) 110 volts   85%
Voltage Source = VD/.03
Voltage Source = 3.3 volts drop/.03 = 110 volts

☆ Challenge Answers

50. (b) temperature coefficient   95%
51. (a) Impedance   97%
52. (c) 0.05 ohm   90%
Chapter 9, Table 9
The alternating current resistance of No. 4 aluminum is .51 ohm in steel conduit for 1,000 feet.
R = (Resistance/1,000 feet) × Conductor Length
R = (.51 ohm/1,000 feet) × (2 wires × 50 feet)
R = .051 ohm

53. (a) No. 6 copper   95%
No calculations are required for this problem. The alternating current resistance of No. 4 aluminum in a steel raceway = .51 ohm. Using Chapter 9, Table 9, the alternating current resistance of No. 6 copper in a steel raceway that has an alternating current resistance of .49 ohm.

<u>54.</u> (b) 232.8 volts   97%

The *NEC®* recommends a maximum 3% voltage drop on the branch circuit conductors, which calculates out to be: 240 volts × .03 = 7.2 volts. The minimum voltage at the load will calculate to be 240 volts less 7.2 volts = 232.8 volts, or 240 volts × .95 = 232.8 volts

<u>55.</u> (b) 5.31 volts dropped   90%

$$CM = \frac{2 \times K \times I \times D}{VD}$$

K = 21.2 ohm, aluminum
I = 55 ampere
D = 95 feet
CM = No. 4, 41,740 Chapter 9, Table 8

$$VD = \frac{2 \times 21.2 \text{ ohm} \times 55 \text{ ampere} \times 95 \text{ feet}}{41,740 \text{ circular mils}}$$

VD = 5.31 volts

<u>56.</u> (a) No. 3, Chapter 9, Table 8   90%

$$CM = \frac{2 \times K \times I \times D}{VD}$$

K = 12.9 ohm, copper
I = 40 ampere
D = 100 feet
VD = 7.2 volts − 4 volts = 3.2 volts

$$CM = \frac{2 \times 12.9 \text{ ohm} \times 40 \text{ ampere} \times 150 \text{ feet}}{3.2 \text{ volts}}$$

CM = 48,375, Chapter 9, Table 8 = No. 3

<u>57.</u> (c) 325 feet   85%

$$D = \frac{CM \times VD \text{ Allowable}}{1.732 \times K \times I}$$

CM = 52,620, Chapter 9, Table 8
VD = 230 volts × 3% = 6.9 volts
K = 12.9 ohm, Copper
I = 50 ampere

$$D = \frac{52,620 \text{ CM} \times 6.9 \text{ volts dropped}}{1.732 \times 12.9 \text{ ohm} \times 50 \text{ ampere}}$$

Distance = 325 feet

<u>58.</u> (c) 10 ampere   90%

$$I = \frac{CM \times VD \text{ Allowable}}{2 \times K \times D}$$

CM = 16,510, Chapter 9, Table 8
VD = 120 volts × 3% = 3.6 volts
K = 12.9 ohm, copper
D = 225 feet

$$I = \frac{16,510 \text{ CM} \times 3.6 \text{ volts dropped}}{2 \times 12.9 \text{ ohm} \times 225 \text{ feet}}$$

I = 10 ampere

<u>59.</u> (a) 14 ampere   90%

To determine the current when you know the volts and resistance, use the formula:
$I = {}^E/_R$
E = 112 volts, R = 8 ohm
I = 112 volts/8 ohm
I = 14 ampere

<u>60.</u> (d) .0180 ohm   90%
R = E/I
E = 7.2 volts dropped, I = 400 ampere
R = 7.2 volts dropped/400 ampere
R = .0180 ohm

<u>61.</u> (a) 2.6%   90%
% of VD = (volts dropped/voltage source) × 100
Volts dropped = 3 volts
Voltage source = 115 volts
% of VD = 3 volts dropped/115 volts source
% of VD = 2.6%

NEC® Answers

62.	(c) 90-4	90%
63.	(a) 100 Accessible, Equipment	75%
64.	(b) 100 Branch Circuit	95%
65.	(c) 100 Building	90%
66.	(b) 100 Demand Factor	60%
67.	(b) 100 Duty, Continuous	90%
68.	(a) 100 Effectively, Grounded	90%
69.	(d) 100 Grounding Electrode Conductor	70%
70.	(d) 100 Hoistway	85%
71.	(d) 100 Qualified Person	90%
72.	(b) 100 Remote-Control Circuit	75%
73.	(c) 100 Service Entrance Cond., Overhead	95%
74.	(b) 100 Voltage, Circuit	90%
75.	(a) 100 Weatherproof	95%
76.	(c) 110-8	80%
77.	(c) 110-14 FPN	80%
78.	(a) 110-22	85%
79.	(c) 110-26(a)(1) Table Condition 3	85%
80.	(a) 110-33(a)	65%

80. Over 6 foot requires two entrances, but a 6-foot control panel requires one entrance.

81.	(a) 200-6(b)	85%
82.	(c) 210-7(f)	90%

83. (b) 210-19(a)

84. (a) 210-21(b)(2), Table   95%

85. (d) 210-52(b)(1)   95%

86. (d) 210-63 FPN and 210-8(b)(2)   75%

87. (a) 210-70(a)(1) Exception No. 1   85%

88. (b) 215-10 Exception No. 3   80%

89. (b) 220-12(a)   90%

90. (a) 220-30(a)   95%

91. (b) 225-6(b)   90%

92. (a) 225-19(a) Exception No. 3   80%

93. (b) 225-32

94. (b) 230-3   95%

95. (d) 230-23(a)   85%

96. (c) 230-31(c) FPN   95%

97. (b) 230-204(c)   90%

98. (d) 240-6(a)   85%

99. (b) 240-20(b)(1)   65%

100. (c) 240-54(c)   95%

101. (b) 250-2(d) and 250-54(c)   95%

102. (a) 250-24(a)(4) Exception   95%

103. (d) 250-32(e)

104. (a) 250-50(a)(2)   85%

105. (b) 250-60

106. (c) 250-68(b)   95%

107. (c) 250-80 Exception

108. (d) 250-96 Exception   85%

109. (c) 250-118(8)   90%

110. (d) 250-130(c)

111. (d) 250-148(a)   95%

112. (d) 250-178   90%

113. (a) 300-4(a)(2)   95%

114. (a) 300-22(d)   75%

115. (c) 305-4(c)

116. (b) 310-4   90%

117. (c) Table 310-13   85%

118. (c) 310-16   65%
Aluminum and copper-clad aluminum have the same ampacities.

119. (d) 318-9(d)   90%

120. (d) 320-8   90%

121. (b) 321-6   90%

122. (a) 328-4   95%

123. (a) 328-17   85%

124. (b) 330-16   90%

125. (b) 331-4(5)   75%

126. (d) 333-12(a)

127. (c) 334-11(a)(2)   70%

128. (a) 336-18   95%

129. (a) 336-21   95%

130. (d) 339-1   95%

131. (d) 342-7(a)(1)   95%

132. (a) 343-1   95%

133. (d) 345-11   90%

134. (a) 346-12(b)(1)   75%
The general rule applies if the condition of the exception (straight runs, threaded couplings, etc.) was given in the question, then the exception would have applied. Note: Table 346-12 is an exception table, but it's not properly labeled as such.

135. (d) 346-16(a)   90%

136. (d) 347-5 and 347-6   80%

137. (b) 348-15   60%
EMT is required to be supported within 3 feet of each termination, but not within 3 feet of each coupling.

138. (c) 350-12 Table Column B   90%

139. (b) 351-8 Exception No. 2   90%

140. (c) 351-23(a)(6)

141. (c) 352-9   80%

142. (c) 354-5   80%

143. (a) 358-7   95%

144. (c) 362-6

145. (c) 363-7   95%

146. (c) 363-16   95%

147. (b) 365-10   75%
The *Code* specifies maximum diameter.

148. (b) 370-16(b)(2)   90%

149. (a) 370-23(h)(1)   90%

150. (d) 373-8   80%

151. (d) 374-5(2)   80%

152. (a) 380-13(a)   80%

153. (c) 384-7   95%

154. (b) 384-16(a) Exception No 1   85%

155. (d) 400-14   70%

156. (d) 410-57(e)   80%

157. (d) 422-16(b)(2)   80%

158. (d) 424-44(g)

159. (b) 426-42   75%

Article 427 – Fixed Electric Heating Equipment for Pipelines And Vessels

160. (a) 430-9(c)   95%

161. (b) 430-52(b)   90%

162. (a) 430-109(a)(1)   75%

163. (b) 430-132   70%

164. (c) 440-22(a)   75%

165. (b) 445-5   80%

166. (b) 450-41   80%

167. (a) 455-8   95%

168. (d) 480-5(b)   95%

169. (b) 500-9(b)   65%

170. (d) 501-5(c)   90%

171. (b) 511-4   80%

172. (b) 515-2 Table   90%

173. (d) 517-13(a) Exception No. 3

174. (d) 520-68(a)(1)   80%

175. (c) 530-14   95%

176. (d) 545-6 and Exception   30%

177. (b) 550-7(a)   70%

178. (d) 553-7(b)   95%

179. (b) 555-4

180. (a) 600-6(a)

181. (c) 610-14(e)(2)   75%

106 ampere + (72 ampere at 50%) = 142 ampere

Note: See Table 610-14(a) for conductor selection.

182. (b) 620-91(c)   75%

183. (b) 630-31(a)(2)   80%

21 ampere $\times$ .39 = 8.19 ampere

184. (c) 650-7   95%

185. (b) 660-9   95%

186. (a) 665-44(a)(1)   90%

187. (b) 680-4   95%

188. (a) 680-6(a)(2)   75%

189. (b) 680-11   75%

The *Code* says normal, not abnormal.

190. (a) 680-21(d)   80%

191. (a) 680-22(b)   95%

192. (b) 680-38

193. (b) 700-12   75%

194. (b) 702-5   85%

195. (a) 725-26(b)   90%

196. (b) 725-54(a)(1)   50%

197. (b) 760-3(a) and 300-21   95%

198. (b) 760-71   95%

199. (a) 810-11 Exception   95%

200. (d) Chapter 9, Table 1, Note 4   80%

Notes:

Unit 9 – Single-Family Dwelling Unit Load Calculations

1. (d) any of these, 220-2   85%

2. (b) 0.5   85%

3. (b) two circuits, 220-11(c)(1)   95%

4. (c) 3,000 VA, 220-16(a)   90%

5. (d) 33 ampere   90%
   8 kW Table 220-19 Column A
   I = P/E
   I = 8,000 watts/240 volts, I = 33 ampere

6. (b) 37 ampere   90%
   Step 1: Table 220-19 Column A = 8 kW
   Step 2: Increase Column A (8 kW) by 10%
   *8 kW × 1.1 = 8.8 kW (8,000 VA+ 800 VA)
   I = P/E = 8,800 VA/240 volts
   I = 37 ampere
   *Increase Column A 5% for each 1 kW or major fractions of a kW (.5 or larger).

7. (b) 25 ampere   75%
   Nameplate = 6,000 watts
   I = P/E
   I = 6,000 watts/240 volts, I = 25 ampere

8. (b) 20 ampere   75%
   Nameplate = 4,800 watts
   I = P/E
   I = 4,800 watts/240 volts, I = 20 ampere

9. (c) 33 ampere   85%
   Step 1: Total connected load: 6 kW + 3 kW = 9 kW
   Step 2: Demand load for one range: 8 kW, Table 220-19
   I = P/E
   I = 8,000 watts/240 volts, I = 33 ampere

10. (d) 37 ampere   95%
    Step 1: The total connected load
    6 kW + 4 kW + 4 kW = 14 kW
    Step 2: Demand load for one range [Table 220-19]
    8 kW
    Step 3: Increase Column A value (8 kW) 5% for each kW, or major fraction (.5 kW), that the range exceeds 12 kW.
    8 kW × 1.1 = 8.8 kW
    I = 8,800 watts/240 volts
    I = 37 ampere

11. (a) True, 210-11(c)(1)   95%

12. (a) 1,500 VA, 220-16(b)   95%

13. (d) all of these, 220-3(a)   95%

14. (a) True, Table 220-3(a) Note   95%

15. (c) 6,300 VA   95%
    (2,100 square foot × 3 VA) No additional load is required for general use receptacles and lighting outlets, Note to Table 220-3(a).

16. (a) True, 210-11(c)(1)   95%

17. (c) 4 Circuits   90%
    Step 1: General lighting VA
    2,340 square feet × 3 VA = 7,020 VA
    Step 2: General lighting ampere: I = VA/E
    I = 7,020 VA/120 volts
    I = 58.5 ampere

    Step 3: Determine the number of circuits:

    $$\text{Circuits} = \frac{\text{General Lighting Ampere}}{\text{Circuit Ampere}}$$

    $$\text{Circuits} = \frac{59 \text{ ampere}}{15 \text{ ampere}}$$

    Circuits = 3.93 or 4

    Note: Use 120 or 120/240 volts, single-phase unless specified otherwise 220-2.

18. (b) 3 Circuits   90%
    Step 1: General lighting VA
    2,100 square feet × 3 VA = 6,300 VA

    Step 2: General lighting ampere: I = VA/E
    I = 6,300 VA/120 volts
    I = 53 ampere

    Step 3: Determine the number of circuits:
    $$\text{Circuits} = \frac{\text{General Lighting Ampere}}{\text{Circuit Ampere}}$$

    $$\text{Circuits} = \frac{53 \text{ ampere}}{20 \text{ ampere}}$$

    Circuits = 2.65 or 3 circuits

19. (a) 2 circuits 95%

Step 1: Determine the general lighting VA load:
Square footage × 3 VA
1,500 square feet × 3 VA = 4,500 VA

Step 2: Determine the general lighting ampere load:
I = P/E
I = 4,500 VA/120 volts
I = 37.5 ampere or 38 ampere

Step 3: Determine the number of circuits:

$$Circuits = \frac{General\ Lighting\ Ampere}{Circuit\ Ampere}$$

$$Circuits = \frac{38\ ampere}{20\ ampere}$$

Circuits = 1.9 or 2 circuits

20. (a) demand factor 90%
Article 100, 220-11

21. (a) True, 220-21 95%

22. (d) four, 220-17 95%

23. (d) the greater of a or b, 220-18 90%

24. (b) False 70%
Section 220-18 requires a dryer load if the dryer is installed in the dwelling unit.

25. (b) False 85%
Section 220-19 applies to cooking appliances (household) over 1 kW.

26. (a) 3-wire, 1-phase, 120/240 volt 95%
systems up to 400 ampere, 310-15(b)(2)(a)

27. (a) 8,100 VA, 220-3(a) 95%
2,700 square feet × 3 VA = 8,100 VA

28. (d) 24,120 VA 90%
General lighting and receptacles
6,540 square feet × 3 VA =   19,620 VA
Small-appliance circuits
1,500 VA × 2 =   3,000 VA
Laundry circuit
1,500 VA × 1 =   1,500 VA
Total connected load   24,120 VA

29. (d) 9,000 VA, 440-34 85%
A/C, 230 volts × 28 ampere × 1.25 = 8,050 VA*
Heat 220-15 3,000 × 3 = 9,000 VA
*Omit the smaller of the two loads [Section 220-21].

30. (a) 10, 1000 VA, 440-34 90%
A/C: Since a 4 horsepower motor is not listed in Table 430-148 we must determine the approximate value of a 4 horsepower 230 volt A/C.
3 horsepower = 230 volts × 17 ampere = 3,910 VA
5 horsepower = 230 volts × 28 ampere = 6,440 VA
10,350 VA/2 = 5,175 VA × 1.25 = 6,469 VA*
Heat: 10,000 VA
*Omit the smaller of the two loads [Section 220-21].

31. (b) 6,690 VA, 220-17 95%
Disposal   940 VA
Dishwasher   1,250 VA
Water heater   + 4,500 VA
    6,690 VA

32. (b) 5,843 VA, 220-17 75%
Disposal   940 VA
Dishwasher   1,250 VA
Trash compactor   1,100 VA
Water heater   + 4,500 VA
    7,790 VA × .75 = 5,843 VA

33. (c) 5,000 VA, 220-18 85%
Dryer load must not be less than 5,000 VA or the nameplate rating if greater than 5 kW (for standard calculation); this does not apply to optional calculations.

34. (d) 5,500 VA, 220-18 95%
Dryer load must not be less than 5,000 VA or the nameplate rating if greater than 5 kW (for standard calculation), this does not apply to optional calculations.

35. (c) 4.5 kW, Table 220-19 75%
Column B: 3 kW × 2 units = 6 kW × .75 = 4.5 kW

36. (b) 4.8 kW, Table 220-19 95%
Column C: 6 kW × .8 = 4.8 kW

37. (d) 9.3 kW, Table 220-19 90%
Column C: 6 kW × .8 =   4.8 kW
Column B: 3 kW × 2 units 6 kW × .75 =   + 4.5 kW
Total demand   9.3 kW

38. (b) 8 kW, Table 220-19 90%
Column A: 8 kW

39. (a) 8.8 kW 95%
Column A value (8 kW) must be increased 5% for each kW or major fraction of a kW (.5 kW or larger over 12 kW).
8 kW × 1.1 = 8.8 kW

40. (b) No. 2/0, 310-15(b)(6) 90%

41. (c) 5,100 VA, Table 220-11   75%
General lighting
1500 sq. ft. × 3 VA = 4,500 VA
Table 220-3(a)
2 small appliances = 3,000 VA, 220-16(a)
Laundry =   1,500 VA, 220-16(b)
                  + 9,000 VA
                  - 3,000 VA at 100% =   3,000 VA
                    6,000 VA at 35% = + 2,100 VA
Net computed demand   5,100 VA
Note: 15- and 20-ampere receptacles are considered part of the general lighting load (3 VA), Table 220-3(a).

42. (b) 5,555 VA, 220-15   90%
A/C = 3 horsepower
230 volts × 17 ampere = 3,910 VA
3,910 VA × 1.25 = 4,888 VA
1/4 horsepower, 230 volts ×2.9 ampere = 667 VA
A/C demand load = 5,555 VA
Heat (4 kW) omitted because it's smaller than air condition [*Section 220-21*].

43. (b) 5.5 kW, 220-17   95%
Water heater   4,000 VA
Dishwasher   + 1,500 VA
Demand load at 100%   5,500 VA/1,000 = 5.5 kW

44. (b) 5 kW, 220-18   95%

45. (c) 8.8 kW, Table 220-19, Note 1   95%
Column A (8 kW) increased 5% for each kW or major fraction of a kW over 12 kW.
8 kW × 1.1 = 8.8 kW

46. (c) No. 2, 310-15(b)(6)   85%

47. (a) No. 8, 250-102 and Table 250-66   90%

48. (a) No. 8, 250-50 and Table 250-66   90%

49. (d) all of these, 220-30   95%

50. (a) No. 1, 220-30   55%
Step 1: Total connected load
(a) General lighting
1,500 sq. ft. × 3 VA   4,500 VA
(b) Small appliance
1,500 VA × 2 circuits   3,000 VA
 Laundry circuit   + 1,500 VA
                               4,500 VA

(c) Appliances (nameplate ratings)
Dishwasher   1,500 VA
Water heater   5,000 VA
Disposal   1,000 VA
Dryer   5,500 VA
Cooktop   6,000 VA
Ovens 3,000 VA × 2   + 6,000 VA
                             25,000 VA

Step 2: Determine the demand load.
General lighting   4,500 VA
Small appliance and laundry   4,500 VA
Appliance nameplate ratings + 25,000 VA
                              34,000 VA
First 10,000 at 100 %   10,000 VA = 10,000 VA
Remainder at 40%   24,000 VA = + 9,600 VA
Demand load   19,600 VA

Step 3: A/C vs. Heat
(a) Air-conditioning or heat pump compressors at 100% vs. 65% of one, two or three separately controlled electric space heating units.
AC: 230 volts ×17 ampere = 3,910 VA, Omit [*Section 220-21*]
Heat: 10,000 VA × .65 = 6,500 VA

Step 4: Determine the total demand load.
Step 2   19,600 VA
Step 3   + 6,500 VA
Total Demand Load =   26,100 VA

Step 5: Feeders and service conductors, 310-15(b)(6)
I = P/E
I = 26,100 VA/240 volts
I = 108 ampere
Feeder and service conductor sized according to 110 ampere, No. 1 AWG aluminum.

51. (d) No. 2, 220-30   85%
Step 1: Total connected load.
General lighting (and receptacles)
(a) 2,330 square foot × 3 VA =   6,990 VA
(b) Small appliance 1,500 × 2 =   3,000 VA
Laundry circuit =   1,500 VA
(c) Appliance (nameplate ratings)
Dishwasher   1,500 VA
Disposal   1,000 VA
Trash compactor   1,500 VA
Water heater   6,000 VA
Range   14,000 VA
Dryer   + 4,500 VA*
                                     28,500 VA

<u>71.</u> (c) 7.8 kW, 220-19 Note 3   90%

Step 1: Determine total connected load.

6 kW + 6 kW = 12 kW

Step 2: Column C demand factor for 2 units, 65%

Step 3: Apply Column C demand factor to the total connected load. 12 kW $\times$ .65 = 7.8 kW

Note: Column A for two units = 11 kW

<u>72.</u> (b) 8 kW   90%

The word "Maximum" in a range question is asking for the larger of Column A or C. Table 220-19, Note 3 permits Column C demand factors to be used.

Step 1: Determine total connected load. 8.5 kW

Step 2: Determine Column C demand factor for 2 units. 80%

Step 3: Apply Column C demand factor to the total connected load, 8.5 kW $\times$ .8 = 6.8 kW.

Note: Column A for one unit = 8 kW

<u>73.</u> (c) 19.5 kW   90%

Table 220-19, Note 3 permits Column C demand factors to be used.

Step 1: Determine the total connected load.

5 units rated 5 kW =	25 kW
2 units rated 4 kW =	8 kW
4 units rated 7 kW =	+ 28 kW
Total connected load	61 kW

Step 2: Column C demand factor for 11 units, 32%.

Step 3: Apply Column C demand factor to the total connected load, 61 kW $\times$ .32 = 19.52 kW.

Note: Column A for 11 units = 26 kW

<u>74.</u> (a) 9.3 kW   90%

Table 220-19, Note 3 permits Column B and Column C demand factors to be used.

Step 1: Determine total connected load:

Column B = 3 kW + 3 kW = 6 kW

Column C = 6 kW

Step 2: Determine Column B and Column C demand factor:

Column B (2 units) = 75%

Column C (1 unit) = 80%

Step 3:

Column B demand factor: 6 kW $\times$ .75% = 4.5 kW

Column C demand factor: 6 kW $\times$ .8% = 4.8 kW

Total demand load = 4.5 kW + 4.8 kW = 9.3 kW

<u>75.</u> (b) No. 2/0, 310-15(b)(6)   75%

<u>76.</u> (b) 100 ampere   90%

Section 220-2 specifies that nominal voltage of 240 volts is to be used:

I = P/E

I = 21,560 VA/240 volts

I = 90 ampere,

Note: Section 220-32 requires a minimum of 100 ampere.

<u>77.</u> (d) 110 ampere, 310-15(b)(6)   95%

Section 220-2 specifies that nominal voltage of 240 volts is to be used:

I = P/E

I = 24,221 VA/240 volts

I = 101 ampere

<u>78.</u> (c) 4,000 watts   95%

Table 220-30(1) and (3)

The larger of:

Air-conditioning at 100% = 4,000 watts

Heat at 65%: 6,000 watts $\times$ .65 = 3,900 watts, Omit

<u>79.</u> (a) 22 kVA, 220-30   95%

Step 1: Apply demand factors:

Total connected load, 25,000 VA	
First 10 kW at 100%	10,000 VA
Remainder 15 kW at 40%	6,000 VA
Net computed load	16,000 VA

Step 2: Larger of air-conditioning vs. heat

A/C at 100%:   = 6,000 VA

Heat at 40%: 10,000 $\times$ .40 = 4,000 VA

Step 3: Total Demand Load (Step 1 and 2), 22,000 VA

Step 4: Total demand load in ampere:

I = P/E

I = 22,000 VA/240 volts

I = 92 ampere

Step 5: Conductor size, 310-15(b)(6), No. 4 THHN

80. (d) 110 ampere   90%

220-30 Optional Method:

Step 1: Small appliance and laundry circuits

2 small-appliance circuits (1,500 VA × 2 circuits)	3,000 VA
1 laundry circuit (1,500 VA × 1 circuit)	1,500 VA

Step 2: General Lighting (1,800 square feet × 3 VA)   5,400 VA

Step 3: Appliances (nameplate rating)

Water heater	4,000 VA
Dishwasher (115 volts × 9.8 ampere)	1,127 VA
Dryer (do not use 5,000 VA)	4,000 VA
Oven	6,000 VA
Cooktop	6,000 VA

Step 4: Motors

Pool pump (115 volts × 13.8 ampere)   + 1,587 VA

Step 5: Totals   32,614 VA

Step 6: Demand load:

Total connected load	32,614 VA	
First 10,000 at 100%	− 10,000 VA at 100% =	10,000 VA
Remainder at 40%	22,614 VA at 40% =	9,046 VA

Step 7: Larger of air-conditioning vs. heat

A/C 100%: VA = Volts × Ampere, 230 volts × 22.5* ampere = 5,175 VA

The full load current for a 4 horsepower motor must be between 3 horsepower
and 5 horsepower.

*3 horsepower = 17 ampere, 5 horsepower = 28 ampere, 4 horsepower = 22.5 ampere   + 5,175 VA

Heat at 65%: 6,000 watts × .65 = 3900 watts (Omit)

Step 8: Total demand load in VA (Step 6 + Step 7)   24,221 VA

Step 9: Total demand load in ampere:

I = P/E

I = 24,221 VA/230 volts

I = 105 ampere

Step 10: Conductor size, 310-15(b)(6), No. 3 THHN

81. (b) 110 ampere, 220-30   90%
    Step 1: Small appliance and laundry circuits
    Small-appliance circuits (1,500 VA × 2 circuits)   3,000 VA
    Laundry circuit (1,500 VA × 1 circuit)   1,500 VA
    Step 2: General lighting (1,800 square feet × 3 VA)   5,400 VA
    Step 3: Appliances (nameplate rating)
    Water heater   14,000 VA
    Dishwasher   1,500 VA
    Dryer   4,500 VA
    Oven   6,000 VA
    Range   + 6,000 VA
    Step 4: Totals   31,900 VA
    Step 5: Demand factors
    Total connected load   31,900 VA
    First 10,000 VA at 100%   − 10,000 VA at 100% = 10,000 VA
    Remainder at 40%   21,900 VA at 40% =   8,760 VA
    Step 6: Larger of air-conditioning vs. heat
    A/C 100%: 6,000 VA   + 6,000 VA
    Heat 40%: 10,000 watts × .40 = 4,000 VA
    Step 7: Total demand load in VA (Step 5 + Step 6) :   24,760 VA
    Step 8: Total demand load in ampere:
    I = P/E
    I = 24,760 VA/240 volts
    I = 103 ampere
    Step 9: Conductor size, 310-15(b)(6), No. 3 THHN

82. (b) No. 3, 220-30   90%
    Step 1: Small appliance and laundry circuits
    2 small-appliance circuits (1,500 VA × 2 circuits)   3,000 VA
    1 laundry circuit (1,500 VA × 1 circuit)   1,500 VA
    Step 2: General lighting (1,800 square feet × 3 VA)   5,400 VA
    Step 3: Appliances (nameplate rating)
    Water heater   4,000 VA
    Dishwasher   1,200 VA
    Dryer at nameplate   4,000 VA
    Range   13,900 VA
    Step 4: Motors
    Pool pump (115 volts × 13.8 ampere or 230 × 6.9 ampere)   + 1,587 VA
    Step 5: Totals   34,587 VA
    Step 6: Demand factors
    Total connected load   34,587 VA
    First 10 kW at 100%   - 10,000 VA at 100% = 10,000 VA
    Remainder at 40%   24,587 VA at 40% =   9,835 VA

    Step 7: Larger of air-conditioning vs. heat
    A/C 100%: 230 volts × 28 ampere = 6,440 VA   + 6,440 VA
    Heat 65%: 6,000 × .65 = 3900 VA Omit
    Step 8: Total demand load in VA (Step 6 and 7)   26,275 VA
    Step 9: Total demand load in ampere:
    I = P/E
    I = 26,275 VA/240 volts
    I = 109 ampere
    Step 10: Conductor size, 310-15(b)(6), No. 3 THHN

NEC® Answers

83. (c) 90-1(a) and (b) — 90%

84. (b) 90-2(b)(2) — 90%

85. (d) 100 Attachment Plug — 70%

86. (d) 100 Connector, Pressure (Solderless) — 85%

87. (c) 100 Exposed, Wiring Methods — 95%

88. (c) 100 Location, Wet — 85%

89. (a) 100 Switchboard — 90%

90. (d) 110-10 — 80%
Careful, read the *NEC®* text completely.

91. (b) 110-26

92. (c) 110-27(a) — 95%

93. (d) 110-34(e) and Table 110-34(e) — 90%

94. (b) 200-3 — 85%

95. (a) 210-4(a) FPN — 85%

96. (c) 210-8(a)(5) — 90%

97. (c) 210-25 — 90%

98. (a) 210-52(c)(5) Exception

99. (a) 215-2(d) — 80%
The 400 ampere requirement can be found in Section 310-15(b)(6).

100. (d) 220-3(b)(9)

101. (b) 220-19 — 75%

102. (d) 225-4 — 85%

103. (c) 225-24 — 90%

104. (a) 230-24(a) — 95%

105. (c) 230-42(a)

106. (d) 230-56 — 75%

107. (a) 230-82(6) — 90%

108. (b) 230-95(c) FPN No. 2 — 95%

109. (a) 240-3(a) — 95%

110. (b) 240-20(a) — 90%

111. (b) 240-51(b) — 90%

112. (b) 240-83(a) — 85%

113. (d) 250-2(a) — 95%

114. (b) 250-24(b) — 90%

115. (d) 250-50(c) — 85%

116. (a) 250-66(b) — 85%

117. (b) 250-94(2) — 85%

118. (b) 250-106 FPN's

119. (d) 250-122(f)(2)

120. (a) 250-146(b) — 90%

121. (a) 280-21 — 90%

122. (b) 300-3(a) — 90%

123. (a) Table 300-5 Column 6 — 95%

124. (a) 300-11(a)(2)

125. (c) 300-22(b) — 90%

126. (d) 305-3(c) — 85%

127. (d) 310-4 — 85%

128. (c) Table 310-13 — 75%

129. (a) 310-70 — 95%

130. (b) 318-3 — 60%

131. (a) 320-6 — 90%

132. (b) 321-1 — 90%

133. (c) 324-6 — 95%

134. (c) 328-1 — 80%

135. (a) 330-3 — 75%

136. (a) 331-3 — 90%

137. (d) 333-3 — 70%
Type AC cable is permitted for exposed work

138. (c) 334-2 — 70%

139. (b) 336-5(a)(1)

140. (d) 338-3(a) — 95%

141. (d) 340-5 — 95%

142. (d) 343-5(b) — 95%

143. (d) 345-3(c) — 95%

144. (a) 346-3(c) — 90%

145. (d) 347-2 FPN — 80%

146. (a) 348-9 — 75%

147. (c) 349-10(b) — 85%

148. (d) 350-5(6) and 350-5(7) — 95%

149. (a) 351-5(a) Exception — 70%

150. (b) 352-4 — 70%

151. (b) 354-8 — 95%

152. (a) 356-4 — 90%

153. (c) 358-5 — 90%

154. (c) 362-1 — 95%

155. (c) 363-4 — 80%

156. (a) 364-6(2)

157. (c) 365-1 — 90%

158. (c) 370-4 — 80%

159. (d) 370-20 — 70%
Be careful, be sure to read the *NEC®* slowly.

160. (d) 370-29 Exception — 95%

161. (d) 373-3 — 95%

162. (b) 374-2 — 95%

163.	(d) 380-7	95%	182.	(c) 430-102(b) Exception No. 1	85%
164.	(b) 384-3(c)	95%	183.	(d) 440-12(a)(1)	95%
165.	(a) 384-36 Table	90%	184.	(a) 450-6	90%
166.	(d) 400-5	75%	185.	(b) 450-43(c)	75%
167.	(b) 400-23	80%	186.	(d) 470-3	95%
168.	(c) 402-6	95%	187.	(a) 500-5(a)(4) FPN No. 2	95%
169.	(c) 410-3 Exception	95%	188.	(b) 500-9	95%
170.	(c) 410-15(b)(1) and (3)	95%	189.	(c) 501-3(b)(1) Exception	75%
171.	(c) 410-31	80%	190.	(a) 501-5(d)(1) Exception	
172.	(b) 410-57(a)	75%	191.	(a) 502-16(a)	80%
173.	(c) 410-77(a)	80%	192.	(c) 511-1	95%
174.	(a) 422-11(a)	90%	193.	(d) 513-3(d)	90%
175.	(b) 422-33	95%	194.	(a) 514-5(a)	
176.	(a) 424-9	90%	195.	(a) 517-18(c)	80%
177.	(a) 424-82	80%	196.	(c) 517-160(b)(2)	95%
178.	(d) 426-20(e)	95%	197.	(b) 518-4(a)	50%
179.	(d) 427-12	95%	198.	(b) 525-13(a)	
180.	(b) 430-22(a)	85%	199.	(c) 540-13	95%
181.	(a) 430-62(a)	55%	200.	(c) 551-71	75%

Notes:

Unit 10 – Multifamily Dwelling Unit Load Calculations

10–1 Multifamily Dwelling-Unit Load Calculations – Standard Method

1. (b) 40,590 VA, Table 220-11   75%
   General lighting 840 sq feet × 3 VA   2,520 VA
   Small appliance 1,500 sq feet × 2 VA   3,000 VA
   Laundry circuit   + 0 VA
   Total connected load   5,520 VA × 20 units = 110,400 VA
   Demand factor, Table 220-11
   Total connected load   110,400 VA
   First 3,000 VA at 100%   − 3,000 VA × 1.00 =   3,000 VA
   Next 117,000 VA at 35%   107,400 VA × .35 =   + 37,590 VA
   Total demand load   40,590 VA
   *Laundry facilities provided

2. (c) 51,300 VA, Table 220-11   75%
   General lighting 990 sq feet × 3 VA   2,970 VA
   Small appliance 1,500 sq feet × 2   3,000 VA
   Laundry circuit 1,500 × 1   + 1,500 VA
   Total connected load   7,470 VA × 20 = 149,400 VA
   Demand factor, Table 220-11
   Total connected load   149,400 VA
   First 3,000 VA at 100%   − 3,000 VA =   3,000 VA
                                                              146,400 VA
   Next 117,000 VA at 35% =   117,000 VA =   40,950 VA
   Remainder VA at 25%   29,400 VA =   + 7,350 VA
   Total demand load   51,300 VA

3. (b) 240 kW, 220-14,   95%
   A/C: 230 volts × 17 ampere = 3,910 VA, 3,910 VA × 1.25 = 4,888 VA
   4,888 VA + (3,910 VA × 39 units) = 157,378 VA, Omit [*Section 220-21*]
   Heat 220-15, 3,000 VA × 2 units = 6,000 × 40 units = 240,000 watts

4. (b) 125 kW, 220-14   95%
   A/C: 230 volts × 17 ampere = 3,910 VA × 1.25 = 4,888 VA
   4,888 VA + (3,910 VA × 24 units) = 98,728 VA, Omit [*Section 220-21*]
   Heat 220-15, 5,000 × 25 = 125,000 watts

5. (c) 80 kW, 220-17   80%
   Disposal   940 VA
   Dishwasher   1,250 VA
   Water heater   + 4,500 VA
   Net connected load   6,690 VA × 16 units = 107,040 VA × 0.75 = 80,280 VA

6. (a) 149 kVA, 220-17   95%
   Disposal   900 VA
   Dishwasher   1,200 VA
   Water Heater   + 5,000 VA
   Net connected load   7,100 VA × 28 units = 198,800 VA × 0.75 = 149,100 VA

  7. (a) 50 kW, 220-18   95%
      5 kW* × 40 units = 200 kW × .25 = 50 kW
      *The minimum load is 5 kW, for standard calculations.

  8. (a) 26 kW, 220-18   95%
      5.25 kW × 10 units = 52.5 kW × .5 = 26.25 kW

  9. (b) 18 kW, Table 220-19   95%
      Column B: 3.25 kW × 12 units × .45 = 17.55 kW

 10. (a) 20 kW, 220-19   95%
      Column C: 8 kW × 7 units × .36 = 20.16 kW

 11. (c) 20 kW, Table 220-19, Column A   95%

 12. (c) 17 kW, Table 220-19, Note 1   75%
      Step 1: Column A demand load = 14 kW (3 units)
      Step 2: The average range (16 kW) exceeds 12 kW by 4 kW, increase Column A demand load (14 kW) by 20%.
      14 kW × 1.2 = 16.8 kW

 13. (d) 22 kW, Table 220-19, Note 2   80%
      Step 1: Determine the total connected load.
      11 kW (use minimum 12 kW): 12 kW × 3 units = 36 kW
      14 kW: 14 kW × 3 units =   42 kW
      Total connected load   78 kW
      Step 2: Determine the average range rating, 78 kW (total connected)/6 (units) =13 kW
      Step 3: Demand load Table 220-19, Column A: 6 ranges = 21 kW
      Step 4: The average range (13 kW) exceeds 12 kW by 1 kW.
      Increase demand load from Column A (21 kW) by 5%, 21 kW × 1.05 = 22.05 kW

 14. (b) 600 kcmil aluminum, 310-15(b)(6)   75%
      I = VA/E
      I = 90,000 VA/230 volts
      I = 391 amperes, 600 kcmil aluminum

 15. (c) 2 – 500 kcmil THHN   85%

$$I = \frac{VA}{E \times 1.732}$$
$$I = \frac{260,000 \ VA}{360 \ volts}$$

I = 722 ampere
Each set must have an ampacity of 722 ampere/2 parallel sets = 361 ampere per parallel set.
500 kcmil THHN has an ampacity of 380 ampere each and two sets (parallel) of 500 kcmil THHN have an ampacity of 760 ampere Table 310-16 at 75°C. These conductors (760 ampere) are permitted to be protected by an 800-ampere protection device [*Section 240-3(b)*].

<u>16.</u> (a) 40 kVA, Table 220-11   75%

     (a) General lighting 1500 square feet × 3 VA =   4,500 VA

     (b) 2 small appliances   3,000 VA

     (c) Laundry   + 1,500 VA

     Total connected load   9,000 VA × 12 units =   108,000 VA

     Demand factor, Table 220-11

     Total connected load   108,000 VA

     First 3,000 VA at 100%   − 3,000 VA × 1.00 =   3,000 VA

     Remaining VA at 35%   105,000 VA × .35 =   + 36,750 VA

     Net computed demand load   39,750 VA

     Note: 15- and 20-ampere receptacles are considered part of the general lighting load (3 VA), see notes to Table 220-3(a).

<u>17.</u> (b) 60 kW, 220-15   85%

     A/C = 3 horsepower, 230 volts × 17 ampere = 3,910 VA, 3,910 VA × 1.25 = 4,888 VA

     4,888 VA + (3,910 VA × 11 units) =   47,898 VA

     1/8 horsepower* 230 volts × 1.45 ampere = 333.5 VA × 12 units =   + 4,002 VA

     Total A/C demand load   51,900 VA, Omit 220-21

     Heat = 5,000 VA × 12 units = 60,000 kW

     * one-half the VA of a $^{1}/_{4}$ horsepower motor

<u>18.</u> (d) 50 kW, 220-17   80%

     Water heater   4,000 VA

     Dishwasher   + 1,500 VA

     Total demand load   5,500 VA × 12 units × .75 = 49,500 VA

     Note: Washing machine on laundry circuit.

<u>19.</u> (b) 27 kW, 220-18   85%

     5,000 watts* × 12 units × .45 demand = 27 kVA

     *5 kW is minimum for dryers.

<u>20.</u> (d) 30 kW, 220-19 Note 1   90%

     Column A = 27 kW × 1.1* = 29.7 kW

     * Column A (27 kW) increased 5% for each kW or major fraction of a kW that the range (14.45 kW) is over 12 kW.

<u>21.</u> (c) 1,000 ampere   80%

     I = VA/E

     I = 206,000 VA/240 volts

     I = 858 ampere

<u>22.</u> (b) No. 1/0, 250-102(c) and Table 250-66   70%

<u>23.</u> (c) No. 3/0, 250-102(c) and Table 250-66   60%

     The parallel equivalent of 400 kcmil × 3 raceways = 1,200 kcmil

<u>24.</u> (d) a or c, 220-32   75%

     100% A/C verses 100% heat

<u>25.</u> (d) 130 kVA, Table 220-32   95%

     (a) General lighting 1500 square feet × 3 VA =   4,500 VA   Table 220-3(a)

     (b) 2 small appliances   3,000 VA   220-16(a)

     (c) Laundry   + 1,500 VA   210-11(c)(2), 220-16(b)

                                    9,000 VA

     9,000 VA × 60 units = 540,000 VA × .24 = 129,600 VA

26. (d) 72 kVA, 220-32   80%
    A/C = 3 horsepower 230 volt × 17 ampere = 3,910 VA
    1/8 horsepower 230 volts × 1.45 ampere =   + 334 VA
    Total demand load   4,244 VA × 60 units × .24 = 61,114 VA, Omit 220-21
    Heat = 5,000 VA × 60 units × .24 = 72,000 VA

27. (a) 80 kVA, 220-32   95%
    Water heater   4,000 VA
    Dishwasher   + 1,500 VA
    Total demand load   5,500 VA × 60 units × .24 = 79,200 VA

28. (d) 58 kW, 220-32   80%
    Note: Use the nameplate rating.
    4 kW × 60 units × .24 derating factor = 57.6 kW

29. (a) 65 kW, 220-32   95%
    4.5 kW × 60 units × .24 = 64.8 kW

30. (d) 200 kW, 220-32   90%
    14 kW × 60 units × .24 = 201.6 kW

31. (b) 800 ampere   80%
    I = VA/(E × 1.732), I = 270,000 VA/(208 volts × 1.732) = 750 ampere

32. (b) 250 kVA, 220-32   60%
    Step 1: Determine the total connected load.
    General lighting 900 sq feet × 3 VA   2,700 VA
    Two small-appliance circuits (required)   3,000 VA
    Laundry circuit (required)   1,500 VA
    Air-conditioning (230 volts × 28 ampere)   6,440 VA*
    Heat 5 kW, omit
    Water heater   5,000 VA
    Range   + 14,000 VA
    Total connected load   32,640 VA × 20 units = 652,800 VA
    Step 2: Net computed load = 652,800 VA × .38 = 248,064 VA
    Step 3: Service size
    I = VA/E
    I = 248,064 VA/240 volts
    I = 1,034 ampere

33. (c) No. 1/0, 250-102(c) and Table 250-66   75%

34. (b) No. 2/0, 250-102(c) and Table 250-66   60%
    Based on 1,000 kcmil

☆ Challenge Answers

General Lighting and Receptacle Calculations Table 220-11

35. (d) 54 kW, Table 220-3(a)   75%
    900 square feet × 3 VA × 20 units = 54,000 VA

<u>36.</u> (b) 41 kVA — 70%

Sections 220-3(a), 220-11, 220-16. Table 220-11 permits a demand factor for the general lighting load (3 VA per square foot of living space), the two small-appliance circuits, and the laundry circuit. See 220-16.

Note: Laundry circuit load not required, see 210-52(f) Exception No. 2

Step 1: Total connected load.

General lighting and receptacles (840 square feet × 3 VA)	2,520 VA
Small-appliance circuits (1,500 VA × 2 circuits)	3,000 VA
Laundry circuit	+ 0 VA
Total connected load	5,520 VA × 20 units = 110,400 VA

Step 2: Demand Factors, Table 220-11

Total load	110,400 VA	
First 3,000 VA at 100%	− 3,000 VA × 1.00 =	3,000 VA
Next 117,000 VA at 35%	107,400 VA × .35 =	+ 37,590 VA
Total demand load		40,590 VA

37. (a) 106,500 VA, 220-17 — 75%

Disposal	900 VA
Dishwasher	1,200 VA
Water heater	+ 5,000 VA
Total demand load	7,100 VA × 20 units × .75 = 106,500 VA

<u>38.</u> (b) 25 kW — 80%

Section 220-18 requires a minimum of 5 kW for each dryer, and the demand factor for 10 units is 50%

5 kW × 10 units × .5 = 25 kW

<u>39.</u> (c) 54 kW, Table 220-19, Note 1 — 75%

30 units = 15 kW + 1 kW for each range in calculation, 15 kW + 30 kW = 45 kW

15.8 exceeds 12 kW by 3.8, ".8" is a major fraction so Column A value must be increased 20%.

Column A: 45 kW × 1.2 = 54 kW

<u>40.</u> (d) 33 kW, Table 220-19, Note 2 — 90%

Step 1: Determine the total connected load (using 12 kW as minimum*).

Five ranges at 10 kW = 5 units × 12 kW* =	60 kW
Five ranges at 14 kW = 5 units × 14 kW =	70 kW
Five ranges at 16 kW = 5 units × 16 kW =	+ 80 kW
Total connected load	210 kW

Step 2: Determine the average range kW, average range = 210 kW/15 units= 14 kW

Step 3: Determine the demand load using Table 220-19, column A, 15 units = 30 kW

Step 4: Increase Column A demand 5% for each kW or major fraction of a kW that the average range exceeds 12 kW. The average range (14 kW) exceeds 12 kW by 2 kW therefore, Column A must be increased by 10%.

30 kW × 1.1 = 33 kW

135. (d) 352-6   90%

136. (a) 356-10   90%

137. (a) 362-9   80%

138. (b) 364-25   85%
A fire seal is not always required. Read the Section more carefully next time.

139. (d) 370-16(a) Table 370-16(b), 370-16(b)(1)(4)   70%
Switch (yoke) counts as two No. 12, cable clamps count as one No. 12, plus six No. 12 = 9 No. 12 conductors. 2.25 cubic inches $\times$ 9 = 20.25 cubic inches

140. (a) 370-23(b)(2)   90%

141. (d) 370-25   90%

142. (c) 370-71(b)(2)   95%

143. (c) 374-7   90%

144. (a) 380-14(c)   95%

145. (a) 384-10   90%

146. (d) 400-8

147. (b) 410-4(c)(1)   80%

148. (a) 410-14(b)   95%

149. (d) 410-23   95%

150. (c) 410-38(a)   80%

151. (a) 410-56(f)(3)

152. (c) 410-67(c)   95%

153. (c) 410-100   95%

154. (d) 422-13   95%

155. (b) 422-48   90%

156. (b) 424-38(a)   90%

157. (b) 430-6(a)(1)   75%

158. (d) 430-31   80%

159. (a) 430-40   75%

160. (a) 430-81(b)   95%

161. (b) 430-110(a)   95%

162. (c) 440-62(c)   80%

163. (d) 450-3(b) Note 1
54 ampere $\times$ 125% = 68 ampere

164. (c) 460-6(a)   95%

165. (b) 500-3(d)   80%

166. (d) 500-7(a)   80%

167. (d) 501-4(b)   75%

168. (a) 501-6(b)(1)   80%

169. (c) 514-5(c)   95%

170. (c) 517-61(a)(5)   80%

171. (a) 525-13(g)

172. (d) 550-2 Definition and FPN   95%

173. (c) 550-23

174. (b) 555-6(a)

175. (d) 600-10(c)(2)   95%

176. (d) 620-51   80%

177. (d) 645-5(d)(5)(c)

178. (a) 660-5   80%

179. (b) 675-2   80%

180. (c) 680-4 Definition, storable pool   95%

181. (b) 680-6(b)(1)   90%

182. (c) 680-20(a)(3)   90%

183. (c) 680-22(a)(1)   65%

184. (d) 680-25(b)(1) Flex is not permitted.

185. (d) 680-42

186. (d) 700-6(a)   90%

187. (a) 700-12(b)(2) and 701-11(b)(2)   95%

188. (b) 701-2   80%

189. (b) 725-2 Definition   80%

190. (b) 725-41(a) FPN   55%
Use Table 11(a) and 11(b) of Chapter 9. Use table formula 100/Vmax 100 volts/25 volts = 4 ampere.

191. (a) 725-54(d) 800-52(a)(1) and 820-52(a)(1)   90%
Note: 300-11(b) Exception No. 2 does permit Class 2 circuits to be supported to a raceway under certain conditions

192. (b) 760-30(a)(1)

193. (d) 770-33   80%

194. (d) 800-52(a)(2)   95%

195. (c) 820-10(f)(3)   95%

196. (d) 830-10(a) and(f)

197. (d) 830-58(a)(2)   95%

198. (b) Chapter 9, Table 1, Note 3   90%

199. (b) Chapter 9, Table 4   90%

200. (b) Appendix C, Table 1   95%

Unit 11 – Commercial Load Calculations

1. (b) Ampacity, Article 100   80%

2. (d) Overcurrent protection, Article 100   80%

3. (a) True, 240-6(a)   95%

4. (d) any of these, 220-2   90%

5. (b) .5, 220-2   90%

6. (b) 24 kVA   90%
   The VA of a single phase 3 horsepower motor rated
   230 volts is 230 volts × 17 ampere = 3,910 VA.
   The feeder demand load is calculated at 125% of the
   largest air conditioner plus 100% of all other air-conditioner units. [(3,910 VA × 1.25) + (3,910 VA × 5] =
   24,438 VA/1,000 = 24.4 kVA.

7. (d) No. 10 with a 30-ampere breaker   90%
   I = VA/E
   I = 6,000 VA/240 volts
   I = 25 ampere
   Note: The minimum overcurrent protection is 30
   ampere and the conductor must not be smaller than a
   No. 10 [*Section 240-3(b)*].

8. (b) 54 kW   80%
   8 dryers × 6.75 kW = Total of 54 kW
   Note: The *NEC®* does not permit a demand factors for
   commercial dryers.

9. (b) No. 6 THHN with 45-ampere protection   65%
   Electric heating equipment must have conductor and
   overcurrent protection device sized no less than 125%
   of the total rating of the equipment (including the
   motors) [*Section 424-3(b)*]
   Total heating = VA/E, 15,000 VA/480 volts = 31.25
   ampere + 1.6 ampere (motor) = 32.85 ampere
   Conductor and overcurrent not to be less than 32.85 ×
   1.25 = 41.06 ampere.
   Overcurrent protection device cannot be less than 41
   ampere [*Section 240-6(a)*] = 45 ampere and the conductor must be sized at a minimum of 41 ampere, at
   60°C which is a No. 6.

10. (d) 85 kVA, 220-15   90%
    Heat = [20,000 + (230 volts × 5.4 ampere)] = 21,242
    VA × 4 units = 84,968 VA/1,000 = 84.97 kVA

11. (c) 48 ampere   95%
    I = VA/E
    I = 11,400 VA/240 volts
    I = 47.5 ampere

12. (b) 21 kW, 220-20   95%
Water heater*	9 kW
Booster heater*	12 kW
Mixer	3 kW
Oven	2 kW
Dishwasher	1.5 kW
Disposal	+ 1 kW
Total connected	28.5 kW

    28.5 kW × .65 = 18.53 kW
    *Feeder demand load cannot be less than the two
    largest appliances (12 kW + 9 kW) = 21 kW

13. (b) 30 kW, 220-20   90%
Water heater*	14 kW
Booster heater *	11 kW
Mixer	7 kW
Oven	9 kW
Dishwasher	1.5 kW
Disposal	+ 3 kW
Total connected	45.5 kW

    45.5 kW × .65 = 29.58 kW
    *Feeder demand load cannot be less than the two
    largest appliances (14 kW + 11 kW) = 25 kW

14. (a) 10,500 VA   80%
    7 washing machines × 1,500 VA = 10,500 VA
    The *NEC*® does not permit a demand factors for commercial laundry circuits.

15. (b) 15 kVA, Table 220-3(a) and 220-11   90%
    24 motel rooms × 685 square feet = 16,440 square feet
    16,440 square feet × 2 VA =   32,880 VA
    First 20,000 VA at 50%   − 20,000 VA × .50 =   10,000 VA
    Next 80,000 VA at 40%   12,880 VA × .40 = + 5,152 VA
    Total demand load   15,152 VA

16. (b) 55 kVA, Table 220-3(a) and 220-11   85%
    General lighting (250,000 square feet × .25 VA)   62,500 VA
    *Receptacles (200 × 180 VA)   + 36,000 VA
    Total connected load   98,500 VA
    First 12,500 VA at 100%   12,500 VA × 1.00 =   12,500 VA
    Remainder at 50%   86,000 VA × .50 =   + 43,000 VA
    Total demand load   155,500 VA
    * The receptacle load (180 VA for each receptacle) is permitted to be included with the general lighting for storage ware-
    houses, see Section 220-13 of the *NEC*® for details.

17. (c) 8,000 VA, 220-3(a), 215-2(a), 230-42(a)   90%
    General lighting load (3,200 square feet × 2 VA × 1.25) = 8,000 VA

18. (b) 338 kVA, 220-3(a), 215-2(a), 230-42(a)   90%
    General lighting (90,000 square feet × 3 VA × 1.25) = 337,500 VA

19. (d) 20 fixtures, 210-19(a)   95%
    Maximum load on an overcurrent device is limited to 80% of the overcurrent device rating [*Section 210-19(a)* and *384-
    16(a)*].
    20 ampere × .8 = 16 ampere maximum load. The maximum number of .8 ampere continuous load lighting fixtures per-
    mitted on a 20-ampere circuit is: 20 ampere × .8 (continuous load) = 16 ampere/.8 ampere = 20 fixtures per circuit.
    Note: 220-4(b) requires that we use total ampere rating of unit (.8 ampere), not voltage of the lamps.

20. (a) 5 fixtures, 220-4(b)   90%
    20 ampere × .8 = 16 ampere × 120 volts = 1,920 VA/360 watts = 5.33 fixtures per circuit or a maximum of 5 fixtures

21. (b) 33 kVA, 220-12   75%
    130 feet × 200 VA per foot = 26,000 VA × 1.25 (continuous load) = 32,500 VA
    See Example D3 in Appendix D.

22. (a) 3,600 VA, 220-3(b)(8)   90%
    Multi-outlet receptacle assembly
    50 feet/ 5 feet per section:
    180 VA = 10 sections × 180 VA =   1,800 VA
    10 feet/ 1 foot per section:
    180 VA = 10 sections × 180 VA =   + 1,800 VA
    Total multi-outlet receptacle assembly load   3,600 VA

23. (b) 13 receptacles, 220-3(c)(9)   80%
    The total circuit VA for a 20-ampere, 120-volt circuit is 120 volts × 20 ampere = 2,400 VA
    The number of receptacle outlets permitted per circuit = 2,400 VA/180 VA = 13 receptacles.

24. (c) 15 kVA, 220-13   85%
    110 receptacles × 180 VA   19,800 VA
    First 10,000 VA at 100%   − 10,000 VA × 1.00 =   10,000 VA
    Remainder at 50%   9,800 VA × .50 =   + 4,900 VA
    Total receptacle demand load   14,900 VA

<u>25.</u> (b) 152 kVA, 220-3(a), 215-2(a) and 220-13   80%

General lighting (30,000 square feet × 3.5 VA × 1.25)   132,250 VA

Receptacle load, Table 220-3(a) Note and 220-13

1 VA per square foot	30,000 VA	
First 10,000 VA at 100%	− 10,000 VA × 1.00 =	10,000 VA
Remainder at 50%	20,000 VA × .50 =	+ 10,000 VA
Total receptacle demand load		20,000 VA

Total general lighting and receptacle demand load = 132,250 VA + 20,000 VA = 152,000 VA

<u>26.</u> (d) 55 kVA, 220-3(a), 215-2(a),

230-42(a) and 220-13   75%

General lighting (10,000 square feet × 3.5 VA × 1.25)   43,750 VA

Receptacles (75 receptacles × 180 VA)	13,500 VA	
First 10,000 VA at 100%	− 10,000 VA × 1.00 =	10,000 VA
Remainder at 50%	3,500 VA × .50 =	+ 1,750 VA
Receptacle demand load		11,750 VA

Total general lighting and receptacle demand load (43,750 VA + 11,750 VA) =   55,500 VA

<u>27.</u> (b) 1,500 VA, 220-3(c)(6) and 600-6(b)(3)   80%

The demand load for the sign is 1,200 VA × 1.25 (continuous load) = 1,500 VA

<u>28.</u> (b) 150 ampere   90%

The following formula can be used to determine the neutral current of three-wire circuit from a four-wire wye system.

$$I_n = \sqrt{\text{Line } 1^2 + \text{Line } 2^2) - (\text{Line } 1 \times \text{Line } 2)}$$

$$I_n = \sqrt{(150^2 + 150^2) - (150 \times 150)}$$

$$I_n = \sqrt{45,000 - 22,500} \quad I_n = \sqrt{22,500} \quad I_n = 150 \text{ ampere}$$

<u>29.</u> (b) 550 ampere

90% Section 555-5 permits a demand factor according to the number of receptacles.

24 receptacles × 20 ampere =	480 ampere
30 receptacles × 30 ampere =	900 ampere
1,380 ampere × .4 (demand factor) =	552 ampere

<u>30.</u> (c) 300 kcmil   80%

Two 300 kcmil conductors have a rating of 285 ampere each × 2 conductors = 570 ampere combined [*Section 240-3(b)* and *Table 310-16*]. Because the load is less than 800 ampere, the conductor must:

(1) have an ampacity of at least 570 ampere, and

(2) be protected by a 600 ampere protection device, and

(3) must be sized according to Table 310-16, 300 kcmil, using the 75°C terminal rating [*Section 110-14(c)(2)*].

Note: 250 kcmil THHN conductor has an ampacity of 290 ampere each, but we cannot use the 90°C ampacity rating for sizing conductors [*Section 110-14(c)*].

<u>31.</u> (a) 644 ampere, Table 550-22   80%

The feeder and service for mobile home parks is sized using the demand factors of Table 550-22.

Be careful. The demand factors of Table 550-22 applies to the larger of:

(1) 16,000 VA for each mobile home lot, or

(2) the calculate load of a one family dwelling unit [*Section 550-13*].

In this question 16,000 VA must be used for the calculation because we don't know the calculated load.

16,000 VA per site × 42 sites × .23 (demand factor) = 154,560 VA

I = VA/(E × 1.732)

I = 154,560 VA/240 volt

I = 644 ampere

32. (b) 450 ampere, 551-73(a)   85%

Recreational vehicles parks are calculated using the demand factor of Table 551-73.

Step 1: Each 20-ampere receptacle site is rate a minimum of 2,400 VA, each 20/30-ampere site is rated 3,600 VA and each 50-ampere site is rated 9,600 VA [*Section 551-73(a)*].

Step 2: The total connected load is:

Total load = (17 sites × 2,400 VA) + (35 sites × 3,600 VA) + (10 sites × 9,600 VA)

Total load = 40,800 VA + 126,000 VA + 96,000 VA = 262,800 VA

Step 3: The demand factors for 62 sites listed in Table 551-73(a) is .41

Step 4: Total demand load: 262,800 VA × .41 = 107,748 VA

Step 5: The total ampere rating is calculated by:

I = VA/E

I = 107,748 VA/240 volts

I = 449 ampere.

33. (a) 200 kVA, 220-36   95%

400 kVA × .5 = 200 kVA

34. (a) 260 kVA, 220-36   95%

400 kVA × .65 = 260 kVA

35. (c) 450 kVA, 220-34   95%

Step 1: Determine the average VA per square foot:

590,000 VA/28,000 square feet = 21.07 VA per square foot.

Step 2: Apply the demand factors from Table 220-34 to the average VA per square foot:

Average VA per square foot =	21.07 VA	
First 3 VA at 100%	− 3.00 VA × 1.00 =	3.00 VA
	18.07 VA	
Next 17 VA at 75%	− 17.00 VA × .75 =	12.75 VA
Remainder at 25%	1.07 VA × .25 =	+ .27 VA
Net VA per square foot		16.02 VA

Step 3: Determine the school net VA (16.02 VA per square foot × 28,000 square feet) = 448,560 VA

CHALLENGE QUESTIONS

36. (d) No. 8 THW   75%

The only 3-phase loads that can be connected to the high-leg is the air-conditioning or heat load. Section 220-21 permits us to omit the smallest load.

Air-conditioning: The conductor size would be calculated at 125% of the load [*Section 440-32*]. The ampere rating would be determined by the use of Table 430-150 and is 28 ampere.

28 ampere × 1.25 = 35 ampere would be a No. 8 [Table 310-16 at 60°C and 110-14(c)].

Heat: The conductor for heat would be calculated at 125% of the total heating load [*Section 424-3(b)*]. The ampere rating of the heat would be calculated as:

I = P/(E × 1.732)

I = 10,000 watts/(230 volts × 1.732)

I = 25 ampere × 1.25

I = 31 ampere

Omit the heat because it's smaller than the air-conditioning [*Section 220-21*].

Note: The question specified service conductors, Sections 230-23(b) and 230-31(b) require a minimum No. 8 copper. Since all of the multiple choice answers are smaller than the No. 8 THW, answer "(d)" is the only possible choice.

<u>37.</u> (b) 7 circuits   75%

To determine the number of circuits required for an installation, we must:

Step 1: Determine the ampere load for each fixture:

I = P/E

I = 250 VA/120 volts

I = 2.083 ampere each

Step 2: Determine the maximum load on the protection device, [*Section 210-20(a)* and *384-16(d)*].

20 ampere circuit × .8 = 16 ampere

Step 3: Determine the maximum number of fixtures per circuit

16 ampere/2.1 ampere = 7 fixtures per circuit

Step 4: Determine the number of circuits, 46 fixtures/7 per circuit = 7 circuits required.

Conductor Sizing

<u>38.</u> (c) No. 1/0, Table 250-66, 250-102(c)   75%

<u>39.</u> (b) No. 2, Table 250-66, 250-102(b)   75%

Kitchen Equipment Calculations [*Section 220-20*]

<u>40.</u> (a) 17 kW, 220-20   85%

The demand load is determined by applying the Table 220-20 demand factor to all kitchen equipment loads that operate intermittently; but, not air-conditioning, heat, or ventilation.

Note: The demand load cannot be less than the sum of the two largest appliances.

Step 1: Determine the total connected load:

Range =	14.00 kW
W/H =	1.50 kW
Mixer =	.75 kW
D/W =	2.00 kW
Booster =	3.00 kW
Coffee =	+ 3.00 kW
	24.25 kW

Step 2: Determine the demand factor for six units Table 220-20, 65%.

Step 3: Determine the demand load: 24.25 kW × .65 = 15.76 kW.

But, the sum of the two largest appliances equals: 14 kW + 3 kW = 17 kW.

<u>41.</u> (c) 23 kVA, Table 220-3(a) and Table 220-11   65%

Hotel:

12 × 15 feet = 180 square feet × 2 VA × 100 rooms	36,000 VA
First 20,000 VA at 50%	20,000 VA × .50 = 10,000 VA
Remainder at 40%	16,000 VA × .40 = 6,400 VA

Office:

60 × 20 feet = 1,200 square feet × 3.5 VA × 1.25 (continuous load)	5,250 VA
Office receptacles: 1,200 square feet × 1 VA*	1,200 VA
Halls: 120 square feet × .5 VA × 1.25 (continuous)	+ 75 VA
	22,925 VA/1,000 = 22.93 kVA

* Table 220-3(a) note b, add 1 VA per square foot for unknown receptacle count

Receptacle Demand Load Calculations [*Section 220-13*]

<u>42.</u> (c) 14,000 VA, 220-13   80%

Table 220-13, Nondwelling receptacle at 180 VA each

100 receptacles × 180 VA	18,000 VA	
First 10,000 VA at 100%	− 10,000 VA × 1.00 =	10,000 VA
Remainder at 50%	8,000 VA × .50 =	+ 4,000 VA
Total demand load		14,000 VA

<u>43.</u> (d) 15,000 VA, Table 220-3(a)   80%
However, the demand factors of Table 220-13 still apply
General use receptacles:

20,000 VA × 1 VA[b] per square foot	20,000 VA	
First 10,000 VA at 100%	− 10,000 VA × 1.00 =	10,000 VA
Remainder at 50%	10,000 VA × .50 =	+ 5,000 VA
Total receptacle demand load		15,000 VA

Note[b], Table 220-3(a): 1 VA per square foot for receptacles when the number of receptacles is unknown in banks and office buildings.

General Lighting Load Calculations [Table 220-3(a) and 220-11]

<u>44.</u> (d) 151 kVA, Table 220-3(a) and Table 220-13   80%

General lighting (30,000 square feet × 3.5 VA × 1.25)		131,250 VA
General use receptacles:		
30,000 VA × 1 VA per square foot	30,000 VA	
First 10,000 VA at 100%	− 10,000 VA × 1.00 =	10,000 VA
Remainder VA at 50%	20,000 VA × .50 =	+ 10,000 VA
Total receptacle demand load =		20,000 VA

Lighting, 131,250 VA + 20,000 VA receptacle = 151,250 VA/1,000 = 151.25 kVA

Neutral Calculations

<u>45.</u> (c) 745 ampere, 220-22   75%

Total feeder load 1,450 ampere		
Phase to phase loads 600 ampere at 0%		0 ampere
Discharge lighting 300 ampere at 100%	300 ampere × 1.00 =	300 ampere
First 200 ampere at 100%	200 ampere × 1.00 =	200 ampere
Remainder at 70%	350 ampere × .70 =	+ 245 ampere
Total neutral demand load		745 ampere

Marina Calculations [Article 555]

<u>46.</u> (b) 630 ampere, 555-3 and 555-5   85%
Step 1: Determine the total connected receptacle ratings:

20 receptacles rated 20 ampere (20 × 20 ampere)	400 ampere
17 receptacles rated 30 ampere (17 × 30 ampere)	510 ampere
7 receptacles rated 50 ampere (7 × 50 ampere)	+ 350 ampere
Total connected receptacle rating	1,260 ampere

Step 2: Determine demand factor from Section 555-5, 50%
Step 3: Determine net compute demand load, 1,260 ampere × .5 = 630 ampere

Mobile Home Parks [Article 550]

<u>47.</u> (b) 264 kVA, 550-22 and Table 550-22   85%
The calculation is based on 16,000 VA
Step 1: Determine the total connected load: 75 sites × 16,000 VA = 1,200,000 VA
Step 2: Determine the demand factor from Table 550-22, 22%
Step 3: Determine the total demand load; 1,200,000 VA × .22 = 264,000 VA/1,000 = 264 kVA

Motel

<u>48.</u> (b) 202 kVA, Table 220-3(a) and 220-11   95%

Step 1: Determine the general lighting demand load:

40 units × 300 square feet × 2 VA =	24,000 VA
First 20,000 VA at 50%	<u>20,000 VA</u> × .50 = 10,000 VA
Remainder at 40%	− 4,000 VA × .40 = 1,600 VA
Total general lighting demand load	

Step 2: Determine the A/C vs. heat load [*Section 220-21*]

Air-conditioning, 3 kW, omit it because it's smaller than the heat

Heat, 4 kW, 4 kVA × 40 units = 160 kVA × 1,000 =	160,000 VA

Step 3: One water heater, per 2 units,

40 units/2 = 20 water heaters × 1,500 VA =	+ 30,000 VA
Total demand load	201,600 VA

Recreational Vehicle Park Calculations [Article 551]

<u>49.</u> (d) 65 kVA, 551-71, 551-73(a), and Table 551-73 80%

Note: A minimum of 70% of the sites must have a 30/20-ampere facility (3,600 VA per site).

Step 1: Determine the total connected load:

3 sites at 50 ampere (3 sites × 9,600 VA)	28,800 VA
30 sites at 30 ampere (30 sites × 3,600 VA)	108,000 VA
9 sites at 20 ampere (9 sites × 2,400 VA)	+ 21,600 VA
Total	158,400 VA

Step 2: Determine demand factor [Table 551-73]: 41%

Step 3: Determine net computed demand load: 158,400 VA × .41 = 64,944 VA/1,000 = 65 kVA

Restaurant Calculations

<u>50.</u> (a) 50 ampere   75%

I = P/E

I = 12,000 VA/240 volts

I = 50 ampere

<u>51.</u> (a) 64 kVA, 220-36   75%

Step 1: Determine the connected load:

General lighting	30 kVA
Dishwasher	5 kW
Coffer makers (2 kW × 2 units)	4 kW
Kitchen appliances (2 kW × 5 units)	10 kVA
Small-appliances circuit (1.5 kVA × 10 units)	+ 15 kVA
Total connected load	64 kVA

Step 2: Apply 220-36 demand factor: 64 kVA at 100% = 64 kVA demand load

Optional School Load Calculation

<u>52.</u> (b) 18.75 VA, Table 220-34   75%

Step 1: Determine the average VA per square foot:

320,000 VA/10,000 square feet = 32 VA

Step 2: Determine the demand VA per square foot:

Total VA per square foot	32.00 VA
First 3 VA at 100%	− 3.00 VA × 1.00 = 3.00 VA per sq/ft
	29.00 VA
Next 17 VA at 75%	− 17.00 VA × .75 = 12.75 VA per sq/ft
Remainder at 25%	12.00 VA × .25 = + 3.00 VA per sq/ft
Net computed VA per square foot	18.75 VA per sq/ft

<u>53.</u> (a) 135 kVA, 220-34   80%

Step 1: Determine the average VA per square foot:
160,000 VA/20,000 square feet = 8 VA

Step 2: Determine the demand VA per square foot:
Total VA per square foot:

	8.00 VA	
First 3 VA	− 3.00 VA × 1.00 =	3.00 VA
Next 17 VA	5.00 VA × 0.75 =	<u>3.75 VA</u>
Net computed VA per square foot		6.75 VA

Step 3: Determine the total demand load:
6.75 VA × 20,000 square feet = 135,000 VA
135,000 VA/1,000 = 135kVA

NEC® Answers

54. (b) Chapter 9, Table 8, 6th Column   80%

55. (d) 645-5(d)(5)   80%

56. (a) 90-3   80%

57. (c) 810-52 Table   75%

58. (c) 400-9   80%

59. (b) 680-70

60. (d) 305-6(a)

61. (d) 318-5(f)   80%

62. (c) Appendix C, Table C1   80%

63. (d) 645-16   95%

64. (d) 250-100, 501-16, 502-16, 503-16   65%

65. (b) 660-2 Definition   95%

66. (d) 100 Switch, Isolating   85%

67. (b) 760-71(h)   85%

68. (d) 680-1   90%

69. (c) 410-6   95%

70. (b) 680-6(a)(1)

71. (d) 760-1 FPN No. 1   95%

72. (a) 760-3(b), 300-22(c) and FPN

73. (d) 410-70   85%

74. (b) 680-6(b)(1) and 680-40   80%

75. (d) 424-34   90%

76. (c) 680-8   95%

77. (d) 339-3(b)   95%

78. (b) Chapter 9, Table 1   95%

79. (d) 680-10   95%

80. (d) 680-21(a)

81. (b) 430-35(b)   80%

82. (d) 680-41(b)(1)   95%

83. (d) 345-12(b)(3)   80%

84. (b) 500-3(b)

85. (b) 680-52(a)

86. (a) 365-3(c)   90%

87. (c) 680-21(a)(5)   95%

88. (d) 700-7(a)(b)(c)   95%

89. (d) 700-9(c)   85%

90. (b) 700-26   95%

91. (a) 670-3(b)   85%

92. (c) 680-20(a)(1)   85%

93. (a) 702-2 FPN   90%

94. (b) 720-6   90%

95. (d) 725-5, 760-5, 800-5, 820-5

96. (d) 725-21(b)   95%

97. (d) 725-27(b)   80%

98. (d) 725-41 FPN, Table 11 of Chapter 9   95%

99. (c) 680-4 definition   95%

100. (b) 680-5(b)   80%
GFCI receptacles are permitted.

101. (a) 380-9(b)

102. (b) 230-92   95%

103. (a) 370-28(a)(1)   90%

104. (c) 230-70(b)

105. (d) 500-5(c) FPN No. 1   90%

106. (c) 450-21(b)   90%

107. (b) 370-16(c)(1)   75%
Careful, the Code states smaller than No. 6, not larger.

108. (c) 424-19(a)(2)(b)   85%

109. (c) 240-24(a)(1)   85%

110. (b) 310-15(a)(1) FPN   85%

111. (a) 410-101(d)   70%

112. (c) 665-25   80%

113. (a) 760-27(b)   90%

114. (c) Chapter 9, Table 8   95%

115. (d) 760-30(a)(2)(3)   90%

116. (a) 760-54(a)(1)   95%

117. (c) Chapter 9, Table 1   80%

118. (d) 770-50 Table and 770-51(d)   85%

119. (d) 800-10(a)(2) and (b)   75%

120. (c) 800-40(d)   95%

121. (a) 300-5 Table Column 2   80%

122. (b) 800-51(b)

123. (b) 620-12(a)(1)   60%

124. (a) 230-52   80%

125. (b) 501-5(a)(4)   95%

126. (c) 517-30(b)(5)

127. (b) 230-26   90%

128. (b) 501-4(a) Exception No. 3

129. (c) 230-72(c) Exception   90%

130. (c) 725-54(a)(1)   90%

131. (a) 300-19(a) Table   75%

132. (d) 725-54(a)(2)   95%

133. (c) 660-6(a)   75%

134. (b) 250-52(d)   90%

135. (d) 680-22(a)   60%

136. (c) 430-32(a)(2)   90%

137. (a) 680-22(a)(1)

138. (b) 230-54(a)   80%

139. (d) 680-22(b)(3)   90%

140. (b) 230-46   70%

141. (d) 680-25(b)(1)   85%

142. (d) 220-34   90%

143. (a) 384-14   90%

144. (a) 225-18   90%

145. (b) 410-65(a)   95%

146. (b) 410-8(d)   80%

147. (b) 695-6(c)

148. (a) 410-16(a)

149. (d) 700-9(b)   90%

150. (d) 410-16(c)

151. (c) 725-21(a)   85%

152. (a) 410-28(c)   75%

153. (a) 800-13   85%

154. (c) 240-61   75%

155. (d) 110-12(b)   90%

156. (d) 520-25(a)   80%

157. (a) 511-3(b)   80%

158. (d) 410-28(e)   95%

159. (a) 110-14(c)(1)(c)   95%

160. (a) 800-48

161. (a) 110-26(b)   75%

162. (a) Chapter 9, Table 8, 6th Column   80%

163. (b) 760-54(a)(1)   85%

164. (b) 110-26(f)

165. (b) 680-25(d)

166. (c) 422-18(b) Exception

167. (d) 680-25(c)   75%
      Flex is not permitted, but liquidtight is.

168. (a) 328-12(a)   90%

169. (d) 680-31   90%

170. (b) 680-72

171. (a) 230-79(a)   60%

172. (d) 250-104(a)   85%

173. (b) 300-6(c) FPN   90%

174. (a) 300-16(b)   90%

175. (c) 331-10   95%

176. (a) 810-11 Exception   90%

177. (a) 230-95   95%

178. (d) 330-22   90%

179. (d) 240-21(b)(4)   90%

180. (d) 310-8(a)   70%

181. (d) 430-101   70%

182. (d) 810-16(a) Table   90%

183. (c) 830-40(d)   95%

184. (b) Chapter 9, Table 1, Note 4   80%

185. (d) Chapter 9, Table 1, Note 7   95%

186. (c) Chapter 9, Table 8, 5th Column   95%

187. (b) 300-5(d)

188. (c) Chapter 9, Table 9, 6th and 9th Columns   95%

189. (c) 300-6   85%

190. (b) 300-14

191. (b) Chapter 9, Table 9, 9th Column   90%

192. (a) 695-7

193. (c) 810-71(c)   75%

194. (b) 820-52(a)(1)(b) Exception No. 1   80%

195. (b) 430-14   95%

196. (d) 430-72(c)(4)   90%

197. (a) 500-5(b)(3)   80%

198. (c) 680-9   90%

199. (c) Chapter 9, Table 1, Note 4 and Table 4   85%
      2.071 square inches $\times$ 0.6 = 1.243 square inches

200. (a) Chapter 9, Table 1, Note 4, and 310-15(b)(2) 95%

Unit 12 – Delta/Delta and Delta/Wye Transformer Calculations

Introduction

1. (a) series   95%
2. (a) delta   95%
3. (b) line   90%
4. (b) line   95%
5. (d) wye, delta   75%
6. (c) phase   95%
7. (d) delta, wye   95%
8. (d) all of these   90%
9. (d) all of these   95%
10. (d) delta, wye   95%
11. (b) 2:1, 4:1   90%
12. (a) 4:1.   90%
    480 primary phase volts, 120 secondary phase volts
13. (b) wye   95%
14. (a) delta   70%
15. (b) wye   95%
16. (a) True   95%
17. (b) False   85%
18. (a) True   95%
19. (a) 190 volts   90%

    110 volts $\times$ 1.732 = 190 volts

20. (b) False   90%
21. (c) 54 ampere   95%
    $I_{Line} = VA_{Line}/(E_{Line} \times \sqrt{3})$
    I = 45,000 VA/(480 volts $\times$ 1.732)
    I = 54 ampere
22. (b) 108 ampere   95%
    $I_{Line} = VA_{Line}/(E_{Line} \times \sqrt{3})$
    I = 45,000 VA/(240 volts $\times$ 1.732)
    I = 108 ampere
23. (a) True   95%
24. (d) 31 ampere   95%
    $I_{Phase} = VA_{Phase}/E_{Phase}$
    I = 15,000 VA/480 volts
    I = 31 ampere
25. (b) 62 ampere   95%
    $I_{Phase} = VA_{Phase}/E_{Phase}$
    I = 15,000 VA/240 volts
    I = 62 ampere
26. (d) all of the above   90%
27. (d) all of the above   70%
28. (d) a and b   65%
29. (c) a and b   90%
30. (d) all of these   85%

31. (d) none of these   75%

The minimum phase must be at least 14 kVA because of the C phase = 5 kVA + 7 kVA.

Phase A = 11 kVA, B = 11 kVA, C = 14 kVA

	Phase A (L1 and L2)	Phase B (L2 and L3)	C1 L1	C2 L3	Line Load
(1) 18 kVA, 240 volts 3Ø	6 kVA	6 kVA	3 kVA	3 kVA	18 kVA
(2) 5 kVA, 240 volts 1Ø	5 kVA	5 kVA			10 kVA
(3) 2 kVA, 120 volts 1Ø	____	____	2 kVA*	4 kVA*	6 kVA
Totals	11 kVA	11 kVA	5 kVA	7 kVA	34 kVA

*Indicates neutral (120-volt) loads.

Once you balance the transformer you can size the transformer according to the load of each phase. The "C" transformer must be sized using; two times the highest of C1 or C2. The "C" transformer is actually a single transformer, if one side has a greater load than the other, then that side determines the transformer size.

32. (a) True   95%

33. (d) all of these   55%

	Line 1	Line 2	Line 3	Load
18 kVA, 240 volts 3Ø	6.0 kVA	6.0 kVA	6.0 kVA	18 kVA
5 kVA, 240 volts 1Ø	2.5 kVA	2.5 kVA		5 kVA
5 kVA, 240 volts 1Ø		2.5 kVA	2.5 kVA	5 kVA
(3) 2 kVA, 120 volts 1Ø	2.0 kVA*	_____	4.0 kVA*	6 kVA
	10.5 kVA	11 kVA	12.5 kVA	34 kVA

*Indicates neutral (120-volt) loads.

34. (d) a and b   90%

	Line 1	Line 2	Line 3	Ampere Calculation
18 kVA, 240 volts 3Ø	43 ampere	43 ampere	43 ampere	18,000 VA/(240 volts × 1.732)
5 kVA, 240 volts 1Ø	21 ampere	21 ampere		5,000 VA/240 volts
5 kVA, 240 volts 1Ø		21 ampere	21 ampere	5,000 VA/240 volts
(3) 2 kVA, 120 volts 1Ø	17 ampere*	____	34 ampere*	3,000 VA/120 volts
	81 ampere	85 ampere	98 ampere	

*Indicates neutral (120-volt) loads.

35. (a) 17 ampere   90%

Neutral current = Line 3 less Line 1

Line 1 = 17 ampere, Line 2 = 34 ampere

Neutral current = 34 ampere – 17 ampere

Neutral current = 17 ampere

	Line 1	Line 2	Line 3	Ampere Calculation
18 kVA, 240 volts 3Ø	43 ampere	43 ampere	43 ampere	18,000 VA/(240 volts × 1.732)
5 kVA, 240 volts 1Ø	21 ampere	21 ampere		5,000 VA/240 volts
5 kVA, 240 volts 1Ø		21 ampere	21 ampere	5,000 VA/240 volts
(3) 2 kVA, 120 volts 1Ø	34 ampere*		17 ampere*	2,000 VA/120 volts

*Indicates neutral (120-volt) loads.

36. (b) 34 ampere   60%

	Line 1	Line 2	Line 3	Ampere Calculation
18 kVA, 240 volts 3Ø	43 ampere	43 ampere	43 ampere	18,000 VA/(240 volts × 1.732)
5 kVA, 240 volts 1Ø	21 ampere	21 ampere		5,000 VA/240 volts
5 kVA, 240 volts 1Ø	_____	21 ampere	21 ampere	5,000 VA/240 volts
2 kVA, 120 volts 1Ø	17 ampere*		34 ampere*	3,000 VA/120 volts

*Indicates neutral (120-volt) loads.

37. (c) 6 kVA   95%

38. (a) 5 kVA   60%

Can you believe you missed this?

39. (a) 29 ampere   75%

$I = VA/(E \times \sqrt{3})$

$I = 12{,}000 \text{ VA}/(240 \text{ volts} \times 1.732)$

$I = 29 \text{ ampere}$

40. (d) 2:1   90%

480 primary phase volts to 240 secondary phase volts

41. (b) False   95%

42. (a) True   90%

43. (a) True   85%

44. (c) 26 ampere   95%

$I_{Line} = VA_{Line}/(E_{Line} \times \sqrt{3})$

$I = 22{,}000 \text{ VA}/(480 \text{ volts} \times 1.732)$

$I = 26 \text{ ampere}$

45. (a) 61 ampere   95%

$I_{Line} = VA_{Line}/(E_{Line} \times \sqrt{3})$

$I = 22{,}000 \text{ VA}/(208 \text{ volts} \times 1.732)$

$I = 61 \text{ ampere}$

46. (a) True   95%

47. (a) 15 ampere   75%

$I_{Phase} = VA_{Phase}/E_{Phase}$

$I = 7{,}333 \text{ VA}/480 \text{ volts}$

$I = 15 \text{ ampere}$

48. (d) 61 ampere   70%

$I_{Phase} = VA_{Phase}/E_{Phase}$

$I = 7{,}333 \text{ VA}/120 \text{ volts}$

$I = 61 \text{ ampere}$

49. (d) all of these   95%

50. (d) all of these   90%

51. (d) all of these   95%

52. (a) 1,945 VA   60%

On a wye system, a 208-volt 1Ø load is connected to two transformer phases.

Motor VA (single-phase) = Volts × Ampere

Motor VA = 208 volts × 18.7 ampere

Motor VA = 3,890 VA

The load on each transformer phase is 3,890 VA/2 phases = 1,945 VA

53. (a) 2,906 VA   75%

On a delta or wye system a three phase load is distributed on all three phases.

Motor VA (three phase) = Volts × Ampere × 1.732

Motor VA = 208 volts × 24.2 ampere × 1.732

Motor VA = 8,718 VA

The load on each transformer phase is 8,718 VA/3 phases = 2,906 VA

54. (d) all of these   95%

55. (c) a or b   70%

	Phase A (Line 1)	Phase B (Line 2)	Phase C (Line 3)	Load
18 kVA, 208 volts 3Ø	6.0 kVA	6.0 kVA	6.0 kVA	18 kVA
5 kVA, 208 volts 1Ø	2.5 kVA	2.5 kVA		5 kVA
5 kVA, 208 volts 1Ø		2.5 kVA	2.5 kVA	5 kVA
(3) 2 kVA, 120 volts 1Ø	4.0 kVA*		2.0 kVA*	6 kVA
	12.5 kVA	11 kVA	10.5 kVA	34 kVA

*Indicates neutral (120-volt) loads.

56. (d) a, b, and d   95%

	Line 1	Line 2	Line 3	Load
18 kVA, 208 volts 3Ø	6.0 kVA	6.0 kVA	6.0 kVA	18 kVA
5 kVA, 208 volts 1Ø	2.5 kVA	2.5 kVA		5 kVA
5 kVA, 208 volts 1Ø		2.5 kVA	2.5 kVA	5 kVA
(3) 2 kVA, 120 volts 1Ø	4.0 kVA*		2.0 kVA*	6 kVA
	12.5 kVA	11 kVA	10.5 kVA	34 kVA

*Indicates neutral (120-volt) loads.

57. (c) A and B   65%

	Line 1	Line 2	Line 3	Ampere Calculation
18 kVA, 208 volts 3Ø	50 ampere	50 ampere	50 ampere	18,000 VA/(208 volts × 1.732)
5 kVA, 208 volts 1Ø	24 ampere	24 ampere		5,000 VA/208 volts
5 kVA, 208 volts 1Ø		24 ampere	24 ampere	5,000 VA/208 volts
(3) 2 kVA, 208 volts 1Ø	34 ampere*		17 ampere*	2,000 VA/120 volts
	108 ampere	98 ampere	91 ampere	

*Indicates neutral (120-volt) loads.

58. (a) True   95%

59. (b) 29 ampere   60%

Neutral current = Line 1 less Line 3

Neutral current = 34 ampere – 17 ampere

Neutral current = 17 ampere

	Line 1	Line 2	Line 3	Ampere Calculation
(1) 18 kVA, 240 volts 3Ø	50 ampere	50 ampere	50 ampere	18,000 VA/(208 volts × 1.732)
(1) 5 kVA, 240 volts 1Ø	24 ampere	24 ampere		5,000 VA/208 volts
(1) 5 kVA, 240 volts 1Ø		24 ampere	24 ampere	5,000 VA/208 volts
(3) 2 kVA, 120 volts 1Ø	34 ampere*		17 ampere*	2,000 VA/120 volts

*Indicates neutral (120-volt) loads.

60. (b) 34 ampere   65%

	Line 1	Line 2	Line 3	Ampere Calculation
(1) 18 kVA 240 volts 3Ø	50 ampere	50 ampere	50 ampere	18,000 VA/(208 volts × 1.732)
(1) 5 kVA 240 volts 1Ø	24 ampere		24 ampere	5,000 VA/208 volts
(1) 5 kVA 240 volts 1Ø		24 ampere	24 ampere	5,000 VA/208 volts
(3) 2 kVA 120 volts 1Ø	34 ampere*		17 ampere*	2,000 VA/120 volts

*Indicates neutral (120-volt) loads.

61. (d) a and c   95%

62. (d) a and c   95%

CHALLENGE QUESTIONS

Note: The following calculations are the basis
to answer questions 63–69.

Motor VA (3Ø) = Volts × FLC × 1.732

Motor VA (1Ø) = Volts × FLC

Note: Use the Table volts and Table amperes, however, some exams use the actual volts times table ampere.

25 Horsepower, 230 volt, 3Ø, Synchronous, Table 430-150:

Motor VA = 230 volt × 53 ampere × 1.732

Motor VA = 21,113 VA/3 = 7,038 per phase

1 Horsepower, 230 volt, 1Ø, Table 430-148:

Motor VA = 230 volt × 10 ampere

Motor VA = 2,300 VA

2 Horsepower, 115 volt 1Ø, Table 430-148:

Motor VA = 115 volt × 24 ampere

Motor VA = 2,760 VA

	Phase A	Phase B	Phase C1	C2	Line VA
25 Horsepower, 230 volt, 3Ø	7,038 VA	7,038 VA	3,519 VA	3,519 VA	21,114 VA
1 Horsepower, 230 volts, 1Ø	2,300 VA				2,300 VA
1 Horsepower, 230 volts, 1Ø		2,300 VA			2,300 VA
2 Horsepower 115 volts, 1Ø			2,760 VA*		2,760 VA
2 Horsepower 115 volts, 1Ø			2,760 VA*		2,760 VA
1900 VA, 115 volts				1,900 VA*	
1900 VA, 115 volts				1,900 VA*	
1900 VA, 115 volts	_____	_____	_____	1,900 VA*	5,700 VA
Total	9,338 VA	9,338 VA	9,039 VA	9,219 VA	36,934 VA

Note: Heat was omitted as a noncoincident load since it is smaller than the air-conditioning load. See 220-21.
*Indicates neutral (120-volt) loads.

63. (c) 10/10/20 kVA   70%

Phase A = 9,338 VA

Phase B = 9,338 VA

Phase C = 9,039 VA + 9,219 VA = 18,258 VA

64. (b) 21,113 VA   40%

Motor VA (3Ø) = Volts × FLC × 1.732

25 horsepower, 240 volt 3Ø, synchronous, Table 430-150:

Motor VA = 230 volt × 53 ampere × 1.732

Motor VA = 21,113 VA

Note: Some exams use 240 volts for calculations: Motor VA = 240 volt × 53 ampere × 1.732 = 22,031 VA

65. (c) 48 ampere   50%

Phase C1 = 2,760 VA + 2,760 VA = 5,520 VA/120 volt = 46 ampere

Phase C2 = 1,900 VA × 3 loads = 5,700 VA/120 volts = 47.5 ampere

66. (c) 2,300 VA   70%

67. (c) 2,300 VA   65%

Motor VA = 230 volts × 10 ampere = 2,300 VA

68. (b) 7,038 VA   50%

Motor VA (3Ø) = Volts × FLC × 1.732

25 horsepower, 240 volt 3Ø, synchronous, Table 430-150:

Motor VA = 230 volt × 53 ampere × 1.732

Motor VA = 21,113 VA/3 = 7,038 VA per phase

69. (b) 66.25 ampere, 430-22(a)   50%

53 ampere × 1.25 = 66.25 ampere

70. (a) 45 ampere   70%

$I_{Primary} = VA_{Primary}/(E_{Primary} \times 3)$

$I_{Primary} = (35.7 \text{ kVA} \times 1,000)/(480 \text{ volts} \times 1.732)$

$I_{Primary} = 45.1$ ampere

71. (d) 60 ampere device   70%

450-3(b) not more than 125% of primary current,
45.1 ampere × 1.25 = 56.4 ampere, 450-3(b) Exception No. 1, next size up, 240-6(a), 60-ampere device

72. (c) 80–90 kW   70%

    30 kW + 40 kW + (7 loads $\times$ 1.5 kW) = 80.5 kW

73. (d) 333 ampere   50%

    100 kW less 20 kW = 80 kW of neutral loads, since the question specified "balanced," then 40 kW is on C1, 40 kW is on C2.I of C1 = 40,000 VA/120 volts = 333.3 ampere

74. (c) 42 ampere   65%

    Phase Current = Phase Power/Phase Voltage

    Phase power = 30,000 VA/3 phases = 10,000 VA per phase

    Phase voltage = 240 volt

    Phase current = 10,000 VA/240 volt = 41.7 ampere per phase or. . .

    Phase current = Line current/1.732

    Line current = Line power/(line volts $\times$ 1.732)

    Line current = 30,000 VA/(240 volts $\times$ 1.732) = 72.2 ampere per line

    Phase current = 72.2 ampere/1.732 = 41.7 ampere per phase

75. (a) 113 ampere   60%

    There are two easy ways to calculate this problem.

    (1) Calculate the primary line current, then use the ratio to determine the secondary line current, or

    (2) Use the ratio to determine the secondary voltage, then calculate the secondary line current.

    Primary Line Current = Primary Phase Power/(Primary Phase Volts $\times$ 1.732)

    Primary line current = 45,000 VA/(460 volts $\times$ 1.732) = 56.5 ampere.

    Voltage ratio is 2:1, therefore secondary voltage is $^1/_2$ of primary voltage and the current will be twice the primary line current, secondary line current; 56.5 ampere $\times$ 2 = 113 ampere

76. (d) 30 ampere   60%

    The primary line current is calculated by dividing the primary line power by the primary line voltage.

    The primary line power is the same as the secondary line power (at 100% efficiency).

    The secondary line power is calculated as: VA = Secondary Line Volts $\times$ Secondary Line Ampere $\times$ 1.732.

    Secondary line power = 220 volts $\times$ 300 ampere $\times$ 1.732 = 114,312 VA

    Primary line current = Primary line power/(Primary line voltage $\times$ 1.732)

    Primary line current = 114,312 VA/(2,200 volts $\times$ 1.732)

    Primary line current = 90 ampere

77. (b) delta primary, delta secondary   70%

Delta/Wye Answers

78. (c) 87 ampere   55%

    Step 1: Calculate transformer VA load

    Transformer VA = 208 volts $\times$ 200 amperes $\times$ 1.732

    Trnsformer VA = 72,000 VA

    Step 2: Primary line current

    Primary line current = 72,000 VA/480 volts $\times$ 1.732

    Primary line current = 86.6 amperes

79. (b) 12 ampere   60%

    The line current for a wye system = Line power/(Line volts $\times$ 1.732)

    Line current = 10,000 VA/(480 volts $\times$ 1.732) = 12 ampere

80. (c) 2,500 watts   60%

    2.5 kW $\times$ 1,000 = 2,500 watts load

81. (b) (Volts × Ampere)/$\sqrt{3}$ 55%

Example: 15 horsepower 208-volt, three-phase motor. Table 430-150 Ampere = 46.2 ampere

Line Power = (Volts × Ampere)/1.732 = (208 volts × 46.2 ampere)/1.732 = 5,548 VA, or

Line Power = Volts × Ampere × 1.732

Line power = 208 volts × 46.2 ampere × 1.732

Line power = 16,644 VA/3 phases

Line power = 5,548 VA per line

82. (b) 4,000 VA   70%

12,000 VA line/3 phases = 4,000 VA per phase

83. (a) 50 ampere   60%

Balance 120-volt loads only, the 208-volts loads do not apply

	Phase A	Phase B	Phase C
	2,000 VA*	2,000 VA*	2,000 VA*
	2,000 VA*	2,000 VA*	2,000 VA*
	2,000 VA*	2,000 VA*	2,000 VA*
	6,000 VA*	6,000 VA*	6,000 VA*

Neutral current = 6,000 VA/120 volts = 50 ampere

*Indicates neutral (120-volt) loads.

84. (b)1,250 watts   55%

2,500 watts line load/2 phases = 1,250 watts per phase

85. (a) 3,700 VA, Table 430-150   55%

Motor VA = Motor Volts × Motor Amperes

Motor VA = 208 volts × 30.8 ampere × 1.732 = 11,096 VA

Motor VA per line = 16,096VA/1.732 = 3698 VA

86. (d) 100 ampere   55%

Maximum unbalance

	Line1	Line 2	Line 3
2.5 kW, 208-volt heating unit	12 ampere	12 ampere	
2.5 kW, 208-volt heating unit		12 ampere	12 ampere
(9) 2 kVA, 120-volt lighting loads	50 ampere*	50 ampere*	50 ampere
(5) 3 kVA ,120-volt loads	50 ampere*	25 ampere*	50 ampere
	112 ampere	99 ampere	112 ampere*

NEC® Answers

87. (a) 370-23(d)(2)
88. (a) 318-7(b)(1)   90%
89. (b) 310-15(b)(4)(a)   85%
90. (a) 370-44   95%
91. (b) 410-49   90%
92. (a) 210-19(a) and 210-20(a)   60%
Maximum continuous load permitted on a 20-ampere
circuit breaker is limited to 80% according to 384-
16(d). 20 ampere × 80% = 16 ampere max continuous load permitted. 16 ampere/2 ampere per fixture =
8 lighting fixtures permitted.
93. (d) 760-23   90%
94. 680-22(b)
95. (d) 210-4(d) FPN
96. (d) 680-21(b)
97. (a) 410-56(c) Exception
98. (c) 725-41(a) FPN, Table 11 in Chapter 9   85%
99. (a) 680-5(c)   80%
100. (d) 110-34(b)   90%
101. (c) 600-6
102. (a) 550-8(d)   80%
103. (d) 725-23   75%
104. (d) 240-3(d), (e) and (g)   70%
105. (d) 305-4(h)   95%
106. (b) 110-27(b)   90%
107. (d) 210-8(b)(1)   90%
108. (c) 210-52(c)(3)   75%
109. (c) 300-20(b) FPN   95%
110. (b) 210-52(f) Exception No. 1   95%
111. (b) 410-101(c)(8)   90%
112. (a) 220-16(a)   90%
113. (d) 90-2(a)   70%
114. (c) 810-52 Table   75%
115. (a) 100 Accessible, Equipment   75%
116. (d) 800-10(a)(2) and (b)   75%
117. (a) Article 100 Bonding Jumper, Main   75%
118. (d) 830-10(a) and (f)
119. (d) 720-4   75%
120. (b) 331-4(5)   75%
121. (d) 680-25(c)   75%
Flex is not permitted, but liquidtight is.
122. (d) 100 Attachment Plug   70%
123. (d) 374-5(a)(1) Current-carrying conductors   55%

124. (d) 400-5   75%
125. (c) 680-22(a)(1)   65%
126. (d) 364-12 and 380-8   65%
127. (d) 215-10   60%
128. (c) 501-3(b)(1) Exception   75%
129. (d) 680-22(a)   60%
130. (b) 220-32(c)(3)   95%
131. (d) 517-35(a)   75%
132. (c) 345-7, Chapter 9 Table 4   75%
133. (d) 210-52(c)(1)   75%
134. (d) 100 Branch Circuit, Multiwire   70%
If you missed this question, you probably answered
the question quickly without checking the responses in
(b) and (c). Slow down and read the complete paragraph of the NEC®, not just the first line. Read all of
the answer choices before you pick (a).
135. (b) 331-4(8) and 518-4   75%
136. (c) 100 Conductor, Covered   60%
137. (c) 210-52(c)(3)   75%
138. (b) 680-11   75%
The Code says normal, not abnormal.
139. (d) 100 Enclosure   75%
140. (d) 350-18 Exception No. 3   70%
Read the exception again.
141. (d) 100 Grounding Electrode Conductor   70%
142. (c) 680-10   75%
143. (c) 320-14   60%
144. (b) 110-14(c)(1)(c)   65%
145. (d) 250-50 Table   55%
Three times 500,000 = 1,500,000 which is greater than
1,100 kcmil
146. (d) 362-2   75%
147. (a) 680-6(a)(2)   75%
148. (a) 100 Service Equipment   75%
149. (d) 501-4(b)   75%
150. (a) 110-26(f)(1)(a)   70%
151. (b) 200-7(c)(2)   75%
152. (a) 110-26(b)   75%
153. (b) 500-9(b)   65%
154. (c) 440-22(a)   75%
155. (c) 210-8(a)(2) Exception No. 2   60%
156. (a) 110-33(a)   65%
Over 6 foot requires two entrances, but a 6-foot control panel requires one entrance.
157. (d) 210-63 FPN and 210-8(b)(2)   75%

158. (c) 550-5(a) Exception No. 1   70%

159. (c) 110-34(a) Table   65%

160. (a) 210-8(a)(6)   75%

161. (b) 100 Service Lateral   60%

162. (b) 550-7(a)   70%

163. (a) 430-109(a)(1)   75%

164. (b) 210-52(b)(1)   70%

165. (b) 550-22 Table 6   65%
     units × 16,000 VA3.29 = 27,840 VA

166. (b) 220-3(a)   75%

167. (b) 518-4(a)   75%

168. (d) 210-70(a)(1 and 2)   75%

169. (b) 430-132   70%

170. (b) Table 210-21(b)(2) and Table 210-24   70%
     I know you want to say 20 ampere, but read *Code* carefully.

171. (d) 511-3(b)   60%

172. (d) 450-24   75%

173. (a) 210-4(c)   60%

174. (c) 351-9 Exception   75%

175. (d) 210-21(b)(1)   70%

176. (b) 680-22   70%

177. (d) 210-7(d)(3)   60%
     Nongrounding type receptacles can be replaced with another nongrounding type receptacle or a GFCI receptacle.

178. (b) 100 Dwelling Unit, Two-Family   65%

179. (b) 110-26(a)(1)   75%

180. (b) 100 Overload   75%
     Be sure to read the definition more carefully.

181. (c) 351-9   75%

182. (b) 110-34(c)   65%
     The *NEC*® requires the word "Danger" in warning signs.

183. (b) 100 Remote-Control Circuit   75%

184. (a) 351-5(a) Exception   70%

185. (b) 100 Demand Factor   60%

186. (c) 326-1   70%

187. (a) 330-3   75%

188. (b) 725-54(a)(1)   75%

189. (b) 516-2(e)   75%

190. (d) 430-101   70%

191. (a) 240-3(b), 310-16, 240-6(a), 110-14(c)(1)   70%

192. (c) 660-6(a)   75%

193. (d) 514-8 Exception No. 2   75%

194. (c) 430-24   75%

195. (b) 725-41(a) FPN, Table 11 in Chapter 9   75%

196. (a) 364-6(b)(1)   75%

197. (c) 100 Askarel   60%

198. (d) 90-2(b)(5)   75%

199. (a) 725-54(d) 800-52(a)(1) and 820-52(a)(1)   90%

200. (d) 90-1(b)   60%

Notes:

Index

W